the study of number (includes measure)

GCSE (9-1)
Mathematics
Higher Level

Entirely self-study

AQA EDEXCEL OCR
Exam Boards

4 books

- Arithmetic
- Algebra
- Geometry
- Statistics

Explanations   Examples
worked Answer <u>next to</u> every Question

Joseph Humphreys

JH GCSE (9-1) Mathematics

Higher Level

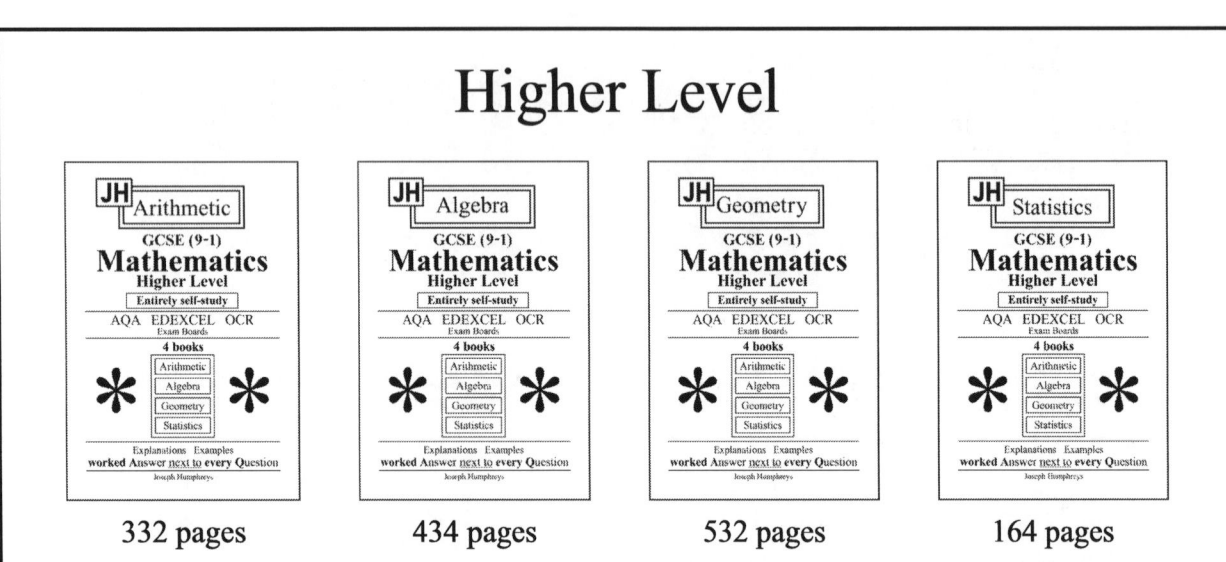

Foundation Level

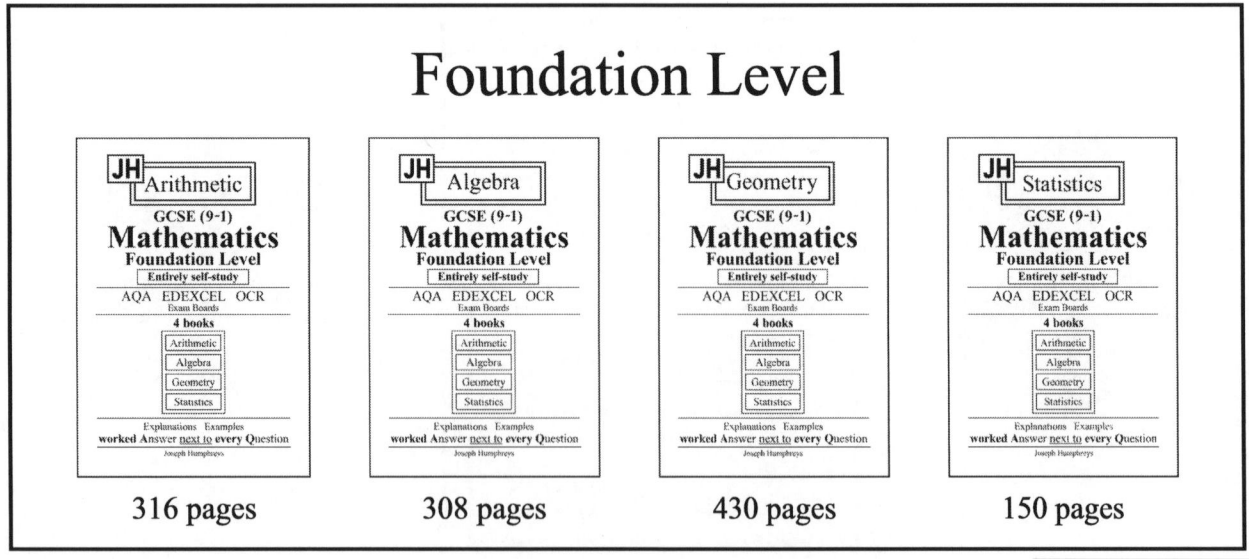

ISBN 9798854090544

© Joseph Humphreys 2023

Manchester
England

JH GCSE (9-1) Mathematics Higher Level
4 books - Arithmetic, Algebra, Geometry, Statistics

To the Student.

The aim of my books is to let the student help themselves in Mathematics.

The books teach the student in '**self-study**' format
the Mathematics which is presented in the GCSE (9-1) Higher Level
and comply with the AQA, Edexcel and OCR Exam Boards.
They are methodical and structured to give the student the opportunity
to work independently.

The strategy throughout is to consider the flow of introductions, explanations
and examples. The Question / Answer (Q/A) format then allows the student
to attempt related questions with the worked answers 'covered up' alongside -
to be checked at any point and progress or re-evaluate.
(Alternative 'working' may be possible for a solution.)

The Higher Level includes the Foundation Level.
The additional Mathematics required for the Higher Level
has Contents / pages marked, ex., ***5**.

The following notes are useful.

1. In general, the topics may be studied in the order presented in each book.
   However, it may be necessary to 'break off' from 1 topic to study
   part of another before returning to complete it.
   It may also be useful to study larger topics in stages.

2. The student can revise regularly knowing that each topic is easily located
   in the Contents pages of the books and any question attempted is easily checked.
   (There are no separate revision sections.)

3. Questions are often written in a compact form down the page -
   this is unusual but necessary for the format of these books.

4. Where an example is not presented, the Q/A which follow act as examples -
   this avoids repetition.

5. An arrow, ex. $\Rightarrow$, mainly indicates development of work.

6. Where more than 1 method of work is shown, a student may already be
   familiar with or not confident with 1 method and can at least consider another.

7. In general, decimal answers are not 'rounded'.

I have been a Mathematics teacher for many years.

I wish you every success.

Joseph Humphreys

Arithmetic Contents Summary

Page		Page Number
Arithmetic 1	Numbers	1
	Addition	3
	Subtraction	4
	Multiplication	8
	Division	14
	Decimals	20
Arithmetic 2	Decimals	36
	Fractions	40
	Types of Numbers	70
Arithmetic 3	Multiples	74
	Factors	75
	Rounding Numbers	82
	Negative Numbers	88
	Money	102
	Ready Reckoner	108
	Magic Squares	109
Arithmetic 4	Percentages	110
Arithmetic 5	Percentages	150
	Number Order	155
	Number Grids	157
	Indices / Powers	158
	BIDMAS	173
	Sequences	176
	Proportion	180
Arithmetic 6	Ratio	186
	Foreign Exchange	198
	Standard Form	202
	Surds	210
	Trial and Improvement	221
Arithmetic 7	Measure	223
	Measure - Metric System	231
	Measure - Imperial System	244
	Measure Conversion	249
Arithmetic 8	Density - Mass - Volume	255
	Pressure - Force - Area	258
	Scale	260
	Time	263
	Time - Speed - Distance	279
	Velocity - Time - Acceleration	285
Arithmetic 9	Calculator	

Arithmetic 1   Contents   *Higher

Page	Title		Details
1	**Numbers**	1	Column Headings / Values.
2		2	
3	**Addition**	1	Addition. 'Problems'.
4	**Subtraction**	1	Method 1 - 'Decomposition'.
5		2	Method 2 - '10 to the top, 10 to the bottom' / 'borrow 10, pay it back'.
6		3	'Problems'.
7		4	Method 3 - 'Add Differences'. 'Problems'.
8	**Multiplication**	1	Multiplication.
9		2	'Problems'.
10		3	Use of Factors.
11		4	Use of Brackets.
12		5	Factorial.
13		6	Multiplication Tables.
14	**Division**	1	Division / Remainder.
15		2	'Problems'.
16		3	'Long Division' - Method 1 (add the dividing number).
17		4	Method 2 (estimate and check).
18		5	(Calculations).
19		6	'Problems'.
20	**Decimals**	1	Column Headings / Values.
21		2	Change a Decimal to a Fraction / Fraction in Lowest Terms.
22		3	Addition.
23		4	Subtraction - Method 1 / 2.
24		5	- Method 3.
25		6	Division - by a Whole Number.
26		7	
27		8	- by 10, 100, 1000, ...
28		9	Multiplication - by 10, 100, 1000, ...
29		10	- by a Whole Number / Decimal.
30		11	
31		12	Division - by a Decimal.
32		13	- 'Fraction Form'.
33		14	Reciprocal. List in Order - Ascending / Descending.
34		15	Powers. Roots.
35		16	Change a Recurring Decimal to a Fraction.
(continued)			

Arithmetic 2   Contents   *Higher

Page	Title		Details
36	**Decimals**	17	Multiplication / Division - by 10, 100, ... - given the result.
37		18	
38		19	
39		20	
40	**Fractions**	1	Types of Fractions - Proper / Improper / Mixed Number.
41		2	
			Change Improper Fraction to Whole Number / Mixed Number.
42		3	Change Whole Number / Mixed Number to Improper Fraction.
43		4	Equal / Equivalent Fractions.
44		5	Cancel Fractions. Lowest Terms.
45		6	Cancel Fraction to Lowest Terms.
46		7	Equivalent Fractions. Change Improper Fraction to Mixed Number.
47		8	Addition - Same Denominators.
48		9	Subtraction - Same Denominators.
49		10	- from a Whole Number. Addition / Subtraction.
50		11	Addition / Subtraction - Different Denominators.
51		12	
52		13	- Negative Fractions (same denominator).
53		14	Subtraction - Mixed Numbers - Method 1.
54		15	- Method 2.
55		16	- Method 3.
56		17	Addition / Subtraction - 'Problems'.
57		18	Compare the Size of Fractions.
58		19	Multiplication.
59		20	Cancel - Methods 1 / 2 / 3.
60		21	
61		22	Mixed Numbers. Negative Fractions.
62		23	Fraction of an 'amount'. Fraction of a Decimal.
63		24	Division - by a Whole Number.
64		25	- by a Whole Number / Fraction.
65		26	- different presentation.
66		27	Multiplication / Division - 'Problems'.
67		28	Reciprocal.
68		29	Change Fraction to Decimal.
69		30	Powers. Roots.
70	**Types of Numbers**	1	Counting. Natural. Whole. Even. Odd.
			Integers. Multiples. Factors. Prime. (Consecutive.)
71		2	Square. Square Root. Cube. Cube Root. Rectangle.
72		3	Powers. Roots. Rational. Irrational. Real. Imaginary. 'Own Name'.
73		4	Prime Numbers - Sieve of Eratosthenes.

Arithmetic 3 — Contents — *Higher

Page	Title		Details
74	**Multiples**	1	Multiples. Common Multiples / Lowest Common Multiple (L.C.M.)
75	**Factors**	1	Factors.
76		2	Common Factors and Highest Common Factor (H.C.F.)
77		3	Prime Factors. A number as a product of its Prime Factors.
78		4	Highest Common Factor (H.C.F.) and (L.C.M.)
79		5	Lowest Common Multiple (L.C.M.)
80		6	H.C.F. / L.C.M. - Index Form.
81		7	Perfect Numbers.
82	**Rounding Numbers**	1	Round to the nearest 10.
83		2	Round to the nearest 100 / nearest 1000.
84		3	Round Decimal to nearest Whole Number.
85		4	Round Decimal to 1 / 2 Decimal Places.
86		5	Round to 1,2,3, ... Significant Figures.
87		6	Estimation.
88	**Negative Numbers**	1	Negative Numbers. Number Line.
89		2	Number Order.
90		3	Increase / Decrease.
91		4	Addition. Subtraction.
92		5	
93		6	
94		7	2 signs next to each other.
95		8	Multiplication.
96		9	Division.
97		10	'Problems'.
98		11	Multiplication (more than 2 numbers).
99		12	Powers (Indices). Roots.
100		13	Combined Operations.
101		14	Absolute Value.
102	**Money**	1	Units. Addition. 'Problems'.
103		2	Subtraction - Method 1 / 2. 'Problems'.
104		3	- Method 3.
105		4	Multiplication. 'Problems'.
106		5	Division. 'Problems'.
107		6	Bank Account. Credit. Debit. Bank Statement.
108	**Ready Reckoner**	1	Ready Reckoner.
109	**Magic Squares**	1	Magic Squares.

Arithmetic 4 — Contents — *Higher

Page	Title	#	Details
110	**Percentages**	1	Change a Percentage to a Fraction. 100% Grid.
111		2	Cancel Fraction.
112		3	Greater than 100%.
113		4	Change Fraction Percentage to Fraction.
114		5	Cancel.
115		6	Change a Fraction to a Percentage.
116		7	
117		8	Change Whole / Mixed Number to a Percentage.
118		9	Change a Percentage to a Decimal. Change a Decimal to a Percentage.
119		10	Change a Fraction Percentage to a Decimal.
120		11	Whole Amount / Parts. Compare Percentage / Fraction / Decimal.
121		12	Percentage of a Value - calculate using ... Fractions - Method 1.
122		13	Decimals - Method 1.
123		14	Fractions - Method 2 / Decimals - Method 2.
124		15	Percentage Increase / Decrease - Method 1.
125		16	- 'Multiplier' - Method 2.
126		17	The Original Value of an Increased / Decreased Value - Method 1.
127		18	- Method 2.
128		19	Express One Value as a Percentage of Another Value.
129		20	Percentage Change - Express Change in a Value as a Percentage.
130		21	Inflation - Deflation.
131		22	Profit / Loss.
132		23	Percentage Profit / Loss.
133		24	Credit / Higher Purchase.
134		25	
135		26	Wage / Salary.
136		27	
137		28	Value Added Tax (V.A.T.).
138		29	Original Value of item that now includes V.A.T.
139		30	
140		31	Simple Interest - Formula.
141		32	
142		33	
143		34	Compound Interest - Method 1.
144		35	
145		*36	Formula / Method 2.
146		*37	
147		*38	
148		39	Appreciation / Depreciation - Method 1.
149		*40	- Formula / Method 2.

(continued)

Arithmetic 5 — Contents *Higher

Page	Title		Details
150	**Percentages**	41	Percentage Increase / Decrease. Combine 'Multipliers'.
151		42	
152		43	Interchange ... Percentages - Fractions - Decimals.
153		44	
154		45	
155	**Number Order**	1	Arrange numbers in Ascending / Descending order.
156		2	Decimals / Fractions.
157	**Number Grids**	1	Number Grids.
158	**Indices / Powers**	1	Numbers in Index Form.
159		2	Decimals / Fractions. Operations.
160		3	Multiplication of Numbers - Addition of Indices.
161		4	Division of Numbers - Subtraction of Indices.
162		5	Multiplication / Division of Numbers - Addition/Subtraction of Indices.
163		6	Raise Power of a Number to a Power - Multiplication of Indices.
164		7	Index 0 / Power 0.
165		8	Negative Indices / Powers $-1, -2, -3, ...$
166		9	
167		10	Division - Cancel Powers.
168		*11	Fraction Indices - Positive.
169		*12	Negative.
170		*13	Positive.
171		*14	Negative.
172		*15	Write number in Index Form.
173	**BIDMAS**	1	Order of Operations.
174		2	
175		3	
176	**Sequences**	1	Find the Next Term in a Sequence.
177		2	
178		3	
179		4	
180	**Proportion**	1	Direct - '1 Value increases ... other Value increases' - Method 1 / 2.
181		2	
182		3	Graph.
183		4	Inverse - '1 Value increases ... other Value decreases' - Method 1 / 2.
184		5	
185		6	Graph.

Arithmetic 6 — Contents   *Higher

Page	Title	#	Details
186	**Ratio**	1	Equivalent Ratios. Simplest Form.
187		2	State Ratio (A:B or B:A).
188		3	Simplest Form.
189		4	Fraction Values.
190		5	
191		6	Decimal Values.
192		7	Unitary Form - 1:n and n:1
193		8	Equivalent Ratios - Find Unknown Value.
194		9	Divide / Share in a Given Ratio. Method 1.
195		10	Method 2.
196		11	Actual difference between Ratio.
197		12	Ratio of 3 values - given 2 values.
198	**Foreign Exchange**	1	Currency Units / Exchange Rate.
199		2	Exchange Pound for other currencies.
200		3	Compare Price of item. Exchange 1 Currency Unit for Pounds.
201		4	Exchange 2 Foreign Currencies (relate each to the Pound).
202	**Standard Form**	1	Write a Number in Standard Form.
203		2	
204		3	Change Standard Form to a Number.
205		4	Addition. Subtraction.
206		5	Multiplication. Division.
207		6	
208		7	'Problems'.
209		8	
210	**Surds**	1	Addition.
211		2	Subtraction.
212		3	Multiplication.
213		4	Division.
214		*5	Factors.
215		*6	Express 'Number × Surd' as a Surd.
216		*7	Expand Brackets.
217		*8	Fraction - Rationalise the Denominator.
218		*9	
219		*10	
220		*11	
221	**Trial and Improvement**	1	Square Root.
222		2	Cube Root.

Arithmetic 7 — Contents [Measure] *Higher

Page	Title	#	Details
223	**Measure**	1	Discrete / Continuous Measure.
224		2	Continuous Measure - Accuracy.
225		3	
226		4	Combining Measures.
227		5	'Problems'.
228		6	
229		7	Error Interval.
230		8	Absolute Error. Percentage Error.
231	**Measure - Metric System**	1	Metric Units - Length / Weight / Capacity.
232		2	Measure / Draw Line (cm / mm).
233		3	Convert Units - Length.
234		4	'Problems'.
235		5	Convert Units - Weight.
236		6	'Problems'.
237		7	Convert Units - Capacity.
238		8	'Problems'.
239		9	List Measures in Ascending / Descending order.
240		10	
241		11	Estimation - Length, Weight, Capacity.
242		12	Convert Units - Area.
243		13	Volume.
244	**Measure - Imperial System**	1	Convert Units - Length.
245		2	Weight.
246		3	Capacity.
247		4	Area.
248		5	Volume.
249	**Measure Conversion**	1	Convert Units - Metric / Imperial - Length.
250		2	
251		3	Weight.
252		4	Capacity.
253		5	Area.
254		6	Volume.

Arithmetic 8   Contents   [Measure]   *Higher

Page	Title		Details
	Density - Mass -		
255	**Volume**	1	⌐ Formulas.
256		2	│
257		3	⌐ Change of Units.
	Pressure - Force -		
258	**Area**	1	⌐ Formulas.
259		2	⌐
260	**Scale**	1	Read a Scale Measurement (on instrument).
261		2	⌐ Estimate a Scale Measurement - Read.
262		3	⌐ - Mark.
263	**Time**	1	Units of Time.
264		2	⌐ 12-Hour Clock - Read Time.
265		3	⌐ Indicate Time.
266		4	⌐
267		5	12 / 24-Hour Clock.
268		6	⌐ Converting Time Units - Higher / Lower.
269		7	⌐ - Decimal / Fraction.
270		8	Addition.
271		9	⌐ Subtraction - Method 1 / 2.
272		10	⌐ - Method 3.
273		11	Multiplication.
274		12	'Problems': $+, -, \times$.
275		13	⌐ Division - no remainder / remainder.
276		14	⌐ - remainder.
277		15	Timetable.
278		16	Calendar. Estimation.
	Time - Speed -		
279	**Distance**	1	⌐ Formulas.
280		2	⌐ Change of Units.
281		3	⌐
282		4	⌐ Average Speed.
283		5	│
284		6	⌐
	Velocity - Time -		
285	**Acceleration**	1	⌐ Formulas.
286		2	│
287		3	⌐ Kinematics Formulas.

Arithmetic 9   Contents

Page	Title		Details
	Calculator		
	Introduction		
1	**Calculator - Direct Input**	1	Direction. Shift. Mode. Clear. Error. +/−.
2		2	Addition. Subtraction. Multiplication. Division.
3		3	Prime Factors. Indices / Powers.
4		4	Roots. Percentages. Factorial.
5		5	Fractions: Input. To Decimal. Equivalent. Improper / Mixed Number.
6		6	+, −, ×.
7		7	÷. Decimal to Fraction.
8		8	Answer Key.
			Combining Operations. Brackets / = Sign.
9		9	
10		10	
11		11	Indices / Powers and Roots.
12		12	π (Pi). Reciprocal. Random Numbers. Trigonometry.
13		13	Standard Form to/from Number.
14		14	Calculations. Multiplication.
15	**Calculator - Indirect Input**	1	Shift. Clear. Error. +/−. Memory.
16		2	Addition. Subtraction. Multiplication. Division.
17		3	Constant. Indices / Powers.
18		4	Roots. Percentages. Factorial.
19		5	Fractions: Input. To/from Decimal. Equivalent. Improper / Mixed Number.
20		6	+, −, ×, ÷.
21		7	Combining Operations. Brackets / = Sign.
22		8	
23		9	Indices / Powers and Roots.
24		10	Standard Form to/from Number. Calculations.
25		11	π (Pi). Reciprocal. Random Numbers. Trigonometry.

Numbers 1
Column Headings / Values.

Numbers

A **number** is made up of one or more **digits / figures**.
The **digits / figures** are 0,1,2,3,4,5,6,7,8,9.

We can refer **generally** to a **specific** number by stating **how many digits / figures** are in the number.
   Ex.   8 has 1 digit / figure   - it is called a '1 **digit** number' or a '1 **figure** number'.
   Ex.   73 has 2 digits / figures - it is called a '2 **digit** number' or a '2 **figure** number'.
   Ex.   502 has 3 digits / figures - it is called a '3 **digit** number' or a '3 **figure** number'.
   Ex.   1,416 has 4 digits / figures - it is called a '4 **digit** number' or a '4 **figure** number'.

The **value** of a digit depends on its **position / place** in a number - this is called its **place value**.

The position of each digit in a number has a **name / heading**, for example, **Hundreds**, **Tens** and **Units**, so we can read / write a number - these are not usually written.

   Ex.   Hundreds   Tens   Units
                             1   The digit 1 has the value 'one Unit'.
                     1   0   The digit 1 has the value 'one Ten' - the 0 must be written.
          1   0   0   The digit 1 has the value 'one Hundred' - the 00 must be written.

The headings are written HTU for short.
The names / headings for numbers up to trillions are shown here -

   Trillions   Billions   Millions   Thousands
   H T U , H T U , H T U , H T U , H T U

There are <u>Hundreds, Tens and Units</u> of these larger numbers -
Ex. ' Hundreds of **Thousands**, Tens of **Thousands**, **Thousands** '

*Note Numbers and names / headings are usually separated by a comma -
         this <u>can</u> be omitted and usually is when calculating.
            (The comma is <u>not</u> entered into a calculator.)

We **read a number** by stating how many **Hundreds**, **Tens** and **Units** there are of <u>each</u> **name** / **heading** starting from the <u>left</u>.

Q	A

1) Write each number **in words**.
   Trillions Billions Millions Thousands
    H T U, H T U, H T U,   H T U , H T U
	1)

a) 1 0 0	a) One hundred.
b) 1 , 0 0 0	b) One thousand.
c) 1 , 0 0 0 , 0 0 0	c) One million.
d) 1 , 0 0 0 , 0 0 0 , 0 0 0	d) One billion.
e) 1 , 0 0 0 , 0 0 0 , 0 0 0 , 0 0 0	e) One trillion.
f) 1 2 3	f) One hundred and twenty three.
g) 4 5 6 , 0 0 0	g) Four hundred and fifty six thousand.
h) 7 8 9 , 0 0 0 , 0 0 0	h) Seven hundred and eighty nine million.
i) 5 0 0 , 0 0 0 , 0 0 0 , 0 0 0	i) Five hundred billion.
j) 9 9 9 , 0 0 0 , 0 0 0 , 0 0 0 , 0 0 0	j) Nine hundred and ninety nine trillion.
k) 7 2 , 3 1 8	k) Seventy two thousand, three hundred and eighteen.
l) 1 0 4 , 0 0 0 , 6 0 5	l) One hundred and four million, six hundred and five.
m) 8 , 0 0 2 , 0 0 7 , 0 0 0 , 0 0 9	m) Eight trillion, two billion, seven million and nine.

Numbers 2

Column Headings / Values.

Q	A
2) Write each number in figures.	2) Trillions Billions Millions Thousands H T U , H T U , H T U , H T U , H T U
a) Two hundred and fifty nine.	a) 2 5 9
b) Thirty three thousand.	b) 3 3 , 0 0 0
c) Forty eight million.	c) 4 8 , 0 0 0 , 0 0 0
d) Twelve billion.	d) 1 2 , 0 0 0 , 0 0 0 , 0 0 0
e) Six hundred trillion.	e) 6 0 0 , 0 0 0 , 0 0 0 , 0 0 0 , 0 0 0
f) Seven thousand and seventy seven.	f) 7 , 0 7 7
g) Nine hundred and four thousand and five.	g) 9 0 4 , 0 0 5
h) Five million, eight hundred thousand.	h) 5 , 8 0 0 , 0 0 0
i) Eighty six billion, twenty million, four hundred.	i) 8 6 , 0 2 0 , 0 0 0 , 4 0 0
j) Fifty trillion, five hundred billion.	j) 5 0 , 5 0 0 , 0 0 0 , 0 0 0 , 0 0 0
k) Three hundred and nineteen million, ten thousand.	k) 3 1 9 , 0 1 0 , 0 0 0
l) Seven hundred and sixty four billion.	l) 7 6 4 , 0 0 0 , 0 0 0 , 0 0 0
m) One thousand, one hundred and ninety two.	m) 1 , 1 9 2

If we are given just one number to read or write, it is useful to write the names / headings to help.
Each of the following Q/A give just one number - and only up to millions, which will be sufficient to help explain.

Q	A
3) Write each number in words.	3) *Note M = Millions Th. = Thousands
A) 123,456,789	M Th. H T U , H T U , H T U A) 1 2 3 , 4 5 6 , 7 8 9 One hundred and twenty three **million**, four hundred and fifty six **thousand**, seven hundred and eighty nine.
B) 30,902,560	M Th. H T U , H T U , H T U B) 3 0 , 9 0 2 , 5 6 0 Thirty **million**, nine hundred and two **thousand**, five hundred and sixty.
C) 7,018,074	M Th. H T U , H T U , H T U C) 7 , 0 1 8 , 0 7 4 Seven **million**, eighteen **thousand** and seventy four.
D) 850,001	Th. H T U , H T U D) 8 5 0 , 0 0 1 Eight hundred and fifty **thousand** and one.

Q	A
4) Write each number in figures.	4) *Note M = Millions Th. = Thousands

A) Five hundred and four **thousand**,
six hundred and eight.
	A) Th.
	     H T U , H T U
	      5 0 4 , 6 0 8

B) Nine **million**,
seventy seven **thousand**,
eight hundred and thirty five.
	B) M Th.
	  H T U , H T U , H T U
	    9 , 0 7 7 , 8 3 5

C) Forty three **million**,
two hundred and ninety **thousand**
and sixteen.
	C) M Th.
	  H T U , H T U , H T U
	  4 3 , 2 9 0 , 0 1 6

D) Nine hundred and eighty seven **million**,
six hundred and fifty four **thousand**,
three hundred and twenty one.
	D) M Th.
	  H T U , H T U , H T U
	  9 8 7 , 6 5 4 , 3 2 1

Addition 1

Addition. 'Problems'.

Arithmetic
Arithmetic includes the 4 main calculations of Addition, Subtraction, Multiplication and Division. These may be presented in many ways which have to be recognised, as shown here.
Calculations may use the = sign or be set out so that numbers are positioned in their columns, ex. HTU, although the headings are usually omitted.

Addition

Add the columns from **right** to **left**. ←

Ex.1 1 + 2

1 + 2 = 3

Ex.2 Add 24 and 49.

$$\begin{array}{r} TU \\ 24 \\ +49 \\ \hline \end{array} \Rightarrow \begin{array}{r} TU \\ 24 \quad \text{4+9=13} \\ +49 \\ \hline 3 \\ 1 \end{array} \Rightarrow \begin{array}{r} TU \\ 24 \\ +49 \\ \hline 2+4+1=7 \quad 73 \\ 1 \end{array}$$

'carry 1'

*Note
When we add the Units column, we '**carry**' any Tens to the Tens column to be **added** there. In the same way we 'carry' from Tens to Hundreds, then from Hundreds to Thousands, ...

Q	A	Q	A
1) **Add** 74 and 12.	1) $\begin{array}{r}74\\+12\\\hline 86\end{array}$	5) 6 + 3	5) 6 + 3 = 9
2) Calculate 38 **+** 55.	2) $\begin{array}{r}38\\+55\\\hline 93\\ _1\end{array}$	6) Find the **total** of 67, 8 and 89.	6) $\begin{array}{r}67\\8\\+\ 89\\\hline 164\\ _{12}\end{array}$
3) Find the **sum** of 486 and 40.	3) $\begin{array}{r}486\\+\ \ 40\\\hline 526\\ _1\end{array}$	7) **Add** the following: 153 + 590 + 174.	7) $\begin{array}{r}153\\590\\+174\\\hline 917\\ _2\end{array}$
4) Work out 98 **plus** 209.	4) $\begin{array}{r}98\\+209\\\hline 307\\ _{11}\end{array}$	8) Work out the answer to the **addition** of 4,600 and 3,700.	8) $\begin{array}{r}4,600\\+3,700\\\hline 8,300\\ _1\end{array}$

The following 'problems' involve addition.

Q	A
1) A team scores 19 points in the first half of a game and 25 points in the second half. How many points did the team score in total?	1) $\begin{array}{r}19\\+25\\\hline 44\\ _1\end{array}$ points
2) In a school there are 526 girls and 479 boys. How many pupils are there altogether?	2) $\begin{array}{r}526\\+\ 479\\\hline 1005\\ _{111}\end{array}$ pupils
3) At the end of a year's production, a bicycle company had sold 2,050 to one store and 1,885 to another. It had a **gross** unsold. Calculate the number of bicycles produced by the company in the year.	3) $\begin{array}{r}2,050\\1,885\\+\ \ 144\quad (=\textbf{gross})\\\hline 4,079\\ _{11}\end{array}$ bicycles
4) The number of patients attending a clinic over a 4-week period was recorded as follows: Week 1 89, Week 2 48, Week 3 66, Week 4 57 Find the total number of patients recorded.	4) $\begin{array}{r}89\\48\\66\\+\ 57\\\hline 260\\ _{23}\end{array}$ patients

Subtraction 1

‘Decomposition’.

Subtraction

Subtract the columns from **right** to **left**. ←

Ex.1 $9-2$

$9-2=7$

Ex.2 $596-120$

$$\begin{array}{r} \text{HTU} \\ 596 \\ -120 \\ \hline 476 \end{array}$$

When a number in a column can not be subtracted from a smaller number, we can use either of the following 3 Methods to allow the subtraction to take place.

Method 1

This method is usually referred to as ‘**decomposition**’ - numbers are ‘**decomposed**’ or ‘**broken up**’ into other numbers.

Ex.1 $80-28$

$$\begin{array}{rr} T & U \\ \mathbf{8} & 0 \\ -2 & 8 \\ \hline \end{array}$$

⇨ **80** becomes **70** + **10**
8 is ‘crossed out’ and replaced by **7**.
1 is placed in front of any units and stands for **10**.

$$\begin{array}{rr} T & U \\ \overset{\mathbf{7}}{\not{8}} & {}^{\mathbf{1}}0 \\ -2 & 8 \\ \hline 5 & 2 \end{array}$$

Ex.2 $753-189$

($\overset{T}{1_4}$ = **14** Tens, so the **1** stands for **10** Tens = **100**)

$$\begin{array}{rrr} H & T & U \\ 7 & \mathbf{5} & 3 \\ -1 & 8 & 9 \\ \hline \end{array}$$

⇨ **50** becomes **40** + **10**

$$\begin{array}{rrr} H & T & U \\ 7 & \overset{\mathbf{4}}{\not{5}} & {}^{\mathbf{1}}3 \\ -1 & 8 & 9 \\ \hline & & 4 \end{array}$$

$13-9=4$

⇨

$$\begin{array}{rrr} H & T & U \\ \mathbf{7} & \overset{4}{\not{5}} & {}^{1}3 \\ -1 & 8 & 9 \\ \hline & & 4 \end{array}$$

⇨ **700** becomes **600** + **100**

$$\begin{array}{rrr} H & T & U \\ \overset{\mathbf{6}}{\not{7}} & \overset{{}^{\mathbf{1}}4}{\not{5}} & {}^{1}3 \\ -1 & 8 & 9 \\ \hline 5 & 6 & 4 \end{array}$$

$14-8=6$

Ex.3 $600-257$

$$\begin{array}{rrr} H & T & U \\ \mathbf{6} & 0 & 0 \\ -2 & 5 & 7 \\ \hline \end{array}$$

⇨ **600** becomes **500** + **90** + **10**

$$\begin{array}{rrr} H & T & U \\ \overset{\mathbf{5}}{\not{6}} & \overset{\mathbf{9}}{\not{0}} & {}^{\mathbf{1}}0 \\ -2 & 5 & 7 \\ \hline 3 & 4 & 3 \end{array}$$

Ex.4 $4{,}002 - 2{,}105$

$$\begin{array}{rrrr} Th & H & T & U \\ \mathbf{4} & 0 & 0 & 2 \\ -2 & 1 & 0 & 5 \\ \hline \end{array}$$

⇨ **4000** becomes **3000** + **900** + **90** + **10**

$$\begin{array}{rrrr} Th & H & T & U \\ \overset{\mathbf{3}}{\not{4}} & \overset{\mathbf{9}}{\not{0}} & \overset{\mathbf{9}}{\not{0}} & {}^{\mathbf{1}}2 \\ -2 & 1 & 0 & 5 \\ \hline 1 & 8 & 9 & 7 \end{array}$$

Ex.5 $8{,}034 - 579$

$$\begin{array}{rrrr} Th & H & T & U \\ 8 & 0 & \mathbf{3} & 4 \\ - & 5 & 7 & 9 \\ \hline \end{array}$$

⇨ **30** becomes **20** + **10**

$$\begin{array}{rrrr} Th & H & T & U \\ 8 & 0 & \overset{\mathbf{2}}{\not{3}} & {}^{\mathbf{1}}4 \\ - & 5 & 7 & 9 \\ \hline & & & 5 \end{array}$$

⇩

$$\begin{array}{rrrr} Th & H & T & U \\ \mathbf{8} & 0 & \overset{2}{\not{3}} & {}^{1}4 \\ - & 5 & 7 & 9 \\ \hline & & & 5 \end{array}$$

⇨ **8000** becomes **7000** + **900** + **100**

$$\begin{array}{rrrr} Th & H & T & U \\ \overset{\mathbf{7}}{\not{8}} & \overset{\mathbf{9}}{\not{0}} & \overset{{}^{\mathbf{1}}2}{\not{3}} & {}^{1}4 \\ - & 5 & 7 & 9 \\ \hline 7 & 4 & 5 & 5 \end{array}$$

Subtraction 2

'10 to the top ...' / 'borrow 10 ...'.

Method 2

This method has no particular name but is usually referred to as '**10 to the top, 10 to the bottom**' or '**borrow 10, pay it back**'.

Ex.1 80 − 28

```
   T U              T U
   8 0      ⇨      8 ¹0       ← 10 is added to the 'top number' in the Units column  and
  -2 8             -2₁8       ← 10 is added to the 'bottom number' in the Tens column.
  ———              ———
            ⇨                        or
   T U              T U
                                10 is 'borrowed' by the 'top number' in the Units column  and
   8 ¹0     ⇨      8 ¹0       10 is 'paid back' to the 'bottom number' in the Tens column.
  -2₁8             -2₁8
     2             5 2
     ↑         8−(2+1) = ↑   We need to mentally add  2+1 = 3
10 − 8 = 2    8 −  3    = 5
```

*Note
Since we have added 10 to <u>both</u> numbers, the subtraction has become 90 − 38 (= 52 also).
In general, when subtracting 2 numbers, adding 10, 100, 1000, ... (or **any** number) to **both** numbers gives the same answer - <u>Method 2</u> uses this idea.

The method is extended to **any column** when a number can not be subtracted from a smaller number.

Ex.2 753 − 189

```
   H T U    ⇨    H T U    ⇨     H T U     ⇨     H T U
   7 5 3         7 5 ¹3         7 ¹5 ¹3          7 ¹5 ¹3
  -1 8 9        -1 8₁9         -1₁8₁9           -1₁8₁9
  —————         —————              6 4           5 6 4
                    4         15−(8+1) = ↑   7−(1+1) = ↑
                              15−  9   = 6   7−  2   = 5
```

Ex.3 600 − 247

```
   H T U    ⇨    H T U    ⇨     H T U     ⇨     H T U
   6 0 0         6 0 ¹0         6 ¹0 ¹0          6 ¹0 ¹0
  -2 5 7        -2 5₁7         -2₁5₁7           -2₁5₁7
  —————         —————              4 3           3 4 3
                    3         10−(5+1) = ↑   6−(2+1) = ↑
                              10−  6   = 4   6−  3   = 3
```

Ex.4 4,002 − 2,105

```
 Th H T U  ⇨  Th H T U  ⇨   Th H T U   ⇨   Th H T U   ⇨   Th H T U
  4 0 0 2     4 0 0 ¹2       4 0 ¹0 ¹2      4 ¹0 ¹0 ¹2      4 ¹0 ¹0 ¹2
 -2 1 0 5    -2 1 0₁5       -2 1₁0₁5       -2₁1₁0₁5        -2₁1₁0₁5
 ———————     ———————            9 7           8 9 7         1 8 9 7
                   7        10−(0+1) = ↑  10−(1+1) = ↑   4−(2+1) = ↑
                            10−  1   = 9  10−  2   = 8   4−  3   = 1
```

Ex.5 8,034 − 579

```
 Th H T U  ⇨  Th H T U  ⇨   Th H T U   ⇨   Th H T U   ⇨   Th H T U
  8 0 3 4     8 0 3 ¹4       8 0 ¹3 ¹4      8 ¹0 ¹3 ¹4      8 ¹0 ¹3 ¹4
 -  5 7 9    -  5 7₁9       -  5₁7₁9       - 1 5₁7₁9       - 1 5₁7₁9
 ———————     ———————            5 5           4 5 5         7 4 5 5
                   5        13−(7+1) = ↑  10−(5+1) = ↑          ↑
                            13−  8   = 5  10−  6   = 4     8 − 1 = 7
```

Subtraction 3

'Problems'.

Q	A	Q	A
1) Subtract 52 from 89.	1) $\begin{array}{r} 89 \\ -52 \\ \hline 37 \end{array}$	5) 23 − 15	5) 23 − 15 = 8
2) Take 103 from 674.	2) $\begin{array}{r} 674 \\ -103 \\ \hline 571 \end{array}$	6) How many is 25 less than 500 ?	6) Method 1 $\quad$ Method 2 $\begin{array}{r} 4\ 9 \\ \cancel{5}\ \cancel{0}\ {}^10 \\ -\ \ 2\ 5 \\ \hline 4\ 7\ 5 \end{array}$ $\begin{array}{r} 5\ {}^10\ {}^10 \\ -\ {}_12\ 5 \\ \hline 4\ 7\ 5 \end{array}$
3) Calculate 91 minus 37.	3) Method 1 $\quad$ Method 2 $\begin{array}{r} \overset{8}{\cancel{9}}\ {}^11 \\ -\ 3\ 7 \\ \hline 5\ 4 \end{array}$ $\begin{array}{r} 9\ {}^11 \\ -{}_13\ 7 \\ \hline 5\ 4 \end{array}$	7) How many is added to 3,309 to make 8,008 ?	7) Method 1 $\quad$ Method 2 $\begin{array}{r} 7\ 9\ 9 \\ \cancel{8}\ \cancel{0}\ \cancel{0}\ {}^18 \\ -3\ 3\ 0\ 9 \\ \hline 4\ 6\ 9\ 9 \end{array}$ $\begin{array}{r} 8\ {}^10\ {}^10\ {}^18 \\ -{}_13\ {}_13\ {}_10\ 9 \\ \hline 4\ 6\ 9\ 9 \end{array}$
4) Find the difference between 645 and 456.	4) Method 1 $\quad$ Method 2 $\begin{array}{r} \overset{5}{\cancel{\overset{1}{3}}} \\ \cancel{6}\ \cancel{4}\ {}^15 \\ -4\ 5\ 6 \\ \hline 1\ 8\ 9 \end{array}$ $\begin{array}{r} 6\ {}^14\ {}^15 \\ -{}_14\ {}_15\ 6 \\ \hline 1\ 8\ 9 \end{array}$	8) 7,067 − 1,798	8) Method 1 $\quad$ Method 2 $\begin{array}{r} 6\ 9\ {}^15 \\ \cancel{7}\ \cancel{0}\ \cancel{6}\ {}^17 \\ -1\ 7\ 9\ 8 \\ \hline 5\ 2\ 6\ 9 \end{array}$ $\begin{array}{r} 7\ {}^10\ {}^16\ {}^17 \\ -{}_11\ {}_17\ {}_19\ 8 \\ \hline 5\ 2\ 6\ 9 \end{array}$

The following 'problems' involve Subtraction.

Q	A
1) Team A scores 278 points and Team B scores 154 points. How many more points did Team A score than Team B?	1) $\begin{array}{r} 278 \\ -154 \\ \hline 124 \end{array}$ points
2) In a box of 85 pens, 58 pens are black. Find the number of pens which are not black.	2) Method 1 $\quad\quad$ Method 2 $\begin{array}{r} \overset{7}{\cancel{8}}\ {}^15 \\ -5\ 8 \\ \hline 2\ 7 \end{array}$ pens $\begin{array}{r} 8\ {}^15 \\ -{}_15\ 8 \\ \hline 2\ 7 \end{array}$ pens
3) A school has 436 boys and 369 girls. What is the difference between the number of girls and the number of boys?	3) Method 1 $\quad\quad$ Method 2 $\begin{array}{r} \overset{3}{\cancel{\overset{1}{2}}} \\ \cancel{4}\ \cancel{3}\ {}^16 \\ -3\ 6\ 9 \\ \hline 6\ 7 \end{array}$ boys $\begin{array}{r} 4\ {}^13\ {}^16 \\ -{}_13\ {}_16\ 9 \\ \hline 6\ 7 \end{array}$ boys
4) There were 902 visitors to a museum on one day. 283 were adults and the rest were children. How many were children?	4) Method 1 $\quad\quad$ Method 2 $\begin{array}{r} 8\ 9 \\ \cancel{9}\ \cancel{0}\ {}^12 \\ -2\ 8\ 3 \\ \hline 6\ 1\ 9 \end{array}$ children $\begin{array}{r} 9\ {}^10\ {}^12 \\ -{}_12\ {}_18\ 3 \\ \hline 6\ 1\ 9 \end{array}$ children
5) A builder needs 8,000 bricks for a wall but only has 3,309. How many more bricks are needed?	5) Method 1 $\quad\quad$ Method 2 $\begin{array}{r} 7\ 9\ 9 \\ \cancel{8}\ \cancel{0}\ \cancel{0}\ {}^10 \\ -3\ 3\ 0\ 9 \\ \hline 4\ 6\ 9\ 1 \end{array}$ bricks $\begin{array}{r} 8\ {}^10\ {}^10\ {}^10 \\ -{}_13\ {}_13\ {}_10\ 9 \\ \hline 4\ 6\ 9\ 1 \end{array}$ bricks
6) Calculate the number of boxes remaining when 4,257 are used from a stock of 6,103.	6) Method 1 $\quad\quad$ Method 2 $\begin{array}{r} 5\ {}^10\ 9 \\ \cancel{6}\ \cancel{1}\ \cancel{0}\ {}^13 \\ -4\ 2\ 5\ 7 \\ \hline 1\ 8\ 4\ 6 \end{array}$ boxes $\begin{array}{r} 6\ {}^11\ {}^10\ {}^13 \\ -{}_14\ {}_12\ {}_15\ 7 \\ \hline 1\ 8\ 4\ 6 \end{array}$ boxes

Subtraction 4   'Add Differences'. 'Problems'.

When **each number** in each column can be subtracted from a larger number, then it is usually faster to just subtract.
The following method is more useful when a number in a column is subtracted from a smaller number - but it is shown here for both possibilities. It is also likely to be used for **mental subtraction**.

Method 3
This method has no particular name but is usually referred to as '**add the differences**'.
Starting from the **lowest number**, find the **differences** (by subtracting or 'adding on') to
the **next 10** (10, 20, 30, ... 100)
the **next 100** (100, 200, 300, ... 1000)
the **next 1000** (1000, 2000, 3000, ... 10000) ...
The **last difference** should include any number '**left over**' - i.e., smaller than the highest place value.

ADD the differences to find the total difference between the given numbers.

The **differences** are usually calculated **mentally** - and only these need be written down.

Ex.1	Ex.2	Ex.3	Ex.4	Ex.5
57 − 24	80 − 15	64 − 37	753 − 189	4261 − 2675
T U	T U	T U	H T U	Th H T U
24 to **30** 6	15 to **20** 5	37 to **40** 3	189 to **190** 1	2675 to **2680** 5
30 to 57 +2 7	20 to 80 +6 0	40 to 64 +2 4	190 to **200** 1 0	2680 to **2700** 2 0
3 3	6 5	2 7	200 to 753 +5 5 3	2700 to **3000** 3 0 0
1			5 6 4	3000 to 4261 +1 2 6 1
				1 5 8 6

Q	A	Q	A

1) Subtract 52 from 89.	1) 52 to **60** 8
	60 to 89 +2 9
	   3 7
	   1
		5) 23 − 15	5) 23 − 15 = 8

2) Take 103 from 674.	2) 103 to **110** 7
	110 to **200** 9 0
	200 to 674 +4 7 4
	   5 7 1
	   1
		6) How many is 25 **less than** 500 ?	6) 25 to **30** 5
			30 to **100** 7 0
			100 to 500 +4 0 0
			   4 7 5

3) Calculate 91 minus 37.	3) 37 to **40** 3
	40 to 91 +5 1
	   5 4
		7) How many is **added to** 3,309 **to make** 8,008 ?	7) 3309 to **3310** 1
			3310 to **3400** 9 0
			3400 to **4000** 6 0 0
			4000 to 8008 +4 0 0 8
			   4 6 9 9

4) Find the **difference** between 645 and 456.	4) 456 to **460** 4
	460 to **500** 4 0
	500 to 645 +1 4 5
	   1 8 9
		8) 7,067 − 1,798	8) 1798 to **1800** 2
			1800 to **2000** 2 0 0
			2000 to 7067 +5 0 6 7
			   5 2 6 9

Q	A
1) In a box of 85 pens, 58 pens are black. Find the number of pens which are not black.	1) 58 to **60** 2 60 to 85 +2 5 2 7 pens
2) There were 902 visitors to a museum on one day. 283 were adults and the rest were children. How many were children?	2) 283 to **290** 7 290 to **300** 1 0 300 to 902 +6 0 2 6 1 9 children
3) Calculate the number of boxes remaining when 4,257 are used from a stock of 6,103.	3) 4257 to **4260** 3 4260 to **4300** 4 0 4300 to **5000** 7 0 0 5000 to 6103 +1 1 0 3 1 8 4 6 boxes

Multiplication 1

Multiplication.

Multiplication

Ex.1   6 × 11

   6 × 11 = 66

Ex.2 **Multiply** 43 **by** 2

```
 TU        TU
 43        43
× 2   ⇨  × 2
  6        86
```

*Note
Multiply **2** by the Units first, then Tens, ...

Ex.3 **Multiply** 43 **by** 20

```
 TU         TU         HTU
 43         43          43
×20   ⇨   ×20    ⇨   ×20
  0         60         860
```

*Note
When multiplying by **Tens** (10, 20, ... 90) we **know** the number ends in a **0** so write this first.

Then multiply **2** by the Units first, then Tens, ...

Ex.4 **Multiply** 43 **by** 22

```
 TU         TU         HTU
 43         43          43
×22   ⇨   ×22    ⇨   ×22
 86         86          86  ⎫
+___       +860        +860 ⎭ Add
                       946
                  'carry' 1
```

Ex. 4 combines Exs. 2 and 3 (above)

Ex.5 Find the **product of** 19 **and** 7.

```
   TU                    HTU
   19                     19
 ×  7    ⇨             ×  7
    3       1×7 +6=13    133
'carry' 6                  6
```

'product' means 'multiply'.

*Note
When multiplying the Units column, we '**carry**' any Tens to the Tens column to be **added** there.

In the same way we 'carry' from Tens to Hundreds, then from Hundreds to Thousands, ...

Ex.6 Find the **product of** 19 **and** 50.

```
   TU         TU         HTU
   19         19          19
 × 50   ⇨  × 50    ⇨   × 50
    0         50         950
        'carry' 4         4
```

Ex.7 Find the **product of** 19 **and** 57.

```
  H T U       H T U       ThH T U
    1 9         1 9           1 9
  ×  5 7  ⇨  ×  5 7   ⇨   ×  5 7
    1 3₆3       1 3₆3         1 3₆3
  + ___       + 9₄5 0       + 9₄5 0
                             1 0 8 3
                                 1
```

Ex. 7 combines Exs. 5 and 6 (above)

*Note
The 'carry' numbers can be written as shown here.

Ex.8 **Times** 784 **by** 36.

```
  ThH T U      T            T
                ThThH T U    ThThH T U
    7 8 4         7 8 4        7 8 4
  ×   3 6   ⇨  ×   3 6    ⇨ ×   3 6
    4 7₅0₂4       4 7₅0₂4      4 7₅0₂4
  + _____       +2 3₂5₁2 0   +2 3₂5₁2 0
                              2 8 2 2 4
                                    1
```

Multiplication 2

Multiplication. 'Problems'.

Q	A	Q	A
1) 12 × 3.	1) $12 \times 3 = 36$ or $\begin{array}{r} 12 \\ \times\ 3 \\ \hline 36 \end{array}$	7) Work out 35 × 26.	7) $\begin{array}{r} 35 \\ \times\ 26 \\ \hline 21_30 \\ +7_100 \\ \hline 910 \end{array}$
2) **Multiply** 21 **by** 4.	2) $21 \times 4 = 84$ or $\begin{array}{r} 21 \\ \times\ 4 \\ \hline 84 \end{array}$	8) Find the **product of** 98 **and** 47.	8) $\begin{array}{r} 98 \\ \times\ 47 \\ \hline 68_56 \\ +39_320 \\ \hline 4606 \\ 1\ 1 \end{array}$
3) Calculate 18 × 5.	3) $18 \times 5 = 90$ or $\begin{array}{r} 18 \\ \times\ 5 \\ \hline 90 \\ 4 \end{array}$	9) **Times** 503 **by** 64.	9) $\begin{array}{r} 503 \\ \times\ 64 \\ \hline 201_22 \\ +301_880 \\ \hline 32192 \end{array}$
4) Find the **product of** 32 **and** 30.	4) $32 \times 30 = 960$ or $\begin{array}{r} 32 \\ \times\ 30 \\ \hline 960 \end{array}$	10) Calculate 318 × 242.	10) $\begin{array}{r} 318 \\ \times\ 242 \\ \hline 63_6 \\ 127_220 \\ +63_8600 \\ \hline 76956 \\ 1 \end{array}$ $318 \Leftarrow \times 200$
5) **Times** 76 **by** 90.	5) $76 \times 90 = 6840$ or $\begin{array}{r} 76 \\ \times\ 90 \\ \hline 6840 \\ 5 \end{array}$		
6) **Multiply** 27 **by** 35.	6) $\begin{array}{r} 27 \\ \times\ 35 \\ \hline 13_55 \\ +8_210 \\ \hline 945 \end{array}$		

The following 'problems' involve multiplication.

Q	A
1) There are 11 teams with 7 players in each team. How many players are there altogether?	1) $11 \times 7 = 77$ players or $\begin{array}{r} 11 \\ \times\ 7 \\ \hline 77 \end{array}$ players
2) A driver travels 85 miles on each of 5 days. What is the total distance travelled.	2) $\begin{array}{r} 85 \\ \times\ 5 \\ \hline 425 \\ 2 \end{array}$ miles
3) If there are 36 tubes of sweets in each box, how many tubes are there altogether in 2 **score** boxes?	3) $\begin{array}{r} 36 \\ \times\ 40 \\ \hline 1440 \\ 2 \end{array}$ tubes (1 **score** = 20, 2 **score** = 40)
4) How many hours are there in February?	4) $\begin{array}{r} 28\ \text{days} \\ \times\ 24\ \text{hours} \\ \hline 11_32 \\ +5_160 \\ \hline 672 \end{array}$ hours (leap year) $\begin{array}{r} 29\ \text{days} \\ \times\ 24\ \text{hours} \\ \hline 11_36 \\ +5_180 \\ \hline 696 \end{array}$ hours
5) A machine can produce 125 plastic bottle tops every minute. Calculate the number of tops the machine can produce in a) 75 minutes. b) 6 hours.	5) b) 6 hours = 6 × 60 minutes = 360 minutes a) $\begin{array}{r} 125 \\ \times\ 75 \\ \hline 6_22_5 \\ +8_73_50 \\ \hline 9375 \\ 1 \end{array}$ tops $\begin{array}{r} 125 \\ \times\ 360 \\ \hline 000 \\ 7_53_00 \\ +37_500 \\ \hline 45000 \\ 1\ 1 \end{array}$ tops $\Leftarrow$ can omit

Multiplication 3

Use of Factors.

Use of Factors

Depending on the numbers being multiplied the following method may be used - it is **a little faster** and reduces the amount of work - it avoids the need to add at the end (as in the 'general method').

It relies on recognising that a number is an **'answer'** in the **'times tables'** -
it has **two factors up to 12** which multiply to give the number.
We **multiply by the factors** - in any order.
There may be a choice of factors. A prime number only has 1 and itself.

This idea may be extended to include factors which are multiples of 10 (20, 30, 40, ...).

Ex.1 Calculate 93×28 We need to recognise that $28 = 4 \times 7$
⇩
$28 = 4 \times 7$ and now multiply $93 \times 4 \times 7$

```
     9 3           or        9 3
  ×    4                  ×    7
    3 7₁2                   6 5₂1
  ×    7                  ×    4
    2 6₅0₁4                 2 6₂0 4
```

Ex.2 72×76
⇩
$72 = 8 \times 9$
or $72 = 6 \times 12$

```
      7 6       or       7 6
   ×    8             ×    6
     6 0₄8              4 5₃6
   ×    9             ×   1 2
     5 4 7₂2           5 4₆7₂2
```

Ex.3 140×13
⇩
$140 = 70 \times 2$
or $140 = 20 \times 7$

```
      1 3       or       1 3
   ×   7 0            ×   2 0
     9₂1 0              2 6 0
   ×    2             ×    7
     1 8 2 0           1 8₄2 0
```

*Note In the following Q/A just one set of calculations is shown.

Q	A	Q	A

1) 27×41	1) $27 = 3 \times 9$
	`` 4 1
	  × 3
	  1 2 3
	  × 9
	1 1₂0₂7```
		6) 36×18	6) $36 = 3 \times 12$
			$36 = 4 \times 9$
			$36 = 6 \times 6$
			$18 = 2 \times 9$
			$18 = 3 \times 6$
			`` 3 6
			  × 2
			    7₁2
			  × 9
			  6 4₁8```

2) 38×49	2) $49 = 7 \times 7$
	`` 3 8
	  × 7
	  2 6₅6
	  × 7
	1 8₄6₄2```
		7) 75×250	7) $250 = 5 \times 50$
			`` 7 5
			   × 5
			     3 7₂5
			   × 5 0
			  1 8₃7₂5 0```

3) 15×62	3) $15 = 3 \times 5$
	`` 6 2
	  × 3
	  1 8 6
	  × 5
	  9₄3 0```
		8) 210×65	8) **$210 = 3 \times 70$**
			$210 = 30 \times 7$
			`` 6 5
			    × 3
			      1 9₁5
			    × 7 0
			  1 3₆6₃5 0```

4) 93×88	4) $88 = 8 \times 11$
	`` 9 3
	  × 8
	  7 4₂4
	  × 1 1
	  8 1₄8₄4```
		9) 540×97	9) $540 = 6 \times 90$
			$540 = 60 \times 9$
			`` 9 7
			    × 6 0
			     5 8₂0
			    × 9
			   5 2₇3₁8 0```

5) 24×56	5) $24 = 2 \times 12$
	$24 = 3 \times 8$
	$24 = 4 \times 6$
	$56 = 7 \times 8$
	`` 5 6
	  × 4
	  2 2₂4
	  × 6
	1 3₁4₂4```
		10) 330×320	10) $330 = 11 \times 30$
			$320 = 4 \times 80$
			$320 = 40 \times 8$
			`` 3 3 0
			    × 4
			     1 3₁2 0
			    × 8 0
			  1 0₂5₆6 0 0```

Multiplication 4 — Use of Brackets.

Use of Brackets
The following shows how 3 examples of **'multiplying brackets'** can be used as **alternative methods** for **specific** calculations - usually when the numbers involved begin to get 'large'.

Multiplying Numbers
Write each given number as the **sum of 2 numbers** - **a multiple of 10 + the units number**.

Ex. Calculate
$24 \times 35 \Rightarrow = (20 + 4)(30 + 5)$

$= (20 + 4)(30 + 5)$ 600
$= 600 + 100 + 120 + 20$ 100
                                    120
                                 + 20
$= 840$ 840

Squaring Numbers
Write the number to be squared as the **sum of 2 numbers** - **a multiple of 10 + the units number**.

Ex. Calculate
$37^2 \Rightarrow = (30 + 7)^2$

$= (30 + 7)(30 + 7)$ 900
$= 900 + 210 + 210 + 49$ 210
                                    210
                                + 49
$= 1369$ 1369

Multiplying is carried out **mentally** - and the results can be written in a **column** instead of a row.

'Difference of 2 squares'
The term is used when 2 square numbers are subtracted.

Write the numbers as (1st. + 2nd.)(1st. − 2nd.)

It is more useful when (1st. − 2nd.) is **small**.

Ex. Calculate
$49^2 - 46^2 \Rightarrow = (49 + 46)(49 - 46)$
             $= (\ 95 \ )(\ 3 \ )$
             $= 285$

Q	A	Q	A

Calculate
1) 53×18
	1) $= (50 + 3)(10 + 8)$
	    $= 954$ 500
	                      400
	                      30
	                   + 24
	                      954
		5) 86^2	5) $= (80 + 6)^2$
			    $= (80 + 6)(80 + 6)$
			    $= 7396$ 6400
			                     480
			                     480
			                  + 36
			                     7396

2) 49×27	2) $= (40 + 9)(20 + 7)$
	    $= 1323$ 800
	                     280
	                     180
	                  + 63
	                    1323
		6) 99^2	6) $= (90 + 9)^2$
			    $= (90 + 9)(90 + 9)$
			    $= 9801$ 8100
			                     810
			                     810
			                  + 81
			                     9801

3) 65×91	3) $= (60 + 5)(90 + 1)$
	    $= 5915$ 5400
	                     60
	                     450
	                  + 5
	                     5915
		7) $38^2 - 37^2$	7) $= (38 + 37)(38 - 37)$
			       $= (\ 75 \ )(\ 1 \ )$
			       $= 75$

		8) $26^2 - 22^2$	8) $= (26 + 22)(26 - 22)$
			       $= (\ 48 \ )(\ 4 \ )$
			       $= 192$

4) 25^2	4) $= (20 + 5)^2$
	    $= (20 + 5)(20 + 5)$
	    $= 625$ 400
	                 100
	                 100
	              + 25
	                625
		9) $75^2 - 65^2$	9) $= (75 + 65)(75 - 65)$
			       $= (\ 140 \ )(\ 10 \ )$
			       $= 1400$

Multiplication 5

Factorial.

Factorial

The term '**Factorial**' is used when **consecutive whole numbers** usually **from 1** are multiplied.
The highest number being multiplied is written with an exclamation mark !.
The numbers are usually written in **descending order**.

*Note This helps explain 0! = 1 which appears 'odd' but is accepted.

Consider the 'pattern' ...

Ex. 4 **factorial** is written 4! $4! = 4 \times 3 \times 2 \times 1 = 24$ $4! = 4 \times 3! = 4 \times 3 \times 2 \times 1$
    3 **factorial** is written 3! $3! = 3 \times 2 \times 1 = 6$ $3! = 3 \times 2! = 3 \times 2 \times 1$
    2 **factorial** is written 2! $2! = 2 \times 1 = 2$ $2! = 2 \times 1! = 2 \times 1$
    1 **factorial** is written 1! $1! = 1$ $1! = 1 \times \mathbf{0!} = 1 \times \mathbf{1}$
    0 **factorial** is written 0! $0! = 1$ so $0! = 1$

Q	A
Find the value of	
1) 5!	1) $5 \times 4 \times 3 \times 2 \times 1 = 120$
2) 6!	2) $6 \times 5 \times 4 \times 3 \times 2 \times 1 = 720$
3) 7!	3) $7 \times 6 \times 5 \times 4 \times 3 \times 2 \times 1 = 5040$
4) 3! + 4!	4) $3 \times 2 \times 1 + 4 \times 3 \times 2 \times 1$ $= 6 + 24 = 30$
5) 2! + 1! + 0!	5) $2 \times 1 + 1 + 1$ $= 2 + 1 + 1 = 4$
6) 5! − 3!	6) $5 \times 4 \times 3 \times 2 \times 1 - 3 \times 2 \times 1$ $= 120 - 6 = 114$
7) 2! × 5!	7) $2 \times 1 \times 5 \times 4 \times 3 \times 2 \times 1$ $= 2 \times 120 = 240$
8) 0! × 1!	8) $1 \times 1 = 1$
9) $\dfrac{6!}{3!}$	9) $\dfrac{6 \times 5 \times 4 \times \cancel{3} \times \cancel{2} \times \cancel{1}}{\cancel{3} \times \cancel{2} \times \cancel{1}} = \dfrac{120}{1} = 120$
10) $\dfrac{7!}{5!}$	10) $\dfrac{7 \times 6 \times \cancel{5} \times \cancel{4} \times \cancel{3} \times \cancel{2} \times \cancel{1}}{\cancel{5} \times \cancel{4} \times \cancel{3} \times \cancel{2} \times \cancel{1}} = \dfrac{42}{1} = 42$
11) $\dfrac{4!}{6!}$	11) $\dfrac{\cancel{4} \times \cancel{3} \times \cancel{2} \times \cancel{1}}{6 \times 5 \times \cancel{4} \times \cancel{3} \times \cancel{2} \times \cancel{1}} = \dfrac{1}{30}$

Multiplication 6

Multiplication Tables.

<u>Multiplication Tables</u>
Time should be taken to learn the multiplication / 'times' tables -
both in 'table order' as they are written here and then in 'any order' from any table.

1 × 2 = 2	1 × 3 = 3	1 × 4 = 4	1 × 5 = 5
2 × 2 = 4	2 × 3 = 6	2 × 4 = 8	2 × 5 = 10
3 × 2 = 6	3 × 3 = 9	3 × 4 = 12	3 × 5 = 15
4 × 2 = 8	4 × 3 = 12	4 × 4 = 16	4 × 5 = 20
5 × 2 = 10	5 × 3 = 15	5 × 4 = 20	5 × 5 = 25
6 × 2 = 12	6 × 3 = 18	6 × 4 = 24	6 × 5 = 30
7 × 2 = 14	7 × 3 = 21	7 × 4 = 28	7 × 5 = 35
8 × 2 = 16	8 × 3 = 24	8 × 4 = 32	8 × 5 = 40
9 × 2 = 18	9 × 3 = 27	9 × 4 = 36	9 × 5 = 45
10 × 2 = 20	10 × 3 = 30	10 × 4 = 40	10 × 5 = 50
11 × 2 = 22	11 × 3 = 33	11 × 4 = 44	11 × 5 = 55
12 × 2 = 24	12 × 3 = 36	12 × 4 = 48	12 × 5 = 60

1 × 6 = 6	1 × 7 = 7	1 × 8 = 8	1 × 9 = 9
2 × 6 = 12	2 × 7 = 14	2 × 8 = 16	2 × 9 = 18
3 × 6 = 18	3 × 7 = 21	3 × 8 = 24	3 × 9 = 27
4 × 6 = 24	4 × 7 = 28	4 × 8 = 32	4 × 9 = 36
5 × 6 = 30	5 × 7 = 35	5 × 8 = 40	5 × 9 = 45
6 × 6 = 36	6 × 7 = 42	6 × 8 = 48	6 × 9 = 54
7 × 6 = 42	7 × 7 = 49	7 × 8 = 56	7 × 9 = 63
8 × 6 = 48	8 × 7 = 56	8 × 8 = 64	8 × 9 = 72
9 × 6 = 54	9 × 7 = 63	9 × 8 = 72	9 × 9 = 81
10 × 6 = 60	10 × 7 = 70	10 × 8 = 80	10 × 9 = 90
11 × 6 = 66	11 × 7 = 77	11 × 8 = 88	11 × 9 = 99
12 × 6 = 72	12 × 7 = 84	12 × 8 = 96	12 × 9 = 108

1 × 10 = 10	1 × 11 = 11	1 × 12 = 12
2 × 10 = 20	2 × 11 = 22	2 × 12 = 24
3 × 10 = 30	3 × 11 = 33	3 × 12 = 36
4 × 10 = 40	4 × 11 = 44	4 × 12 = 48
5 × 10 = 50	5 × 11 = 55	5 × 12 = 60
6 × 10 = 60	6 × 11 = 66	6 × 12 = 72
7 × 10 = 70	7 × 11 = 77	7 × 12 = 84
8 × 10 = 80	8 × 11 = 88	8 × 12 = 96
9 × 10 = 90	9 × 11 = 99	9 × 12 = 108
10 × 10 = 100	10 × 11 = 110	10 × 12 = 120
11 × 10 = 110	11 × 11 = 121	11 × 12 = 132
12 × 10 = 120	12 × 11 = 132	12 × 12 = 144

<u>Multiplication Tables</u>

Division 1

Division

If a number does not divide into a digit of a number exactly, then we '**carry**' the '**remainder**' to the next digit.

After the last digit, any 'remainder' can be written

a) as a 'remainder' - this is shown here.

b) as a fraction - this is shown in **Fractions 2**.

c) as a decimal - this is shown in **Decimals 4-5**.

Ex.1 $6 \div 2$

$6 \div 2 = 3$

Ex.2 **Divide** 9 **by** 5.

$9 \div 5 = 1$ remainder 4

$= 1$ rem. 4

$= 1$ r 4

The answer is usually written in 1 of these 3 ways.

Ex.3 **Share** 86 **by** 2

This is usually set out

$$\begin{array}{r} \text{TU} \\ \mathbf{4} \\ 2\overline{)\mathbf{86}} \end{array} \Rightarrow \begin{array}{r} \text{TU} \\ \mathbf{43} \\ 2\overline{)\mathbf{86}} \end{array}$$

or

$$\frac{\mathbf{86}}{\mathbf{2}} = \mathbf{4} \Rightarrow \frac{\mathbf{86}}{\mathbf{2}} = 43$$

Ex.4 **How many times does** 3 **go into** 462 ?

$$\begin{array}{r} \text{HTU} \\ \mathbf{1} \\ 3\overline{)\mathbf{4}62} \end{array} \Rightarrow \begin{array}{r} \text{HTU} \\ 1\mathbf{5} \\ 3\overline{)4\,{}^{1}\mathbf{6}\,2} \end{array} \Rightarrow \begin{array}{r} \text{HTU} \\ 15\mathbf{4} \\ 3\overline{)4\,{}^{1}6\,{}^{1}\mathbf{2}} \end{array}$$

'**carry**' ↑ remainder   '**carry**' ↑

or

'**carry**' ↓ remainder   '**carry**' ↓

$$\frac{\mathbf{4}\,6\,2}{\mathbf{3}} = \mathbf{1} \Rightarrow \frac{4\,{}^{1}\mathbf{6}\,2}{\mathbf{3}} = 1\mathbf{5} \Rightarrow \frac{4\,{}^{1}6\,{}^{1}\mathbf{2}}{\mathbf{3}} = 154$$

Ex.5 **How many** 4's **are there in** 78 ?

$$\begin{array}{r} \text{TU} \\ \mathbf{1} \\ 4\overline{)\mathbf{7}8} \end{array} \Rightarrow \begin{array}{r} \text{TU} \\ 1\,\mathbf{9}\text{ rem. }\mathbf{2} \\ 4\overline{)7\,{}^{3}\mathbf{8}} \end{array}$$

'**carry**' ↑

or

'**carry**' ↓

$$\frac{\mathbf{7}\,8}{\mathbf{4}} = \mathbf{1} \Rightarrow \frac{7\,{}^{3}\mathbf{8}}{\mathbf{4}} = 1\mathbf{9}\text{ rem. }\mathbf{2}$$

*Note

When we divide into the Hundreds column, we '**carry**' any Hundreds 'left over' to the Tens column to be divided there as Tens.

In the same way we '**carry**' from Tens to Units.

Q	A
1) $48 \div 8$	1) $48 \div 8 = 6$
2) $19 \div 7$	2) $19 \div 7 = 2$ remainder 5
3) **Divide** 54 **by** 10	3) $54 \div 10 = 5$ rem. 4
4) **Share** 96 **by** 3	4) $\begin{array}{r} 32 \\ 3\overline{)96} \end{array}$ or $\frac{96}{3} = 32$
5) **How many times does** 2 **go into** 246 ?	5) $\begin{array}{r} 123 \\ 2\overline{)246} \end{array}$ or $\frac{246}{2} = 123$
6) **How many** 4's **are there in** 4,084 ?	6) $\begin{array}{r} 2021 \\ 4\overline{)8084} \end{array}$ or $\frac{8084}{4} = 2021$
7) Calculate 90 **divided by** 6.	7) $\begin{array}{r} 15 \\ 6\overline{)9\,{}^{3}0} \end{array}$ or $\frac{9\,{}^{3}0}{6} = 15$

Q	A
8) Work out $605 \div 5$	8) $\begin{array}{r} 121 \\ 5\overline{)6\,{}^{1}0\,5} \end{array}$ or $\frac{6\,{}^{1}0\,5}{5} = 121$
9) **Divide** 296 **by** 8	9) $\begin{array}{r} \mathbf{0}\,3\,7 \\ 8\overline{)2\,9\,{}^{5}6} \end{array}$ or $\frac{2\,9\,{}^{5}6}{8} = \mathbf{0}37$ *Note 8 does not divide into the first digit 2 so we divide into 29 - but do not usually write the **0**.
10) **Share** 89 **by** 6	10) $\begin{array}{r} 1\,4\text{ r }5 \\ 6\overline{)8\,{}^{2}9} \end{array}$ or $\frac{8\,{}^{2}9}{6} = 14\text{ r }5$
11) $568 \div 9$	11) $\begin{array}{r} 6\,3\text{ r }1 \\ 9\overline{)5\,6\,{}^{2}8} \end{array}$ or $\frac{5\,6\,{}^{2}8}{9} = 63\text{ r }1$
12) **How many** 11's **are there in** 3,244 ?	12) $\begin{array}{r} 2\;\;9\;\;4\text{ rem. }10 \\ 11\overline{)3\;\;2\,{}^{10}4\,{}^{5}4} \end{array}$ or $\frac{3\;\;2\,{}^{10}4\,{}^{5}4}{11} = 294\text{ rem. }10$

Division 2

'Problems'.

The following 'problems' involve division.

Q	A
1) 32 pupils are divided into 8 equal teams. How many pupils are in each team?	1) $32 \div 8 = 4$ pupils.
2) Crayons are sold in packets of 5. Find the number of packets required for 27 crayons.	2) $27 \div 5 = 5$ packets (r 2 crayons)
3) A machine needs 4 operators. If 84 operators are working, how many machines are being used?	3) $4\overline{)84}^{\,21}$ machines or $\dfrac{84}{4} = 21$ machines
4) Calculate the number of tables that can be made with 936 legs if each table has 3 legs.	4) $3\overline{)936}^{\,312}$ tables or $\dfrac{936}{3} = 312$ tables
5) £2,048 is shared equally between 2 friends. How much does each receive?	5) $2\overline{)2048}^{\,£1024}$ or $\dfrac{2048}{2} = £1024$
6) A child takes 8 cubes from a pile of 848 to make a cuboid. How many times can 8 cubes be taken from the pile?	6) $8\overline{)848}^{\,106}$ times or $\dfrac{848}{8} = 106$ times
7) How many weeks are there in 91 days?	7) $7\overline{)9\,^{2}1}^{\,1\ 3}$ weeks or $\dfrac{9^{2}1}{7} = 13$ weeks
8) If 840 eggs are packed in boxes of 6 eggs, how many boxes are there?	8) $6\overline{)8\,^{2}4\,0}^{\,1\ 4\ 0}$ boxes or $\dfrac{8^{2}4\,0}{6} = 140$ boxes
9) State the number of books that a shop must sell at £9 to receive £522.	9) $9\overline{)5\,2\,^{7}2}^{\,5\ 8}$ books $\dfrac{5\,2^{7}2}{9} = 58$ books
10) Find the minimum number of cars needed to transport 98 people if each car can carry 5 people.	10) $5\overline{)9\,^{4}8}^{\,19}$ rem. 3 or $\dfrac{9^{4}8}{5} = 19$ rem. 3 **20** cars (1 car is needed for **3** people)
11) Calculate the remainder when 306 million is divided by 4 million.	11) $4\overline{)3\,0\,^{2}6}^{\,7\ 6\ \mathbf{r\ 2}}$ or $\dfrac{3\,0^{2}6}{4} = 76\ \mathbf{r\ 2}$ Remainder = **2 million**.
12) How many dozens are there in 4,715?	12) 1 dozen = 12 $12\overline{)4\ 7^{11}1\ ^{3}4}^{\quad 3\ 9\ 2\ \text{dozens r 10}}$ or $\dfrac{4\ 7^{11}1\ ^{3}4}{12} = 392$ dozens r 10
13) A 5600m roll of wire is cut into pieces, each 11m long. How many such pieces are cut?	13) $11\overline{)5\ 6^{1}0\ 0}^{\quad 5\ 0\ 9\ \text{pieces (rem. 1m)}}$ or $\dfrac{5\ 6^{1}0\ 0}{11} = 509$ pieces (rem. 1m)
14) 1,000 £1 coins are shared equally between 12 people. How many £1 coins does each person receive?	14) $12\overline{)1\ 0\ 0\ ^{4}0}^{\qquad 8\ 3\ £1\ \text{coins (r 4 £1 coins)}}$ or $\dfrac{1\ 0\ 0^{4}0}{12} = 83$ £1 coins (r 4 £1 coins)

Division 3   'Long Division'.

Long Division
'Long Division' generally refers to division by a number greater than 12 (not in the 'times tables'), since it generally takes longer to calculate the answer and there is more work involved.
Dividing by a multiple of 10, for ex., 20, 30 ,40, ... is not quite so difficult.

We can set out the work as already shown in **Division 1-2** - **but** it needs **very good mental arithmetic** to do so without writing down any 'working out' apart from a remainder.

The additional 2 Methods shown here are alternatives.

Method 1
We keep **adding** the number we are dividing by until the total is **equal to** or **less than** the number we are dividing into - then we **count** how many times we have **added** the number.
We add 9 numbers at the most - this is like writing out the **table** for the number we are dividing by.

*Note
Ex.1 is 'easy' but helps explain the method.

Ex.1   26 ÷ 13

                                        2   2
13|26   ⇨   13|26   13 ✓   **Tick**✓ and **count**
                  −26   +13 ✓   how many times
                    0   26   13 is added ... **2**.

13 divides into 26 **2 times** with **0 remainder**.

Ex.2   55 ÷ 14

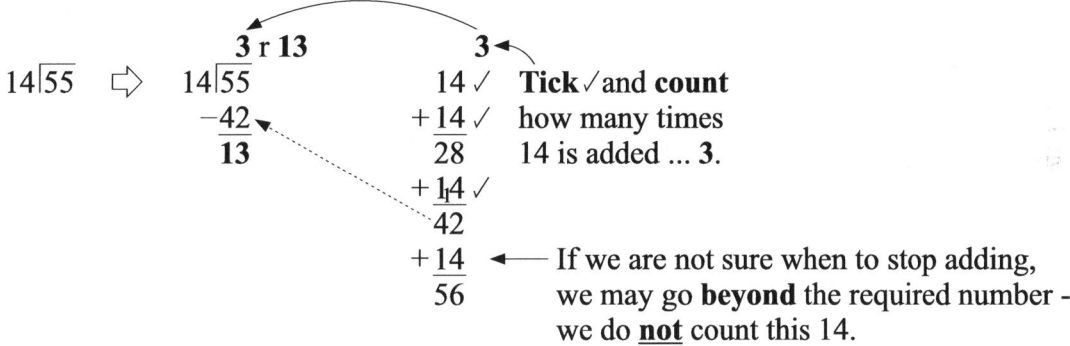

14 divides into 55 **3 times** with **13 remainder**.

Ex.3   789 ÷ 15

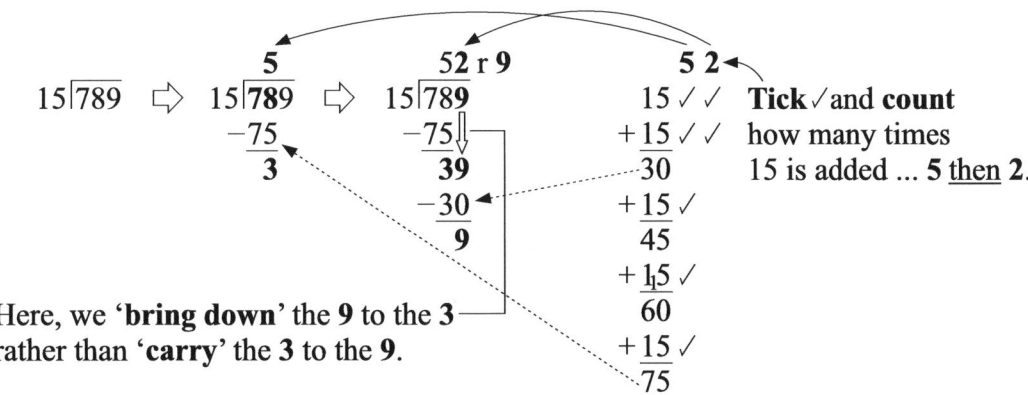

Here, we '**bring down**' the **9** to the **3**
rather than '**carry**' the **3** to the **9**.

15 divides into 789 **52 times** with **9 remainder**.

Division 4   'Long Division'.

Long Division

Method 2
This method relies on **estimation**.

We **multiply** an **estimate** - and if it is not the exact answer, decide how near it is to the answer.
It may be near enough to be accepted - so we calculate the 'remainder' and continue, if necessary.

If it is not near enough, or we are not sure, we try **another estimate**.

*Note
In the following **Examples**, only '**close estimates**' are shown in order to help explain.

In the following Q/A, only the '**correct estimate**' is shown!

*Note
Ex.1 is 'easy' but helps explain the method.

Ex.1   26 ÷ 13   **estimate**
                         (1)   (2)

$$13\overline{)26} \Rightarrow \begin{array}{r}2\\13\overline{)26}\\-26\\\hline 0\end{array} \quad \begin{array}{r}13\\\times\,1\\\hline 13\end{array} \quad \begin{array}{r}13\\\times\,2\\\hline 26\end{array}$$

13 divides into 26 **2 times** with **0 remainder**.

Ex.2   55 ÷ 14   **estimate**
                       (1)   (2)   (3)

$$14\overline{)55} \Rightarrow \begin{array}{r}3\text{ r }13\\14\overline{)55}\\-42\\\hline 13\end{array} \quad \begin{array}{r}14\\\times\,2\\\hline 28\end{array} \quad \begin{array}{r}14\\\times\,3\\\hline 42\end{array} \quad \begin{array}{r}14\\\times\,4\\\hline 56\end{array}$$

14 divides into 55 **3 times** with **13 remainder**.

Ex.3   789 ÷ 15   **estimate**   **estimate**
                       (1)   (2)   (3)

$$15\overline{)789} \Rightarrow \begin{array}{r}5\\15\overline{)789}\\-75\\\hline 3\end{array} \quad \begin{array}{r}15\\\times\,4\\\hline 60\end{array} \quad \begin{array}{r}15\\\times\,5\\\hline 75\end{array} \Rightarrow \begin{array}{r}52\text{ r }9\\15\overline{)789}\\-75\downarrow\\\hline 39\\-30\\\hline 9\end{array} \quad \begin{array}{r}15\\\times\,2\\\hline 30\end{array}$$

Here, we '**bring down**' the **9** to the **3**
rather than '**carry**' the **3** to the **9**.

15 divides into 789 **52 times** with **9 remainder**.

Division 5

'Long Division'.

Q	A	Q	A
1) Calculate 52 ÷ 13	1) $\quad\quad$ **4** 13|52 $\underline{-52}$ $\quad\quad$ **0** Method 1 13 ✓ $\underline{+13}$ ✓ 26 $\underline{+13}$ ✓ 39 $\underline{+1_13}$ ✓ 52 Method 2 estimate 13 $\underline{\times\ \textbf{4}}$ $\underline{52}$ $\quad_1$	6) Calculate 3496 ÷ 46	6) $\quad\quad$ **76** 46|3496 $\underline{-322}\Downarrow$ $\quad$ 276 $\underline{-276}$ $\quad\quad$ **0** Method 1 $\quad\quad$ **7 6** 46 ✓✓ $\underline{+\ 4_16}$ ✓✓ $\quad$ 9 2 $\underline{+_1 4 6}$ ✓✓ 1 3 8 $\underline{+\ 4_16}$ ✓✓ 1 8 4 $\underline{+_1 4_16}$ ✓✓ 2 3 0 $\underline{+\ 4 6}$ ✓✓ 2 7 6 $\underline{+_1 4 6}$ ✓ 3 2 2 Method 2 estimate 46 $\quad$ 46 $\underline{\times\ \textbf{7}}\quad\underline{\times\ \textbf{6}}$ $\underline{322}\quad\underline{276}$ $\quad_4\quad\quad_3$
2) Calculate 49 ÷ 15	2) $\quad\ $ **3 r 4** 15|49 $\underline{-45}$ $\quad\quad$ **4** Method 1 $\ $ **3** 15 ✓ $\underline{+1_15}$ ✓ 30 $\underline{+15}$ ✓ 45 Method 2 estimate 15 $\underline{\times\ \textbf{3}}$ $\underline{45}$ $\ _1$		
3) Calculate 98 ÷ 24	3) $\quad\ $ **4 r 2** 24|98 $\underline{-96}$ $\quad\quad$ **2** Method 1 $\ $ **4** 24 ✓ $\underline{+24}$ ✓ 48 $\underline{+24}$ ✓ 72 $\underline{+24}$ ✓ 96 Method 2 estimate 24 $\underline{\times\ \textbf{4}}$ $\underline{96}$ $\ _1$	7) Calculate 8775 ÷ 65	7) $\quad\ $ **135** 65|8775 $\underline{-65}\Downarrow$ $\ $ 227 $\underline{-195}\Downarrow$ $\ $ 325 $\underline{-325}$ $\quad\quad$ **0** Method 1 $\ $ **1 3 5** 65 ✓✓✓ $\underline{+\ 6_15}$ ✓✓ 1 3 0 $\underline{+\ 6 5}$ ✓✓ 1 9 5 $\underline{+_1 6_15}$ ✓ 2 6 0 $\underline{+_1 6 5}$ ✓ 3 2 5 Method 2 estimate 65 $\ $ 65 $\ $ 65 $\underline{\times\textbf{1}}\underline{\times\ \textbf{3}}\underline{\times\ \textbf{5}}$ $\underline{65}\ \underline{195}\ \underline{325}$ $\quad\ _1\quad\ _2$ $\quad_1$ $2\,^12\,7\quad\quad 2\,^12\,7$ $\underline{-1\ 9\ 5}\quad\underline{-1_19\ 5}$ $\quad 3\,2\quad\quad\ \ 3\,2$ or
4) Calculate 884 ÷ 17	4) $\quad$ **52** 17|884 $\underline{-85}\Downarrow$ $\ $ 34 $\underline{-34}$ $\quad\quad$ **0** Method 1 $\ $ **5 2** 17 ✓✓ $\underline{+17}$ ✓✓ 34 $\underline{+17}$ ✓ 51 $\underline{+17}$ ✓ 68 $\underline{+17}$ ✓ 85 Method 2 estimate 17 $\ $ 17 $\underline{\times\ \textbf{5}}\underline{\times\ \textbf{2}}$ $\underline{85}\quad\underline{34}$ $\ _3\quad\ _1$	8) Calculate 1,500 ÷ 51	8) $\quad$ **29 r 21** 51|1500 $\underline{-102}\Downarrow$ a) $\ $ 480 $\underline{-459}$ $\quad\quad$ **21** b) Method 1 $\quad$ **2 9** 51 ✓✓ $\underline{+\ 51}$ ✓✓ 1 0 2 $\underline{+\ 5 1}$ ✓ 1 5 3 $\underline{+_1 5 1}$ ✓ 2 0 4 $\underline{+\ 5_11}$ ✓ 2 5 5 $\underline{+_1 5 1}$ ✓ 3 0 6 $\underline{+\ 5 1}$ ✓ 3 5 7 $\underline{+_1 5 1}$ ✓ 4 0 8 $\underline{+\ 5 1}$ ✓ 4 5 9 Method 2 estimate 51 $\ $ 51 $\underline{\times\ \textbf{2}}\underline{\times\ \textbf{9}}$ $\underline{102}\ \underline{459}$
5) Calculate 759 ÷ 33	5) $\quad$ **23** 33|759 $\underline{-66}\Downarrow$ $\ $ 99 $\underline{-99}$ $\quad\quad$ **0** $\ \ ^6\!7\,^15$ $\ $ or $\ $ $7\,^15$ $\underline{-6\ 6}\quad\quad\underline{-6_1 6}$ $\quad\ 9\quad\quad\quad\ 9$ *Note Here, the subtraction is shown separately for clarity - but it does not have to be. Method 1 $\ $ **2 3** 33 ✓✓ $\underline{+33}$ ✓✓ 66 $\underline{+33}$ ✓ 99 Method 2 estimate 33 $\ $ 33 $\underline{\times\ \textbf{2}}\underline{\times\ \textbf{3}}$ $\underline{66}\ \underline{99}$		a) $\ \ \ ^4\!1\,\cancel{5}\,^10$ $\ $ or $\ $ $1\,5\,^10$ $\underline{-1\ 0\ 2}\quad\underline{-1\ 0_12}$ $\quad\ 4\,8\quad\quad\quad 4\,8$ b) $\ \ \ ^7\!4\,\cancel{8}\,^10$ $\ $ or $\ $ $4\,8\,^10$ $\underline{-4\ 5\ 9}\quad\underline{-4\ 5_19}$ $\quad\ 2\,1\quad\quad\quad 2\,1$

Division 6

'Long division' - 'Problems'.

The following 'problems' involve 'long division'.

Q	A
1) Calculate the number of teams that can be formed with 90 pupils if each team needs to have 18 pupils.	1) Method 1 Method 2 5 5 estimate 18|90 18 ✓ 18 −90 + 1,8 ✓ × 5 0 36 90 + 1,8 ✓ 4 54 + 1,8 ✓ 72 + 1,8 ✓ 90 Answer = **5 teams**
2) A box holds 36 books. How many boxes are needed to hold 828 books?	2) Method 1 Method 2 **23** 2 3 estimate 36|828 36 ✓✓ 36 36 −72⇓ + 3,6 ✓✓ × 2 × 3 108 7 2 72 108 −108 + 3 6 ✓ 1 1 **0** 1 0 8 Answer = **23 boxes**
3) An aircraft, which holds up to 82 passengers, carried 3680 passengers in 1 month. Calculate the minimum number of flights that the aircraft made in the month.	3) Method 1 Method 2 **44 r 72** 4 4 estimate 82|3680 82 ✓✓ 82 −328⇓ + 82 ✓✓ × 4 400 1 6 4 328 −328 + 1 82 ✓✓ **72** 2 4 6 + 1 82 ✓✓ 3 2 8 Answer = **45 flights** 3 9 or 4⁄0¹0 4¹0¹0 3 2 8 −3₁2₁8 7 2 7 2
4) A garage carwash uses 77 litres of water per car. How many cars are washed if 69,377 litres of water are used in a given month?	4) Note Method 1 Method 2 **901** 9 1 estimate 77|69377 77 ✓✓ 77 77 −693⇓⇓ + 7,7 ✓ × 9 × 1 0 77 1 5 4 693 77 −77 + 1 7,7 ✓ 6 **0** 2 3 1 + 1 7 7 ✓ 3 0 8 + 7,7 ✓ 3 8 5 + 1 7,7 ✓ 4 6 2 + 1 7 7 ✓ 5 3 9 + 1 7,7 ✓ 6 1 6 + 7,7 ✓ 6 9 3 Answer = **901 cars**

Decimals 1   Column Headings / Values.

*Note
To help explain, a decimal is first considered as a positive number - it may be a negative number.

Decimals
A **decimal** or **decimal number** involves a number between 0 and 1 -
this is **part of one unit** and can also be called a **decimal fraction**.

A decimal can have **any number of digits / figures - the decimal fraction part** it is written to the **right** of any whole number and separated by the **decimal point**.

   Exs.   A)   0·**1**   B)   5·**7**   C)   23·**849**   D)   15,476·**04**   E)   0·**555**...   F)   3·**142**...

The **value** of a digit depends on its **position** in the decimal fraction.
Each position has a **column heading** so we can read / write a number - these are not usually written.

For short, the headings (when written) are usually written as **fractions** and are shown here from **Tenths** down to **Millionths** - **but** there is no limit as to how **small** a decimal can be.

Each column heading is also referred to in terms of a '**decimal place**' -
ex. **1st. decimal place**. (This is the position of the 1st. digit of the decimal on the right of the point.)

whole number				decimal					
						decimal place			
				1st.	2nd.	3rd.	4th.	5th.	6th.
			decimal point				Ten	Hundred	
H	T	U		Tenths	Hundredths	Thousandths	Thousandths	Thousandths	Millionths
100	10	1	·	$\frac{1}{10}$	$\frac{1}{100}$	$\frac{1}{1000}$	$\frac{1}{10000}$	$\frac{1}{100000}$	$\frac{1}{1000000}$

Exs.
		0	·	1					
		2	·	3	1				
		4	·	7	5	9			
	2	9	·	0	8	2	6		
4	1	5	·	6	9	3	0	3	
		3	·	2	2	4	5	1	9

As with whole numbers - each column heading is **one tenth** of the value of the heading on its **left**
          or   - each column heading is **ten times** the value of the heading on its **right**.

We **read the decimal fraction part** by stating **each digit** starting from the **left**.

   Exs.   A)   0·**1**   is read   'nought point **one**' or 'zero point **one**'
          B)   25·**25**   is read   'twenty five point **two five**'
          C)   14·**678** is read   'fourteen point **six seven eight**'

*Note It is usual and clearer to write, for example, 0·1 rather than ·1 when there is no whole number.

Unlike whole numbers, any **0's on the <u>RIGHT</u>** of a decimal / point do <u>not</u> change its value.

   Exs.   A)   0·3   =   0·30   =   0·300   =   0·3000   = ...
          B)   0·03   =   0·030   =   0·0300   =   0·03000   = ...
          C)   9·87   =   9·870   =   9·8700   =   9·87000   = ...
          D)   62·405   =   62·4050   =   62·40500   =   62·405000   = ...
          E)   1   =   1·0   =   1·00   =   1·000   = ...

*Note
When using a **calculator**, if we input **0's** on the **right** of a decimal, they will be **ignored** and not show at the end of a calculation. They will **disappear** if the = sign or a calculation sign is pressed.

   Exs.   0.30   = 0.3   The decimal point is usually at the base of the digits on a calculator / computer.
          9.8700 = 9.87
          4.0   = 4

Decimals 2 — Change Decimal to Fraction.

Change Decimal to Fraction

We should recognise that the value of a decimal is really understood in terms of fractions.

Each decimal is the sum of ... **Tenths** + **Hundredths** + **Thousandths** + ...

Ex. $0.12 = 0 \cdot 1\ 2$
$= 0 + \dfrac{1}{10} + \dfrac{2}{100}$

and simplified $= \dfrac{10}{100} + \dfrac{2}{100} = \dfrac{12}{100}$

Ex. $79.345 = 79 \cdot 3\ 4\ 5$
$= 79 + \dfrac{3}{10} + \dfrac{4}{100} + \dfrac{5}{1000}$

and simplified $= 79 + \dfrac{300}{1000} + \dfrac{40}{1000} + \dfrac{5}{1000} = 79\dfrac{345}{1000}$

From the above Exs., it can be seen that to change a decimal to a fraction -
the decimal part of the number is written over **the place value of the last digit**.
It can also be seen that the **denominator** is **1 followed by a 0 for each digit** in the decimal part.

Ex. Change each decimal to a fraction.

a) 9·4	b) 0·58	c) 2·017	d) 0·3333	e) 1·00894	f) 7·246172
a) $9\dfrac{4}{10}$	b) $\dfrac{58}{100}$	c) $2\dfrac{017}{1000}$ $= 2\dfrac{17}{1000}$	d) $\dfrac{3333}{10000}$	e) $1\dfrac{00894}{100000}$ $= 1\dfrac{894}{100000}$	f) $7\dfrac{246172}{1000000}$

Any 0's at the front of the decimal part of a number are omitted from the fraction - as in c), e).

Q	A	Q	A
Change each decimal to a fraction. 1) 0·7	1) $\dfrac{7}{10}$	4) 8·0029	4) $8\dfrac{29}{10000}$
2) 5·04	2) $5\dfrac{4}{100}$	5) 0·12345	5) $\dfrac{12345}{100000}$
3) 0·618	3) $\dfrac{618}{1000}$	6) 9·806043	6) $9\dfrac{806043}{1000000}$

Change Decimal to Fraction in its Lowest Terms / Simplest Form (see also **Fractions 5-6**)

('Divide the numerator and denominator by the highest common factor to obtain an equivalent fraction'.)

Ex. Change the decimal to a fraction 'in its lowest terms'.

a) 0·4   b) 5·36   c) 7·05   d) 0·175

a) $\dfrac{4}{10} \overset{\div 2}{\underset{\div 2}{=}} \dfrac{2}{5}$ b) $5\dfrac{36}{100} \overset{\div 4}{\underset{\div 4}{=}} 5\dfrac{9}{25}$ c) $7\dfrac{5}{100} \overset{\div 5}{\underset{\div 5}{=}} 7\dfrac{1}{20}$ d) $\dfrac{175}{1000} \overset{\div 25}{\underset{\div 25}{=}} \dfrac{7}{40}$

Q	A	Q	A
Change each decimal to a fraction in its lowest terms. 1) 0·5	1) $\dfrac{5}{10} = \dfrac{1}{2}$ (÷5)	3) 1·08	3) $1\dfrac{8}{100} = 1\dfrac{2}{25}$ (÷4)
2) 8·25	2) $8\dfrac{25}{100} = 8\dfrac{1}{4}$ (÷25)	4) 0·222	4) $\dfrac{222}{1000} = \dfrac{111}{500}$ (÷2)

Decimals 3

Addition.

Operations
We can add, subtract, multiply and divide decimals.
We will **accept** that the **operations are carried out in the same manner as whole numbers** - since the value of each column heading is ten times the value of the heading on its right, as with whole numbers.
(A good understanding of fractions is needed to show this - and it would also need a lot of work!)

Generally, we do not write the column headings.

Addition
We make sure that the **decimal points are under each other** and the digits are in the correct columns.

Ex.1 0·3 + 1·4	Ex.2 5·86 + 0·09	Ex.3 27·45 + 6·581	Ex.4 3 + 0·7 + 2·8 + 1·9
0·3 +1·4 ── 1·7	5·86 +0·09 ──── 5·95 'carry' 1	27·450 ← + 6·581 ───── 34·031 'carry' 11 1 We can write 0 here	3·0 ← 0·7 2·8 +1·9 ──── 8·4 'carry' 2 Always write ·0 here

Q	A	Q	A

1) **Add** 2·5 and 0·1	1) 2·5
	   +0·1
	   ───
	   2·6
		5) Find the **total** of 123·4 and 55·007	5) 123·400
			   + 55·007
			   ──────
			   178·407

2) Calculate 3·43 **+** 6·39	2) 3·43
	   +6·39
	   ────
	   9·82
	      1
		6) **Add** the following:
		    2 + 0·2 + 0·02 + 0·002
			6) 2·0
			   0·2
			   0·02
			   +0·002
			   ─────
			   2·222

3) Find the **sum** of 7·654 and 80·97	3) 7·654
	   +80·970
	   ──────
	   88·624
	      1 1
		7) 1·0101 **+** 800 **+** 88	7) 1·0101
			   800·0
			   + 88·0
			   ──────
			   889·0101

4) Work out
    4·2 **plus** 5 **plus** 0·8
	4) 4·2
	   5·0
	   + 0·8
	   ────
	   10·0 = 10
	      1		

The following 'problems' involve addition.

Q	A

1) The distance from A to B is 3·9 km and from B to C is 0·4 km. What is the total distance from A to C ?	1) 3·5
	   +0·4
	   ────
	   3·9 km

2) Three parcels weigh 8 kg, 5·79 kg and 2·462 kg. How much do the parcels weigh altogether ?	2) 8·0
	   5·79
	   + 2·462
	   ──────
	   16·252 kg
	      1 1

3) A tank has 141·8 litres of liquid in it. 33·95 litres are added and then a further 20·65 litres. How many litres of liquid are now in the tank ?	3) 141·8
	   33·95
	   + 20·65
	   ──────
	   196·40 = 196·4 litres
	      2 1

Decimals 4

Subtraction.

Subtraction

We make sure that the **decimal points are under each other** and the digits are in the correct columns.

Ex.1 7·4 − 2·3

```
  7·4
− 2·3
─────
  5·1
```

Ex.2 6·859 − 5·01

```
  6·859
− 5·01
──────
  1·849
```

Ex.3 75·72 − 3·49

Method 1
```
     6
  7 5·7¹2
−    5·4 9
──────────
  7 0·2 3
```

Method 2
```
  7 5·7¹2
−    5·4₁9
──────────
  7 0·2 3
```

Ex.4 8·6 − 2·13

Method 1
```
     5
  8·6¹0 ←
− 2·1 3
────────
  6·4 7
```

Method 2
```
  8·6¹0 ←
− 2·1₁3
────────
  6·4 7
```

<u>Always</u> write **0** here

Q	A	Q	A

1) **Subtract** 4·4 **from** 9·6	1) 9·6
	   − 4·4
	   ─────
	     5·2
		5) How much is 0·57 **less than** 5·7 ?	5) Method 1
			     6
			 5·7¹0
			−0·5 7
			─────
			 5·1 3
			
			Method 2
			 5·7¹0
			−0·5₁7
			─────
			 5·1 3

2) **Take** 1·02 **from** 7·385	2) 7·385
	   − 1·02
	   ──────
	     6·365		

		6) How much is **added to** 0·9 **to make** 6 ?	6) Method 1 5 6·¹0 −0·9 ───── 5·1 Method 2 6·¹0 −0₁·9 ───── 5·1
3) Calculate 69·53 **minus** 2·17	3) Method 1 4 6 9·5̸¹3 − 2·1 7 ──────── 6 7·3 6 Method 2 6 9·5¹3 − 2·1₁7 ──────── 6 7·3 6		
		7) Calculate 3 − 1·23	7) Method 1 2 9 3̸·0̸¹0 ← −1·2 3 ──────── 1·7 7 Method 2 3·¹0 ¹0 ← −1₁·2₁3 ──────── 1·7 7 write **0 0** here
4) Find the **difference** between 8·46 and 0·89	4) Method 1 7 ¹3 8̸·4̸¹6 −0·8 9 ─────── 7·5 7 Method 2 8·⁴4¹6 −0₁·8₁9 ─────── 7·5 7		

The following 'problems' involve subtraction.

Q	A

1) A car has travelled 5·2 miles of an 8·7 mile journey. How far has the car still to travel?	1) 8·7
	   − 5·2
	   ─────
	     3·5 miles

2) The cost of a watch was £46·95. It is reduced by £4·69 in a sale. What is the new price of the watch?	2) Method 1 8 £ 4 6·9̸¹5 − 4·6 9 ────────── £ 4 2·2 6 Method 2 £ 4 6·9¹5 − 4·6₁9 ────────── £ 4 2·2 6
3) Parcel A weighs 9·17 kg and parcel B weighs 3·015 kg. How much heavier is parcel A than parcel B?	3) Method 1 6 9·1 7̸¹0 −3·0 1 5 ───────── 6·1 5 5 kg Method 2 9·1 7¹0 −3·0 1₁5 ───────── 6·1 5 5 kg
4) Calculate the rise in temperature from 10·8 °F to 25 °F.	4) Method 1 4 2 5̸·¹0 −1 0·8 ──────── 1 4·2 °F Method 2 2 5·¹0 −1 0₁·8 ──────── 1 4·2 °F

Decimals 5

Subtraction.

Subtraction

*Note Method 3 (below) is explained up to 3 decimal places - it can be extended to more.
It is explained in a) **Subtraction 4** for whole numbers only.
b) **Money 3** which is similar to decimals, up to 2 decimal places.

When **each number** in each column can be subtracted from a larger number, then it is usually faster to just subtract.
The following method is more useful when a number in a column is subtraction from a smaller number - but it is shown here for both possibilities.

Method 3

Starting from the **lowest number**, find the **differences** (by subtracting or 'adding on') to
the **next 0·010** (0·010, 0·020, 0·030, ... 0·100)
the **next 0·100** (0·100, 0·200, 0·300, ... 1·000)   } it is useful to write the 'end 0's '
the **next 1·000** (1·000, 2·000, 3·000, ... 10·000) ...
The **last difference** should include any number **'left over'** - i.e., smaller than the highest place value.

ADD the differences to find the total difference between the given numbers.

The **differences** are usually calculated **mentally** - and only these need be written down.

Ex.1	Ex.2	Ex.3
7·4 − 2·3	8·6 − 2·13	9·045 − 5·487
2·3 to **3·0** 0·7	2·13 to **2·20** 0·07	5·487 to **5·490** 0·003
3·0 to 7·4 +4·4	2·20 to **3·00** 0·80	5·490 to **5·500** 0·010
5·1	3·00 to 8·60 +5·60	5·500 to **6·000** 0·500
1	6·47	6·000 to 9·045 +3·045
	1	3·558

Q	A	Q	A
1) **Subtract** 4·4 from 9·6	1) 4·4 to **5·0** 0·6 5·0 to 9·6 +4·6 5·2 1	5) How much is 0·81 **less than** 5·5 ?	5) 0·81 to **0·90** 0·09 0·90 to **1·00** 0·10 1·00 to **5·50** +4·50 4·69
2) Find 5·3 **take away** 2·7	2) 2·7 to **3·0** 0·3 3·0 to 5·3 +2·3 2·6	6) How much is **added to** 0·222 **to make** 2 ?	6) 0·222 to **0·230** 0·008 0·230 to **0·300** 0·070 0·300 to **1·000** 0·700 1·000 to **2·000** +1·000 1·778
3) Take 3·68 **from** 6·31	3) 3·68 to **3·70** 0·02 3·70 to **4·00** 0·30 4·00 to 6·31 +2·31 2·63		
4) Calculate 8·125 **minus** 1·749	4) 1·749 to **1·750** 0·001 1·750 to **1·800** 0·050 1·800 to **2·000** 0·200 2·000 to 8·125 +6·125 6·376	7) Find 70·04 − 15·6	7) 15·60 to **16·00** 0·40 16·00 to **20·00** 4·00 20·00 to **70·04** +50·04 54·44

The following 'problems' involve subtraction.

Q	A
1) Calculate the rise in temperature from 10·8 °F to 15·1 °F.	1) 10·8 to **11·0** 0·2 11·0 to 15·1 +4·1 4·3 °F
2) Parcel A weighs 9·17 kg and parcel B weighs 3·015 kg. How much heavier is parcel A than parcel B?	2) 3·015 to **3·020** 0·005 3·020 to **3·100** 0·080 3·100 to **4·000** 0·900 4·000 to **9·170** +5·170 6·155 kg 1 1

Decimals 6   Division by a Whole Number.

Division by a Whole Number
The **decimal points are under each other** in the number being divided into and the answer.
There are 5 possibilities to consider - A, B i), B ii), B iii), C.

A) A number divides **exactly** into a number **as it is presented**.

Ex.1 $6·8 \div 2$ $\begin{array}{r}3·4\\2\overline{)6·8}\end{array}$ Ex.2 $0·0276 \div 4$ $\begin{array}{r}0·0069\\4\overline{)0·0\,2\,7^3 6}\end{array}$

Q	A	Q	A
1) $4·2 \div 2$	1) $\begin{array}{r}2·1\\2\overline{)4·2}\end{array}$	4) $8·0 \div 5$	4) $\begin{array}{r}1·6\\5\overline{)8·^30}\end{array}$
2) $36·09 \div 3$	2) $\begin{array}{r}12·03\\3\overline{)36·09}\end{array}$	5) $7·68 \div 6$	5) $\begin{array}{r}1·2\,8\\6\overline{)7·^16\,^48}\end{array}$
3) $0·864 \div 4$	3) $\begin{array}{r}0·2\,1\,6\\4\overline{)0·8\,6^2 4}\end{array}$	6) $0·35 \div 7$	6) $\begin{array}{r}0·05\\7\overline{)0·3\,5}\end{array}$

B) We divide a number but we **change how it is presented to deal with a remainder**.

 We use the idea shown in **Decimals 1** ... 0's on the RIGHT of a decimal do **not** change its value.

 After dividing the given number, there is a **remainder** -
 we '**add**' a **0** to the number, 'carry' the remainder to it and divide again.
 We can repeat this as necessary, depending on the 'problem' / how the answer is to be written.

i) A number divides **exactly** into a number - eventually!

Ex.1 $6·9 \div 2$ $\begin{array}{r}3·4\,5\\2\overline{)6·9^10}\end{array}$ Ex.2 $0·5 \div 4$ $\begin{array}{r}0·1\,2\,5\\4\overline{)0·5\,^10\,^20}\end{array}$

Q	A	Q	A
1) $8·1 \div 2$	1) $\begin{array}{r}4·05\\2\overline{)8·1\,0}\end{array}$	4) $3 \div 5$	4) $\begin{array}{r}0·6\\5\overline{)3·0}\end{array}$
2) $0·55 \div 2$	2) $\begin{array}{r}0·2\,7\,5\\2\overline{)0·5\,^15\,^10}\end{array}$	5) $31 \div 4$	5) $\begin{array}{r}7·75\\4\overline{)3\,1·^30\,^20}\end{array}$
3) $6·49 \div 10$	3) $\begin{array}{r}0·6\,4\,9\\10\overline{)6·4\,^49\,^90}\end{array}$	6) $0·1 \div 8$	6) $\begin{array}{r}0·0\,1\,2\,5\\8\overline{)0·1\,0\,^20\,^40}\end{array}$

ii) A number does **not** divide **exactly** into a number - the remainder and the digit in the answer **recurs** / **repeats** itself **infinitely**. The number in the answer is called a **recurring decimal** - Ex.

Ex.1 $6·4 \div 3$ $\begin{array}{r}2·1\,3\,3\,3...\\3\overline{)6·4\,^10\,^10\,^10...}\end{array}$ or $\begin{array}{r}2·1\,\dot{3}\\3\overline{)6·4\,^10\,^1}\end{array}$ | Ex.2 $0·52 \div 6$ $\begin{array}{r}0·0\,8\,6\,6...\\6\overline{)0·5\,2\,^40\,^40...}\end{array}$ or $\begin{array}{r}0·0\,8\,\dot{6}\\6\overline{)0·5\,2\,^40\,^4}\end{array}$

*Note We can show the digit repeating itself and 3 dots **or** the digit with just 1 dot.
 The **digit repeats** itself when the **remainder repeats** itself.

Q	A	Q	A
1) $0·5 \div 3$	1) $\begin{array}{r}0·1\,6\,6\,6...\\3\overline{)0·5\,^20\,^20\,^20...}\end{array}$ or $\begin{array}{r}0·1\,\dot{6}\\3\overline{)0·5\,^20\,^2}\end{array}$	4) $3·2 \div 9$	4) $\begin{array}{r}0·3\,5\,5\,5...\\9\overline{)3·2\,^50\,^50\,^50...}\end{array}$ or $\begin{array}{r}0·3\,\dot{5}\\9\overline{)3·2\,^50\,^5}\end{array}$
2) $2·2 \div 3$	2) $\begin{array}{r}0·7\,3\,3\,3...\\3\overline{)2·2\,^10\,^10\,^10...}\end{array}$ or $\begin{array}{r}0·7\,\dot{3}\\6\overline{)2·2\,^10\,^1}\end{array}$	5) $0·8 \div 9$	5) $\begin{array}{r}0·0\,8\,8\,8...\\9\overline{)0·8\,^80\,^80...}\end{array}$ or $\begin{array}{r}0·0\,\dot{8}\\9\overline{)0·8\,^80\,^8}\end{array}$
3) $8·6 \div 6$	3) $\begin{array}{r}1·4\,3\,3\,3...\\6\overline{)8·^26\,^20\,^20\,^20...}\end{array}$ or $\begin{array}{r}1·4\,\dot{3}\\6\overline{)8·^26\,^20\,^2}\end{array}$	6) $5 \div 12$	6) $\begin{array}{r}0·4\,1\,6\,6...\\12\overline{)5·0\,^20\,^80\,^80...}\end{array}$ or $\begin{array}{r}0·4\,1\,\dot{6}\\12\overline{)5·0\,^20\,^80\,^8}\end{array}$

*Note For example, writing 1·4333... is usually preferred to 1·4$\dot{3}$ both for clarity and because it is easier to continue to use the number, which we often do.
 (Apart from this topic, it is preferred and used in other topics.)

Decimals 7 — Division by a Whole Number.

Division by a Whole Number

iii) A number does **not** divide **exactly** into a number - more than 1 remainder and more than 1 digit in the answer **recurs** / **repeats** itself **infinitely**.
The number in the answer is a **recurring decimal**.

Ex.1 $3 \cdot 6 \div 11$

$$11 \overline{) 3 \cdot 6\,^3 0\,^8 0\,^3 0\,^8 0 \ldots} \quad \text{giving} \quad 0 \cdot 3\,2\,7\,2\,7\ldots$$

or

$$11 \overline{) 3 \cdot 6\,^3 0\,^8 0\,^3 0} \quad \text{giving} \quad 0 \cdot 3\,\dot{2}\,\dot{7}$$

*Note 1 We can show the repeating digits and 3 dots **or** 1 dot over the **first** and **last** repeating digits.

*Note 2
To obtain such recurring decimals, the numbers we **divide by** usually involve using a calculator.
The following Q/A are possible without a calculator.

Q	A	Q	A
1) $5 \cdot 7 \div 11$	1) $0 \cdot 5\,1\,8\,1\,8\ldots$; $11\overline{)5 \cdot 7\,^2 0\,^9 0\,^2 0\,^9 0 \ldots}$ or $0 \cdot 5\,\dot{1}\,\dot{8}$; $11\overline{)5 \cdot 7\,^2 0\,^9 0\,^2}$	3) $1 \div 111$	3) $0 \cdot 0\,0\,9\,0\,0\,9\,0\,0\,9 \ldots$; $111\overline{)1 \cdot 0\,0\,0\,^1 0\,0\,0\,^1 0\,0\,0 \ldots}$ or $0 \cdot \dot{0}\,0\,\dot{9}$; $111\overline{)1 \cdot 0\,0\,0\,^1}$
2) $4 \cdot 8 \div 7$	2) $0 \cdot 6\,8\,5\,7\,1\,4\,2\,8\,5\,7 \ldots$; $7\overline{)4 \cdot 8\,^6 0\,^4 0\,^5 0\,^1 0\,^3 0\,^2 0\,^6 0\,^4 0\,^5 0 \ldots}$ or $0 \cdot 6\,\dot{8}\,5\,7\,1\,4\,\dot{2}$; $7\overline{)4 \cdot 8\,^6 0\,^4 0\,^5 0\,^1 0\,^3 0\,^2 0\,^6}$	4) $1 \div 101$	4) $0 \cdot 0\,0\,9\,9\,0\,0\,9\,9\,0 \ldots$; $101\overline{)1 \cdot 0\,0\,0\,^{91} 0\,^1 0\,0\,0\,^{91} 0\,^1 0 \ldots}$ or $0 \cdot \dot{0}\,0\,9\,\dot{9}$; $101\overline{)1 \cdot 0\,0\,0\,^{91} 0\,^1}$

C) We divide a number **as it is presented - and the remainder** is needed.

*Note We are not usually **asked** to calculate the remainder as shown here but it makes sense in some 'problems' to do so.

Ex.1 Calculate $6 \cdot 9 \div 2$
State the remainder.

$$2\overline{)6 \cdot 9} \quad = \quad 3 \cdot 4 \text{ r } 0 \cdot 1$$

*Note
The remainder is from this column - it is **not** 1.

Ex.2 Calculate £$0 \cdot 19 \div 7$
State the remainder.

$$7\overline{)£0 \cdot 1\,9} \quad = \quad £0 \cdot 0\,2 \text{ r } £0 \cdot 05$$

*Note
The remainder is from this column - it is **5p**.

Q	A	Q	A
1) Calculate $8 \cdot 7 \div 2$ State the remainder.	1) $2\overline{)8 \cdot 7} = 4 \cdot 3 \text{ r } 0 \cdot 1$	3) Calculate $0 \cdot 283 \div 4$ State the remainder.	3) $4\overline{)0 \cdot 2\,8\,3} = 0 \cdot 0\,7\,0 \text{ r } 0 \cdot 003$
2) Calculate £$3 \cdot 10 \div 6$ State the remainder.	2) $6\overline{)£3 \cdot 1\,^1 0} = £0 \cdot 5\,1 \text{ r } £0 \cdot 04$	4) Calculate $0 \cdot 0548 \div 5$ State the remainder.	4) $5\overline{)0 \cdot 0\,5\,4\,8} = 0 \cdot 0\,1\,0\,9 \text{ r } 0 \cdot 0003$

*Note
We can refer to a decimal by using the word '**recurring**' immediately after the number(s) that recur or 'more loosely' by stating **some** of the digits followed by '**dot dot dot**' or something similar!
Ex. We can read 2·333333...   as '2·3 **recurring**'   or '2·33 **dot dot dot**'
                 0·6969...   as '0·69 **recurring**'   or '0·69 **dot dot dot**'
                 7·14285714... as '7·142857 **recurring**' or '7·142 **dot dot dot**'

Decimals 8

Division by 10, 100, 1000, ...

Division by 10, 100, 1000, ...
Consider the following examples.

Ex.1 $23.4 \div 10$

$$\begin{array}{r} 2.34 \\ 10\overline{)23.^34\ ^40} \end{array}$$

Ex.2 $23.4 \div 100$

$$\begin{array}{r} 0.234 \\ 100\overline{)23.4^{34}0^{40}0} \end{array}$$

Ex.3 $23.4 \div 1000$

$$\begin{array}{r} 0.0234 \\ 1000\overline{)23.4\ 0^{340}0\ ^{4000}0} \end{array}$$

Divide by **10** and the digits move **1 place** to the right in the answer.

Divide by **100** and the digits move **2 places** to the right in the answer.

Divide by **1000** and the digits move **3 places** to the right in the answer.

Since the digits we start with just move columns, we can **write down the answer** without showing the above work **but** ...

rather than think of moving the digits, **it is easier to think of moving the decimal point**.

$\div 10$ Move the point **1 place LEFT**
$\div 100$ Move the point **2 places LEFT** } Move the point **1 place LEFT for each 0**.
$\div 1000$ Move the point **3 places LEFT** ...

Ex.1 $23.4 \div 10$
= 2.34

Ex.2 $23.4 \div 100$
= 0.234

Ex.3 $23.4 \div 1000$
= 0.0234
⇧
0's may be needed

Ex.4 $55.555... \div 10$
= 5.5555...

*Note We do not need to show the arrows!

Q	A	Q	A
1) $34.5 \div 10$	1) $= 3.45$	13) $2 \div 10000$	13) $= 0.0002$
2) $34.5 \div 100$	2) $= 0.345$	14) $50.13 \div 100$	14) $= 0.5013$
3) $34.5 \div 1000$	3) $= 0.0345$	15) $4.807 \div 1000$	15) $= 0.004807$
4) $9.76 \div 10$	4) $= 0.976$	16) $0.006 \div 10$	16) $= 0.0006$
5) $9.76 \div 100$	5) $= 0.0976$	17) $1515 \div 100$	17) $= 15.15$
6) $9.76 \div 1000$	6) $= 0.00976$	18) $932.9 \div 100000$	18) $= 0.009329$
7) $0.8 \div 10$	7) $= 0.08$	19) $77.77... \div 10$	19) $= 7.777...$
8) $0.8 \div 100$	8) $= 0.008$	20) $8.6222... \div 100$	20) $= 0.086222...$
9) $0.8 \div 1000$	9) $= 0.0008$	21) $0.999... \div 1000$	21) $= 0.000999...$
10) $2 \div 10$	10) $= 0.2$ ($2 = 2.$)	22) $45.454... \div 10$	22) $= 4.5454...$
11) $2 \div 100$	11) $= 0.02$	23) $3.142... \div 1000$	23) $= 0.003142...$
12) $2 \div 1000$	12) $= 0.002$	24) $202.02... \div 100$	24) $= 2.0202...$

Decimals 9

Multiplication by 10, 100, 1000, ...

Multiplication by 10, 100, 1000, ...
Multiplication is the **reverse** of division - this is used to help explain.

*Note **Decimals 10** explains a direct method of multiplication.

Consider the following examples shown in **Decimals 8**.

Ex.1 $23·4 \div 10$ Ex.2 $23·4 \div 100$ Ex.3 $23·4 \div 1000$

$$10\overline{)23·4}\ \to\ 2·34$$

$$100\overline{)23·4}\ \to\ 0·234$$

$$1000\overline{)23·4}\ \to\ 0·0234$$

Now we **reverse** the work ...

Multiply by **10** and the digits move **1 place** to the left in the answer.

Multiply by **100** and the digits move **2 places** to the left in the answer.

Multiply by **1000** and the digits move **3 places** to the left in the answer.

Since the digits we start with just move columns, we can **write down the answer** without showing any work ... but ...

rather than think of moving the digits, **it is easier to think of moving the decimal point**.

× 10   Move the point **1 place RIGHT**
× 100   Move the point **2 places RIGHT**   } Move the point **1 place RIGHT for each 0**.
× 1000   Move the point **3 places RIGHT** ...

Ex.1 $2·34 \times 10$ Ex.2 $0·234 \times 100$ Ex.3 $0·0234 \times 1000$
    $= 23·4$ $= 23·4$ $= 23·4$

0's are not needed before a whole number.

Ex.4 $5·555... \times 10$ Ex.5 $6·7 \times 100$
    $= 55·55...$ $= 670· = 670$

0's are needed   *Note We do not need to show the arrows!

Q	A	Q	A
1) $0·456 \times 10$	1) $= 4·56$	10) $0·0003 \times 10000$	10) $= 3· = 3$
2) $0·456 \times 100$	2) $= 45·6$	11) $0·015 \times 100$	11) $= 1·5$
3) $0·456 \times 1000$	3) $= 456· = 456$	12) $8·4 \times 10$	12) $= 84· = 84$
4) $7·89 \times 10$	4) $= 78·9$	13) $9·7 \times 100$	13) $= 970· = 970$
5) $7·89 \times 100$	5) $= 789· = 789$	14) $0·6 \times 1000$	14) $= 600· = 600$
6) $7·89 \times 1000$	6) $= 7890· = 7890$	15) $5·555... \times 1000$	15) $= 5555·5...$
7) $0·0002 \times 10$	7) $= 0·002$	16) $0·777... \times 10$	16) $= 7·77...$
8) $0·0002 \times 100$	8) $= 0·02$	17) $90·9090... \times 100$	17) $= 9090·90...$
9) $0·0002 \times 1000$	9) $= 0·2$	18) $3·1415... \times 10$	18) $= 31·415...$

Decimals 10   Multiplication.

Multiplication
Division is used to help explain.

Unlike addition, subtraction and division of decimals, the decimal points are **not** just placed under each other in a column.

Each of the following examples has part a) and b) which are directly related.

Consider the **numbers being multiplied** in a) ... we divide by 10's (each is shown as ÷10) to obtain the **numbers being multiplied** in b).

We can divide the **answer** in a) **by the same** 10's to obtain the **answer** in b).

This shows that ...   **we multiply decimals in the same way as whole numbers** ...
                      then position the decimal point in the answer.

To do this, notice there are the **same number of digits in decimal places** in the 'question' and 'answer'.
(These are **bold underlined** for emphasis.)

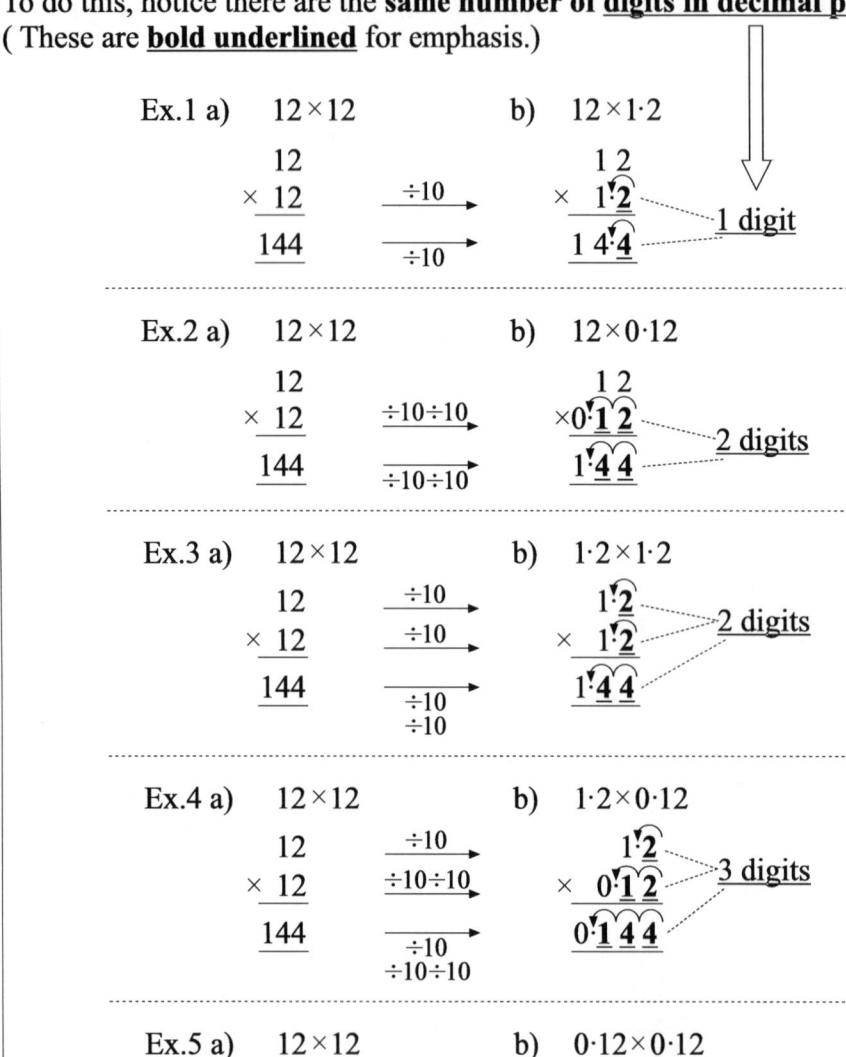

Ex.1 a)   12 × 12   b)   12 × 1·2
            12   1 2
          × 12   ÷10→   × 1·**2**   ⟶ 1 digit
          ‾‾‾‾   ÷10→   ‾‾‾‾‾‾
           144   14·**4**

Ex.2 a)   12 × 12   b)   12 × 0·12
            12   1 2
          × 12   ÷10÷10→   × 0·**1 2** ⟶ 2 digits
          ‾‾‾‾   ÷10÷10→   ‾‾‾‾‾‾‾
           144   1·**4 4**

Ex.3 a)   12 × 12   b)   1·2 × 1·2
            12   ÷10→   1·**2**
          × 12   ÷10→   × 1·**2**   ⟶ 2 digits
          ‾‾‾‾   ‾‾‾‾‾‾
           144   ÷10→   1·**4 4**
                    ÷10

Ex.4 a)   12 × 12   b)   1·2 × 0·12
            12   ÷10→   1·**2**
          × 12   ÷10÷10→   × 0·**1 2** ⟶ 3 digits
          ‾‾‾‾   ‾‾‾‾‾‾‾
           144   ÷10→   0·**1 4 4**
                   ÷10÷10

Ex.5 a)   12 × 12   b)   0·12 × 0·12
            12   ÷10÷10→   0·**1 2**
          × 12   ÷10÷10→   × 0·**1 2** ⟶ 4 digits
          ‾‾‾‾   ‾‾‾‾‾‾‾‾
           144   ÷10÷10→   0·**0 1 4 4**
                  ÷10÷10   ⇧
                                0 is needed

*Note
Since we multiply decimals as whole numbers, they are usually set out as whole numbers and **not** in their decimal columns.

Decimals 11

Multiplication.

Multiplication

When working, we can show the arrows ─÷10→ as in the Examples, but we usually just ...

... **count the digits** in decimal places in the **numbers being multiplied** -

and **count the same number of digits** in decimal places in the **answer** to **locate the decimal point**.

Here, the **counted digits** are underlined for emphasis.

Ex.1 4·3 × 2
```
   4·3  ⟩ 1 digit
×    2
   8·6
```

Ex.2 0·4 × 0·2
```
   0·4  ⟩ 2 digits
×  0·2
  0·08
    ⇑
  0 is needed
```

Ex.3 5·9 × 0·7
```
   5·9  ⟩ 2 digits
×  0·7
  4·13
    6
```

Ex.4 0·65 × 0·1
```
   0·65  ⟩ 3 digits
×  0·1
  0·065
```

Ex.5 0·32 × 0·03
```
   0·32  ⟩ 4 digits
×  0·03
  0·0096
```
Do **not** place decimal points here

Ex.6 2·7 × 3·5
```
    2·7
×   3·5
   13·5    ⟩ 2 digits
+  821·0
    9·45
```

Ex.7 9·8 × 0·46
```
    9·8
×  0·46
   58·8    ⟩ 3 digits
+ 39·320
   4·508
    1 1
```

Q	A	Q	A

1) 3·4 × 2	1)
	    3·4
	 × 2
	    6·8
		11) 0·001 × 0·001	11)
			     0·001
			 × 0·001
			 0·000001

2) 0·34 × 2	2)
	   0·34
	 × 2
	   0·68
		12) 1·2 × 1·1	12)
			   1·2
			 × 1·1
			   1·32

3) 3·4 × 0·2	3)
	   3·4
	 × 0·2
	   0·68
		13) 1·5 × 1·3	13)
			    1·5
			 × 1·3
			    4·5
			 + 1·50
			    1·95

4) 0·34 × 0·2	4)
	   0·34
	 × 0·2
	 0·068
		14) 8·9 × 0·77	14)
			     8·9
			 × 0·77
			    6·23
			 + 6·230
			    6·853

5) 0·34 × 0·02	5)
	   0·34
	 × 0·02
	 0·0068
		15) 0·45 × 32	15)
			   0·45
			 × 32
			    9·0
			 + 13·50
			   14·40 = 14·4

6) 34 × 0·2	6)
	    34
	 × 0·2
	   6·8
		16) 0·34 × 0·021	16)
			     0·34
			 × 0·021
			    34
			 + 680
			 0·00718

7) 7·8 × 0·9	7)
	   7·8
	 × 0·9
	   7·02		

8) 41·6 × 0·05	8)
	   41·6
	 × 0·05
	 2·080 = 2·08
		17) 5·78 × 6·9	17)
			    5·78
			 × 6·9
			   52·02
			 + 34·680
			   39·882

9) 0·3 × 0·3	9) 0·3 × 0·3 0·09		
10) 8·097 × 10	10) 8·097 × 10 80·970 = 80·97		

30

Decimals 12

Division by a Decimal.

<u>Division by a Decimal</u>
We first multiply **both** numbers of the division by 10, 100, 1000, ... to **change the decimal** that we are **dividing by** into a **whole number**.
This ensures that the **decimal point** is placed correctly in the answer - (as explained in **Decimals 6-7**).

Consider the following examples.

Ex.1 $6 \div 2$ $2\overline{)6}$ quotient 3

⇩ ×10

$60 \div 20$ $20\overline{)60}$ quotient 3

⇩ ×100

$600 \div 200$ $200\overline{)600}$ quotient 3

⇩ ×1000

$6000 \div 2000$ $2000\overline{)6000}$ quotient 3

Multiplying **both** numbers of the division by 10, 100, 1000, ... **gives the same answer**.

The same method is used for decimals ...

Change into a whole number

Ex.2 $0.6 \div 0.2$ $0.2\overline{)0.6}$

⇩ ×10

$06. \div 02.$ $2.\overline{)06.}$ quotient 3.

Ex.3 $0.8642 \div 0.02$ $0.02\overline{)0.8642}$

⇩ ×100

$086.42 \div 002.$ $002.\overline{)086.42}$ quotient 43.21

Ex.4 $0.0084 \div 0.002$ $0.002\overline{)0.0084}$

⇩ ×1000

$0008.4 \div 0002.$ $0002.\overline{)0008.4}$ quotient 4.2

*Note The following Q/A are presented with the 'division box' only.

Q	A	Q	A
1) $0.3\overline{)0.6}$	1) ×10 $03.\overline{)06.}$ quotient 2.	5) $0.6\overline{)0.078}$	5) $06.\overline{)00.78}$ quotient 0.13
2) $0.4\overline{)8.4}$	2) $04.\overline{)84.}$ quotient 21.	6) $0.5\overline{)0.0045}$	6) $05.\overline{)00.045}$ quotient 0.009
3) $0.2\overline{)0.26}$	3) $02.\overline{)02.6}$ quotient 1.3	7) $0.3\overline{)2.56}$	7) $03.\overline{)26.3²0²0}$ quotient 8.766...
4) $0.9\overline{)99}$	4) $09.\overline{)990.}$ quotient 110.	8) $0.8\overline{)80.4}$	8) $08.\overline{)804.0}$ quotient 100.5

Q	A	Q	A
1) $0.02\overline{)0.08}$	1) ×100 $002.\overline{)008.}$ quotient 4.	4) $0.05\overline{)55.82}$	4) $005.\overline{)558²2⁰}$ quotient 1116.4
2) $0.03\overline{)3.69}$	2) $003.\overline{)369.}$ quotient 123.	5) $0.02\overline{)6}$	5) $002.\overline{)600.}$ quotient 300.
3) $0.04\overline{)0.008}$	3) $004.\overline{)000.8}$ quotient 0.2	6) $0.07\overline{)0.00049}$	6) $007.\overline{)000.049}$ quotient 0.007

Q	A	Q	A
1) $0.002\overline{)0.004}$	1) ×1000 $0002.\overline{)0004.}$ quotient 2.	4) $0.008\overline{)72.0008}$	4) $0008.\overline{)72000.8}$ quotient 9000.1
2) $0.006\overline{)2.454}$	2) $0006.\overline{)2454.}$ quotient 409.	5) $0.003\overline{)0.00009}$	5) $0003.\overline{)00000.09}$ quotient 0.03
3) $0.007\overline{)50}$	3) $0007.\overline{)50 0¹0²0⁶0⁴0...}$ quotient 7142.85...	6) $0.001\overline{)1.2345}$	6) $0001.\overline{)1234.5}$ quotient 1234.5

Decimals 13

Division - 'Fraction Form'.

Division - 'Fraction Form'
Division may be set out in 'fraction form'.
The methods already explained are the same.
The decimal point is positioned in the 'answer' as it is passed in the 'question'.

Ex.1 $\dfrac{6\cdot 4}{2}$ $\dfrac{6\cdot 4}{2} = 3\cdot 2$

Ex.2 $\dfrac{5\cdot 1}{3}$ $\dfrac{5\cdot {}^2 1}{3} = 1\cdot 7$

Ex.3 $\dfrac{9\cdot 8}{4}$ $\dfrac{9\cdot {}^1 8\, {}^2 0}{4} = 2\cdot 45$

Ex.4 $\dfrac{8}{5}$ $\dfrac{8\, {}^3 0}{5} = 1\cdot 6$

Ex.5 $\dfrac{4\cdot 4}{6}$ $\dfrac{4\cdot {}^4 2\, {}^2 0\, {}^2 0\,\ldots}{6} = 0\cdot 733\ldots$

Ex.6 $\dfrac{0\cdot 8}{0\cdot 2}$ $\dfrac{0\,8\cdot}{0\,2\cdot} = 4$ (×10)

Ex.7 $\dfrac{6\cdot 3\,9\,6}{0\cdot 0\,3}$ $\dfrac{6\,3\,9\cdot 6}{0\,0\,3\cdot} = 213\cdot 2$ (×100)

Ex.8 $\dfrac{4\,0\cdot 0\,0\,7\,6}{0\cdot 0\,0\,4}$ $\dfrac{4\,0\,0\,0\,7\cdot {}^3 6}{0\,0\,0\,4\cdot} = 10001\cdot 9$ (×1000)

Q	A	Q	A
1.		2.	
1) $\dfrac{6\cdot 9}{3}$	1) $\dfrac{6\cdot 9}{3} = 2\cdot 3$	1) $\dfrac{0\cdot 6}{0\cdot 3}$	1) $\dfrac{0\,6\cdot}{0\,3\cdot} = 2$
2) $\dfrac{9\cdot 1}{7}$	2) $\dfrac{9\cdot {}^2 1}{7} = 1\cdot 3$	2) $\dfrac{0\cdot 0\,0\,8}{0\cdot 4}$	2) $\dfrac{0\,0\cdot 0\,8}{0\,4\cdot} = 0\cdot 0\,2$
3) $\dfrac{1\cdot 8}{5}$	3) $\dfrac{1\cdot 8\,{}^3 0}{5} = 0\cdot 36$	3) $\dfrac{1\cdot 9}{0\cdot 2}$	3) $\dfrac{1\,9\cdot {}^1 0}{0\,2\cdot} = 9\cdot 5$
4) $\dfrac{7}{4}$	4) $\dfrac{7\,{}^3 0\,{}^2 0}{4} = 1\cdot 75$	4) $\dfrac{5\,0}{0\cdot 6}$	4) $\dfrac{5\,0\,{}^2 0\,{}^2 0\,{}^2 0\,\ldots}{0\,6\cdot} = 83\cdot 33\ldots$
5) $\dfrac{0\cdot 7}{9}$	5) $\dfrac{0\cdot 7\,{}^7 0\,{}^7 0\,\ldots}{9} = 0\cdot 0777\ldots$	5) $\dfrac{0\cdot 0\,7}{0\cdot 0\,1}$	5) $\dfrac{0\,0\,7\cdot}{0\,0\,1\cdot} = 7$
6) $\dfrac{44\cdot 6}{8}$	6) $\dfrac{44\cdot {}^4 6\,{}^6 0\,{}^4 0}{8} = 5\cdot 575$	6) $\dfrac{4\cdot 0\,6\,2}{0\cdot 0\,2}$	6) $\dfrac{4\,0\,6\cdot 2}{0\,0\,2\cdot} = 203\cdot 1$
7) $\dfrac{23\cdot 4\,5}{5}$	7) $\dfrac{23\cdot {}^3 4\,{}^4 5}{5} = 4\cdot 69$	7) $\dfrac{5\,9\cdot 7}{0\cdot 0\,4}$	7) $\dfrac{5\,{}^1 9\,{}^3 7\,{}^1 0\,{}^2 0}{0\,0\,4\cdot} = 1492\cdot 5$
8) $\dfrac{65\cdot 0\,1}{11}$	8) $\dfrac{65\,{}^{10}0\,{}^1 1}{11} = 5\cdot 91$	8) $\dfrac{0\cdot 8\,6\,4}{0\cdot 0\,0\,2}$	8) $\dfrac{0\,8\,6\,4\cdot}{0\,0\,0\,2\cdot} = 432$
9) $\dfrac{1}{3}$	9) $\dfrac{1\cdot {}^1 0\,{}^1 0\,{}^1 0\,\ldots}{3} = 0\cdot 333\ldots$	9) $\dfrac{0\cdot 0\,0\,0\,3}{0\cdot 0\,0\,6}$	9) $\dfrac{0\,0\,0\,0\cdot 3\,0}{0\,0\,0\,6\cdot} = 0\cdot 5$
10) $\dfrac{3}{8}$	10) $\dfrac{3\cdot {}^3 0\,{}^6 0\,{}^4 0}{8} = 0\cdot 375$	10) $\dfrac{1\,0\,0\,0}{0\cdot 0\,0\,7}$	10) $\dfrac{1\,0\,{}^3 0\,{}^2 0\,{}^6 0\,{}^4 0\,{}^5 0\,{}^1 0\,\ldots}{0\,0\,0\,7\cdot} = 142857\cdot 1\ldots$

Decimals 14

Reciprocal. Order - Ascending / Descending.

Reciprocal
(see also **Fractions 28**)

The **reciprocal** of a number is the number which it is **multiplied by** to give the value **1**.

In general, the reciprocal of a number is found by **dividing** the number into 1, usually written in the form

$$\text{reciprocal of } x = \frac{1}{x}$$

Ex.1 The reciprocal of 2 is $\frac{1}{2} = 0.5$
since $2 \times 0.5 = 1$

Ex.2 The reciprocal of 0.5 is $\frac{1}{0.5} = 2$
since $0.5 \times 2 = 1$

Q	A	Q	A
State the **reciprocal** of 1) 4	1) $\frac{1}{4} = 0.25$	4) 0.8	4) $\frac{1}{0.8} = 1.25$
2) 3	2) $\frac{1}{3} = 0.333...$	5) -0.02	5) $\frac{1}{-0.02} = -50$
3) -5	3) $\frac{1}{-5} = -0.2$	6) $\frac{1}{9.7}$	6) 9.7

Decimal Order

Decimals can be arranged in **ascending** order (from smallest to largest)
or **descending** order (from largest to smallest).

Given several numbers to arrange in order, it is useful to list them in a **column** -
with the decimal points **directly under each other**.

It is easy to then list the numbers in a **row**, in **either** order, if required.

The decimal point is on the **right** of a whole number - it is useful to show it.

It can be helpful to insert **0** where a decimal place has no value - as in Exs. part b).
If we then ignore the decimal point, the numbers can be seen and compared as whole numbers - a slightly easier task!

Ex.1
List the following decimals in **ascending** order.
5.07 3.2 0.44 8.6 0.091

ascending order
a)   or b)
0.091   0.091
0.44   0.440
3.2   3.200
5.07   5.070
8.6   8.600

Ex.2
List the following decimals in **descending** order.
0.8 3.5 1.11 2 0.74

descending order
a)   or b)
3.5   3.50
2.   2.00
1.11   1.11
0.8   0.80
0.74   0.74

Q	A	Q	A
1) List in **ascending** order 1.3 0.59 0.6 2.2 0.77	1) 0.59 or 0.59 0.6 0.60 0.77 0.77 1.3 1.30 2.2 2.20	4) List in **descending** order 24 21.4 25.07 23.01 22	4) 25.07 or 25.07 24. 24.00 23.01 23.01 22. 22.00 21.4 21.40
2) List in **descending** order 0.08 0.315 0.8 4 0.42	2) 4. or 4.000 0.8 0.800 0.42 0.420 0.315 0.315 0.08 0.080	5) List in **ascending** order 0.3333 0.333 0.333... 0.3	5) 0.3 or 0.3000 0.333 0.3330 0.3333 0.3333 0.333... 0.33333...
3) List in **ascending** order 5.11 6.03 5.067 6.2 6	3) 5.067 or 5.067 5.11 5.110 6. 6.000 6.03 6.030 6.2 6.200	6) List in **descending** order 7.8 12 9.7 8 7.79 8.7	6) 12. or 12.00 9.7 9.70 8.7 8.70 8. 8.00 7.8 7.80 7.79 7.79

Decimals 15

Powers. Roots.

Powers

A decimal can be 'raised to a power' - it is usually written in **brackets** ( ) for clarity.
(It is expected to use a calculator for some calculations.)

Ex.1 Calculate

a) $(0.1)^2$ b) $(-0.4)^3$

a) $(0.1)^2 = 0.1 \times 0.1 = 0.01$ b) $(-0.4)^3 = -0.4 \times -0.4 \times -0.4 = -0.064$

Q	A	Q	A
Calculate			
1) $(0.3)^2$	1) $0.3 \times 0.3 = 0.09$	8) $(6.7)^4$	8) $6.7 \times 6.7 \times 6.7 \times 6.7 = 2015.1121$
2) $(-0.1)^2$	2) $-0.1 \times -0.1 = 0.01$	9) $(-0.2)^4$	9) $-0.2 \times -0.2 \times -0.2 \times -0.2 = 0.0016$
3) $(0.04)^2$	3) $0.04 \times 0.04 = 0.0016$	10) $(0.8)^5$	10) $0.8 \times 0.8 \times 0.8 \times 0.8 \times 0.8 = 0.32768$
4) $(1.2)^2$	4) $1.2 \times 1.2 = 1.44$	11) $(0.333...)^2$	11) $0.333... \times 0.333... = 0.111...$ [input $\frac{1}{3}$ ($= 0.333...$) into calculator.]
5) $(0.5)^3$	5) $0.5 \times 0.5 \times 0.5 = 0.125$		
6) $(-0.1)^3$	6) $-0.1 \times -0.1 \times -0.1 = -0.001$	12) $(0.111...)^3$	12) $0.111... \times 0.111... \times 0.111... = 0.00137...$ [input $\frac{1}{9}$ ($= 0.111...$) into calculator.]
7) $(1.25)^2$	7) $1.25 \times 1.25 = 1.5625$		

Roots

The root of some decimals may be found without the use of a calculator.

*Note

To help explain, the above Exs. for 'Powers' are reversed in the following Exs.

Ex.1 Calculate

    2 is usually omitted.

a) $\sqrt[2]{0.01}$ or $\sqrt{0.01}$ b) $\sqrt[3]{-0.064}$

a) $\sqrt{0.01} = 0.1$ or -0.1 ($= \pm 0.1$) b) $\sqrt[3]{-0.064} = -0.4$

Q	A	Q	A
Calculate			
1) $\sqrt{0.04}$	1) 0.2 or -0.2 ($= \pm 0.2$)	6) $\sqrt[3]{0.008}$	6) 0.2
2) $\sqrt{0.25}$	2) 0.5 or -0.5 ($= \pm 0.5$)	7) $\sqrt[3]{-0.008}$	7) -0.2
3) $\sqrt{1.21}$	3) 1.1 or -1.1 ($= \pm 1.1$)	8) $\sqrt[3]{2.197}$	8) 1.3
4) $\sqrt[2]{5.76}$	4) 2.4 or -2.4 ($= \pm 2.4$)	9) $\sqrt[3]{40.8}$	9) $3.442...$
5) $\sqrt{0.9}$	5) $0.948...$ or $-0.948...$ ($= \pm 0.948...$)	10) $\sqrt[4]{6.5536}$	10) 1.6 or -1.6 ($= \pm 1.6$)
		11) $\sqrt[5]{0.9}$	11) $0.979...$

Decimals 16 — Change Recurring Decimal to Fraction.

Change A Recurring Decimal To A Fraction
To do this we 'set up' and solve an **equation**.

Ex.1 Change 0·333... to a fraction.

Method
A. Let x = decimal.
B. Multiply both x and decimal by 10, or 100, 1000, ...
  so all / most of the decimal part of the numbers are the same.
C. Subtract values at A. to eliminate all / most decimal parts
D. ... and so leave **equation**.
E. Solve equation.

A. Let $x = 0·333...$
B. $10x = 3·333...$ ×10: $0·333... = 3·33...$
C. $- x = 0·333...$
D. $9x = 3$
E. $x = \frac{3}{9} = \frac{1}{3}$ (÷3)

*Note Multiplying by a larger number than necessary at B. just leads to larger numbers.

Let $x = 0·333...$
$100x = 33·333...$
$- x = 0·333...$
$99x = 33$
$x = \frac{33}{99} = \frac{1}{3}$ (÷33)

Ex.2 Change 0·51222... to a fraction.

Let $x = 0·51222...$
$10x = 5·12222...$
$- x = 0·51222...$
$9x = 4·61$
$x = \frac{4·61}{9} = \frac{461}{900}$

Q	A	Q	A
Change to a fraction.		4) 0·5656...	4) Let $x = 0·5656...$ $100x = 56·5656...$ $- x = 0·5656...$ $99x = 56$ $x = \frac{56}{99}$
1) 0·777...	1) Let $x = 0·777...$ $10x = 7·777...$ $- x = 0·777...$ $9x = 7$ $x = \frac{7}{9}$	5) 0·234234...	5) Let $x = 0·234234...$ $1000x = 234·234234...$ $- x = 0·234234...$ $999x = 234$ $x = \frac{234}{999} = \frac{26}{111}$ (÷9)
2) 0·999...	2) Let $x = 0·999...$ $10x = 9·999...$ $- x = 0·999...$ $9x = 9$ $x = \frac{9}{9} = 1$	6) 0·30555...	6) Let $x = 0·30555...$ $10x = 3·05555...$ $- x = 0·30555...$ $9x = 2·75$ $x = \frac{2·75}{9} = \frac{275}{900} = \frac{11}{36}$ (÷25)
3) 0·1818...	3) Let $x = 0·1818...$ $100x = 18·1818...$ $- x = 0·1818...$ $99x = 18$ $x = \frac{18}{99} = \frac{2}{11}$ (÷9) *Note Multiplying x by, say, 10 gives $10x = 1·8181...$ and is no use.	7) 1·4̇8̇	7) Let $x = 1·4888...$ $10x = 14·8888...$ $- x = 1·4888...$ $9x = 13·4$ $x = \frac{13·4}{9} = \frac{134}{90} = \frac{67}{45}$

Decimals 17 — Multiplication / Division by 10, 100, ...

*Note 1 The Exs. only show whole numbers - for clarity.
           The Q/A only show positive values.
*Note 2 The Method of calculating explained here should be used in the Q/A.

Multiplication / Division by 10, 100, ...

Given the MULTIPLICATION of 2 or more values AND THE RESULT –
A) if a value is **multiplied** by 10, 100, ... then the result is **multiplied** by 10, 100, ...
B) if a value is **divided** by 10, 100, ... then the result is **divided** by 10, 100, ...

Ex.1 Given $4 \times 2 = 8$
Calculate 400×20
(with ×100, ×10 ⇒ ×100 ×10)
$400 \times 20 = 8000$

Ex.2 Given $400 \times 20 = 8000$
Calculate 4×2
(with ÷100, ÷10 ⇒ ÷100 ÷10)
$(4 \times 2 = 8)$

Ex.3 Given $400 \times 20 = 8000$
Calculate 4×200
(÷100, ×10 ⇒ ÷100 ×10; also ÷10 on result)
$(4 \times 200 = 800)$

Ex.4 Given $400 \times 20 = 8000$
Calculate 4000×2
(×10, ÷10 ⇒ ×10; also ÷10 on result)
$(4000 \times 2 = 8000)$

Q	A	Q	A
1) Given $2 \times 3 = 6$ Calculate 20×300	1) $2 \times 3 = 6$; ×10 ×100 ⇒ ×10 ×100; $20 \times 300 = 6000$ $(20 \times 300 = 6000)$	5) Given $1.23 \times 0.45 = 0.5535$ Calculate 123×4.5	5) $1.23 \times 0.45 = 0.5535$; ×100 ×10 ⇒ ×100 ×10; $123 \times 4.5 = 553.5$ $(123 \times 4.5 = 553.5)$
2) Given $2000 \times 30 = 60000$ Calculate 20×3	2) $2000 \times 30 = 60000$; ÷100 ÷10 ⇒ ÷100 ÷10; $20 \times 3 = 60$ $(20 \times 3 = 60)$	6) Given $7 \times 50.6 = 354.2$ Calculate 0.7×0.506	6) $7 \times 50.6 = 354.2$; ÷10 ÷100 ⇒ ÷10 ÷100; $0.7 \times 0.506 = 0.3542$ $(0.7 \times 0.506 = 0.3542)$
3) Given $20 \times 300 = 6000$ Calculate 2000×30	3) $20 \times 300 = 6000$; ×100 ÷10 ⇒ ×100 (÷10 on result becomes ×10); $2000 \times 30 = 60000$ $(2000 \times 30 = 60000)$	7) Given $0.08 \times 92 = 7.36$ Calculate 0.8×0.92	7) $0.08 \times 92 = 7.36$; ×10 ÷100 ⇒ ÷100 (×10 on result); $0.8 \times 0.92 = 0.736$ $(0.8 \times 0.92 = 0.736)$
4) Given $200 \times 3 = 600$ Calculate 20×30	4) $200 \times 3 = 600$; ÷10 ×10 ⇒ ×10 (÷10 on result); $20 \times 30 = 600$ $(20 \times 30 = 600)$	8) Given $4.2 \times 0.5 = 2.1$ Calculate 0.0042×5	8) $4.2 \times 0.5 = 2.1$; ÷1000 ×10 ⇒ ÷1000 (×10 on result); $0.0042 \times 5 = 0.021$ $(0.0042 \times 5 = 0.021)$

Decimals 18 — Multiplication / Division by 10, 100, ...

Q	A
9) Given $2.743 \times 0.608 = 1.667744$ a) Calculate 274.3×6.08	9) a) $\quad 2.743 \times 0.608 = 1.667744$ $\qquad \times 100 \quad \times 10 \;\Rightarrow\; \times 100 \times 10$ $\qquad 274.3 \times 06.08 = 1667.744$ $\qquad (274.3 \times 6.08 = 1667.744)$
b) Calculate 0.02743×0.0608	b) $\quad 2.743 \times 0.608 = 1.667744$ $\qquad \div 100 \quad \div 10 \;\Rightarrow\; \div 100 \div 10$ $\qquad 0.02743 \times 0.00608 = 0.001667744$ $\qquad (0.02743 \times 0.00608 = 0.001667744)$
10) Given $71.65 \times 53.4 = 3826.11$ a) Calculate 7.165×534	10) a) $\quad 71.65 \times 53.4 = 3826.11$ $\qquad \qquad\qquad\qquad\qquad \div 10$ $\qquad \div 10 \quad \times 10 \;\Rightarrow\; \times 10$ $\qquad 7.165 \times 534. = 3826.11$ $\qquad (7.165 \times 534 = 3826.11)$
b) Calculate 716.5×0.534	b) $\quad 71.65 \times 53.4 = 3826.11$ $\qquad\qquad\qquad\qquad\qquad \times 10$ $\qquad \times 10 \quad \div 100 \;\Rightarrow\; \div 100$ $\qquad 716.5 \times 0.534 = 382.611$ $\qquad (716.5 \times 0.534 = 382.611)$
11) Given $0.034 \times 1.292 = 0.043928$ a) Calculate 34×129.2	11) a) $\quad 0.034 \times 1.292 = 0.043928$ $\qquad \times 1000 \; \times 100 \;\Rightarrow\; \times 1000 \times 100$ $\qquad 0034. \times 129.2 = 04392.8$ $\qquad (34 \times 129.2 = 4392.8)$
b) Calculate 0.0034×0.01292	b) $\quad 0.034 \times 1.292 = 0.043928$ $\qquad \div 10 \quad \div 100 \;\Rightarrow\; \div 10 \div 100$ $\qquad 0.0034 \times 0.01292 = 0.0000 43928$ $\qquad (0.0034 \times 0.01292 = 0.000043928)$
12) Given $444.4 \times 55 = 24442$ a) Calculate 4.444×550	12) a) $\quad 444.4 \times 55. = 24442.$ $\qquad\qquad\qquad\qquad\qquad \div 100$ $\qquad \div 100 \; \times 10 \;\Rightarrow\; \times 10$ $\qquad 4.444 \times 550. = 2444.2$ $\qquad (4.444 \times 550 = 2444.2)$
b) Calculate 4444×0.055	b) $\quad 444.4 \times 55. = 24442.$ $\qquad\qquad\qquad\qquad\qquad \div 1000$ $\qquad \times 10 \; \div 1000 \;\Rightarrow\; \times 10$ $\qquad 4444. \times 0.055 = 244.42$ $\qquad (4444 \times 0.055 = 244.42)$

37

Decimals 19   Multiplication / Division by 10, 100, ...

*Note 1 The Division of decimals is set out in **fraction form** so calculations are more easily seen -
the numerator and denominator are referred to.

*Note 2 The Exs. only show whole numbers - for clarity.
The Q/A only show positive values.

*Note 3 The Method of calculating explained here should be used in the Q/A.

Multiplication / Division by 10, 100, ...
Given the DIVISION of 2 or more values AND THE RESULT -
A) if a numerator is **multiplied** by 10, 100, ... then the result is **multiplied** by 10, 100, ...
B) if a numerator is **divided** by 10, 100, ... then the result is **divided** by 10, 100, ...
C) if a denominator is **multiplied** by 10, 100, ... then the result is **divided** by 10, 100, ...
(since division of a number by a larger number gives a smaller result)
D) if a denominator is **divided** by 10, 100, ... then the result is **multiplied** by 10, 100, ...
(since division of a number by a smaller number gives a larger result)

Ex.1 Given $\frac{8}{4} = 2$

Calculate $\frac{800}{40}$ with numerator ×100, denominator ×10 ⇒ result ×100, ÷10

$\left(\frac{800}{40} = 20\right)$

Ex.2 Given $\frac{8000}{400} = 20$

Calculate $\frac{800}{4}$ with numerator ÷10, denominator ÷100 ⇒ result ÷10, ×100

$\left(\frac{800}{4} = 200\right)$

Ex.3 Given $\frac{800}{400} = 2$

Calculate $\frac{8000}{40}$ with numerator ×10, denominator ÷10 ⇒ result ×10, ×10

$\left(\frac{8000}{40} = 200\right)$

Ex.4 Given $\frac{8000}{4} = 2000$

Calculate $\frac{800}{40}$ with numerator ÷10, denominator ×10 ⇒ result ÷10, ÷10

$\left(\frac{800}{40} = 20\right)$

Q	A	Q	A
1) Given $\frac{6}{2} = 3$ Calculate $\frac{60}{20}$	1) $\frac{6}{2} = 3$ numerator ×10, denominator ×10 ⇒ result ×10, ÷10 $\left(\frac{60}{20} = 3\right)$	3) Given $\frac{6000}{200} = 30$ Calculate $\frac{60000}{20}$	3) $\frac{6000}{200} = 30$ numerator ×10, denominator ÷10 ⇒ result ×10, ×10 $\left(\frac{60000}{20} = 3000\right)$
2) Given $\frac{600}{200} = 3$ Calculate $\frac{60}{2}$	2) $\frac{600}{200} = 3$ numerator ÷10, denominator ÷100 ⇒ result ÷10, ×100 $\left(\frac{60}{2} = 30\right)$	4) Given $\frac{6000}{2} = 3000$ Calculate $\frac{600}{20}$	4) $\frac{6000}{2} = 3000$ numerator ÷10, denominator ×10 ⇒ result ÷10, ÷10 $\left(\frac{600}{20} = 30\right)$

Decimals 20 — Multiplication / Division by 10, 100, ...

Q	A	Q	A
5) Given $\dfrac{39.68}{1.24} = 32$ Calculate $\dfrac{3968}{12.4}$	5) $\dfrac{39.68}{1.24} = 32$ $\dfrac{3968.}{12.4} \xrightarrow{\times 100}_{\times 10} = 320. \xrightarrow{\times 100}_{\div 10}$ $\left(\dfrac{3968}{12.4} = 320\right)$	10) Given $\dfrac{0.0007}{0.14} = 0.005$ Calculate $\dfrac{0.00007}{0.00014}$	10) $\dfrac{0.0007}{0.14} = 0.005$ $\dfrac{0.00007}{0.00014} \xrightarrow{\div 10}_{\div 1000} = 0.005 \xrightarrow{\div 10}_{\times 1000}$ $\left(\dfrac{0.00007}{0.00014} = 0.5\right)$
6) Given $\dfrac{26.6}{5.6} = 4.75$ Calculate $\dfrac{0.266}{0.56}$	6) $\dfrac{26.6}{5.6} = 4.75$ $\dfrac{0.266}{0.56} \xrightarrow{\div 100}_{\div 10} = 0.475 \xrightarrow{\div 100}_{\times 10}$ $\left(\dfrac{0.266}{0.56} = 0.475\right)$	11) Given $\dfrac{739.26}{33.3} = 22.2$ Calculate $\dfrac{0.73926}{333}$	11) $\dfrac{739.26}{33.3} = 22.2$ $\dfrac{0.73926}{333.} \xrightarrow{\div 1000}_{\times 10} = 0.00222 \xrightarrow{\div 1000}_{\div 10}$ $\left(\dfrac{0.73926}{333} = 0.00222\right)$
7) Given $\dfrac{5.7477}{0.0714} = 80.5$ Calculate $\dfrac{0.57477}{7.14}$	7) $\dfrac{5.7477}{0.0714} = 80.5$ $\dfrac{0.57477}{007.14} \xrightarrow{\div 10}_{\times 100} = 0.0805 \xrightarrow{\div 10}_{\div 100}$ $\left(\dfrac{0.57477}{7.14} = 0.0805\right)$	12) Given $\dfrac{0.08}{0.625} = 0.128$ Calculate $\dfrac{0.8}{0.00625}$	12) $\dfrac{0.08}{0.625} = 0.128$ $\dfrac{00.8}{0.00625} \xrightarrow{\times 10}_{\div 100} = 01.28 \xrightarrow{\times 10}_{\times 100}$ $\left(\dfrac{0.8}{0.00625} = 128\right)$
8) Given $\dfrac{7.1921}{2.95} = 2.438$ Calculate $\dfrac{71.921}{0.295}$	8) $\dfrac{7.1921}{2.95} = 2.438$ $\dfrac{71.921}{0.295} \xrightarrow{\times 10}_{\div 10} = 243.8 \xrightarrow{\times 10}_{\times 10}$ $\left(\dfrac{71.921}{0.295} = 243.8\right)$	13) Given $\dfrac{0.1053}{23.4} = 0.0045$ Calculate $\dfrac{1.053}{2340}$	13) $\dfrac{0.1053}{23.4} = 0.0045$ $\dfrac{01.053}{2340.} \xrightarrow{\times 10}_{\times 100} = 0.00045 \xrightarrow{\times 10}_{\div 100}$ $\left(\dfrac{1.053}{2340} = 0.00045\right)$
9) Given $\dfrac{0.04698}{0.0009} = 52.2$ Calculate $\dfrac{0.4698}{0.009}$	9) $\dfrac{0.04698}{0.0009} = 52.2$ $\dfrac{00.4698}{00.009} \xrightarrow{\times 10}_{\times 10} = 52.2 \xrightarrow{\times 10}_{\div 10}$ $\left(\dfrac{0.4698}{0.009} = 52.2\right)$	14) Given $\dfrac{62000}{50} = 1240$ Calculate $\dfrac{620}{0.05}$	14) $\dfrac{62000.}{50.} = 1240.$ $\dfrac{620.0}{0.05} \xrightarrow{\div 100}_{\div 1000} = 12400. \xrightarrow{\div 100}_{\times 1000}$ $\left(\dfrac{620}{0.05} = 12400\right)$

Fractions 1 — Types of Fractions.

Fractions
A **fraction** is a part of a **whole** - it is presented in the form of $\frac{numerator}{denominator}$ ← Division line

The **numerator** indicates how many parts we are concerned with.
The **denominator** indicates how many parts there are altogether.
The **parts** must all be equal in size.

Ex. $\frac{1}{2}$ is read as 'one half'
or 'one out of two' / 'one part out of two parts'
or 'one divided by two'.

Types of Fractions
Names are given to the different types of fractions -

A. Common / Proper / Vulgar Fraction.
B. Improper Fraction.
C. Mixed Number.

A. Common / Proper / Vulgar Fraction
This is a fraction with the numerator **less than** the denominator.

Exs. $\frac{1}{2}$ $\frac{6}{8}$ $\frac{11}{15}$ $\frac{227}{404}$

*Note In the following Q/A 1 3, each diagram represents **1 whole**.

Q	A	Q	A
1) State the fraction shaded in each diagram. a) b) c) d)	1) a) $\frac{3}{5}$ b) $\frac{5}{6}$ c) $\frac{2}{11}$ d) $\frac{4}{8} = \frac{1}{2}$	3) Shade the given fraction of each diagram. a) $\frac{1}{4}$ b) $\frac{2}{5}$ c) $\frac{3}{4}$	3) *Note Alternative parts may be shaded. a) (shade 1 part for each 4) b) (shade 2 parts for each 5) c) (shade 3 parts for each 4)
2) Shade the given fraction of each diagram. a) $\frac{1}{2}$ b) $\frac{2}{3}$ c) $\frac{5}{10}$ d) $\frac{7}{12}$	2) *Note Alternative parts may be shaded. a) b) c) d)	4) In a class of 30 pupils 17 are girls. What fraction of the class are girls?	4) $\frac{17}{30}$
		5) On 1 day, 5 buses are late and 9 are on time. What fraction of the buses are late?	5) $\frac{5}{5+9} = \frac{5}{14}$
		6) In May it rains on 12 days. What fraction of May does it not rain?	6) $\frac{31-12}{31} = \frac{19}{31}$

Fractions 2   Change Improper Fraction to Whole / Mixed Number.

B. <u>Improper Fraction</u>
This is a fraction with the numerator **equal to** or **greater than** the denominator.
If the numerator is **equal to** the denominator - the fraction is always **equal to 1**.

Exs. $\frac{2}{2} = 1$ $\frac{5}{5} = 1$ $\frac{64}{64} = 1$

If the numerator is **greater than** the numerator -
<u>either</u> the fraction is equal to a **whole number**

Exs. $\frac{7}{1} = 7$ $\frac{6}{2} = 3$ $\frac{8}{4} = 2$ $\frac{50}{10} = 5$

<u>or</u> the fraction is equal to a **whole number and a common fraction** together - **a Mixed Number**.

Exs. $\frac{3}{2} = 1\frac{1}{2}$ $\frac{8}{3} = 2\frac{2}{3}$ $\frac{97}{20} = 4\frac{17}{20}$

C. <u>Mixed Number</u>
This is a **whole number <u>and</u> a common fraction** together.

Exs. $7\frac{1}{2}$ $9\frac{3}{4}$ $8\frac{11}{16}$ $15\frac{59}{100}$

<u>Change an Improper Fraction to a Whole Number or Mixed Number</u>
We **divide** the **numerator** by the **denominator** to give the **whole number** -
write any **remainder** over the **denominator** to give the **fraction**.
(The denominator shows how many parts are needed to make a whole number -
by dividing we find how many whole numbers there are and how many parts, if any, are left over.)

Exs. 1) $\frac{4}{1} = 4$ 2) $\frac{6}{2} = 3$ 3) $\frac{9}{2} = 4\frac{1}{2}$ 4) $\frac{77}{9} = 8\frac{5}{7}$

Q	A	Q	A
1) State the value of each diagram - A) - if each square ☐ = $\frac{1}{2}$ a) ☐ b) ☐ c) ☐ d) ☐ e) ☐	1) A) a) $\frac{1}{2}$ b) $\frac{2}{2} = 1$ c) $\frac{3}{2} = 1\frac{1}{2}$ d) $\frac{4}{2} = 2$ e) $\frac{9}{2} = 4\frac{1}{2}$	2) Change each **improper fraction** to a **whole number** or **mixed number**. a) $\frac{3}{3}$ b) $\frac{9}{1}$ c) $\frac{20}{5}$ d) $\frac{11}{2}$ e) $\frac{7}{4}$ f) $\frac{32}{5}$ g) $\frac{52}{21}$	2) a) $= 1$ b) $= 9$ c) $= 4$ d) $= 5\frac{1}{2}$ e) $= 1\frac{3}{4}$ f) $= 6\frac{2}{5}$ g) $= 2\frac{10}{21}$
B) - if each square ☐ = $\frac{1}{3}$ a) ☐ b) ☐ c) ☐ d) ☐ e) ☐ f) ☐	B) a) $\frac{1}{3}$ b) $\frac{2}{3}$ c) $\frac{3}{3} = 1$ d) $\frac{4}{3} = 1\frac{1}{3}$ e) $\frac{5}{3} = 1\frac{2}{3}$ f) $\frac{17}{3} = 5\frac{2}{3}$	3) How many teams of 8 can be formed from 29 players? 4) 10 pens are in a set. How many sets are possible from 87 pens?	3) $\frac{29}{8} = 3\frac{5}{8}$ teams 4) $\frac{87}{10} = 8\frac{7}{10}$ sets

Fractions 3 — Change Whole / Mixed Number to Improper Fraction.

Change a Whole Number to an Improper Fraction

We simply write the **whole number as the numerator** and the **denominator is always 1**.
(write the **whole number 'over 1'**)

Exs. 1) $1 = \dfrac{1}{1}$ 2) $5 = \dfrac{5}{1}$ 3) $49 = \dfrac{49}{1}$ ('49 divided by 1 = 49')

Change a Mixed Number to an Improper Fraction

We **multiply** the **whole number** by the **denominator**, **add** the **numerator** and write the **total** over the **denominator**.   } whole number $\dfrac{+\text{ numerator}}{\text{denominator}}$

(we change each **whole number** into the parts shown by the **denominator** and **add** the extra parts)

Ex. Change each Mixed Number to an Improper Fraction.

1) $4\dfrac{1}{2}$ ⇩ $4\dfrac{+1}{2} = \dfrac{9}{2}$ ('9 halves')

$4 \times 2 = 8$
$\quad\quad +1$
$\quad\quad\ \ 9$

Each whole number has 2 parts.

2) $3\dfrac{2}{5}$ ⇩ $3\dfrac{+2}{5} = \dfrac{17}{5}$ ('17 fifths')

$3 \times 5 = 15$
$\quad\quad\ + 2$
$\quad\quad\ \ 17$

Each whole number has 5 parts.

3) $1\dfrac{3}{7}$ ⇩ $1\dfrac{+3}{7} = \dfrac{10}{7}$ ('10 sevenths')

$1 \times 7 = 7$
$\quad\quad + 3$
$\quad\quad\ \ 10$

Each whole number has 7 parts.

The calculations are usually done mentally.

Q	A	Q	A
1) Change each **whole number** or **mixed number** to an **improper fraction**.	1)		
a) 4	a) $\dfrac{4}{1}$	h) $3\dfrac{21}{25}$	h) $3\dfrac{+21}{25} = \dfrac{96}{25}$
b) 7	b) $\dfrac{7}{1}$	i) $2\dfrac{46}{111}$	i) $2\dfrac{+46}{111} = \dfrac{268}{111}$
c) 68	c) $\dfrac{68}{1}$	j) $10\dfrac{2}{7}$	j) $10\dfrac{+2}{7} = \dfrac{72}{7}$
d) $2\dfrac{1}{2}$	d) $2\dfrac{+1}{2} = \dfrac{5}{2}$	k) $450\dfrac{1}{6}$	k) $450\dfrac{+1}{6} = \dfrac{2701}{6}$
e) $1\dfrac{3}{4}$	e) $1\dfrac{+3}{4} = \dfrac{7}{4}$	2) A month has $4\dfrac{3}{7}$ weeks. How many days are in the month?	2) $4\dfrac{+3}{7} = \dfrac{31}{7}$ 31 days
f) $5\dfrac{5}{9}$	f) $5\dfrac{+5}{9} = \dfrac{50}{9}$		
g) $8\dfrac{7}{10}$	g) $8\dfrac{+7}{10} = \dfrac{87}{10}$	3) There are $6\dfrac{5}{8}$ teams. If there are 8 in a team, how many players are there?	3) $6\dfrac{+5}{8} = \dfrac{53}{8}$ 53 players

Fractions 4

Equal / Equivalent Fractions.

Equal / Equivalent fractions

We can obtain **equal** or **equivalent** fractions if we
a) 'break' parts into smaller parts   or   b) 'group' parts into larger parts.

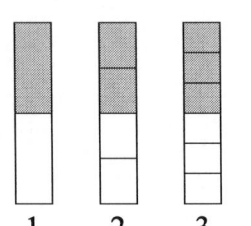

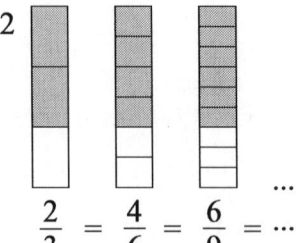

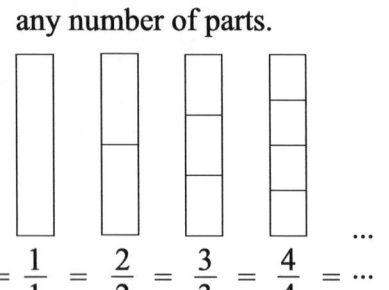

A whole number can have any number of parts.

Ex.1 $\dfrac{1}{2} = \dfrac{2}{4} = \dfrac{3}{6} = \cdots$

Ex.2 $\dfrac{2}{3} = \dfrac{4}{6} = \dfrac{6}{9} = \cdots$

Ex.3 $1 = \dfrac{1}{1} = \dfrac{2}{2} = \dfrac{3}{3} = \dfrac{4}{4} = \cdots$

Q	A	Q	A
1) Shade the given fraction of each diagram a) and b). State the equivalent fractions for diagram b).	**1)** *Note Alternative parts may be shaded.	**2)** Shade the given fraction of each diagram a) and b). State the equivalent fractions for diagram b).	**2)** *Note Alternative parts may be shaded.
1. a) $\dfrac{1}{5}$ b) $\dfrac{1}{5}$	1. a) $\dfrac{1}{5}$ b) $\dfrac{1}{5} = \dfrac{2}{10}$	1. a) $\dfrac{2}{8}$ b)	1. a) $\dfrac{2}{8}$ b) $\dfrac{2}{8} = \dfrac{1}{4}$
2. a) $\dfrac{2}{5}$ b) $\dfrac{2}{5}$	2. a) $\dfrac{2}{5}$ b) $\dfrac{2}{5} = \dfrac{6}{15}$	2. a) $\dfrac{10}{12}$ b)	2. a) $\dfrac{10}{12}$ b) $\dfrac{10}{12} = \dfrac{5}{6}$
3. a) $\dfrac{3}{4}$ b) $\dfrac{3}{4}$	3. a) $\dfrac{3}{4}$ b) $\dfrac{3}{4} = \dfrac{6}{8} = \dfrac{12}{16}$	3. a) $\dfrac{10}{15}$ b)	3. a) $\dfrac{10}{15}$ b) $\dfrac{10}{15} = \dfrac{2}{3}$
4. a) $\dfrac{2}{3}$ b) $\dfrac{2}{3}$	4. a) $\dfrac{2}{3}$ b) $\dfrac{2}{3} = \dfrac{4}{6} = \dfrac{8}{12} = \dfrac{16}{24}$	4. a) $\dfrac{9}{21}$ b)	4. a) $\dfrac{9}{21}$ b) $\dfrac{9}{21} = \dfrac{3}{7}$

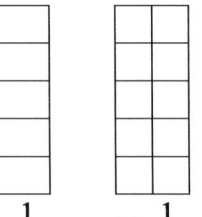

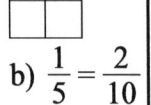

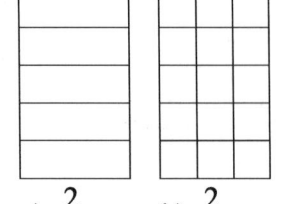

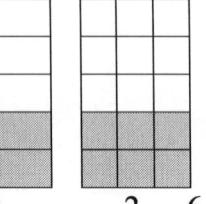

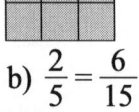

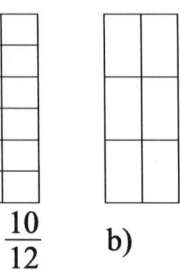

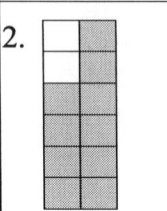

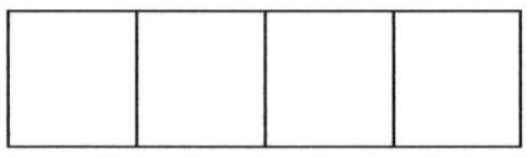

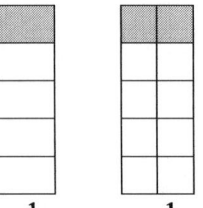

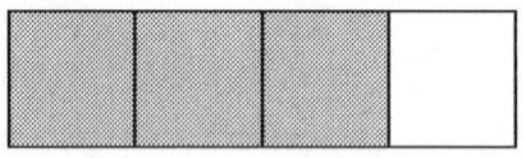

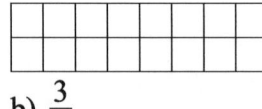

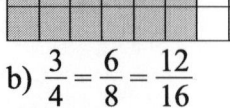

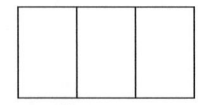

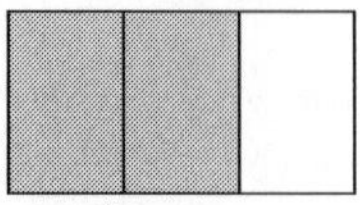

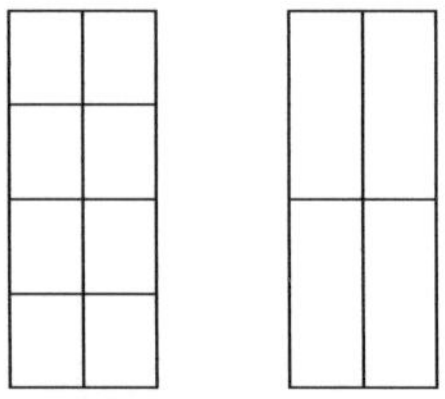

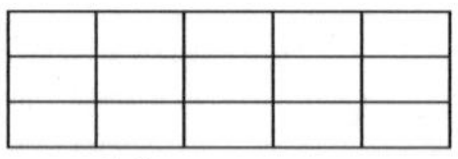

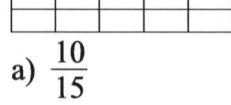

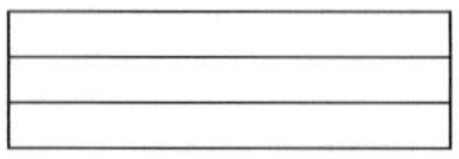

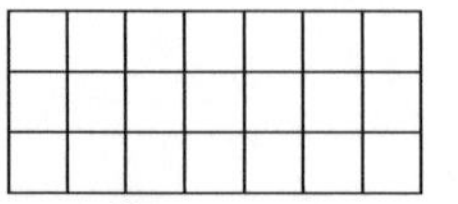

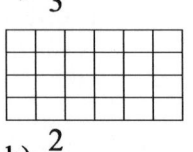

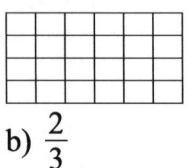

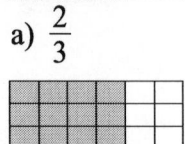

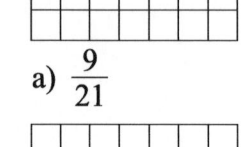

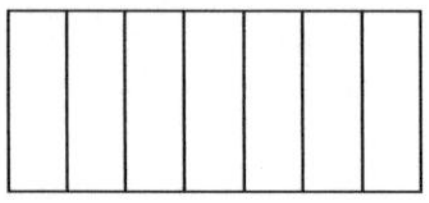

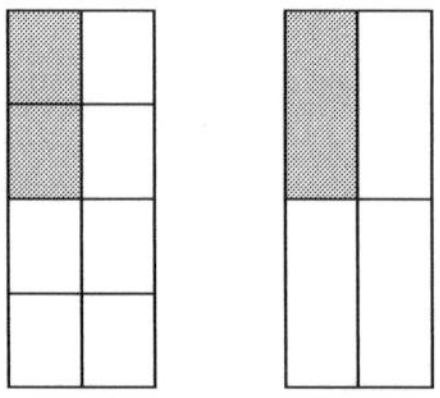

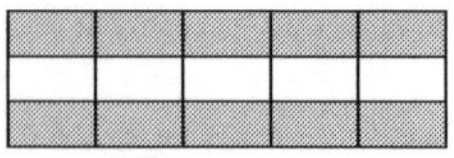

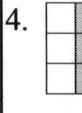

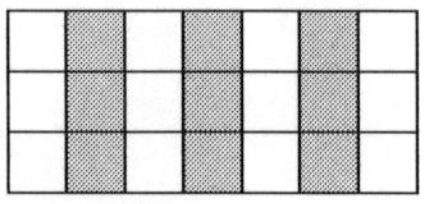

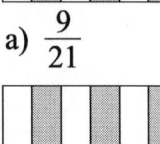

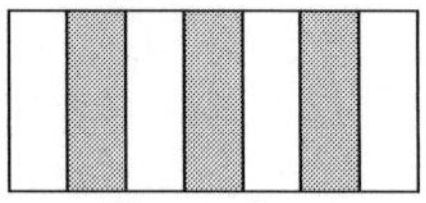

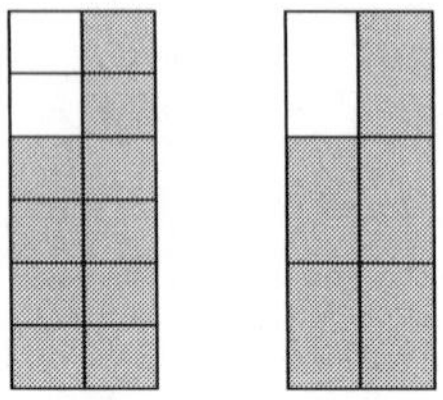

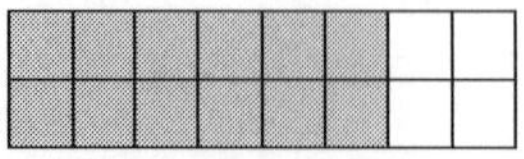

Fractions 5

Equal Fractions. Simplify / Cancel. Lowest Terms.

Equal / Equivalent Fractions

If we **multiply** or **divide** both the **numerator** and **denominator** of a fraction by the **same number** then we obtain equal / equivalent fractions.

Ex.1 $\dfrac{1}{2} = \dfrac{2}{4}$ (×2 top and bottom)

Ex.2 $\dfrac{6}{9} = \dfrac{2}{3}$ (÷3 top and bottom)

Q	A	Q	A
Complete each of the following to form an **equivalent** fraction.			
1) $\dfrac{1}{2} = \dfrac{3}{\_\_}$	1) $\dfrac{1}{2} = \dfrac{3}{6}$ (×3)	4) $\dfrac{4}{6} = \dfrac{2}{\_\_}$	4) $\dfrac{4}{6} = \dfrac{2}{3}$ (÷2)
2) $\dfrac{3}{4} = \dfrac{\_\_}{8}$	2) $\dfrac{3}{4} = \dfrac{6}{8}$ (×2)	5) $\dfrac{12}{16} = \dfrac{\_\_}{4}$	5) $\dfrac{12}{16} = \dfrac{3}{4}$ (÷4)
3) $\dfrac{7}{9} = \dfrac{35}{\_\_}$	3) $\dfrac{7}{9} = \dfrac{35}{45}$ (×5)	6) $\dfrac{14}{42} = \dfrac{\_\_}{6}$	6) $\dfrac{14}{42} = \dfrac{2}{6}$ (÷7)

'Simplify / Cancel Fractions'

Fractions can be **simplified** if the numerator and denominator have a common factor (except 1).
The terms **'simplify'**, **'cancel'** or **'cancel down'** a fraction are used to mean
'**divide** the numerator and denominator by the same number (except 1) to obtain an equivalent fraction'.

'Lowest Terms' / 'Simplest Form'

If a fraction can **not be cancelled any further**, (numerator and denominator can not be divided further) then the fraction is said to be either '**in its lowest terms**' or '**in its simplest form**'.

We may **cancel** a fraction more than once to make sure it is '**in its lowest terms**' - but we usually aim to find the highest number to divide by (the **highest common factor**) in order to keep work to a minimum.

Ex.1 **Cancel** the fraction to its '**lowest terms**'.

$\dfrac{6}{8}$ ⇒ $\dfrac{6}{8} = \dfrac{3}{4}$ (÷2)

1 is the only common factor of 3 and 4 ...
we can not cancel further ...
the fraction is **in its lowest terms**.

Ex.2 Write the fraction '**in its simplest form**'.

$\dfrac{12}{18}$ ⇒ $\dfrac{12}{18} = \dfrac{6}{9} = \dfrac{2}{3}$ (÷2, then ÷3)

or

$\dfrac{12}{18} = \dfrac{4}{6} = \dfrac{2}{3}$ (÷3, then ÷2)

or

6 is the **highest common factor**.

$\dfrac{12}{18} = \dfrac{2}{3}$ (÷6)

1 is the only common factor of 2 and 3 ...
the fraction is **in its simplest form**.

Fractions 6

Cancel Fraction to Lowest Terms / Simplest Form.

Q	A	Q	A
Cancel each fraction to its **lowest terms**.		Write each fraction in its '**simplest form**'.	
1) $\dfrac{3}{6}$	1) $\dfrac{3}{6} \xrightarrow{\div 3} \dfrac{1}{2}$ (÷3 top and bottom)	1) $\dfrac{44}{99}$	1) $\dfrac{44}{99} \xrightarrow{\div 11} \dfrac{4}{9}$
2) $\dfrac{6}{8}$	2) $\dfrac{6}{8} \xrightarrow{\div 2} \dfrac{3}{4}$	2) $\dfrac{13}{26}$	2) $\dfrac{13}{26} \xrightarrow{\div 13} \dfrac{1}{2}$
3) $\dfrac{15}{25}$	3) $\dfrac{15}{25} \xrightarrow{\div 5} \dfrac{3}{5}$	3) $\dfrac{20}{32}$	3) $\dfrac{20}{32} \xrightarrow{\div 2} \dfrac{10}{16} \xrightarrow{\div 2} \dfrac{5}{8}$ or $\dfrac{20}{32} \xrightarrow{\div 4} \dfrac{5}{8}$
4) $\dfrac{21}{49}$	4) $\dfrac{21}{49} \xrightarrow{\div 7} \dfrac{3}{7}$		
5) $\dfrac{18}{24}$	5) $\dfrac{18}{24} \xrightarrow{\div 2} \dfrac{9}{12} \xrightarrow{\div 3} \dfrac{3}{4}$ or $\dfrac{18}{24} \xrightarrow{\div 3} \dfrac{6}{8} \xrightarrow{\div 2} \dfrac{3}{4}$ or $\dfrac{18}{24} \xrightarrow{\div 6} \dfrac{3}{4}$	4) $\dfrac{24}{60}$	4) $\dfrac{24}{60} \xrightarrow{\div 2} \dfrac{12}{30} \xrightarrow{\div 2} \dfrac{6}{15} \xrightarrow{\div 3} \dfrac{2}{5}$ or $\dfrac{24}{60} \xrightarrow{\div 2} \dfrac{12}{30} \xrightarrow{\div 6} \dfrac{2}{5}$ or $\dfrac{24}{60} \xrightarrow{\div 12} \dfrac{2}{5}$
6) $\dfrac{30}{45}$	6) $\dfrac{30}{45} \xrightarrow{\div 3} \dfrac{10}{15} \xrightarrow{\div 5} \dfrac{2}{3}$ or $\dfrac{30}{45} \xrightarrow{\div 5} \dfrac{6}{9} \xrightarrow{\div 3} \dfrac{2}{3}$ or $\dfrac{30}{45} \xrightarrow{\div 15} \dfrac{2}{3}$	*Note Other ways are also possible. 5) $\dfrac{49}{147}$	5) $\dfrac{49}{147} \xrightarrow{\div 7} \dfrac{7}{21} \xrightarrow{\div 7} \dfrac{1}{3}$ or $\dfrac{49}{147} \xrightarrow{\div 49} \dfrac{1}{3}$

Fractions 7

Equal Fractions. Change Improper Fraction to Mixed Number.

Equal / Equivalent Fractions - Improper Fractions

If we **multiply** or **divide** both the **numerator and denominator** of an **improper fraction** by the **same number** then we obtain equal / equivalent fractions.

Ex.1 $\dfrac{3}{2} = \dfrac{6}{4}$ (×2)

Ex.2 $\dfrac{9}{6} = \dfrac{3}{2}$ (÷3)

Q	A	Q	A
Complete each of the following to form an **equivalent** fraction. 1) $\dfrac{4}{3} = \dfrac{20}{}$	1) $\dfrac{4}{3} = \dfrac{20}{15}$ (×5)	3) $\dfrac{8}{6} = \dfrac{4}{}$	3) $\dfrac{8}{6} = \dfrac{4}{3}$ (÷2)
2) $\dfrac{25}{7} = \dfrac{250}{}$	2) $\dfrac{25}{7} = \dfrac{250}{70}$ (×10)	4) $\dfrac{30}{18} = \dfrac{}{3}$	4) $\dfrac{30}{18} = \dfrac{5}{3}$ (÷6)

Change an Improper Fraction to a Mixed Number - 'Cancel'

If an Improper Fraction can be **cancelled**, we can **choose** to do so **before** or **after** changing it to a Mixed Number.

Ex. Change $\dfrac{8}{6}$ to a mixed number.

cancel **before** changing to a mixed number ...

$\dfrac{8}{6} = \dfrac{4}{3} = 1\dfrac{1}{3}$ (÷2)

cancel **after** changing to a mixed number ...

$\dfrac{8}{6} = 1\dfrac{2}{6} = 1\dfrac{1}{3}$ (÷2)

Q	A
Cancel each improper fraction a) **before** changing it to a mixed number. b) **after** changing it to a mixed number. 1) $\dfrac{15}{10}$	1) a) $\dfrac{15}{10} = \dfrac{3}{2} = 1\dfrac{1}{2}$ (÷5) b) $\dfrac{15}{10} = 1\dfrac{5}{10} = 1\dfrac{1}{2}$ (÷5)
2) $\dfrac{30}{8}$	2) a) $\dfrac{30}{8} = \dfrac{15}{4} = 3\dfrac{3}{4}$ (÷2) b) $\dfrac{30}{8} = 3\dfrac{6}{8} = 3\dfrac{3}{4}$ (÷2)
3) $\dfrac{49}{21}$	3) a) $\dfrac{49}{21} = \dfrac{7}{3} = 2\dfrac{1}{3}$ (÷7) b) $\dfrac{49}{21} = 2\dfrac{7}{21} = 2\dfrac{1}{3}$ (÷7)

Fractions 8

Addition - Same Denominators.

Addition - Same Denominators

To add fractions with the **same denominators** (parts are the same size) - we add the **numerators** and write the answer over the **denominator**.

Whole numbers / whole numbers of mixed numbers are added as usual.

For the answer, we usually write fractions in their lowest terms / simplest form and change improper fractions to whole numbers / mixed numbers.

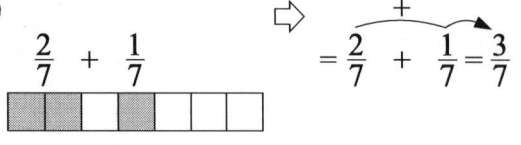

Ex. Calculate
a) $\frac{2}{7} + \frac{1}{7} \Rightarrow = \frac{2}{7} + \frac{1}{7} = \frac{3}{7}$

b) $2\frac{4}{9} + 4\frac{3}{9} \Rightarrow = 2\frac{4}{9} + 4\frac{3}{9} = 6\frac{7}{9}$

Q	A	Q	A
Calculate 1) $\frac{3}{5} + \frac{1}{5}$	1) $\frac{3}{5} + \frac{1}{5} = \frac{4}{5}$	3) $\frac{5}{6} + \frac{5}{6}$	3) $\frac{5}{6} + \frac{5}{6} = \frac{10}{6} = 1\frac{4}{6} = 1\frac{2}{3}$ or $= \frac{5}{3} = 1\frac{2}{3}$
2) $\frac{1}{2} + \frac{1}{2}$	2) $\frac{1}{2} + \frac{1}{2} = \frac{2}{2} = \frac{1}{1} = 1$		

Q	A	Q	A
Calculate 1) $\frac{2}{7} + \frac{3}{7}$	1) $\frac{5}{7}$	13) $\frac{2}{20} + \frac{3}{20} + \frac{7}{20}$	13) $\frac{12}{20} = \frac{3}{5}$
2) $\frac{5}{39} + \frac{14}{39}$	2) $\frac{19}{39}$	14) $\frac{9}{8} + \frac{5}{8}$	14) $\frac{14}{8} = \frac{7}{4} = 1\frac{3}{4}$
3) $\frac{4}{9} + \frac{2}{9} + \frac{1}{9}$	3) $\frac{7}{9}$	15) $\frac{2}{7} + 6$	15) $6\frac{2}{7}$
4) $\frac{1}{4} + \frac{3}{4}$	4) $\frac{4}{4} = 1$	16) $5 + 5\frac{81}{100}$	16) $10\frac{81}{100}$
5) $\frac{5}{5} + \frac{10}{5} + \frac{15}{5}$	5) $\frac{30}{5} = 6$	17) $2\frac{1}{5} + 1\frac{3}{5}$	17) $3\frac{4}{5}$
6) $\frac{8}{3} + \frac{4}{3}$	6) $\frac{12}{3} = 4$	18) $3\frac{2}{9} + 3\frac{2}{9} + 1\frac{2}{9}$	18) $7\frac{6}{9} = 7\frac{2}{3}$
7) $\frac{3}{2} + \frac{5}{2} + \frac{7}{2}$	7) $\frac{15}{2} = 7\frac{1}{2}$	19) $4\frac{1}{6} + 4\frac{5}{6}$	19) $8 + \frac{6}{6}$ $= 8 + 1 = 9$
8) $\frac{5}{9} + \frac{8}{9}$	8) $\frac{13}{9} = 1\frac{4}{9}$	20) $1\frac{5}{7} + 2\frac{3}{7}$	20) $3 + \frac{8}{7}$ $= 3 + 1\frac{1}{7} = 4\frac{1}{7}$
9) $\frac{21}{10} + \frac{27}{10} + \frac{11}{10}$	9) $\frac{59}{10} = 5\frac{9}{10}$	21) $5\frac{7}{8} + 2\frac{7}{8}$	21) $7 + \frac{14}{8}$ $= 7 + 1\frac{6}{8}$ $= 8\frac{6}{8} = 8\frac{3}{4}$
10) $\frac{1}{6} + \frac{1}{6}$	10) $\frac{2}{6} = \frac{1}{3}$		
11) $\frac{1}{6} + \frac{1}{6} + \frac{1}{6}$	11) $\frac{3}{6} = \frac{1}{2}$		
12) $\frac{8}{55} + \frac{12}{55}$	12) $\frac{20}{55} = \frac{4}{11}$		

Fractions 9

Subtraction - Same Denominators.

Subtraction - Same Denominators

To subtract fractions with the **same denominators** (parts are the same size) - we subtract the **numerators** and write the answer over the **denominator**.

Whole numbers / whole numbers of mixed numbers are subtracted as usual.

For the answer, we usually write fractions in their lowest terms / simplest form and change improper fractions to whole numbers / mixed numbers.

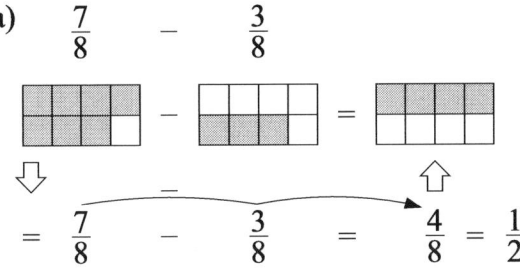

Ex. Calculate a) $\frac{7}{8} - \frac{3}{8} = \frac{7}{8} - \frac{3}{8} = \frac{4}{8} = \frac{1}{2}$

b) $3\frac{4}{5} - 1\frac{2}{5} = 3\frac{4}{5} - 1\frac{2}{5} = 2\frac{2}{5}$

Q	A	Q	A
Calculate 1) $\frac{2}{3} - \frac{1}{3}$	1) $\frac{2}{3} - \frac{1}{3} = \frac{1}{3}$	3) $1\frac{5}{7} - \frac{3}{7}$	3) $1\frac{5}{7} - \frac{3}{7} = 1\frac{2}{7}$
2) $\frac{9}{10} - \frac{4}{10}$	2) $\frac{9}{10} - \frac{4}{10} = \frac{5}{10} = \frac{1}{2}$		

Q	A	Q	A
Calculate 1) $\frac{1}{2} - \frac{1}{2}$	1) $\frac{0}{2} = 0$	10) $9\frac{9}{10} - \frac{3}{10}$	10) $9\frac{6}{10} = 9\frac{3}{5}$
2) $\frac{8}{9} - \frac{3}{9}$	2) $\frac{5}{9}$	11) $6\frac{7}{8} - 6\frac{7}{8}$	11) 0
3) $\frac{7}{7} - \frac{6}{7}$	3) $\frac{1}{7}$	12) $4\frac{1}{2} - 2\frac{1}{2}$	12) 2
4) $\frac{33}{5} - \frac{3}{5}$	4) $\frac{30}{5} = 6$	13) $5\frac{6}{7} - 1\frac{4}{7}$	13) $4\frac{2}{7}$
5) $\frac{5}{6} - \frac{1}{6}$	5) $\frac{4}{6} = \frac{2}{3}$	14) $\frac{7}{9} - \frac{5}{9} - \frac{2}{9}$	14) $\frac{0}{9} = 0$
6) $\frac{29}{30} - \frac{11}{30}$	4) $\frac{18}{30} = \frac{3}{5}$	15) $\frac{6}{7} - \frac{3}{7} - \frac{1}{7}$	15) $\frac{2}{7}$
7) $\frac{16}{3} - \frac{8}{3}$	7) $\frac{8}{3} = 2\frac{2}{3}$	16) $5\frac{9}{11} - \frac{2}{11} - \frac{3}{11}$	16) $5\frac{4}{11}$
8) $\frac{9}{4} - \frac{3}{4}$	8) $\frac{6}{4} = 1\frac{2}{4} = 1\frac{1}{2}$	17) $8\frac{4}{5} - 2\frac{1}{5} - 5\frac{2}{5}$	17) $1\frac{1}{5}$
9) $8\frac{2}{5} - 3$	9) $5\frac{2}{5}$	18) $9\frac{7}{8} - 1\frac{3}{8} - 6\frac{4}{8}$	18) $2\frac{0}{8} = 2$

Fractions 10 — Subtraction - from a Whole Number. Addition / Subtraction.

Subtraction - from a Whole Number

To subtract a **fraction from a whole number** - change **1 unit of the whole number** to an improper fraction with the **same denominator** as the given fraction - the numerators are then subtracted as already seen.

*Note This is usually done **mentally**.

Ex. Calculate

a) $1 - \frac{3}{7} \Rightarrow = \frac{7}{7} - \frac{3}{7} = \frac{4}{7}$

b) $6 - \frac{2}{3} \Rightarrow = 5\frac{3}{3} - \frac{2}{3} = 5\frac{1}{3}$

b) $8 - 4\frac{1}{6} \Rightarrow = 7\frac{6}{6} - 4\frac{1}{6} = 3\frac{5}{6}$

Q Calculate	A	Q	A
1) $1 - \frac{1}{2}$	1) $\frac{2}{2} - \frac{1}{2} = \frac{1}{2}$	8) $6 - \frac{7}{9}$	8) $5\frac{9}{9} - \frac{7}{9} = 5\frac{2}{9}$
2) $1 - \frac{3}{8}$	2) $\frac{8}{8} - \frac{3}{8} = \frac{5}{8}$	9) $2 - \frac{18}{53}$	9) $1\frac{53}{53} - \frac{18}{53} = 1\frac{35}{53}$
3) $1 - \frac{7}{9}$	3) $\frac{9}{9} - \frac{7}{9} = \frac{2}{9}$	10) $4 - \frac{101}{200}$	10) $3\frac{200}{200} - \frac{101}{200} = 3\frac{99}{200}$
4) $1 - \frac{11}{21}$	4) $\frac{21}{21} - \frac{11}{21} = \frac{10}{21}$	11) $7 - 1\frac{3}{4}$	11) $6\frac{4}{4} - 1\frac{3}{4} = 5\frac{1}{4}$
5) $1 - \frac{33}{100}$	5) $\frac{100}{100} - \frac{33}{100} = \frac{67}{100}$	12) $5 - 2\frac{1}{6}$	12) $4\frac{6}{6} - 2\frac{1}{6} = 2\frac{5}{6}$
6) $8 - \frac{1}{4}$	6) $7\frac{4}{4} - \frac{1}{4} = 7\frac{3}{4}$	13) $9 - 8\frac{7}{8}$	13) $8\frac{8}{8} - 8\frac{7}{8} = \frac{1}{8}$
7) $3 - \frac{3}{8}$	7) $2\frac{8}{8} - \frac{3}{8} = 2\frac{5}{8}$	14) $3 - 1\frac{25}{44}$	14) $2\frac{44}{44} - 1\frac{25}{44} = 1\frac{19}{44}$

Addition and Subtraction

Addition and Subtraction may be combined.

Q Calculate	A	Q	A
1) $\frac{5}{7} - \frac{4}{7} + \frac{2}{7}$	1) $\frac{3}{7}$	7) $8\frac{4}{5} - 4\frac{2}{5} + 8\frac{1}{5} - 9\frac{3}{5}$	7) 3
2) $\frac{8}{9} + \frac{1}{9} - \frac{4}{9}$	2) $\frac{5}{9}$	8) $2\frac{6}{7} + 3\frac{4}{7} - 1\frac{1}{7}$	8) $4\frac{9}{7} = 4 + 1\frac{2}{7} = 5\frac{2}{7}$
3) $\frac{1}{5} + \frac{2}{5} - \frac{3}{5}$	3) $\frac{0}{5} = 0$	9) $4 - 1\frac{5}{6} + 5\frac{1}{6}$	9) $= 3\frac{6}{6} - 1\frac{5}{6} + 5\frac{1}{6}$ $= 7\frac{2}{6} = 7\frac{1}{3}$
4) $\frac{7}{8} - \frac{3}{8} + \frac{5}{8}$	4) $\frac{9}{8} = 1\frac{1}{8}$		
5) $7\frac{11}{12} + \frac{5}{12} - 1\frac{7}{12}$	5) $6\frac{9}{12} = 6\frac{3}{4}$	10) $7\frac{2}{9} + 6 - 8\frac{4}{9}$	10) $= 7\frac{2}{9} + 5\frac{9}{9} - 8\frac{4}{9}$ $= 4\frac{7}{9}$
6) $6\frac{2}{3} - 2\frac{1}{3} + 4\frac{1}{3}$	6) $8\frac{2}{3}$		

Fractions 11

Addition / Subtraction - Different Denominators.

Addition / Subtraction - Different Denominators

To add or subtract fractions with **different denominators** -
we find **equivalent fractions** with the **same denominators** (usually the smallest possible number) ...
then add or subtract the **numerators** and write the answer over the **denominator**.
(In diagram form, think to 'break' the denominator parts so they are the same size.)

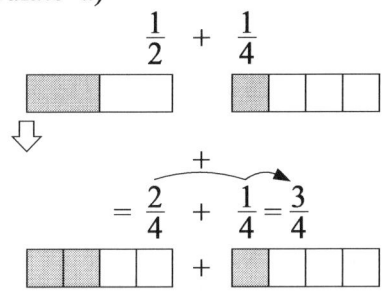

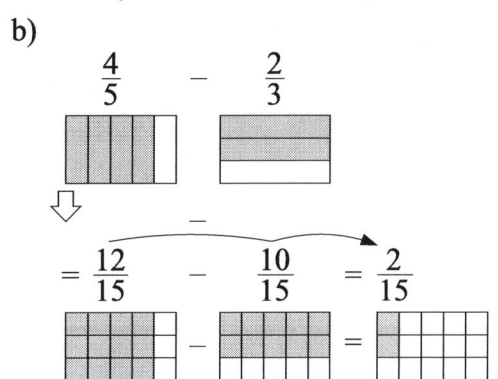

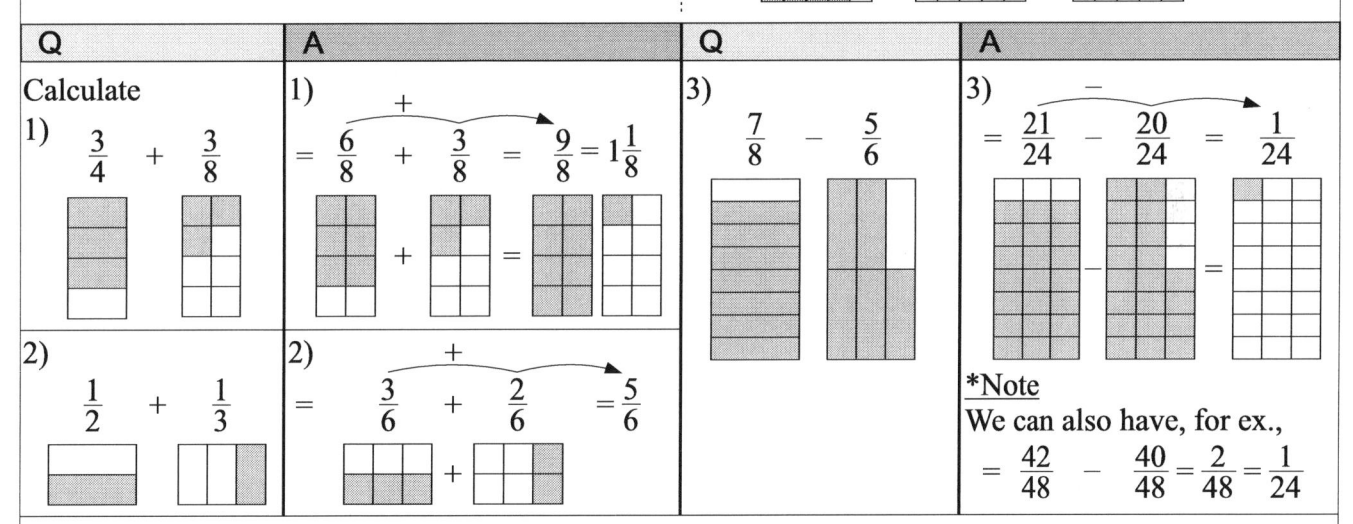

Ex. Calculate a) $\frac{1}{2} + \frac{1}{4}$

$= \frac{2}{4} + \frac{1}{4} = \frac{3}{4}$

b) $\frac{4}{5} - \frac{2}{3}$

$= \frac{12}{15} - \frac{10}{15} = \frac{2}{15}$

Q	A	Q	A
Calculate 1) $\frac{3}{4} + \frac{3}{8}$	1) $= \frac{6}{8} + \frac{3}{8} = \frac{9}{8} = 1\frac{1}{8}$	3) $\frac{7}{8} - \frac{5}{6}$	3) $= \frac{21}{24} - \frac{20}{24} = \frac{1}{24}$ *Note We can also have, for ex., $= \frac{42}{48} - \frac{40}{48} = \frac{2}{48} = \frac{1}{24}$
2) $\frac{1}{2} + \frac{1}{3}$	2) $= \frac{3}{6} + \frac{2}{6} = \frac{5}{6}$		

To find Equivalent Fractions with the Same Denominators

Multiply each denominator by **a number** that makes **both denominators equal**.
Multiply each numerator by the **same number** to give **equivalent fractions**.

*Note
We usually find the **smallest** number that makes both denominators equal -
this is also called '**finding the Lowest Common Denominator**' -
otherwise we have to cancel at the end ... but this is acceptable.

Ex.
Calculate a) $\frac{1}{2} + \frac{1}{4}$ ⇒ $\times 2 \left(\frac{1}{2} + \frac{1}{4} \right) \times 1 = \frac{2}{4} + \frac{1}{4} = \frac{3}{4}$

or (for ex.) $\times 4 \left(\frac{1}{2} + \frac{1}{4} \right) \times 2 = \frac{4}{8} + \frac{2}{8} = \frac{6}{8} = \frac{3}{4}$

b) $\frac{4}{5} - \frac{2}{3}$ ⇒ $\times 3 \left(\frac{4}{5} - \frac{2}{3} \right) \times 5 = \frac{12}{15} - \frac{10}{15} = \frac{2}{15}$

or (for ex.) $\times 6 \left(\frac{4}{5} - \frac{2}{3} \right) \times 10 = \frac{24}{30} - \frac{20}{30} = \frac{4}{15} = \frac{2}{15}$

c) $\frac{5}{6} + \frac{7}{8}$ ⇒ $\times 4 \left(\frac{5}{6} + \frac{7}{8} \right) \times 3 = \frac{20}{24} + \frac{21}{24} = \frac{41}{24} = 1\frac{17}{24}$

or (for ex.) $\times 8 \left(\frac{5}{6} - \frac{7}{8} \right) \times 6 = \frac{40}{48} - \frac{42}{48} = \frac{82}{48} = \frac{41}{24} = 1\frac{17}{24}$

Fractions 12

Addition / Subtraction - Different Denominators.

*Note
The numerators may be written over the common denominator as follows (to avoid writing it more than once) -

Ex. $\times 1\left(\frac{1}{6} + \frac{1}{3}\right)\times 2 = \frac{1 + 2}{6} = \frac{3}{6} = \frac{1}{2}$

Q	A	Q	A
Calculate 1) $\frac{1}{6} + \frac{1}{3}$	1) $\times 1\left(\frac{1}{6} + \frac{1}{3}\right)\times 2$ $= \frac{1}{6} + \frac{2}{6} = \frac{3}{6} = \frac{1}{2}$	11) $\frac{1}{2} - \frac{1}{4}$	11) $\times 2\left(\frac{1}{2} - \frac{1}{4}\right)\times 1$ $= \frac{2}{4} - \frac{1}{4} = \frac{1}{4}$
2) $\frac{2}{5} + \frac{9}{20}$	2) $\times 4\left(\frac{2}{5} + \frac{9}{20}\right)\times 1$ $= \frac{8}{20} + \frac{9}{20} = \frac{17}{20}$	12) $\frac{33}{35} - \frac{3}{7}$	12) $\times 1\left(\frac{33}{35} - \frac{3}{7}\right)\times 5$ $= \frac{33}{35} - \frac{15}{35} = \frac{18}{35}$
3) $\frac{1}{4} + \frac{7}{8}$	3) $\times 2\left(\frac{1}{4} + \frac{7}{8}\right)\times 1$ $= \frac{2}{8} + \frac{7}{8} = \frac{9}{8} = 1\frac{1}{8}$	13) $\frac{5}{7} - \frac{1}{2}$	13) $\times 2\left(\frac{5}{7} - \frac{1}{2}\right)\times 7$ $= \frac{10}{14} - \frac{7}{14} = \frac{3}{14}$
4) $\frac{2}{5} + \frac{1}{2}$	4) $\times 2\left(\frac{2}{5} + \frac{1}{2}\right)\times 5$ $= \frac{4}{10} + \frac{5}{10} = \frac{9}{10}$	14) $\frac{3}{4} - \frac{7}{10}$	14) $\times 5\left(\frac{3}{4} - \frac{7}{10}\right)\times 2$ $= \frac{15}{20} - \frac{14}{20} = \frac{1}{14}$
5) $\frac{3}{4} + \frac{2}{3}$	5) $\times 3\left(\frac{3}{4} + \frac{2}{3}\right)\times 4$ $= \frac{9}{12} + \frac{8}{12} = \frac{17}{12} = 1\frac{5}{12}$	15) $\frac{1}{2} - \frac{1}{3} - \frac{1}{6}$	15) $\times 3\left(\frac{1}{2} - \frac{1}{3}\right)\times 2 - \frac{1}{6}\times 1$ $= \frac{3}{6} - \frac{2}{6} - \frac{1}{6} = \frac{0}{6} = 0$
6) $\frac{1}{6} + \frac{1}{4}$	6) $\times 2\left(\frac{1}{6} + \frac{1}{4}\right)\times 3$ $= \frac{2}{12} + \frac{3}{12} = \frac{5}{12}$	16) $\frac{2}{3} + \frac{1}{4} - \frac{3}{5}$	16) $\times 20\left(\frac{2}{3} + \frac{1}{4}\right)\times 15 - \frac{3}{5}\times 12$ $= \frac{40}{60} + \frac{15}{60} - \frac{36}{60} = \frac{19}{60}$
7) $\frac{5}{8} + \frac{5}{6}$	7) $\times 3\left(\frac{5}{8} + \frac{5}{6}\right)\times 4$ $= \frac{15}{24} + \frac{20}{24} = \frac{35}{24} = 1\frac{11}{24}$	17) $\frac{3}{4} - \frac{5}{8} + \frac{13}{16}$	17) $\times 4\left(\frac{3}{4} - \frac{5}{8}\right)\times 2 + \frac{13}{16}\times 1$ $= \frac{12}{16} - \frac{10}{16} + \frac{13}{16} = \frac{15}{16}$
8) $\frac{11}{25} + \frac{9}{20}$	8) $\times 4\left(\frac{11}{25} + \frac{9}{20}\right)\times 5$ $= \frac{44}{100} + \frac{45}{100} = \frac{89}{100}$	18) $8\frac{9}{10} - \frac{2}{5}$	18) $8\frac{9}{10}\times 1 - \frac{2}{5}\times 2$ $= 8\frac{9}{10} - \frac{4}{10} = 8\frac{5}{10} = 8\frac{1}{2}$
9) $\frac{1}{2} + \frac{1}{4} + \frac{1}{8}$	9) $\times 4\left(\frac{1}{2} + \frac{1}{4}\right)\times 2 + \frac{1}{8}\times 1$ $= \frac{4}{8} + \frac{2}{8} + \frac{1}{8} = \frac{7}{8}$	19) $5\frac{2}{3} - 1\frac{5}{9}$	19) $4+\left[\times 3\left(\frac{2}{3} - \frac{5}{9}\right)\times 1 = \frac{6}{9} - \frac{5}{9} = \frac{1}{9}\right] = 4\frac{1}{9}$
10) $1\frac{7}{9} + 3\frac{5}{6}$	10) $4+\left[\times 2\left(\frac{7}{9} + \frac{5}{6}\right)\times 3 = \frac{14}{18} + \frac{15}{18} = \frac{29}{18} = 1\frac{11}{18}\right]$ $= 5\frac{11}{18}$	20) $7\frac{1}{6} - 6\frac{1}{7}$	20) $1+\left[\times 7\left(\frac{1}{6} - \frac{1}{7}\right)\times 6 = \frac{7}{42} - \frac{6}{42} = \frac{1}{42}\right] = 1\frac{1}{42}$

Fractions 13

Addition / Subtraction - Negative Fractions

Addition / subtraction of fractions can involve **negative fractions**.
(The following Exs. and Q/A just consider fractions with the **same denominator**.)

Ex. Calculate

a) $\frac{1}{7} - \frac{3}{7} \Rightarrow = \frac{1}{7} - \frac{3}{7} = -\frac{2}{7}$

b) $-\frac{1}{7} + \frac{3}{7} \Rightarrow = -\frac{1}{7} + \frac{3}{7} = \frac{2}{7}$

c) $-\frac{1}{7} - \frac{3}{7} \Rightarrow = -\frac{1}{7} - \frac{3}{7} = -\frac{4}{7}$

d) $\frac{1}{7} + -\frac{3}{7} \Rightarrow = \frac{1}{7} - \frac{3}{7} = -\frac{2}{7}$

e) $\frac{1}{7} - -\frac{3}{7} \Rightarrow = \frac{1}{7} + \frac{3}{7} = \frac{4}{7}$

Q Calculate	A	Q	A
1) $\frac{1}{5} - \frac{4}{5}$	1) $-\frac{3}{5}$	7) $\frac{4}{9} - -\frac{4}{9}$	7) $\frac{4}{9} + \frac{4}{9} = \frac{8}{9}$
2) $-\frac{3}{8} + \frac{3}{8}$	2) $\frac{0}{8} = 0$	8) $-\frac{9}{10} + \frac{3}{10}$	8) $-\frac{6}{10} = -\frac{3}{5}$
3) $-\frac{2}{9} + \frac{7}{9}$	3) $\frac{5}{9}$	9) $\frac{7}{8} + -\frac{2}{8}$	9) $\frac{7}{8} - \frac{2}{8} = \frac{5}{8}$
4) $-\frac{1}{3} - \frac{1}{3}$	4) $-\frac{2}{3}$	10) $-\frac{3}{4} + -\frac{3}{4}$	10) $-\frac{3}{4} - \frac{3}{4} = -\frac{6}{4} = -1\frac{2}{4} = -1\frac{1}{2}$
5) $-\frac{1}{6} - \frac{5}{6}$	5) $-\frac{6}{6} = -1$	11) $\frac{5}{6} - -\frac{2}{6}$	11) $\frac{5}{6} + \frac{2}{6} = \frac{7}{6} = 1\frac{1}{6}$
6) $\frac{5}{7} + -\frac{6}{7}$	6) $\frac{5}{7} - \frac{6}{7} = -\frac{1}{7}$	12) $-\frac{19}{20} - -\frac{7}{20}$	12) $-\frac{19}{20} + \frac{7}{20} = -\frac{12}{20} = -\frac{3}{5}$

In the following Ex.1 and Q/A 1-6, it is useful to consider **mentally** reversing the values - subtract the **smaller value** <u>from</u> the **larger value** - and then place a minus sign before the answer value to compensate for this.

Ex.1 Calculate

a) $2\frac{1}{5} - 9\frac{4}{5} = -7\frac{3}{5}$

'think' $9\frac{4}{5} - 2\frac{1}{5} = 7\frac{3}{5}$

Ex.2 Calculate

b) $-2\frac{1}{5} - 9\frac{3}{5} = -11\frac{4}{5}$

(add together both negative values)

Q Calculate	A	Q	A
1) $\frac{1}{6} - 3$	1) $-2\frac{5}{6}$	7) $-\frac{3}{4} - 8$	7) $-8\frac{3}{4}$
2) $1\frac{2}{9} - 6$	2) $-4\frac{7}{9}$	8) $-5 - \frac{1}{5}$	8) $-5\frac{1}{5}$
3) $4 - 7\frac{5}{8}$	3) $-3\frac{5}{8}$	9) $-2\frac{4}{7} - 6\frac{2}{7}$	9) $-8\frac{6}{7}$
4) $2 - 9\frac{6}{7}$	4) $-7\frac{6}{7}$	10) $-4\frac{1}{6} - 4\frac{1}{6}$	10) $-8\frac{2}{6} = -8\frac{1}{3}$
5) $3\frac{1}{3} - 4\frac{2}{3}$	5) $-1\frac{1}{3}$	11) $-1\frac{2}{3} - 2\frac{1}{3}$	11) $-3\frac{3}{3} = -(3+1) = -4$
6) $5\frac{3}{10} - 8\frac{7}{10}$	6) $-3\frac{4}{10} = -3\frac{2}{5}$	12) $-3\frac{4}{5} - 1\frac{3}{5}$	12) $-4\frac{7}{5} = -(4+1\frac{2}{5}) = -5\frac{2}{5}$

Fractions 14

Subtraction - Mixed Numbers.

<u>Subtraction - Mixed Numbers</u>

Subtracting a **larger fraction part** from a **smaller fraction part** of mixed numbers leads to extra work ... 3 Methods are considered.

<u>Method 1</u>

We subtract the whole numbers and make sure **both fractions** have the **same denominator**.

1 whole unit is then written as a fraction with the **same denominator** as those being subtracted to allow the subtraction to take place - its numerator is **added** to the smaller numerator.

Ex.1 $8\frac{2}{7} - 3\frac{6}{7}$

$5 + \left[\frac{2}{7} - \frac{6}{7} \right] =$

let $1 = \frac{7}{7}$ to allow the subtraction.

$4\cancel{5} + \left[\cancel{\frac{2}{7}}^{(+7)9} - \frac{6}{7} = \frac{3}{7} \right] = 4\frac{3}{7}$

Ex.2 $9\frac{1}{3} - 1\frac{1}{2}$ (×2, ×3)

$8 + \left[\frac{2}{6} - \frac{3}{6} \right] =$

let $1 = \frac{6}{6}$ to allow the subtraction.

$7\cancel{8} + \left[\cancel{\frac{2}{6}}^{(+6)8} - \frac{3}{6} = \frac{5}{6} \right] = 7\frac{5}{6}$

Q	A	Q	A
Calculate 1) $7\frac{1}{4} - 5\frac{3}{4}$	1) $7\frac{1}{4} - 5\frac{3}{4} =$ $2 + \left[\frac{1}{4} - \frac{3}{4} \right] =$ let $1 = \frac{4}{4}$ to allow the subtraction. $1\cancel{2} + \left[\cancel{\frac{1}{4}}^{(+4)5} - \frac{3}{4} = \frac{2}{4} = \frac{1}{2} \right] = 1\frac{1}{2}$	4) $2\frac{1}{3} - 1\frac{8}{9}$	4) $2\frac{1}{3} - 1\frac{8}{9} =$ (×3, ×1) $1 + \left[\frac{3}{9} - \frac{8}{9} \right] =$ let $1 = \frac{9}{9}$ to allow the subtraction. $0\cancel{1} + \left[\cancel{\frac{3}{9}}^{(+9)12} - \frac{8}{9} = \frac{4}{9} \right] = \frac{4}{9}$
2) $6\frac{2}{5} - 2\frac{4}{5}$	2) $6\frac{2}{5} - 2\frac{4}{5} =$ $4 + \left[\frac{2}{5} - \frac{4}{5} \right] =$ let $1 = \frac{5}{5}$ to allow the subtraction. $3\cancel{4} + \left[\cancel{\frac{2}{5}}^{(+5)7} - \frac{4}{5} = \frac{3}{5} \right] = 3\frac{3}{5}$	5) $9\frac{1}{2} - 7\frac{4}{5}$	5) $9\frac{1}{2} - 7\frac{4}{5} =$ (×5, ×2) $2 + \left[\frac{5}{10} - \frac{8}{10} \right] =$ let $1 = \frac{10}{10}$ to allow the subtraction. $1\cancel{2} + \left[\cancel{\frac{5}{10}}^{(+10)15} - \frac{8}{10} = \frac{7}{10} \right] = 1\frac{7}{10}$
3) $7\frac{3}{4} - \frac{7}{8}$	3) $7\frac{3}{4} - \frac{7}{8} =$ (×2, ×1) $7 + \left[\frac{6}{8} - \frac{7}{8} \right] =$ let $1 = \frac{8}{8}$ to allow the subtraction. $6\cancel{7} + \left[\cancel{\frac{6}{8}}^{(+8)14} - \frac{7}{8} = \frac{7}{8} \right] = 6\frac{7}{8}$	6) $6\frac{5}{8} - 3\frac{5}{6}$	6) $6\frac{5}{8} - 3\frac{5}{6} =$ (×3, ×4) $3 + \left[\frac{15}{24} - \frac{20}{24} \right] =$ let $1 = \frac{24}{24}$ to allow the subtraction. $2\cancel{3} + \left[\cancel{\frac{15}{24}}^{(+24)39} - \frac{20}{24} = \frac{19}{24} \right] = 2\frac{19}{24}$

Fractions 15 — Subtraction - Mixed Numbers.

54

Method 2
We make sure **both fractions** have the **same denominator**, then break the subtraction into **2 parts** - calculate **mentally** the difference a) from the **smallest fraction** to the **next whole number**
                    and b) from that **whole number** to the **largest fraction**.
We then **add the 2 answers**.

Ex.1 $8\frac{2}{7} - 3\frac{6}{7}$

$3\frac{6}{7}$ to $4\ = \frac{1}{7}$
4 to $8\frac{2}{7} = 4\frac{2}{7}$ $\Big\}+$
Total $= 4\frac{3}{7}$

Ex.2 $9\frac{1}{3} - 1\frac{1}{2}$ (×2, ×3)

$9\frac{2}{6} - 1\frac{3}{6}$

$1\frac{3}{6}$ to $2 = \frac{3}{6}$
2 to $9\frac{2}{6} = 7\frac{2}{6}$ $\Big\}+$
Total $= 7\frac{5}{6}$

Q	A	Q	A
Calculate 1) $7\frac{1}{4} - 5\frac{3}{4}$	1) $5\frac{3}{4}$ to $6 = \frac{1}{4}$ 6 to $7\frac{1}{4} = 1\frac{1}{4}$ $\Big\}+$ Total $= 1\frac{2}{4}$ $= 1\frac{1}{2}$	4) $2\frac{1}{3} - 1\frac{8}{9}$	4) $2\frac{1}{3} - 1\frac{8}{9}$ (×3, ×1) $= 2\frac{3}{9} - 1\frac{8}{9}$ $1\frac{8}{9}$ to $2 = \frac{1}{9}$ 2 to $2\frac{3}{9} = \frac{3}{9}$ $\Big\}+$ Total $= \frac{4}{9}$
2) $6\frac{2}{5} - 2\frac{4}{5}$	2) $2\frac{4}{5}$ to $3 = \frac{1}{5}$ 3 to $6\frac{2}{5} = 3\frac{2}{5}$ $\Big\}+$ Total $= 3\frac{3}{5}$	5) $9\frac{1}{2} - 7\frac{4}{5}$	5) $9\frac{1}{2} - 7\frac{4}{5}$ (×5, ×2) $= 9\frac{5}{10} - 7\frac{8}{10}$ $7\frac{8}{10}$ to $8 = \frac{2}{10}$ 8 to $9\frac{5}{10} = 1\frac{5}{10}$ $\Big\}+$ Total $= 1\frac{7}{10}$
3) $7\frac{3}{4} - \frac{7}{8}$	3) $7\frac{3}{4} - \frac{7}{8}$ (×2, ×1) $= 7\frac{6}{8} - \frac{7}{8}$ $\frac{7}{8}$ to $1 = \frac{1}{8}$ 1 to $7\frac{6}{8} = 6\frac{6}{8}$ $\Big\}+$ Total $= 6\frac{7}{8}$	6) $6\frac{5}{8} - 3\frac{5}{6}$	6) $6\frac{5}{8} - 3\frac{5}{6}$ (×3, ×4) $= 6\frac{15}{24} - 3\frac{20}{24}$ $3\frac{20}{24}$ to $4 = \frac{4}{24}$ 4 to $6\frac{15}{24} = 2\frac{15}{24}$ $\Big\}+$ Total $= 2\frac{19}{24}$

Fractions 16 — Subtraction - Mixed Numbers.

Method 3
We subtract the whole numbers and make sure **both fractions** have the **same denominator**.
Subtract the **numerators** to give a **negative fraction**.
Add the **whole numbers** to this **negative fraction**.

Ex.1 $8\frac{2}{7} - 3\frac{6}{7}$
⇩
$5 + \frac{-4}{7} =$
$5 + \frac{-4}{7} = 4\frac{3}{7}$

Ex.2 $9\frac{1}{3} - 1\frac{1}{2}$ (×2, ×3)
⇩
$9\frac{2}{6} - 1\frac{3}{6}$
$8 + \frac{-1}{6} =$
$8 + \frac{-1}{6} = 7\frac{5}{6}$

Q	A	Q	A
Calculate 1) $7\frac{1}{4} - 5\frac{3}{4}$	1) $2 + \frac{-2}{4} = 1\frac{2}{4}$ $= 1\frac{1}{2}$	4) $2\frac{1}{3} - 1\frac{8}{9}$	4) $2\frac{1}{3} - 1\frac{8}{9}$ (×3, ×1) $= 2\frac{3}{9} - 1\frac{8}{9}$ $1 + \frac{-5}{9} = \frac{4}{9}$
2) $6\frac{2}{5} - 2\frac{4}{5}$	2) $4 + \frac{-2}{5} = 3\frac{3}{5}$	5) $9\frac{1}{2} - 7\frac{4}{5}$	5) $9\frac{1}{2} - 7\frac{4}{5}$ (×5, ×2) $= 9\frac{5}{10} - 7\frac{8}{10}$ $2 + \frac{-3}{10} = 1\frac{7}{10}$
3) $7\frac{3}{4} - \frac{7}{8}$	3) $7\frac{3}{4} - \frac{7}{8}$ (×2, ×1) $= 7\frac{6}{8} - \frac{7}{8}$ $7 + \frac{-1}{8} = 6\frac{7}{8}$	6) $6\frac{5}{8} - 3\frac{5}{6}$	6) $6\frac{5}{8} - 3\frac{5}{6}$ (×3, ×4) $= 6\frac{15}{24} - 3\frac{20}{24}$ $3 + \frac{-5}{24} = 2\frac{19}{24}$

Fractions 17 Addition / Subtraction - 'Problems'.

Addition / Subtraction - 'Problems'

Q	A
1) A boy walks $\frac{5}{8}$ of a mile and then runs $\frac{1}{8}$ of a mile. How far has he travelled?	1) $\frac{5}{8} + \frac{1}{8} = \frac{6}{8} = \frac{3}{4}$ mile (÷2)
2) $\frac{4}{5}$ of a can of paint is mixed with $\frac{3}{5}$ of another can. The cans are the same size. How much paint is mixed?	2) $\frac{4}{5} + \frac{3}{5} = \frac{7}{5} = 1\frac{2}{5}$ cans
3) A box weighs $\frac{3}{4}$ kg and its contents weigh $\frac{2}{5}$ kg. What is the total weight of the box and its contents?	3) ×5$\left(\frac{3}{4} + \frac{2}{5}\right)$×4 $= \frac{15}{20} + \frac{8}{20} = \frac{23}{20} = 1\frac{3}{20}$ kg
4) Find the total weight when $2\frac{8}{9}$g and $1\frac{1}{6}$g of silver are added together.	4) $2\frac{8}{9} + 1\frac{1}{6} =$ $3 + \left[×2\left(\frac{8}{9} + \frac{1}{6}\right)×3\right]$ $= \frac{16}{18} + \frac{3}{18} = \frac{19}{18} = 1\frac{1}{18} = 4\frac{1}{18}$ g
5) A container holds $\frac{7}{9}$ litres of water. Find the fraction remaining if $\frac{2}{9}$ litres of water is poured out.	5) $\frac{7}{9} - \frac{2}{9} = \frac{5}{9}$ litres
6) The temperature drops from 10°C by $\frac{11}{20}$°C. What is the new temperature?	6) $10 - \frac{11}{20} =$ $9\frac{20}{20} - \frac{11}{20} = 9\frac{9}{20}$ °C
7) The length of a tube is $\frac{6}{7}$m. If $\frac{32}{49}$m is cut from it, how long is the remaining tube?	7) ×7$\left(\frac{6}{7} - \frac{32}{49}\right)$×1 $\frac{42}{49} - \frac{32}{49} = \frac{10}{49}$ m
8) A car travels from X to Y to Z. X to Z is $6\frac{5}{8}$km. X to Y is $1\frac{1}{3}$km. Calculate the distance Y to Z.	8) $6\frac{5}{8} - 1\frac{1}{3} =$ $5 + \left[×3\left(\frac{5}{8} - \frac{1}{3}\right)×8\right]$ $= \frac{15}{24} - \frac{8}{24} = \frac{7}{24} = 5\frac{7}{24}$ km
9) A family spends $\frac{2}{5}$ of its income on food, $\frac{1}{6}$ on bills, and $\frac{3}{10}$ on 'other items'. The rest is saved. What fraction of the family's income is a) spent? b) saved?	9) a) ×6$\left(\frac{2}{5} + \frac{1}{6}\right)$×5 $+ \frac{3}{10}$ ×3 $\frac{12}{30} + \frac{5}{30} + \frac{9}{30} = \frac{26}{30} = \frac{13}{15}$ b) $1 - \frac{13}{15} =$ $\frac{15}{15} - \frac{13}{15} = \frac{2}{15}$

Fractions 18

Compare the Size of Fractions.

Compare the Size of Fractions
To compare the size of fractions -
we make **equivalent fractions** with the **same denominators** and then just **compare the numerators**.

*****Note**
Fractions may be **converted to decimals** - see **Fractions 29** - and then compared.

Ex. State the largest fraction

$\frac{3}{5}, \frac{2}{3} \Rightarrow$ ×3$\left(\frac{3}{5}, \frac{2}{3}\right)$×5
$\frac{9}{15}, \frac{10}{15}$

largest $\frac{2}{3}$

Q	A	Q	A
State the largest fraction			
1) $\frac{7}{8}, \frac{3}{4}$	1) ×1$\left(\frac{7}{8}, \frac{3}{4}\right)$×2 $\frac{7}{8}, \frac{6}{8}$ $\frac{7}{8}$ largest	6) $\frac{6}{7}, \frac{9}{11}$	6) ×11$\left(\frac{6}{7}, \frac{9}{11}\right)$×7 $\frac{66}{77}, \frac{63}{77}$ $\frac{6}{7}$ largest
2) $\frac{9}{10}, \frac{47}{50}$	2) ×5$\left(\frac{9}{10}, \frac{47}{50}\right)$×1 $\frac{45}{50}, \frac{47}{50}$ largest $\frac{47}{50}$	7) $\frac{5}{6}, \frac{7}{9}$	7) ×3$\left(\frac{5}{6}, \frac{7}{9}\right)$×2 $\frac{15}{18}, \frac{14}{18}$ $\frac{5}{6}$ largest
3) $\frac{5}{36}, \frac{2}{9}$	3) ×1$\left(\frac{5}{36}, \frac{2}{9}\right)$×4 $\frac{5}{36}, \frac{8}{36}$ largest $\frac{2}{9}$	8) $\frac{3}{8}, \frac{5}{12}$	8) ×3$\left(\frac{3}{8}, \frac{5}{12}\right)$×2 $\frac{9}{24}, \frac{10}{24}$ largest $\frac{5}{12}$
4) $\frac{4}{5}, \frac{3}{4}$	4) ×4$\left(\frac{4}{5}, \frac{3}{4}\right)$×5 $\frac{16}{20}, \frac{15}{20}$ $\frac{4}{5}$ largest	9) $\frac{2}{25}, \frac{1}{10}$	9) ×2$\left(\frac{2}{25}, \frac{1}{10}\right)$×5 $\frac{4}{50}, \frac{5}{50}$ largest $\frac{1}{10}$
5) $\frac{5}{9}, \frac{4}{7}$	5) ×7$\left(\frac{5}{9}, \frac{4}{7}\right)$×9 $\frac{35}{63}, \frac{36}{63}$ largest $\frac{4}{7}$	10) $\frac{17}{40}, \frac{21}{50}$	10) ×5$\left(\frac{17}{40}, \frac{21}{50}\right)$×4 $\frac{85}{200}, \frac{84}{200}$ $\frac{17}{40}$ largest

Fractions 19

Multiplication.

Multiplication

To multiply fractions we **multiply the numerators** and then **multiply the denominators**.

*Note The word '**of**' is replaced by the **multiplication × sign** ... Ex. 2 lots **of** 3 = 2×3 = 6.
Ex. 7 boxes **of** 5 dice = 7×5 = 35 dice.

Ex.1 Calculate $\frac{1}{2}$ of $\frac{1}{3}$

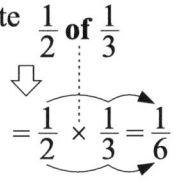

$= \frac{1}{2} \times \frac{1}{3} = \frac{1}{6}$

Ex.2 Calculate - and draw a diagram to help explain the calculation.

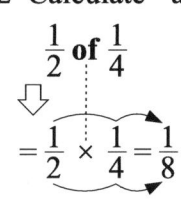

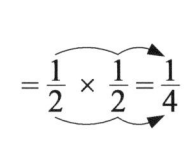

$\frac{1}{2}$ of $\frac{1}{4}$

$= \frac{1}{2} \times \frac{1}{4} = \frac{1}{8}$

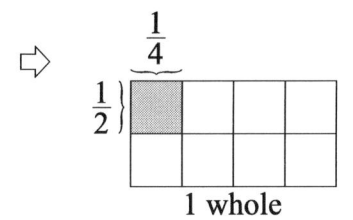

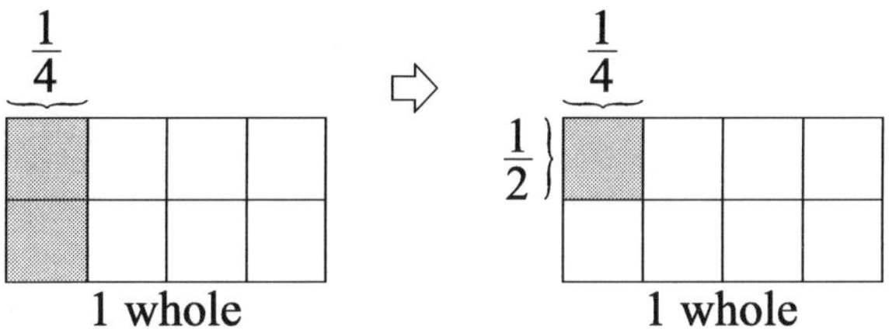

Q	A
Calculate - and draw a diagram to help explain the calculation. 1) $\frac{1}{2}$ of $\frac{1}{2}$	1) $= \frac{1}{2} \times \frac{1}{2} = \frac{1}{4}$
2) $\frac{1}{5}$ of $\frac{1}{3}$	2) $= \frac{1}{5} \times \frac{1}{3} = \frac{1}{15}$
3) $\frac{1}{3}$ of $\frac{2}{3}$	3) $= \frac{1}{3} \times \frac{2}{3} = \frac{2}{9}$
4) $\frac{3}{4}$ of $\frac{3}{5}$	4) $= \frac{3}{4} \times \frac{3}{5} = \frac{9}{20}$
5) $\frac{7}{8}$ of $\frac{5}{6}$	5) $= \frac{7}{8} \times \frac{5}{6} = \frac{35}{48}$

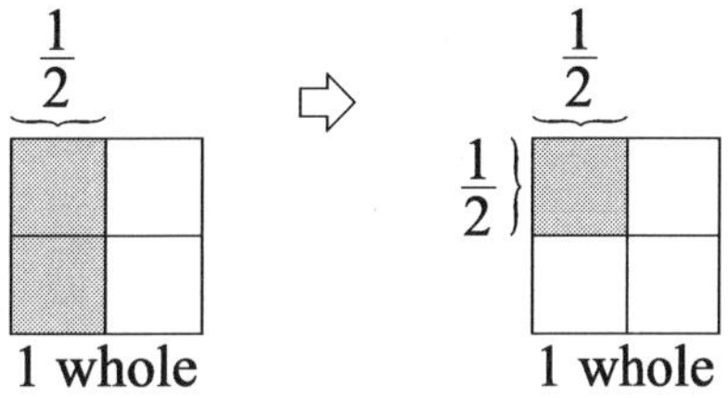

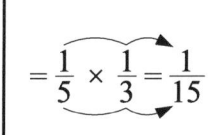

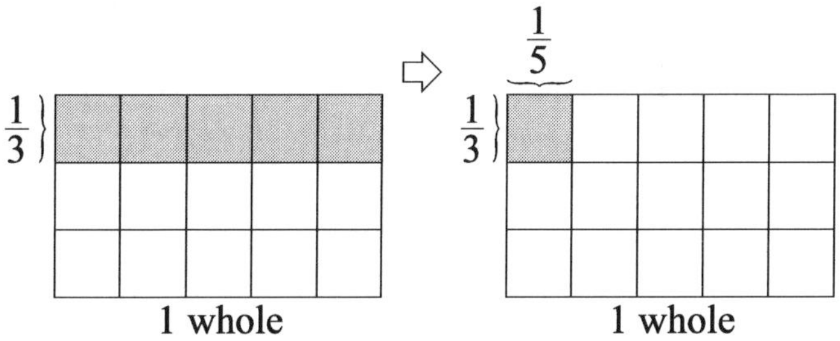

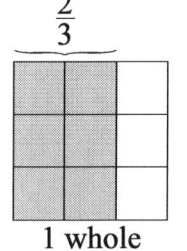

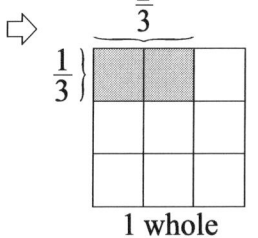

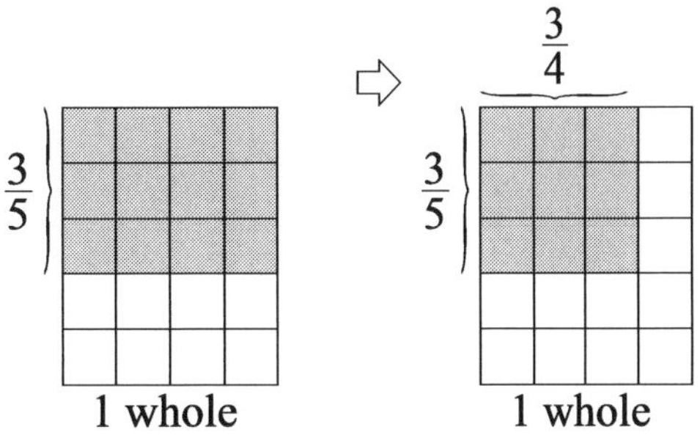

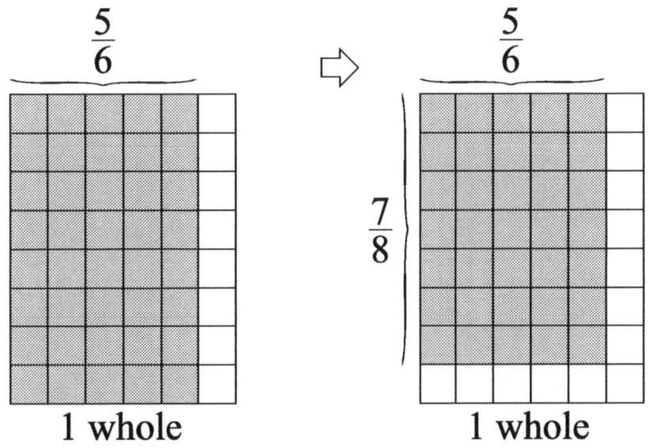

Fractions 20 — Multiplication. Cancel.

Multiplication

A **whole number by** can be written '**over 1**' to make it look like a fraction.

Ex. $\frac{1}{2}$ of 7 $\Rightarrow$ $\frac{1}{2} \times \frac{7}{1} = \frac{7}{2} = 3\frac{1}{2}$ or just $\frac{1}{2} \times 7 = \frac{7}{2} = 3\frac{1}{2}$

For the answer, we usually change improper fractions to whole numbers / mixed numbers.

Q	A	Q	A
Calculate			
1) $\frac{1}{3}$ of $\frac{1}{2}$	1) $= \frac{1}{3} \times \frac{1}{2} = \frac{1}{6}$	5) $\frac{1}{2}$ of 9	5) $= \frac{1}{2} \times \frac{9}{1} = \frac{9}{2} = 4\frac{1}{2}$
2) $\frac{2}{3} \times \frac{1}{3}$	2) $\frac{2}{9}$	6) $5 \times \frac{4}{7}$	6) $= \frac{5}{1} \times \frac{4}{7} = \frac{20}{7} = 2\frac{6}{7}$
3) $\frac{3}{5}$ of $\frac{4}{5}$	3) $= \frac{3}{5} \times \frac{4}{5} = \frac{12}{25}$	7) $\frac{1}{2} \times \frac{3}{4} \times \frac{5}{8}$	7) $\frac{15}{64}$
4) $\frac{6}{1} \times \frac{8}{1}$	4) $\frac{48}{1} = 48$	8) $\frac{4}{5} \times \frac{8}{1} \times \frac{1}{3}$	8) $\frac{32}{15} = 2\frac{2}{15}$

Multiplication - Cancel

When multiplying, fractions are usually presented in their **lowest form** and can not be cancelled.
However, we may be able to **cancel** the **numerator of one fraction** with the **denominator of another**.
Consider the following Example.

Ex. $\frac{1}{2} \times \frac{2}{1}$

Method 1

$\frac{1}{\cancel{2}} \times \frac{\cancel{2}^{1}}{1} = \frac{1}{1} = 1$

cancel then multiply

Method 2

$\frac{1}{2} \times \frac{2}{1} = \frac{\cancel{2}^{1}}{\cancel{2}_{1}} = 1$

multiply then cancel

Method 3

$\frac{1}{2} \times \frac{2}{1} = \frac{\cancel{2}^{1}}{\cancel{2}_{1}} \times \frac{1}{1} = \frac{1}{1} = 1$

exchange numerators / denominators
- then cancel and multiply
This is allowed since we can multiply in any order.

*Note To show that we have **cancelled** -
    cross out the given values and **write the answer** when we have **divided by the common factor**.

Method 1 is generally used - it is 'easier' to recognize that numbers cancel before multiplying and also ensures that smaller values are multiplied.
We should aim to cancel by the **highest common factor** - the highest number that divides into each.

*Note Method 1 is used in the following Q/A.

Q	A	Q	A
Calculate			
1) $\frac{1}{3} \times \frac{3}{1}$	1) $\frac{1}{\cancel{3}} \times \frac{\cancel{3}^{1}}{1} = \frac{1}{1} = 1$	4) $\frac{8}{1} \times \frac{1}{4}$	4) $\frac{\cancel{8}^{2}}{1} \times \frac{1}{\cancel{4}_{1}} = \frac{2}{1} = 2$
2) $\frac{7}{1} \times \frac{1}{7}$	2) $\frac{\cancel{7}^{1}}{1} \times \frac{1}{\cancel{7}_{1}} = \frac{1}{1} = 1$	5) $\frac{1}{8} \times \frac{6}{1}$	5) (cancel by 2) $\frac{1}{\cancel{8}_{4}} \times \frac{\cancel{6}^{3}}{1} = \frac{3}{4}$
3) $\frac{1}{6} \times \frac{2}{1}$	3) $\frac{1}{\cancel{6}_{3}} \times \frac{\cancel{2}^{1}}{1} = \frac{1}{3}$	6) $\frac{15}{1} \times \frac{1}{10}$	6) (cancel by 5) $\frac{\cancel{15}^{3}}{1} \times \frac{1}{\cancel{10}_{2}} = \frac{3}{2} = 1\frac{1}{2}$

Fractions 21

Multiplication. Cancel.

Q	A	Q	A
Calculate			
1) $\dfrac{7}{8} \times \dfrac{1}{7}$	1) $\dfrac{\cancel{7}^{1}}{8} \times \dfrac{1}{\cancel{7}_{1}} = \dfrac{1}{8}$	5) $\dfrac{6}{11} \times \dfrac{4}{9}$	5) $\dfrac{\cancel{6}^{2}}{11} \times \dfrac{4}{\cancel{9}_{3}} = \dfrac{8}{33}$
2) $\dfrac{1}{9} \times \dfrac{3}{2}$	2) $\dfrac{1}{\cancel{9}_{3}} \times \dfrac{\cancel{3}^{1}}{2} = \dfrac{1}{6}$	6) $\dfrac{7}{25} \times \dfrac{20}{3}$	6) $\dfrac{7}{\cancel{25}_{5}} \times \dfrac{\cancel{20}^{4}}{3} = \dfrac{28}{15} = 1\dfrac{13}{15}$
3) $\dfrac{4}{1} \times \dfrac{5}{8}$	3) $\dfrac{\cancel{4}^{1}}{1} \times \dfrac{5}{\cancel{8}_{2}} = \dfrac{5}{2} = 2\dfrac{1}{2}$	7) $\dfrac{50}{7} \times \dfrac{9}{80}$	7) $\dfrac{\cancel{50}^{5}}{7} \times \dfrac{9}{\cancel{80}_{8}} = \dfrac{45}{56}$
4) $\dfrac{5}{8} \times \dfrac{6}{7}$	4) $\dfrac{5}{\cancel{8}_{4}} \times \dfrac{\cancel{6}^{3}}{7} = \dfrac{15}{28}$	8) $\dfrac{13}{100} \times \dfrac{15}{16}$	8) $\dfrac{13}{\cancel{100}_{20}} \times \dfrac{\cancel{15}^{3}}{16} = \dfrac{39}{320}$

Multiplication - Cancel

It may be possible to **cancel** the **numerator of each fraction** with the **denominator of the other**. Consider the following Example.

Ex. $\dfrac{3}{2} \times \dfrac{2}{9}$

Method 1

$\dfrac{\cancel{3}^{1}}{\cancel{2}_{1}} \times \dfrac{\cancel{2}^{1}}{\cancel{9}_{3}} = \dfrac{1}{3}$ '2 and 2 cancel'
'3 and 9 cancel'

cancel then multiply

Method 2

$\dfrac{3}{2} \times \dfrac{2}{9} = \dfrac{\cancel{6}^{1}}{\cancel{18}_{3}} = \dfrac{1}{3}$

multiply then cancel

Method 3

$\dfrac{3}{9} \times \dfrac{2}{2} = \dfrac{\cancel{3}^{1}}{\cancel{9}_{3}} \times \dfrac{\cancel{2}^{1}}{\cancel{2}_{1}} = \dfrac{1}{3}$ '2 and 2 cancel'
'3 and 9 cancel'

exchange numerators / denominators - then cancel and multiply

*Note Method 1 is used in the following Q/A.

Q	A	Q	A
Calculate			
1) $\dfrac{2}{9} \times \dfrac{9}{2}$	1) $\dfrac{\cancel{2}^{1}}{\cancel{9}_{1}} \times \dfrac{\cancel{9}^{1}}{\cancel{2}_{1}} = \dfrac{1}{1} = 1$	6) $\dfrac{6}{25} \times \dfrac{10}{27}$	6) $\dfrac{\cancel{6}^{2}}{\cancel{25}_{5}} \times \dfrac{\cancel{10}^{2}}{\cancel{27}_{9}} = \dfrac{4}{45}$
2) $\dfrac{7}{8} \times \dfrac{4}{7}$	2) $\dfrac{\cancel{7}^{1}}{\cancel{8}_{2}} \times \dfrac{\cancel{4}^{1}}{\cancel{7}_{1}} = \dfrac{1}{2}$	7) $\dfrac{25}{6} \times \dfrac{27}{10}$	7) $\dfrac{\cancel{25}^{5}}{\cancel{6}_{2}} \times \dfrac{\cancel{27}^{9}}{\cancel{10}_{2}} = \dfrac{45}{4} = 11\dfrac{1}{4}$
3) $\dfrac{3}{4} \times \dfrac{8}{9}$	3) $\dfrac{\cancel{3}^{1}}{\cancel{4}_{1}} \times \dfrac{\cancel{8}^{2}}{\cancel{9}_{3}} = \dfrac{2}{3}$	8) $\dfrac{14}{33} \times \dfrac{11}{16}$	8) $\dfrac{\cancel{14}^{7}}{\cancel{33}_{3}} \times \dfrac{\cancel{11}^{1}}{\cancel{16}_{8}} = \dfrac{7}{24}$
4) $\dfrac{6}{5} \times \dfrac{35}{36}$	4) $\dfrac{\cancel{6}^{1}}{\cancel{5}_{1}} \times \dfrac{\cancel{35}^{7}}{\cancel{36}_{6}} = \dfrac{7}{6} = 1\dfrac{1}{6}$	9) $\dfrac{55}{7} \times \dfrac{28}{15}$	9) $\dfrac{\cancel{55}^{11}}{\cancel{7}_{1}} \times \dfrac{\cancel{28}^{4}}{\cancel{15}_{3}} = \dfrac{44}{3} = 14\dfrac{2}{3}$
5) $\dfrac{5}{12} \times \dfrac{9}{10}$	5) $\dfrac{\cancel{5}^{1}}{\cancel{12}_{4}} \times \dfrac{\cancel{9}^{3}}{\cancel{10}_{2}} = \dfrac{3}{8}$	10) $\dfrac{90}{77} \times \dfrac{99}{100}$	10) $\dfrac{\cancel{90}^{9}}{\cancel{77}_{7}} \times \dfrac{\cancel{99}^{9}}{\cancel{100}_{10}} = \dfrac{81}{70} = 1\dfrac{11}{70}$

Fractions 22

Multiplication - Mixed Numbers / Negative Fractions.

Q	A	Q	A
Calculate 1) $\frac{4}{5} \times \frac{1}{3} \times \frac{1}{4}$	1) $\frac{\overset{1}{\cancel{4}}}{5} \times \frac{1}{3} \times \frac{1}{\cancel{4}} = \frac{1}{15}$	4) $\frac{8}{7} \times \frac{7}{5} \times \frac{5}{8}$	4) $\frac{\overset{1}{\cancel{8}}}{\underset{1}{\cancel{7}}} \times \frac{\overset{1}{\cancel{7}}}{\underset{1}{\cancel{5}}} \times \frac{\overset{1}{\cancel{5}}}{\underset{1}{\cancel{8}}} = \frac{1}{1} = 1$
2) $\frac{3}{8} \times \frac{6}{1} \times \frac{3}{5}$	2) $\frac{3}{\underset{4}{\cancel{8}}} \times \frac{\overset{3}{\cancel{6}}}{1} \times \frac{3}{5} = \frac{27}{20} = 1\frac{7}{20}$	5) $\frac{10}{9} \times \frac{7}{20} \times \frac{27}{49}$	5) $\frac{\overset{1}{\cancel{10}}}{\underset{1}{\cancel{9}}} \times \frac{\overset{1}{\cancel{7}}}{\underset{2}{\cancel{20}}} \times \frac{\overset{3}{\cancel{27}}}{\underset{7}{\cancel{49}}} = \frac{3}{14}$
3) $\frac{4}{7} \times \frac{5}{6} \times \frac{7}{9}$	3) $\frac{\overset{2}{\cancel{4}}}{\underset{1}{\cancel{7}}} \times \frac{5}{\underset{3}{\cancel{6}}} \times \frac{\overset{1}{\cancel{7}}}{9} = \frac{10}{27}$	6) $\frac{63}{44} \times \frac{16}{55} \times \frac{25}{81}$	6) $\frac{\overset{7}{\cancel{63}}}{\underset{11}{\cancel{44}}} \times \frac{\overset{4}{\cancel{16}}}{\underset{11}{\cancel{55}}} \times \frac{\overset{5}{\cancel{25}}}{\underset{9}{\cancel{81}}} = \frac{140}{1089}$

Multiplication - Mixed Numbers

We change **Mixed Numbers to Improper Fractions** in order to multiply - (this is the simplest method).

Ex.1 $\frac{1}{2} \times 1\frac{1}{2}$ ⇩ $= \frac{1}{2} \times \frac{3}{2} = \frac{3}{4}$

Ex.2 $3\frac{1}{4} \times 2\frac{2}{3}$ ⇩ $= \frac{13}{\underset{1}{\cancel{4}}} \times \frac{\overset{2}{\cancel{8}}}{3} = \frac{26}{3} = 8\frac{2}{3}$

Ex.3 $5\frac{2}{5} \times 4\frac{1}{6}$ ⇩ $= \frac{\overset{9}{\cancel{27}}}{\underset{1}{\cancel{5}}} \times \frac{\overset{5}{\cancel{25}}}{\underset{2}{\cancel{6}}} = \frac{45}{2} = 22\frac{1}{2}$

Q	A	Q	A
Calculate 1) $1\frac{2}{3} \times \frac{1}{2}$	1) $\frac{5}{3} \times \frac{1}{2} = \frac{5}{6}$	5) $1\frac{5}{9} \times 1\frac{5}{7}$	5) $\frac{\overset{2}{\cancel{14}}}{\underset{3}{\cancel{9}}} \times \frac{\overset{4}{\cancel{12}}}{\underset{1}{\cancel{7}}} = \frac{8}{3} = 2\frac{2}{3}$
2) $\frac{5}{9} \times 5\frac{1}{4}$	2) $\frac{5}{\underset{3}{\cancel{9}}} \times \frac{\overset{7}{\cancel{21}}}{4} = \frac{35}{12} = 2\frac{11}{12}$	6) $9\frac{1}{6} \times 3\frac{3}{10}$	6) $\frac{\overset{11}{\cancel{55}}}{\underset{2}{\cancel{6}}} \times \frac{\overset{11}{\cancel{33}}}{\underset{2}{\cancel{10}}} = \frac{121}{4} = 30\frac{1}{4}$
3) $1\frac{1}{2} \times 2\frac{1}{2}$	3) $\frac{3}{2} \times \frac{5}{2} = \frac{15}{4} = 3\frac{3}{4}$	7) $1\frac{47}{100} \times 5\frac{5}{7}$	7) $\frac{\overset{21}{\cancel{147}}}{\underset{5}{\cancel{100}}} \times \frac{\overset{2}{\cancel{40}}}{\underset{1}{\cancel{7}}} = \frac{42}{5} = 8\frac{2}{5}$
4) $2\frac{3}{4} \times 3\frac{1}{5}$	4) $\frac{11}{\underset{1}{\cancel{4}}} \times \frac{\overset{4}{\cancel{16}}}{5} = \frac{44}{5} = 8\frac{4}{5}$	8) $1\frac{2}{7} \times 2\frac{7}{9} \times 1\frac{3}{25}$	8) $\frac{\overset{1}{\cancel{9}}}{\underset{1}{\cancel{7}}} \times \frac{\overset{1}{\cancel{25}}}{\underset{1}{\cancel{9}}} \times \frac{\overset{4}{\cancel{28}}}{\underset{1}{\cancel{25}}} = \frac{4}{1} = 4$

Multiplication - Negative Fractions

Negative fractions can be multiplied - the positive / negative signs in front of the fractions follow the **same rules** as when multiplying other numbers.

Ex.1 $-\frac{1}{2} \times \frac{1}{3}$ ⇨ $-\frac{1}{2} \times \frac{1}{3} = -\frac{1}{6}$

Ex.2 $-\frac{3}{5} \times -\frac{4}{7}$ ⇨ $-\frac{3}{5} \times -\frac{4}{7} = \frac{12}{35}$

Q	A	Q	A
Calculate 1) $-\frac{2}{3} \times \frac{4}{5}$	1) $-\frac{2}{3} \times \frac{4}{5} = -\frac{8}{15}$	3) $-6\frac{2}{3} \times -4\frac{7}{8}$	3) $-\frac{\overset{5}{\cancel{20}}}{\underset{1}{\cancel{3}}} \times -\frac{\overset{13}{\cancel{39}}}{\underset{2}{\cancel{8}}} = \frac{65}{2} = 32\frac{1}{2}$
2) $\frac{5}{6} \times -\frac{9}{35}$	2) $\frac{\overset{1}{\cancel{5}}}{\underset{2}{\cancel{6}}} \times -\frac{\overset{3}{\cancel{9}}}{\underset{7}{\cancel{35}}} = -\frac{3}{14}$	4) $-\frac{8}{9} \times -\frac{1}{5} \times -2\frac{1}{4}$	4) $-\frac{\overset{2}{\cancel{8}}}{\underset{1}{\cancel{9}}} \times -\frac{1}{5} \times -\frac{\overset{1}{\cancel{9}}}{\underset{1}{\cancel{4}}} = -\frac{2}{5}$

Fractions 23

Multiplication - Fraction of an 'amount' / Decimal.

Multiplication - Fraction of an 'amount'

A fraction of an 'amount' (having a unit of measure) is calculated in the same way as already explained.

Ex. $\frac{1}{2}$ of £7 ⇨ $\frac{1}{2} \times \frac{7}{1} = \frac{7}{2} = £3\frac{1}{2}$ or just $\frac{1}{2} \times 7 = \frac{7}{2} = £3\frac{1}{2}$

Q	A	Q	A
Calculate			
1) $\frac{1}{2}$ of 9g	1) $= \frac{1}{2} \times \frac{9}{1} = \frac{9}{2} = 4\frac{1}{2}$ g	7) $\frac{1}{2}$ of $3\frac{1}{2}$ kg	7) $= \frac{1}{2} \times \frac{7}{2} = \frac{7}{4} = 1\frac{3}{4}$ kg
2) $\frac{1}{2}$ of 80p	2) $= \frac{1}{2} \times \frac{\cancel{80}^{40}}{1} = \frac{40}{1} = 40$p	8) $\frac{1}{5}$ of $1\frac{2}{3}$ cm	8) $= \frac{1}{\cancel{5}} \times \frac{\cancel{5}^{1}}{3} = \frac{1}{3}$ cm
3) $\frac{1}{3}$ of 5m	3) $= \frac{1}{3} \times \frac{5}{1} = \frac{5}{3} = 1\frac{2}{3}$ m	9) $8 \times \frac{6}{7}$ cm²	9) $= \frac{8}{1} \times \frac{6}{7} = \frac{48}{7} = 6\frac{6}{7}$ cm²
4) $\frac{1}{3}$ of 60 l	4) $= \frac{1}{\cancel{2}} \times \frac{\cancel{60}^{20}}{1} = \frac{20}{1} = 20$ l	10) $\frac{11}{50}$ m³ × 90	10) $= \frac{11}{\cancel{50}_{5}} \times \frac{\cancel{90}^{9}}{1} = \frac{99}{5} = 19\frac{4}{5}$ m³
5) $\frac{3}{4}$ of 7km	5) $= \frac{3}{4} \times \frac{7}{1} = \frac{21}{4} = 5\frac{1}{4}$ km	11) $2\frac{7}{9}$ mins × 2	11) $= \frac{25}{9} \times \frac{2}{1} = \frac{50}{9} = 5\frac{5}{9}$ mins
6) $\frac{5}{8}$ of £36	6) $= \frac{5}{\cancel{8}_{2}} \times \frac{\cancel{36}^{9}}{1} = \frac{45}{2} = £22\frac{1}{2}$	12) $1\frac{3}{10} \times 4\frac{1}{6}$ °C	12) $= \frac{13}{\cancel{10}_{2}} \times \frac{\cancel{25}^{5}}{6} = \frac{65}{12} = 5\frac{5}{12}$ °C

Multiplication - Fraction of a Decimal

A fraction of a decimal is calculated in the same way as a fraction of a whole number -
a decimal can be written '**over 1**' to make it look like a fraction.
A decimal may also 'cancel' with the denominator of the fraction - by dividing by a common factor.
(to divide a decimal by a whole number - see **Decimals 5 and 12**)

Ex. $\frac{1}{2}$ of 8·3 ⇨ $\frac{1}{2} \times \frac{8\cdot3}{1} = \frac{8\cdot3}{2} = 4\cdot15$ or just $\frac{1}{2} \times 8\cdot3 = \frac{8\cdot3}{2} = 4\cdot15$

Q	A	Q	A
Calculate			
1) $\frac{1}{2}$ of 4·5	1) $= \frac{1}{2} \times \frac{4\cdot5}{1} = \frac{4\cdot5}{2} = 2\cdot25$	6) $\frac{3}{4}$ of 1·5ml	6) $= \frac{3}{4} \times \frac{1\cdot5}{1} = \frac{4\cdot5}{4} = 1\cdot125$ ml
2) $\frac{1}{2}$ of 6·8m	2) $= \frac{1}{\cancel{2}} \times \frac{\cancel{6\cdot8}^{3\cdot4}}{1} = \frac{3\cdot4}{1} = 3\cdot4$ m	7) $\frac{7}{8}$ of 2·8	7) $= \frac{7}{\cancel{8}_{2}} \times \frac{\cancel{2\cdot8}^{0\cdot7}}{1} = \frac{4\cdot9}{2} = 2\cdot45$
3) $\frac{1}{3}$ of £3·69	3) $= \frac{1}{\cancel{3}} \times \frac{\cancel{3\cdot69}^{1\cdot23}}{1} = \frac{1\cdot23}{1} = £1\cdot23$	8) $\frac{5}{6}$ of 7·2m²	8) $= \frac{5}{\cancel{6}} \times \frac{\cancel{7\cdot2}^{1\cdot2}}{1} = \frac{6\cdot0}{1} = 6$ m²
4) $\frac{3}{5}$ of 0·2g	4) $= \frac{3}{5} \times \frac{0\cdot2}{1} = \frac{0\cdot6}{5} = 0\cdot12$ g	9) $\frac{2}{3}$ of 0·4	9) $= \frac{2}{3} \times \frac{0\cdot4}{1} = \frac{0\cdot8}{3} = 0\cdot222...$
5) $\frac{5}{8}$ of 2·4	5) $= \frac{5}{\cancel{8}_{1}} \times \frac{\cancel{2\cdot4}^{0\cdot3}}{1} = \frac{1\cdot5}{1} = 1\cdot5$	10) $\frac{9}{50}$ of 0·5	10) $= \frac{9}{\cancel{50}_{10}} \times \frac{\cancel{0\cdot5}^{0\cdot1}}{1} = \frac{0\cdot9}{10} = 0\cdot09$

Fractions 24 — Fraction as Division of Whole Numbers.

Division

We can also think of a fraction as being a division of whole numbers - i.e., a 'sharing process'.
The **numerator** is the whole number to be shared and the **denominator** is how many we are sharing between.
The **result** of the sharing is the **same fraction** that we start with.

Ex.1 Share 1 between 2 or 1 divided by 2 ⇨ one 'divided by' two ⟶ $\frac{1}{2}$ = $\frac{1}{2}$

Ex.2 Share 2 between 3 or 2 divided by 3 ⇨ two 'divided by' three ⟶ $\frac{2}{3}$ = $\frac{2}{3}$

Each whole is divided by 3.

Q	A	Q	A
What is the result of			
1) 1 divided by 4	1) $\frac{1}{4}$	6) 4 divided by 6	6) $\frac{4}{6}$; $\frac{4}{6} = \frac{2}{3}$
2) 2 divided by 4	2) $\frac{2}{4}$; $\frac{2}{4} = \frac{1}{2}$		
3) 3 divided by 4	3) $\frac{3}{4}$	7) 5 divided by 6	7) $\frac{5}{6}$
4) 1 divided by 5	4) $\frac{1}{5}$	8) 3 divided by 7	8) $\frac{3}{7}$
5) 2 divided by 5	5) $\frac{2}{5}$		

Q	A	Q	A
What is the result of			
1) 1 divided by 8	1) $\frac{1}{8}$	7) 4 divided by 4	7) $\frac{4}{4} = 1$
2) 2 divided by 8	2) $\frac{2}{8} = \frac{1}{4}$	8) 6 divided by 3	8) $\frac{6}{3} = 2$
3) 3 divided by 9	3) $\frac{3}{9} = \frac{1}{3}$	9) 9 divided by 2	9) $\frac{9}{2} = 4\frac{1}{2}$
4) 6 divided by 8	4) $\frac{6}{8} = \frac{3}{4}$	10) 7 divided by 4	10) $\frac{7}{4} = 1\frac{3}{4}$
5) 5 divided by 7	5) $\frac{5}{7}$	11) 25 divided by 9	11) $\frac{25}{9} = 2\frac{7}{9}$
6) 4 divided by 10	6) $\frac{4}{10} = \frac{2}{5}$	12) 68 divided by 10	12) $\frac{68}{10} = 6\frac{8}{10} = 6\frac{4}{5}$

Fractions 25 — Division.

Division

Consider the following Examples.

Ex.1

$6 \div 2$ (meaning 'how many 2's in 6')

This can be written in fraction form as

$\frac{6}{1} \div \frac{2}{1}$

Dividing by 2 is the same as **multiplying by $\frac{1}{2}$**
since we want half **of** the number.

So we can replace $\div \frac{2}{1}$ by $\times \frac{1}{2}$ and calculate

$6 \div 2 = \frac{6}{1} \div \frac{2}{1}$

$= \frac{6}{1} \times \frac{1}{2}$ $\Rightarrow$ $\frac{\cancel{6}^3}{1} \times \frac{1}{\cancel{2}_1} = \frac{3}{1} = 3$

Ex.2

$6 \div \frac{1}{2}$ (meaning 'how many $\frac{1}{2}$'s in 6')

This can be written in fraction form as

$\frac{6}{1} \div \frac{1}{2}$

Dividing by $\frac{1}{2}$ is the same as **multiplying by 2**
since there are two $\frac{1}{2}$'s in each whole number.

So we can replace $\div \frac{1}{2}$ by $\times \frac{2}{1}$ and calculate

$6 \div \frac{1}{2}$

$= \frac{6}{1} \times \frac{2}{1} = \frac{12}{1} = 12$

From these 2 Examples, we can see that **to divide fractions** -
change the divide $\div$ sign to multiply $\times$ AND invert (turn upside down) the following fraction.

A **Whole Number** is first written **'over 1'**.
A **Mixed Number** is first changed to an **Improper Fraction**.

Q	A	Q	A
Calculate 1) $2 \div \frac{1}{4}$	1) $= \frac{2}{1} \div \frac{1}{4} = \frac{2}{1} \times \frac{4}{1} = \frac{8}{1} = 8$	10) $\frac{3}{5} \div \frac{7}{8}$	10) $= \frac{3}{5} \times \frac{8}{7} = \frac{24}{35}$
2) $\frac{1}{2} \div 3$	2) $= \frac{1}{2} \div \frac{3}{1} = \frac{1}{2} \times \frac{1}{3} = \frac{1}{6}$	11) $\frac{25}{32} \div \frac{35}{36}$	11) $= \frac{\cancel{25}^5}{\cancel{32}_8} \times \frac{\cancel{36}^9}{\cancel{35}_7} = \frac{45}{56}$
3) $8 \div \frac{2}{5}$	3) $= \frac{8}{1} \div \frac{2}{5} = \frac{\cancel{8}^4}{1} \times \frac{5}{\cancel{2}_1} = \frac{20}{1} = 20$	12) $\frac{11}{14} \div 1\frac{4}{7}$	12) $= \frac{11}{14} \div \frac{11}{7}$ $= \frac{\cancel{11}^1}{\cancel{14}_2} \times \frac{\cancel{7}^1}{\cancel{11}_1} = \frac{1}{2}$
4) $\frac{3}{7} \div \frac{3}{7}$	4) $= \frac{\cancel{3}^1}{\cancel{7}_1} \times \frac{\cancel{7}^1}{\cancel{3}_1} = \frac{1}{1} = 1$		
5) $\frac{1}{4} \div \frac{4}{5}$	5) $= \frac{1}{4} \times \frac{5}{4} = \frac{5}{16}$	13) $5\frac{5}{8} \div 4\frac{1}{6}$	13) $= \frac{45}{8} \div \frac{25}{6}$ $= \frac{\cancel{45}^9}{\cancel{8}_4} \times \frac{\cancel{6}^3}{\cancel{25}_5} = \frac{27}{20} = 1\frac{7}{20}$
6) $\frac{1}{9} \div \frac{1}{6}$	6) $= \frac{1}{\cancel{9}_3} \times \frac{\cancel{6}^2}{1} = \frac{2}{3}$		
7) $\frac{1}{6} \div \frac{1}{9}$	7) $= \frac{1}{\cancel{6}_2} \times \frac{\cancel{9}^3}{1} = \frac{3}{2} = 1\frac{1}{2}$	14) $-\frac{3}{5} \div \frac{3}{4}$	14) $= -\frac{\cancel{3}^1}{5} \times \frac{4}{\cancel{3}_1} = -\frac{4}{5}$
8) $\frac{7}{8} \div 4$	8) $= \frac{7}{8} \div \frac{4}{1} = \frac{7}{8} \times \frac{1}{4} = \frac{7}{32}$	15) $-2\frac{1}{2} \div -1\frac{3}{5}$	15) $= -\frac{5}{2} \div -\frac{8}{5}$ $= -\frac{5}{2} \times -\frac{5}{8} = \frac{25}{16} = 1\frac{9}{16}$
9) $5 \div 1\frac{2}{3}$	9) $= \frac{5}{1} \div \frac{5}{3} = \frac{\cancel{5}^1}{1} \times \frac{3}{\cancel{5}_1} = \frac{3}{1} = 3$		

Fractions 26 Division.

Division
Division of fractions may be presented as follows - (the longer division line separates the 2 values)

Ex.1 $\dfrac{3}{\frac{1}{2}} \Rightarrow = 3 \div \dfrac{1}{2}$

Ex.2 $\dfrac{\frac{4}{5}}{6} \Rightarrow = \dfrac{4}{5} \div 6$

Ex.3 $\dfrac{\frac{2}{7}}{\frac{5}{8}} \Rightarrow = \dfrac{2}{7} \div \dfrac{5}{8}$

Ex.4 $\dfrac{9}{4\frac{5}{7}} \Rightarrow = 9 \div 4\dfrac{5}{7}$

Ex.5 $\dfrac{2\frac{1}{3}}{8} \Rightarrow = 2\dfrac{1}{3} \div 8$

Ex.6 $\dfrac{1\frac{4}{9}}{5\frac{3}{4}} \Rightarrow = 1\dfrac{4}{9} \div 5\dfrac{3}{4}$

The division is then calculated as in **Fractions 25** - this is shown as Method 1 in the following Q/A.

*Note
The line of work showing the division ÷ sign can be omitted - shown as Method 2 in the following Q/A.

Q	A	
Calculate	Method 1	Method 2
1) $\dfrac{8}{\frac{1}{3}}$	1) $= 8 \div \dfrac{1}{3}$ $= \dfrac{8}{1} \times \dfrac{3}{1} = \dfrac{24}{1} = 24$	$= \dfrac{8}{1} \times \dfrac{3}{1} = \dfrac{24}{1} = 24$
2) $\dfrac{\frac{5}{7}}{4}$	2) $= \dfrac{5}{7} \div \dfrac{4}{1}$ $= \dfrac{5}{7} \times \dfrac{1}{4} = \dfrac{5}{28}$	$= \dfrac{5}{7} \times \dfrac{1}{4} = \dfrac{5}{28}$ or $= \dfrac{5}{7 \times 4} = \dfrac{5}{28}$
3) $\dfrac{\frac{7}{8}}{\frac{3}{4}}$	3) $= \dfrac{7}{8} \div \dfrac{3}{4}$ $= \dfrac{7}{\underset{2}{8}} \times \dfrac{\overset{1}{4}}{3} = \dfrac{7}{6} = 1\dfrac{1}{6}$	$= \dfrac{7}{\underset{2}{8}} \times \dfrac{\overset{1}{4}}{3} = \dfrac{7}{6} = 1\dfrac{1}{6}$
4) $\dfrac{2}{3\frac{4}{5}}$	4) $= 2 \div 3\dfrac{4}{5} = \dfrac{2}{1} \div \dfrac{19}{5}$ $= \dfrac{2}{1} \times \dfrac{5}{19} = \dfrac{10}{19}$	$= \dfrac{2}{\frac{19}{5}}$ $= \dfrac{2}{1} \times \dfrac{5}{19} = \dfrac{10}{19}$
5) $\dfrac{4\frac{5}{9}}{2}$	5) $= 4\dfrac{5}{9} \div 2 = \dfrac{41}{9} \div \dfrac{2}{1}$ $= \dfrac{41}{9} \times \dfrac{1}{2} = \dfrac{41}{18} = 2\dfrac{5}{18}$	$= \dfrac{41}{9} \times \dfrac{1}{2} = \dfrac{41}{18} = 2\dfrac{5}{18}$ or $= \dfrac{41}{9 \times 2} = \dfrac{41}{18} = 2\dfrac{5}{18}$
6) $\dfrac{9\frac{3}{4}}{2\frac{1}{6}}$	6) $= 9\dfrac{3}{4} \div 2\dfrac{1}{6} = \dfrac{39}{4} \div \dfrac{13}{6}$ $= \dfrac{\overset{3}{\cancel{39}}}{\underset{2}{\cancel{4}}} \times \dfrac{\overset{3}{\cancel{6}}}{\underset{1}{\cancel{13}}} = \dfrac{9}{2} = 4\dfrac{1}{2}$	$= \dfrac{\frac{39}{4}}{\frac{13}{6}}$ $= \dfrac{\overset{3}{\cancel{39}}}{\underset{2}{\cancel{4}}} \times \dfrac{\overset{3}{\cancel{6}}}{\underset{1}{\cancel{13}}} = \dfrac{9}{2} = 4\dfrac{1}{2}$

Fractions 27

Multiplication / Division - 'Problems'

Q	A
1) A girl walks $\frac{3}{4}$ km. Find $\frac{1}{2}$ of this distance.	1) $\frac{1}{2}$ of $\frac{3}{4}$ = $\frac{1}{2} \times \frac{3}{4} = \frac{3}{8}$ km
2) A container holds $\frac{7}{9}$ litres of water. Find the amount poured out if $\frac{2}{9}$ of the water is poured out.	2) $\frac{2}{9}$ of $\frac{7}{9}$ = $\frac{2}{9} \times \frac{7}{9} = \frac{14}{81}$ l
3) $\frac{8}{9}$ of a cake weighing $\frac{5}{6}$ kg is eaten. What is the weight of the cake eaten?	3) $\frac{8}{9}$ of $\frac{5}{6}$ = $\frac{8}{9} \times \frac{5}{6} = \frac{20}{27}$ kg
4) From a batch of 480 cans, $\frac{3}{80}$ are damaged. How many cans are damaged?	4) $\frac{3}{80}$ of 480 = $\frac{3}{80} \times \frac{480}{1} = \frac{18}{1} = 18$ cans
5) Calculate the area of a rectangle $5\frac{1}{5}$ cm by $3\frac{3}{4}$ cm.	5) A = $l \times b$ A = $5\frac{1}{5} \times 3\frac{3}{4} = \frac{26}{5} \times \frac{15}{4} = \frac{39}{2} = 19\frac{1}{2}$ cm²
6) The time taken to assemble a switch is $2\frac{5}{8}$ mins. How long does it take to assemble 36 switches?	6) $36 \times 2\frac{5}{8} = \frac{36}{1} \times \frac{21}{8} = \frac{189}{2} = 94\frac{1}{2}$ mins
7) $4\frac{9}{10}$ m³ of concrete is used to make 21 posts. What is the volume of each post?	7) $4\frac{9}{10} \div 21$ $\frac{49}{10} \div \frac{21}{1} = \frac{49}{10} \times \frac{1}{21} = \frac{7}{30}$ m³
8) A boy runs $\frac{3}{4}$ km around a track. How many laps of $\frac{1}{8}$ km does he run?	8) $\frac{3}{4} \div \frac{1}{8} = \frac{3}{4} \times \frac{8}{1} = \frac{6}{1} = 6$ laps
9) A container holds $\frac{8}{9}$ litres of oil. How many times can $\frac{2}{45}$ litres be poured out?	9) $\frac{8}{9} \div \frac{2}{45} = \frac{8}{9} \times \frac{45}{2} = \frac{20}{1} = 20$ times
10) 5kg of sweets are used to fill boxes. How many boxes of $\frac{7}{20}$ kg can be filled?	10) $5 \div \frac{7}{20} = \frac{5}{1} \times \frac{20}{7} = \frac{100}{7} = 14\frac{2}{7}$ boxes
11) What fraction of 5m is $\frac{4}{7}$ m?	11) $\frac{\frac{4}{7}}{5} = \frac{4}{7} \div \frac{5}{1} = \frac{4}{7} \times \frac{1}{5} = \frac{4}{35}$
12) How many pieces of wire of length $1\frac{2}{3}$ m can be cut from a piece of length $10\frac{1}{5}$ m?	12) $10\frac{1}{5} \div 1\frac{2}{3}$ $\frac{51}{5} \div \frac{5}{3} = \frac{51}{5} \times \frac{3}{5} = \frac{153}{25} = 6\frac{3}{25}$ pieces

Fractions 28

Reciprocal.

Reciprocal
(see also **Decimals 13**)

The **reciprocal** of a number is the number which it is **multiplied by** to give the value **1**.

In general, the reciprocal of a number is found by **dividing** the number into 1, usually written in the form

$$\text{reciprocal of } x = \frac{1}{x}$$

Ex.1   The reciprocal of 2 is $\frac{1}{2}$ since $2 \times \frac{1}{2} = 1$

Ex.2   The reciprocal of $\frac{1}{2}$ is 2 since $\frac{1}{2} \times 2 = 1$ or $\frac{1}{\frac{1}{2}} = \frac{1}{1} \times \frac{2}{1} = \frac{2}{1} = 2$
   invert and multiply

Ex.3   The reciprocal of $\frac{5}{8}$ is $\frac{8}{5}$ since $\frac{5}{8} \times \frac{8}{5} = \frac{40}{40} = 1$ or $\frac{1}{\frac{5}{8}} = \frac{1}{1} \times \frac{8}{5} = \frac{8}{5}$
   $\left(\frac{8}{5} = 1\frac{3}{5}\right)$ invert and multiply

Ex.4   The reciprocal of $2\frac{1}{4} \left(= \frac{9}{4}\right)$ is $\frac{4}{9}$ since $\frac{9}{4} \times \frac{4}{9} = \frac{36}{36} = 1$ or $\frac{1}{2\frac{1}{4}} = \frac{1}{\frac{9}{4}} = \frac{1}{1} \times \frac{4}{9} = \frac{4}{9}$
   (change mixed number to improper fraction)   invert and multiply

Ex.5   The reciprocal of -7 is $\frac{1}{-7}$ since $-7 \times \frac{1}{-7} = \frac{-7}{-7} = 1$

From the Exs. it can be seen that -

The reciprocal of a **positive integer** $=$ $\dfrac{1}{\text{positive integer}}$
$(1, 2, 3, 4 ...)$

The reciprocal of a **negative integer** $=$ $\dfrac{1}{\text{negative integer}}$
$(... -4, -3, -2, -1)$

The reciprocal of a fraction is the **same fraction inverted** (turned upside down).

Q	A	Q	A
State the **reciprocal** of			
1) 5	1) $\frac{1}{5}$	8) $8\frac{3}{5}$	8) $\left(8\frac{3}{5} = \frac{43}{5}\right)$ $\frac{5}{43}$
2) 88	2) $\frac{1}{88}$	9) -4	9) $\frac{1}{-4} = -\frac{1}{4}$
3) $\frac{1}{6}$	3) $\frac{6}{1} = 6$	10) $-\frac{10}{39}$	10) $-\frac{39}{10} = -3\frac{9}{10}$
4) $\frac{1}{41}$	4) $\frac{41}{1} = 41$	11) -6	11) $\frac{1}{-6} = -\frac{1}{6}$
5) $\frac{2}{3}$	5) $\frac{3}{2} = 1\frac{1}{2}$	12) $-4\frac{1}{2}$	12) $\left(-4\frac{1}{2} = -\frac{9}{2}\right)$ $-\frac{2}{9}$
6) $-\frac{9}{7}$	6) $-\frac{7}{9}$	13) $\frac{0.2}{1.5}$	13) $\frac{1.5}{0.2} = 7.5$
7) $-\frac{1}{3}$	7) $-\frac{3}{1} = -3$	14) $-\frac{5.89}{3.04}$	14) $-\frac{3.04}{5.89} = -0.516...$

Fractions 29 — Change Fraction to Decimal.

Change a Fraction to a Decimal
To change / convert a fraction to a decimal **divide** the numerator by the denominator.
(**Decimals 6-7** explain the <u>method of division</u> needed here.)

Ex. Change each fraction to a decimal.

Means '1 divided by 2'

A) $\dfrac{1}{2}$ $\quad 2\overline{)1 \cdot 0}\;\;\; = 0.5$

$1 = 1 \cdot 0 = 1 \cdot 00 = \ldots$

Means '5 divided by 8'

B) $\dfrac{5}{8}$ $\quad 8\overline{)5 \cdot 0\,^20\,^40}\;\;\; = 0.625$

$5 = 5 \cdot 0 = 5 \cdot 00 = \ldots$

Means '2 divided by 3'

C) $5\dfrac{2}{3}$ $\quad 3\overline{)2 \cdot 0\,^20\,^20 \ldots}\;\;\; = 0.666\ldots$
$= 5.666\ldots$

$2 = 2 \cdot 0 = 2 \cdot 00 = \ldots$

*Note The following Q/A are possible without a calculator.

Q	A	Q	A
Change each fraction to a decimal.		17) $99\dfrac{1}{12}$	17) $12\overline{)1 \cdot 0\,0\,^40\,^40 \ldots}\;\;=0.0833\ldots$ ⇨ $99 \cdot 0833\ldots$
1) $\dfrac{1}{4}$	1) $4\overline{)1 \cdot 0\,^20}\;\;=0.25$		
2) $\dfrac{3}{4}$	2) $4\overline{)3 \cdot 0\,^20}\;\;=0.75$	18) $\dfrac{7}{12}$	18) $12\overline{)7 \cdot 0\,^{10}0\,^40\,^40 \ldots}\;\;=0.5833\ldots$
3) $\dfrac{1}{5}$	3) $5\overline{)1 \cdot 0}\;\;=0.2$	19) $\dfrac{1}{20}$	19) $20\overline{)1 \cdot 0\,0}\;\;=0.05$
4) $2\dfrac{3}{5}$	4) $5\overline{)3 \cdot 0}\;\;=0.6$ ⇨ 2.6	20) $\dfrac{9}{20}$	20) $20\overline{)9 \cdot 0\,^{10}0}\;\;=0.45$
5) $\dfrac{1}{6}$	5) $6\overline{)1 \cdot 0\,^40\,^40\,^40 \ldots}\;\;=0.1666\ldots$	21) $\dfrac{1}{25}$	21) $25\overline{)1 \cdot 0\,0}\;\;=0.04$
6) $\dfrac{5}{6}$	6) $6\overline{)5 \cdot 0\,^20\,^20\,^20 \ldots}\;\;=0.8333\ldots$	22) $5\dfrac{9}{25}$	22) $25\overline{)9 \cdot 0\,^{10}0}\;\;=0.45$ ⇨ 5.45
7) $\dfrac{1}{7}$	7) $7\overline{)1 \cdot 0\,^30\,^20\,^60\,^40\,^50\,^10 \ldots}\;\;=0.142857 1\ldots$	23) $\dfrac{1}{30}$	23) $30\overline{)1 \cdot 0\,0\,^{10}0\,^{10}0 \ldots}\;\;=0.0333\ldots$
8) $\dfrac{4}{7}$	8) $7\overline{)4 \cdot 0\,^50\,^10\,^30\,^20\,^60\,^40 \ldots}\;\;=0.571428 5\ldots$	24) $\dfrac{7}{30}$	24) $30\overline{)7 \cdot 0\,^{10}0\,^{10}0\,^{10}0}\;\;=0.2333\ldots$
9) $3\dfrac{1}{8}$	9) $8\overline{)1 \cdot 0\,^20\,^40}\;\;=0.125$ ⇨ 3.125	25) $\dfrac{1}{40}$	25) $40\overline{)1 \cdot 0\,0\,^{20}0}\;\;=0.025$
10) $\dfrac{7}{8}$	10) $8\overline{)7 \cdot 0\,^60\,^40}\;\;=0.875$	26) $\dfrac{9}{40}$	26) $40\overline{)9 \cdot 0\,^{10}0\,^{20}0}\;\;=0.225$
11) $\dfrac{1}{9}$	11) $9\overline{)1 \cdot 0\,^10\,^10 \ldots}\;\;=0.111\ldots$	27) $\dfrac{1}{50}$	27) $50\overline{)1 \cdot 0\,0}\;\;=0.02$
12) $7\dfrac{2}{9}$	12) $9\overline{)2 \cdot 0\,^20\,^2 \ldots}\;\;=0.22\ldots$ ⇨ $7.22\ldots$	28) $\dfrac{43}{50}$	28) $50\overline{)43 \cdot 0\,^{30}0}\;\;=0.86$
13) $\dfrac{1}{10}$	13) $10\overline{)1 \cdot 0}\;\;=0.1$	29) $4\dfrac{1}{100}$	29) $100\overline{)1 \cdot 0\,0}\;\;=0.01$ ⇨ 4.01
14) $\dfrac{3}{10}$	14) $10\overline{)3 \cdot 0}\;\;=0.3$	30) $\dfrac{79}{100}$	30) $100\overline{)79 \cdot 0\,^{90}0}\;\;=0.79$
15) $\dfrac{1}{11}$	15) $11\overline{)1 \cdot 0\,0\,^10\,0 \ldots}\;\;=0.0909\ldots$	31) $\dfrac{1}{200}$	31) $200\overline{)1 \cdot 0\,0\,0}\;\;=0.005$
16) $\dfrac{6}{11}$	16) $11\overline{)6 \cdot 0\,^50\,^60\,^50 \ldots}\;\;=0.5454\ldots$	32) $6\dfrac{1}{1000}$	32) $1000\overline{)1 \cdot 0\,0\,0}\;\;=0.001$ ⇨ 6.001

Fractions 30

Powers. Roots.

Powers

A fraction can be 'raised to a power'.

Ex.1 Calculate

a) $\left(\frac{2}{7}\right)^3$

b) $\left(-1\frac{1}{4}\right)^2$

a) $\left(\frac{2}{7}\right)^3 = \frac{2}{7} \times \frac{2}{7} \times \frac{2}{7} = \frac{8}{343}$

b) $\left(-1\frac{1}{4}\right)^2 = -1\frac{1}{4} \times -1\frac{1}{4}$

$-\frac{5}{4} \times -\frac{5}{4} = \frac{25}{16} = 1\frac{9}{16}$

Q	A	Q	A
Calculate			
1) $\left(\frac{1}{2}\right)^2$	1) $\frac{1}{2} \times \frac{1}{2} = \frac{1}{4}$	5) $\left(\frac{1}{5}\right)^4$	5) $\frac{1}{5} \times \frac{1}{5} \times \frac{1}{5} \times \frac{1}{5} = \frac{1}{625}$
2) $\left(-\frac{1}{2}\right)^2$	2) $-\frac{1}{2} \times -\frac{1}{2} = \frac{1}{4}$	6) $\left(1\frac{2}{9}\right)^2$	6) $1\frac{2}{9} \times 1\frac{2}{9} =$ $\frac{11}{9} \times \frac{11}{9} = \frac{121}{81} = 1\frac{40}{81}$
3) $\left(\frac{3}{4}\right)^3$	3) $\frac{3}{4} \times \frac{3}{4} \times \frac{3}{4} = \frac{27}{64}$	7) $\left(-2\frac{1}{2}\right)^3$	7) $-2\frac{1}{2} \times -2\frac{1}{2} \times -2\frac{1}{2} =$ $-\frac{5}{2} \times -\frac{5}{2} \times -\frac{5}{2} = -\frac{125}{8} = -15\frac{5}{8}$
4) $\left(-\frac{2}{3}\right)^3$	4) $-\frac{2}{3} \times -\frac{2}{3} \times -\frac{2}{3} = -\frac{8}{27}$		

Roots

The root of some fractions may be found without the use of a calculator (not using decimals).
The root of both the **numerator** and **denominator** are calculated separately.

*Note
To help explain, the above Exs. for 'Powers' are reversed in the following Exs.

Ex.1 Calculate

a) $\sqrt[3]{\frac{8}{343}}$

b) 2 is usually omitted. $\sqrt[2]{\frac{25}{16}}$ or $\sqrt{\frac{25}{16}}$

a) $\sqrt[3]{\frac{8}{343}} = \left(\frac{\sqrt[3]{8}}{\sqrt[3]{343}}\right) = \frac{2}{7}$

(omitted)

b) $\sqrt{\frac{25}{16}} = \left(\frac{\sqrt{25}}{\sqrt{16}}\right) = \frac{5}{4}$ or $-\frac{5}{4}$ $\left(=\pm\frac{5}{4}\right)$

(omitted) $= 1\frac{1}{4}$ or $-1\frac{1}{4}$ $\left(=\pm 1\frac{1}{4}\right)$

Q	A	Q	A
Calculate			
1) $\sqrt{\frac{1}{4}}$	1) $\frac{1}{2}$ or $-\frac{1}{2}$ $\left(=\pm\frac{1}{2}\right)$	6) $\sqrt[5]{\frac{1}{32}}$	6) $\frac{1}{2}$
2) $\sqrt{\frac{4}{9}}$	2) $\frac{2}{3}$ or $-\frac{2}{3}$ $\left(=\pm\frac{2}{3}\right)$	7) $\sqrt[5]{-\frac{1}{32}}$	7) $-\frac{1}{2}$
3) $\sqrt[3]{\frac{8}{27}}$	3) $\frac{2}{3}$	8) $\sqrt{\frac{16}{81}}$	8) $\frac{4}{9}$ or $-\frac{4}{9}$ $\left(=\pm\frac{4}{9}\right)$
4) $\sqrt[3]{-\frac{27}{64}}$	4) $-\frac{3}{4}$	9) $\sqrt{\frac{49}{100}}$	9) $\frac{7}{10}$ or $-\frac{7}{10}$ $\left(=\pm\frac{7}{10}\right)$
5) $\sqrt[4]{\frac{1}{625}}$	5) $\frac{1}{5}$ or $-\frac{1}{5}$ $\left(=\pm\frac{1}{5}\right)$	10) $\sqrt{6\frac{1}{4}}$	10) $= \sqrt{\frac{25}{4}} = \frac{5}{2}$ or $-\frac{5}{2}$ $\left(=\pm\frac{5}{2}\right)$ $= 2\frac{1}{2}$ or $-2\frac{1}{2}$ $\left(=\pm 2\frac{1}{2}\right)$

Types of Numbers 1

Counting / Natural /...

Types of Numbers

Numbers can be grouped together in different ways and accordingly given a particular name.

For example, the 'Counting Numbers' are 1, 2, 3, 4, 5, ... ('Starting from 1, increase each number by 1')

The 3 dots ... are used to show that the numbers continue infinitely.

(The numbers may be described in different ways than shown here.)

Q	A
Explain each type of numbers. 1) Natural Numbers (referred to by the letter N)	1) 1, 2, 3, 4, 5, 6, ... These are the same as the 'Counting Numbers' - 'Starting from 1, increase each number by 1'.
2) Whole Numbers (referred to by the letter W)	2) 0, 1, 2, 3, 4, 5, 6, ... The Counting / Natural Numbers, including 0.
3) Even Numbers	3) 0, 2, 4, 6, 8, 10, ... Whole Numbers that divide exactly by 2.
4) Odd Numbers	4) 1, 3, 5, 7, 9, 11, ... Whole Numbers that do not divide exactly by 2.
5) Integers (referred to by the letter Z)	5) ... -4, -3, -2, -1, 0, 1, 2, 3, 4, ... Negative and Positive Whole Numbers and 0.
6) Multiples Ex. The multiples of 2.	6) The result of multiplying a number by the Counting Numbers. Ex. 1×2 = **2**, 2×2 = **4**, 3×2 = **6**, 4×2 = **8**, 5×2 = **10** ... so the multiples of 2 are **2, 4, 6, 8, 10,** ...
7) Factors Ex. The factors of 6.	7) Numbers which are multiplied to give a number. (Numbers which divide into a number without remainder) Ex. 1×**6** = 6, 2×**3** = 6. so the factors of 6 are **1, 2, 3, 6**.

8) Prime Numbers.
 	
 Ex. Prime numbers up to 100.
	8) Numbers that have just 2 different factors - 1 and the number itself.
	
	 2 3 5 7
	 11 13 17 19
	     23 29
	 31 37
	 41 43 47
	     53 59
	 61 67
	 71 73 79
	     83 89
	         97
	
	 Ex. 1×**2** = 2 - the **only factors** of 2 are **1, 2**.
	 Ex. 1×**97** = 97 - the **only factors** of 97 are **1, 97**.

*Note **Consecutive numbers** follow each other in regular order.

Ex. 3 consecutive **whole** numbers are   0,1,2   or   54,55,56.
Ex. 4 consecutive **odd** numbers are   1,3,5,7   or   15,17,19,21.
Ex. 5 consecutive **even** numbers are   2,4,6,8,10   or   100,102,104,106,108.
Ex. 3 consecutive **integers** are   2,3,4   or   -4,-3,-2.
Ex. 3 consecutive **prime numbers** are   5,7,11   or   23, 29,31.

Types of Numbers 2

Square / Root. Cube / Root. Rectangle.

Q	A
9) Square Numbers (also called Perfect Squares) Ex. The Square Numbers up to 100.	9) The results of multiplying a Whole Number by itself. The 2 numbers are the same - as are the sides of a square. $0^2 \quad 1^2 \quad 2^2 \quad 3^2 \quad 4^2 \quad 5^2$ ----- Read $= 0 \times 0 \quad 1 \times 1 \quad 2 \times 2 \quad 3 \times 3 \quad 4 \times 4 \quad 5 \times 5$ '5 squared or $= 0 \quad 1 \quad 4 \quad 9 \quad 16 \quad 25$ 5 to the power 2' $6^2 \quad 7^2 \quad 8^2 \quad 9^2 \quad 10^2$ $= 6 \times 6 \quad 7 \times 7 \quad 8 \times 8 \quad 9 \times 9 \quad 10 \times 10$ $= 36 \quad 49 \quad 64 \quad 81 \quad 100$

10) Square Roots.
 	
Ex. The Square Root of
0, 1, 4, 9, 16, 25.
	10) The opposite of the square of a number.
	The Square Root of a number **n** is the number which is multiplied by itself to give **n**.
	*Note
	This 2 is usually omitted as the Square Root is used so much.
	↓ ↓ 'Root' sign Read
	$\sqrt[2]{0} \quad \sqrt[2]{1} \quad \sqrt[2]{4} \quad \sqrt[2]{9} \quad \sqrt[2]{16} \quad \sqrt[2]{25}$ 'square root of 25'
	⇓
	$= \sqrt{0} \quad \sqrt{1} \quad \sqrt{4} \quad \sqrt{9} \quad \sqrt{16} \quad \sqrt{25}$
	$= 0 \quad 1 \quad 2 \quad 3 \quad 4 \quad 5$
	   or
	      -1 -2 -3 -4 -5
	since also -1×-1=1 -2×-2=4 -3×-3=9 -4×-4=16 -5×-5=25

11) Cube Numbers. Ex. The Cube Numbers up to 1,000.	11) The results of multiplying a Whole Number by itself and multiplying the answer by itself again. The 3 numbers are the same - as are the sides of a cube. $0^3 \quad 1^3 \quad 2^3 \quad 3^3 \quad 4^3 \quad 5^3$ $= 0 \times 0 \times 0 \quad 1 \times 1 \times 1 \quad 2 \times 2 \times 2 \quad 3 \times 3 \times 3 \quad 4 \times 4 \times 4 \quad 5 \times 5 \times 5$ $= 0 \quad 1 \quad 8 \quad 27 \quad 64 \quad 125$ Read '5 cubed or $6^3 \quad 7^3 \quad 8^3 \quad 9^3 \quad 10^3$ 5 to the power 3' $= 6 \times 6 \times 6 \quad 7 \times 7 \times 7 \quad 8 \times 8 \times 8 \quad 9 \times 9 \times 9 \quad 10 \times 10 \times 10$ $= 216 \quad 343 \quad 512 \quad 729 \quad 1,000$
12) Cube Roots. Ex. The Cube Root of 0, 1, 8, 27, 64, 125.	12) The opposite of the cube of a number. The Cube Root of a number **n** is the number which Read is multiplied by itself and itself again to give **n**. 'Cube root of 125' $\sqrt[3]{0} \quad \sqrt[3]{1} \quad \sqrt[3]{8} \quad \sqrt[3]{27} \quad \sqrt[3]{64} \quad \sqrt[3]{125}$ $= 0 \quad 1 \quad 2 \quad 3 \quad 4 \quad 5$ (since 0×0×0 1×1×1 2×2×2 3×3×3 4×4×4 5×5×5 = 0 = 1 = 8 = 27 = 64 = 125)

13) Rectangle Numbers
 	
Ex.
The Rectangle Numbers up to 16.
	13) Numbers which have more than 2 factors.
	Numbers can be represented by **dots** arranged in a **Rectangle** - but **not** a single row or column - (so ... **not** 1, **not** prime).
	Square Numbers are included (a square is a <u>special</u> rectangle).
	
	                                       1×12 1×16
	     1×4 1×6 1×8 1×9 1×10 **2×6** 1×14 1×15 **2×8**
	     2×2 2×3 2×4 3×3 2×5 **3×4** 2×7 3×5 **4×4**
	  = 4 6 8 9 10 12 14 15 16

Types of Numbers 3

Powers. Roots. Rational. Irrational. Real.

Q	A
14) Powers. (including Square Numbers and Cube Numbers) Calculate Exs. a) 5^1 b) 7^2 c) $(0.2)^3$ d) $\left(\frac{1}{2}\right)^4$ e) $(-3)^5$	14) The result of multiplying a number by itself the 'power' number of times. a) $5^1 = 5$ Read '5 to the power 1' b) $7^2 = 7 \times 7 = 49$ c) $(0.2)^3 = 0.2 \times 0.2 \times 0.2 = 0.008$ d) $\left(\frac{1}{2}\right)^4 = \frac{1}{2} \times \frac{1}{2} \times \frac{1}{2} \times \frac{1}{2} = \frac{1}{16}$ $\frac{1}{2}$ 'to the power 4' or 'to the 4th power' e) $(-3)^5 = -3 \times -3 \times -3 \times -3 \times -3 = -243$ -3 'to the power 5' or 'to the 5th power'

15) Roots.
(including Square Roots and Cube Roots)
Calculate
Exs. a) $\sqrt[2]{2}$
b) $\sqrt[3]{64}$
c) $\sqrt[4]{81}$
d) $\sqrt[5]{-1}$
	15) The opposite of Powers.
	The Root of a number **n** is the number which is multiplied by itself the 'root' times to give **n**.
	
	Since -
	a) $\sqrt[2]{2} = \begin{cases} 1.4142... \\ \text{or } -1.4142... \end{cases}$ $1.4142... \times 1.4142... = 2$
	                 or $-1.4142... \times -1.4142... = 2$
	b) $\sqrt[3]{64} = 4$ $4 \times 4 \times 4 = 64$
	c) $\sqrt[4]{81} = \pm 3$ $3 \times 3 \times 3 \times 3 = 81$
	   Read 'the 4th root of 81' or $-3 \times -3 \times -3 \times -3 = 81$
	d) $\sqrt[5]{-1} = -1$ $-1 \times -1 \times -1 \times -1 \times -1 = -1$
	   Read 'the 5th root of -1'

16) Rational Numbers.	16) Numbers which can be expressed as a fraction or terminating decimal (the decimal ends / has finite digits) or recurring decimal (a digit / number of digits repeat infinitely). Exs. 2, $\frac{1}{2}$, $\frac{999}{1111}$, -6, $-5\frac{3}{7}$, 0.8, -2.345, $0.333...$, $7.6767...$
17) Irrational Numbers. (referred to by the letter Q)	17) Numbers which can **not** be expressed as a fraction and are a non-terminating **and** non-recurring decimal. Exs. $\sqrt{2} = \begin{cases} 1.4142... \\ \text{or } -1.4142... \end{cases}$, $\sqrt[3]{500} = 7.9370...$, $\pi = 3.14159...$
18) Real Numbers. (referred to by the letter R)	18) The Rational and Irrational Numbers. Exs. 3, $\frac{4}{5}$, -9.7, $0.111...$, $\sqrt{8}$, π
19) Imaginary Numbers.	19) Numbers which include $\sqrt{-\text{ve}}$ and the letter *i*. Ex. $\sqrt{-1} = i$ so $i^2 = (\sqrt{-1})^2 = \sqrt{-1} \times \sqrt{-1} = -1$ Ex. $\sqrt{-4} = \sqrt{4} \times \sqrt{-1} = \pm 2i$
20) 'Own Name' Numbers. a) Dozen b) Score c) Gross d) Grand e) k	20) These numbers have their 'own name' (others may have also). a) 12 b) 20 c) 144 (12 dozen = 12×12) d) £1,000 (word is in common use). $1,000 in U.S.A. e) 1,000 (ex. when used with the £ sign. £5k = £5,000)

Types of Numbers 4

Prime Numbers - Sieve of Eratosthenes.

Prime Numbers - Sieve of Eratosthenes

There is no pattern or sequence to the Prime Numbers.
Eratosthenes (275 - 195 B.C.) was a Greek Mathematician who devised a method to **separate** or <u>sieve</u> the Prime Numbers in a list of numbers (hence the name).   Here, a list of just 50 numbers is used.

The 1st. 4 prime numbers 2,3,5,7 are kept in the list.

In turn, 'cross out' 1 and then the numbers which have at least 1 of these 4 prime numbers as a factor...

        1 - this is not prime.
        the multiples of 2 (every 2nd number).
        the multiples of 3 (every 3rd number).
        the multiples of 5 (every 5th number).
        the multiples of 7 (every 7th number).

Some numbers are 'crossed out' more than once.   The remaining numbers are prime.

Q	A

1) In order, 'cross out' 1 and the multiples of the 1st. 4 prime numbers to find the prime numbers less than 50.
 	
1 2 3 4 5 6 7 8 9 10
11 12 13 14 15 16 17 18 19 20
21 22 23 24 25 26 27 28 29 30
31 32 33 34 35 36 37 38 39 40
41 42 43 44 45 46 47 48 49 50
	1)
	
	'cross out' 1̶ and multiples of 2 — (but not 2).
	
	1̶ 2 3 4̶ 5 6̶ 7 8̶ 9 1̶0̶
	11 1̶2̶ 13 1̶4̶ 15 1̶6̶ 17 1̶8̶ 19 2̶0̶
	21 2̶2̶ 23 2̶4̶ 25 2̶6̶ 27 2̶8̶ 29 3̶0̶
	31 3̶2̶ 33 3̶4̶ 35 3̶6̶ 37 3̶8̶ 39 4̶0̶
	41 4̶2̶ 43 4̶4̶ 45 4̶6̶ 47 4̶8̶ 49 5̶0̶
	⇩
	'cross out' multiples of 3 ▓ (but not 3).
	
	1̶ 2 3 4̶ 5 6̶ 7 8̶ 9 1̶0̶
	11 1̶2̶ 13 1̶4̶ 15 1̶6̶ 17 1̶8̶ 19 2̶0̶
	21 2̶2̶ 23 2̶4̶ 25 2̶6̶ 27 2̶8̶ 29 3̶0̶
	31 3̶2̶ 33 3̶4̶ 35 3̶6̶ 37 3̶8̶ 39 4̶0̶
	41 4̶2̶ 43 4̶4̶ 45 4̶6̶ 47 4̶8̶ 49 5̶0̶
	⇩
	'cross out' multiples of 5 ╱ (but not 5).
	
	1̶ 2 3 4̶ 5 6̶ 7 8̶ 9 1̶0̶
	11 1̶2̶ 13 1̶4̶ 1̸5̸ 1̶6̶ 17 1̶8̶ 19 2̸0̸
	21 2̶2̶ 23 2̶4̶ 2̸5̸ 2̶6̶ 27 2̶8̶ 29 3̸0̸
	31 3̶2̶ 33 3̶4̶ 3̸5̸ 3̶6̶ 37 3̶8̶ 39 4̸0̸
	41 4̶2̶ 43 4̶4̶ 4̸5̸ 4̶6̶ 47 4̶8̶ 49 5̸0̸
	⇩
	'cross out' multiples of 7 ◯ (but not 7).
	
	1̶ 2 3 4̶ 5 6̶ 7 8̶ 9 1̶0̶
	11 1̶2̶ 13 ⑭ 1̸5̸ 1̶6̶ 17 1̶8̶ 19 2̸0̸
	㉑ 2̶2̶ 23 2̶4̶ 2̸5̸ 2̶6̶ 27 ㉘ 29 3̸0̸
	31 3̶2̶ 33 3̶4̶ ㉟ 3̶6̶ 37 3̶8̶ 39 4̸0̸
	41 ㊷ 43 4̶4̶ 4̸5̸ 4̶6̶ 47 4̶8̶ ㊾ 5̸0̸
	⇩
	Only prime numbers are left in the list.
	
	    2 3 5 7
	11 13 17 19
	       23 29
	31 37
	41 43 47

*<u>Note</u> Prime Numbers up to 120 ($11^2 - 1$) may be found in the same way- just using multiples of 2,3,5,7.
     Prime Numbers up to 168 ($13^2 - 1$) may be found by including Prime Number 11 and its multiples.

Multiples 1

Multiples. L.C.M.

Multiples
The Multiples of a number are the results of multiplying the number by positive whole numbers - as in the 'Times Tables'.
The Multiples of a number are **infinite**.

Ex.1 List the **multiples** of 2.

$1 \times 2 = 2$
$2 \times 2 = 4$ or   **2, 4, 6, 8, 10,** ...
$3 \times 2 = 6$
$4 \times 2 = 8$
$5 \times 2 = 10$
⋮   ⋮   ⋮

Ex.2 List the **multiples** of 6 between 20 and 40.

$4 \times 6 = 24$
$5 \times 6 = 30$ or   **24, 30, 36**.
$6 \times 6 = 36$

Q	A	Q	A
1) List the multiples of :		2) List the multiples of 4 less than 24.	2) 4, 8, 12, 16, 20.
a) 3	a) 3, 6, 9, 12, 15, ...		
b) 5	b) 5, 10, 15, 20, 25, ...	3) List the multiples of 7 between 20 and 50.	3) 21, 28, 35, 42, 49.
c) 8	c) 8, 16, 24, 32, 40, ...	4) List the multiples of 9 between 35 and 90.	4) 36, 45, 54, 63, 72, 81.
d) 25	d) 25, 50, 75, 100, ...	5) List the multiples of 13 less than 70.	5) 13, 26, 39, 52, 65.

Common Multiples and Lowest Common Multiple (L.C.M.)

The multiples of two or more numbers have **common multiples**.
It is useful to be able to recognise - usually mentally - the **common multiples** and in particular, the **Lowest Common Multiple (L.C.M.)** ...
(... when for example, finding a common denominator so as to add / subtract fractions.)

The **L.C.M.** can also be thought of as the **smallest number** that 2 or more numbers divide into (without remainder).

Ex.   State the **Lowest Common Multiple** of 8 and 10.

The multiples of 8 are 8, 16, 24, 32, **40**, 48, 56, 64, 72, **80,** ...

The multiples of 10 are   10, 20, 30, **40**, 50, 60, 70, **80,** ...

The **common multiples** are **40, 80,** ...
The **Lowest Common Multiple = 40**. (only the answer is required)

Q	A	Q	A
State the Lowest Common Multiple of the given numbers.		4) 10, 6	4) 10, 20, **30**, 40, ... 6, 12, 18, 24, **30**, 36, ... L.C.M. = **30**.
1) 2, 5	1) 2,4,6,8,**10**,12,14,16,18,**20**, ... 5, **10**, 15, **20**, ... L.C.M. = **10**.	5) 12, 9	5) 12, 24, **36**, 48, 60, ... 9, 18, 27, **36**, 45, ... L.C.M. = **36**.
2) 3, 4	2) 3, 6, 9, **12**, 15, 18, 21, **24**, ... 4, 8, **12**, 16, 20, **24**, ... L.C.M. = **12**.	6) 5, 10, 15	6) 5, 10, 15, 20, 25, **30**, ... 10, 20, **30**, ... 15, **30**, ... L.C.M. = **30**.
3) 6, 8	3) 6, 12, 18, **24**, 30, 36, 42, **48**, ... 8, 16, **24**, 32, 40, **48**, ... L.C.M. = **24**.		

Factors 1

Factors.

Factors

The **factors** of a number are the whole numbers which divide into it exactly.
(When numbers are **multiplied**, they are **factors** of the number which results.)

Ex.1 List the factors of 2.

$$\frac{2}{1 \times 2}$$

The factors are 1, 2.
(2 is a Prime Number)

Ex.2 List the factors of 9.

$$\frac{9}{\begin{array}{c} 1 \times 9 \\ \cancel{2\times} \\ 3 \times 3 \end{array}}$$

'mental' check — 2 is **not** a factor

The factors are 1, 3, 9
(9 is a Square Number)

Ex.3 List the factors of 28.

$$\frac{28}{\begin{array}{c} 1 \times 28 \\ 2 \times 14 \\ \cancel{3\times} \\ 4 \times 7 \\ \cancel{5\times} \\ \cancel{6\times} \end{array}}$$

'mental' check
'mental' check
'mental' check

The factors are 1, 2, 4, 7, 14, 28

*Note

It is useful to consider the factors by multiplying pairs of factors, always starting with 1.
We should check each **possible** factor on the left side, 2, 3, 4, 5, 6, 7, ... at least mentally.
[The pairs of (possible) factors will become closer to each other and may equal each other.]

The work can be reduced to the following, (omit the × sign), to give the list of factors ...

```
    2              9             28
  1   2          1   9         1   28
                   3   'omit' 3  2   14
                                 4    7
```

... and is presented in this way in the Q/A (below).

Prime Number

A Prime Number is defined as having just 2 factors, **1** and the **number itself**.
(The number 1 has only 1 factor and is **not** a prime number.)

Square Number

In terms of factors, a Square Number is the result of **multiplying a factor by itself**.
It has an odd number of factors.

Q	A	Q	A

List the factors of each number. State if any number is Prime or Square.
 	
 1) 6
	1) 6
	    1 6
	    2 3
		6) 33	6) 33
			    1 33
			    3 11

2) 7	2) 7
	    1 7 Prime
		7) 24	7) 24
			    1 24
			    2 12
			    3 8
			    4 6

3) 18	3) 18
	    1 18
	    2 9
	    3 6
		8) 47
		
		 9) 91
			8) 47
			    1 47 Prime
			
			 9) 91
			    1 91
			    7 13

4) 25	4) 25
	    1 25
	    5 Square
		10) 36	10) 36
			    1 36
			    2 18
			    3 12
			    4 9
			    6 Square

5) 13	5) 13 1 13 Prime		

Factors 2 — Common Factors. H.C.F.

Common Factors and Highest Common Factor (H.C.F.)
Two or more numbers may have one or more **common factors**.
It is useful to be able to recognise - usually mentally - the **common factors** and in particular,
the **Highest Common Factor (H.C.F.)** ...
(... for example, when we cancel fractions or factorise in Algebra)

*Note
In the following Exs. and Q/A, the factors are presented as shown in **Factors 1**.

Ex.1
State the **Highest Common Factor** of 2 and 8.

```
   2          8
 1  2       1  8
            2  4
```

The **common factors** are **1, 2**.
The **Highest Common Factor = 2**.
(only the answer is required)

Ex.2
State the **Highest Common Factor** of 12 and 30.

```
   12         30
 1  12      1  30
 2  6       2  15
 3  4       3  10
            5  6
```

The **common factors** are **1, 2, 3, 6**.
The **Highest Common Factor = 6**.

Q	A	Q	A

State the Highest Common Factor of the given numbers.		7) 14, 42	7) 14 42
			1 14 1 42
			2 7 2 21
			           3 14
			           6 7
			H.C.F. = 14

1) 5, 15	1) 5 15 1 5 1 15 3 5 H.C.F. = 5	8) 60, 48	8) 60 48 1 60 1 48 2 30 2 24 3 20 3 16 4 15 4 12 5 12 6 8 6 10 H.C.F. = 12
2) 8, 20	2) 8 20 1 8 1 20 2 4 2 10 4 5 H.C.F. = 4	9) 4, 6, 8	9) 4 6 8 1 4 1 6 1 8 2 2 3 2 4 H.C.F. = 2
3) 18, 27	3) 18 27 1 18 2 9 3 6 H.C.F. = 9	10) 50, 10, 25	10) 50 10 25 1 50 1 10 1 25 2 25 2 5 5 5 10 H.C.F. = 5

4) 21, 35	4) 21 35
	1 21 1 35
	3 7 5 7
	H.C.F. = 7
		11) 49, 56, 63	11) 49 56 63
			1 49 1 56 1 63
			7 2 28 3 21
			        4 14 7 9
			        7 8
			H.C.F. = 7

5) 9, 4	5) 9 4 1 9 1 4 3 2 H.C.F. = 1	12) 11, 13, 17	12) 11 13 17 1 11 1 13 1 17 H.C.F. = 1
6) 24, 16	6) 24 16 1 24 1 16 2 12 2 8 3 8 4 4 6 H.C.F. = 8		

Factors 3 — Prime Factors.

Prime Factors
A number may be written **as a product of its Prime Factors** ...
the factors are **ALL prime numbers** and **multiply** to give the number.

If a prime number is multiplied more than once, then it can be written in **exponent form** / as a **power**.

Method 1 / 2 are shown or this may be done **mentally**.

Ex. Write 20 as a product of its **prime factors**.

Method 1
Divide by Prime Numbers - in any order - until the last number is Prime.
(it is useful to divide 'downwards' and write the answers underneath each other)

$$\begin{array}{r}2\,|\,20\\10\end{array} \Rightarrow \begin{array}{r}2\,|\,20\\2\,|\,10\\5\end{array}$$
$\Downarrow$
$2\times 2\times 5$
$= 2^2 \times 5$

or

$$\begin{array}{r}2\,|\,20\\10\end{array} \Rightarrow \begin{array}{r}2\,|\,20\\5\,|\,10\\2\end{array}$$
$\Downarrow$
$2\times 2\times 5$
$= 2^2 \times 5$

Method 2
Multiply a pair of factors to give the number. For any factor which is not prime, repeat the above until all the factors are prime.

$$\begin{array}{c}20\\\overline{2\times 10}\end{array} \Rightarrow \begin{array}{c}20\\\overline{2\times 10}\\\overline{2\times 5}\end{array}$$
$\Downarrow$
$2\times 2\times 5$
$= 2^2 \times 5$

or

$$\begin{array}{c}20\\\overline{4\times 5}\end{array} \Rightarrow \begin{array}{c}20\\\overline{4\times 5}\\\overline{2\times 2}\end{array}$$
$\Downarrow$
$2\times 2\times 5$
$= 2^2 \times 5$

---

*Note In the following Q/A,
  Method 1 divides by Prime Numbers in **ascending** order.
  Method 2 may use pairs of factors other than those shown.

Q	A	Q	A
Write as a product of its Prime Factors :	**Method 1** **Method 2**		**Method 1** **Method 2**
1) 6	1) $2\,|\,6$ $\dfrac{6}{2\times 3}$ $\quad\;3$ 2×3	7) 30	7) $2\,|\,30$ $\dfrac{30}{3\times 10}$ $3\,|\,15$ $\overline{2\times 5}$ $\quad\;5$ $2\times 3\times 5$
2) 8	2) $2\,|\,8$ $\dfrac{8}{2\times 4}$ $2\,|\,4$ $\overline{2\times 2}$ $\quad\;2$ $2\times 2\times 2 = 2^3$	8) 36	8) $2\,|\,36$ $\dfrac{36}{6\times 6}$ $2\,|\,18$ $\overline{2\times 3\;\;2\times 3}$ $3\,|\,9$ $\quad\;3$ $2\times 2\times 3\times 3 = 2^2\times 3^2$
3) 9	3) $3\,|\,9$ $\dfrac{9}{3\times 3}$ $\quad\;3$ $3\times 3 = 3^2$	9) 75	9) $3\,|\,75$ $\dfrac{75}{5\times 15}$ $5\,|\,25$ $\overline{3\times 5}$ $\quad\;5$ $3\times 5\times 5 = 3\times 5^2$
4) 12	4) $2\,|\,12$ $\dfrac{12}{3\times 4}$ $2\,|\,6$ $\overline{2\times 2}$ $\quad\;3$ $2\times 2\times 3 = 2^2\times 3$	10) 84	10) $2\,|\,84$ $\dfrac{84}{7\times 12}$ $2\,|\,42$ $\overline{2\times 6}$ $3\,|\,21$ $\overline{2\times 3}$ $\quad\;7$ $2\times 2\times 3\times 7 = 2^2\times 3\times 7$
5) 18	5) $2\,|\,18$ $\dfrac{18}{3\times 6}$ $3\,|\,9$ $\overline{2\times 3}$ $\quad\;3$ $2\times 3\times 3 = 2\times 3^2$	11) 130	11) $2\,|\,130$ $\dfrac{130}{10\times 13}$ $5\,|\,65$ $\overline{2\times 5}$ $\quad\;13$ $2\times 5\times 13$
6) 55	6) $5\,|\,55$ $\dfrac{55}{5\times 11}$ $\quad\;11$ 5×11		

Factors 4

Prime Factors - H.C.F. / L.C.M.

Prime Factors - H.C.F. / L.C.M.

A number may be written **as a product of its Prime Factors** ...
the factors are **ALL prime numbers** and **multiply** to give the number.

When 2 or more numbers are each written **as a product of its Prime Factors**,
the Highest Common Factor (H.C.F.) and Lowest Common Multiple (L.C.M.) can also be found.

*Note
Factors 3 Method 1 / 2 are shown or this may be done mentally.
The factors may be presented in a different order than shown here.

Highest Common Factor (H.C.F.) - the **highest** number that divides into 2 or more numbers.
Write each number **as a product of its prime factors** - MULTIPLY the **common** factors.

Ex. Find the Highest Common Factor (H.C.F.) of 12 and 18.

Method 1

$2 \mid 12$ $2 \mid 18$
$2 \mid 6$ $3 \mid 9$
$\quad 3$ $\quad 3$

⇩   ⇩

$2 \times 2 \times 3$ $2 \times 3 \times 3$

H.C.F. = **2** × **3** = 6

2 and **3** are **common** factors

Method 2

$\dfrac{12}{2 \times 6}$ $\dfrac{18}{2 \times 9}$
2×3 3×3

⇩   ⇩

$2 \times 2 \times 3$ $2 \times 3 \times 3$

H.C.F. = **2** × **3** = 6

2 and **3** are **common** factors

Q	A
1) Find the Highest Common Factor (H.C.F.) of 14 and 42.	1) **Method 1**: $2\mid14$, 7 ; $2\mid42$, $3\mid21$, 7 → 2×7 and $2\times3\times7$ → H.C.F. = **2** × **7** = 14. **Method 2**: $14 = 2\times7$; $42 = 2\times21 = 3\times7$ → 2×7 and $2\times3\times7$ → H.C.F. = **2** × **7** = 14
2) Find the Highest Common Factor (H.C.F.) of 36 and 54.	2) **Method 1**: $2\mid36$, $2\mid18$, $3\mid9$, 3 ; $2\mid54$, $3\mid27$, $3\mid9$, 3 → $2\times2\times3\times3$ and $2\times3\times3\times3$ → H.C.F. = **2** × **3** × **3** = 18. **Method 2**: $36 = 6\times6 = 2\times3\;2\times3$; $54 = 6\times9 = 2\times3\;3\times3$ → $2\times2\times3\times3$ and $2\times3\times3\times3$ → H.C.F. = **2** × **3** × **3** = 18
3) Find the Highest Common Factor (H.C.F.) of 175 and 70.	3) **Method 1**: $5\mid175$, $5\mid35$, 7 ; $2\mid70$, $5\mid35$, 7 → $5\times5\times7$ and $2\times5\times7$ → H.C.F. = **5** × **7** = 35. **Method 2**: $175 = 5\times35 = 5\times7$; $70 = 7\times10 = 2\times5$ → $5\times5\times7$ and $2\times5\times7$ → H.C.F. = **5** × **7** = 35
4) Find the Highest Common Factor (H.C.F.) of 165 and 385.	4) **Method 1**: $3\mid165$, $5\mid55$, 11 ; $5\mid385$, $7\mid77$, 11 → $3\times5\times11$ and $5\times7\times11$ → H.C.F. = **5** × **11** = 55. **Method 2**: $165 = 5\times33 = 3\times11$; $385 = 5\times77 = 7\times11$ → $3\times5\times11$ and $5\times7\times11$ → H.C.F. = **5** × **11** = 55

Factors 5

Prime Factors - L.C.M.

Lowest Common Multiple (L.C.M.) - the **lowest** number that 2 or more numbers divide into.
Write each number **as a product of its prime factors** -
MULTIPLY **all** of the factors which **appear most times** for the given numbers ...
this may be just 1 factor / include the same factor repeated.
(this ensures each of the given numbers divides into the L.C.M. - the factors of each can be 'checked'.)

Ex. Find the Lowest Common Multiple (L.C.M.) of 12 and 18.

Method 1

$$2 | 12 \quad\quad 2 | 18$$
$$2 | 6 \quad\quad 3 | 9$$
$$3 \quad\quad\quad 3$$

⇩ ⇩

$2 \times 2 \times 3 \quad\quad 2 \times 3 \times 3$

L.C.M. = $2 \times 2 \times 3 \times 3 = 36$

2 appears **most times** here.
3 appears **most times** here.

Method 2

$$\frac{12}{2 \times 6} \quad\quad \frac{18}{2 \times 9}$$
$$2 \times 3 \quad\quad 3 \times 3$$

⇩ ⇩

$2 \times 2 \times 3 \quad\quad 2 \times 3 \times 3$

L.C.M. = $2 \times 2 \times 3 \times 3 = 36$

2 appears **most times** here.
3 appears **most times** here.

$1 \times 12 = 12 \quad 1 \times 18 = 18$
$2 \times 12 = 24 \quad 2 \times 18 = \mathbf{36}$
$3 \times 12 = \mathbf{36}$

the **lowest** number that
12 and 18 divide into is 36.

Q	A
1) Find the Lowest Common Multiple (L.C.M.) of 9 and 15.	1) **Method 1** $3|9 \quad 3|15$ $3 \quad\quad 5$ $3 \times 3 \quad 3 \times 5$ L.C.M. = $3 \times 3 \times 5 = 45$ **Method 2** $\frac{9}{3 \times 3} \quad \frac{15}{3 \times 5}$ L.C.M. = $3 \times 3 \times 5 = 45$
2) Find the Lowest Common Multiple (L.C.M.) of 8 and 14.	2) $2|8 \quad 2|14$ $2|4 \quad\quad 7$ 2 $2 \times 2 \times 2 \quad 2 \times 7$ L.C.M. = $2 \times 2 \times 2 \times 7 = 56$ $\frac{8}{2 \times 4} \quad \frac{14}{2 \times 7}$ 2×2 $2 \times 2 \times 2 \quad 2 \times 7$ L.C.M. = $2 \times 2 \times 2 \times 7 = 56$
3) Find the Lowest Common Multiple (L.C.M.) of 20 and 30.	3) $2|20 \quad 2|30$ $2|10 \quad 3|15$ $5 \quad\quad 5$ $2 \times 2 \times 5 \quad 2 \times 3 \times 5$ (either 5) L.C.M. = $2 \times 2 \times 3 \times 5 = 60$ $\frac{20}{2 \times 10} \quad \frac{30}{5 \times 6}$ $2 \times 5 \quad\quad 2 \times 3$ $2 \times 2 \times 5 \quad 2 \times 3 \times 5$ (either 5) L.C.M. = $2 \times 2 \times 3 \times 5 = 60$
4) Find the Lowest Common Multiple (L.C.M.) of 56 and 98.	4) $2|56 \quad 2|98$ $2|28 \quad 7|49$ $2|14 \quad\quad 7$ 7 $2 \times 2 \times 2 \times 7 \quad 2 \times 7 \times 7$ L.C.M. = $2 \times 2 \times 2 \times 7 \times 7 = 392$ $\frac{56}{7 \times 8} \quad \frac{98}{2 \times 49}$ $2 \times 2 \times 2 \quad\quad 7 \times 7$ $2 \times 2 \times 2 \times 7 \quad 2 \times 7 \times 7$ L.C.M. = $2 \times 2 \times 2 \times 7 \times 7 = 392$
5) Find the Lowest Common Multiple (L.C.M.) of 42 and 55.	5) $2|42 \quad 5|55$ $3|21 \quad\quad 11$ 7 $2 \times 3 \times 7 \quad 5 \times 11$ L.C.M. = $2 \times 3 \times 5 \times 7 \times 11 = 2310$ $\frac{42}{2 \times 21} \quad \frac{55}{5 \times 11}$ 3×7 $2 \times 3 \times 7 \quad 5 \times 11$ L.C.M. = $2 \times 3 \times 5 \times 7 \times 11 = 2310$

Factors 6

Prime Factors - H.C.F. / L.C.M. - Index Form.

Prime Factors - H.C.F. / L.C.M.
When 2 or more numbers are each <u>presented</u> as a product of its Prime Factors <u>in index form</u>, the Highest Common Factor (H.C.F.) and Lowest Common Multiple (L.C.M.) can be found and **left in index form**.

The factors may be written as already shown in **Factors 4-5**.
From this it is possible to see that the factors may be left in index form (shown <u>underlined</u>) to find ...

   H.C.F. = multiply **each common number** with its **LOWEST** power.

   L.C.M. = multiply **each number** with its **HIGHEST** power.

(Here, both ways are shown.)

Ex. $A = 4^2 \times 5^3$ $B = 4^3 \times 5 \times 6$

1. Find the Highest Common Factor (H.C.F.) of A and B.
2. Find the Lowest Common Multiple (L.C.M.) of A and B.

1. $\underline{4^2} \times 5^3$ = **4×4×5×5×5**   2. $4^2 \times \underline{5^3}$ = **4×4×5×5×5**

      $4^3 \times \underline{5} \times 6$ = **4×4×4×5×6** $\underline{4^3} \times 5 \times \underline{6}$ = **4×4×4×5×6**

      H.C.F. = **4×4×5**   L.C.M = **4×4×4×5×5×5×6**

           = $\underline{4^2} \times \underline{5}$ = $\underline{4^3} \times \underline{5^3} \times \underline{6}$

Q	A

1) $A = 5^3 \times 7$ $B = 4^2 \times 5^2 \times 7^2$
1.
Find the Highest Common Factor (H.C.F.) of A and B.
2.
Find the Lowest Common Multiple (L.C.M.) of A and B.
	1) 1.
	$5^3 \times \underline{7}$ = **5×5×5×7**
	$4^2 \times \underline{5^2} \times 7^2$ = **4×4×5×5×7×7**
	H.C.F. = **5×5×7**
	        = $\underline{5^2} \times \underline{7}$ 2.
	$\underline{5^3} \times 7$ = **5×5×5×7**
	$4^2 \times 5^2 \times \underline{7^2}$ = **4×4×5×5×7×7**
	L.C.M. = **4×4×5×5×5×7×7**
	        = $\underline{4^2} \times \underline{5^3} \times \underline{7^2}$

2) $A = 2^4 \times 3^2 \times 5$ $B = 2^3 \times 5^2$
1.
Find the Highest Common Factor (H.C.F.) of A and B.
2.
Find the Lowest Common Multiple (L.C.M.) of A and B.
	2) 1.
	$2^4 \times 3^2 \times \underline{5}$ = **2×2×2×2×3×3×5**
	$\underline{2^3} \times 5^2$ = **2×2×2×5×5**
	H.C.F. = **2×2×2×5×5**
	        = $\underline{2^3} \times \underline{5}$ 2.
	$\underline{2^4} \times \underline{3^2} \times 5$ = **2×2×2×2×3×3×5**
	$2^3 \times \underline{5^2}$ = **2×2×2×5×5**
	L.C.M. = **2×2×2×2×3×3×5×5**
	        = $\underline{2^4} \times \underline{3^2} \times \underline{5^2}$

3) $A = 4^2 \times 6^3 \times 8^2$ $B = 4 \times 6^2 \times 8^3$
1.
Find the Highest Common Factor (H.C.F.) of A and B.
2.
Find the Lowest Common Multiple (L.C.M.) of A and B.
	3) 1.
	$4^2 \times 6^3 \times \underline{8^2}$ = **4×4×6×6×6×8×8**
	$\underline{4} \times \underline{6^2} \times 8^3$ = **4×6×6×8×8×8**
	H.C.F. = **4×6×6×8×8**
	            = $\underline{4} \times \underline{6^2} \times \underline{8^2}$ 2.
	$\underline{4^2} \times \underline{6^3} \times 8^2$ = **4×4×6×6×6×8×8**
	$4 \times 6^2 \times \underline{8^3}$ = **4×6×6×8×8×8**
	H.C.F. = **4×4×6×6×6×8×8×8**
	            = $\underline{4^2} \times \underline{6^3} \times \underline{8^3}$

4) $A = 2 \times 7^2 \times 9^4$ $B = 2^2 \times 7^2 \times 9^3$
1.
Find the Highest Common Factor (H.C.F.) of A and B.
2.
Find the Lowest Common Multiple (L.C.M.) of A and B.
	4) 1.
	$\underline{2} \times \underline{7^2} \times 9^4$ = **2×7×7×9×9×9×9**
	$2^2 \times \underline{7^2} \times \underline{9^3}$ = **2×2×7×7×9×9×9**
	H.C.F. = **2×7×7×9×9×9**
	            = $\underline{2} \times \underline{7^2} \times \underline{9^3}$ 2.
	$2 \times \underline{7^2} \times \underline{9^4}$ = **2×7×7×9×9×9×9**
	$\underline{2^2} \times \underline{7^2} \times 9^3$ = **2×2×7×7×9×9×9**
	H.C.F. = **2×2×7×7×9×9×9×9**
	            = $\underline{2^2} \times \underline{7^2} \times \underline{9^4}$

Factors 7   Perfect Numbers.

*Note There are very few Perfect Numbers known and only a few of these are small numbers.
The examples do **not** reveal any and the Q/A are presented so that the 'smaller' numbers are found without too much work!

Perfect Numbers
A Perfect Number is a number which
all of its factors - except the number itself - add to give the SAME number.
The factors are listed in the same way as already shown.

Ex.   Show that the following numbers are **NOT** Perfect Numbers.

a) 2   b) 9   c) 12

'cross out' the number itself

```
        2          9          12
      1  2̶      1  9̶      1  1̶2̶
                    3         2   6
                              3   4
```
number **NOT** the same - so **NOT** Perfect

Total of **required** factors: 1   4   16

Q	A
1) Find which of the following numbers are Perfect Numbers. a) 15 b) 6 c) 4	1) a) 15 / 1 1̶5̶ / 3 5 → Total of required factors 9 b) 6 / 1 6̶ / 2 3 → 6 Perfect c) 4 / 1 4̶ / 2 → 3
2) Find which of the following numbers are Perfect Numbers. a) 33 b) 40 c) 28	2) a) 33 / 1 3̶3̶ / 3 11 → 15 b) 40 / 1 4̶0̶ / 2 20 / 4 10 / 5 8 → 50 c) 28 / 1 2̶8̶ / 2 14 / 4 7 → 28 Perfect
3) Find which of the following numbers are Perfect Numbers. a) 496 b) 100	3) a) 496 / 1 4̶9̶6̶ / 2 248 / 4 124 / 8 62 / 16 31 → 496 Perfect b) 100 / 1 1̶0̶0̶ / 2 50 / 4 25 / 5 20 / 10 → 117
4) Find which of the following numbers are Perfect Numbers. a) 1016 b) 8128	4) a) 1016 / 1 1̶0̶1̶6̶ / 2 508 / 4 254 / 8 127 → 904 b) 8128 / 1 8̶1̶2̶8̶ / 2 4064 / 4 2032 / 8 1016 / 16 508 / 32 254 / 64 127 → 8128 Perfect

*Note Perfect Numbers
So far ... as of 2023 ...only 51 are known ... they are **all even** and end in 6 or 8.
The 51st has 49,724,095 digits.
The 1st. seven are   6   28   496   8128   33,550,336   8,589,869,056   137,438,691,328 .

Rounding Numbers 1

Round to the nearest 10.

Rounding Numbers to the nearest 10, 100, 1000, ...
We may increase or decrease a number to become a number ending in 0.
This is done because we do not need such accuracy or so we can approximate calculations - which is usually easier with numbers ending in 0.

We call this 'rounding' / 'rounding a number' and we usually '**round**' up or down '**to the nearest 10**' or '**to the nearest 100**' or '**to the nearest 1000**' ...

We look at the figure in the column that we are rounding to (the 10's or 100's or 1000's ... column.)
If we **round down** - the figure does **not change**.
If we **round up** - the figure is **increased by 1**. (figures to the **left** may have to change)
Whether we round up or down depends on the number to the **right** of this figure.

Round to the nearest 10

Ex. Round each number to the **nearest 10**.

If the number on the **right** of the 10's column is ...

below 5 - round down.
5 - round up or down.
above 5 - round up.

10's

a) 72   a) 7**2** ⟶ 70   This is nearer to 70 so **round down to 70**.

b) 75   b) 7**5** ⟶ 70 or 80   This is exactly half way between 70 and 80 so **round down to 70 or round up to 80**.

c) 79·3   c) 7**9**·3 ⟶ 80   This is nearer to 80 so **round up to 80**.

Q	A	Q	A
1) Round to the **nearest 10**.	1)		
A) 3	A) **3** ⟶ 0	G) 64·5	G) 6**4**·5 ⟶ 60
B) 5	B) **5** 0 or 10	H) 42·9	H) 4**2**·9 40
C) 6	C) **6** 10	I) 165·22...	I) 16**5**·22... 170
D) 14	D) 1**4** 10	J) 755	J) 75**5** 750 or 760
E) 86	E) 8**6** 90	K) 498	K) 49**8** 500
F) 25	F) 2**5** 20 or 30	L) 999	L) 99**9** 1,000

Rounding Numbers 2

Round to nearest 100 / 1000.

Round to the nearest 100

Ex. Round each number to the **nearest 100**.

If the number on the **right** of the 100's column is ...

below 50 -round down.
50 -round up or down.
above 50 -round up.

100's
⇩

a) 2<u>14</u> → 200   This is nearer to 200 so **round down to 200**.

b) 2<u>50</u> → 250 or 300   This is exactly half way between 250 and 300 so **round down to 200 or round up to 300**.

c) 2<u>62</u> → 300   This is nearer to 300 so **round up to 300**.

Q	A	Q	A
1) Round to the nearest 100.	1)		
A) 29·6	A) <u>29·6</u> → 0	G) 750·01	G) 7<u>50·01</u> → 800
B) 50	B) <u>50</u> 0 or 100	H) 950	H) 9<u>50</u> 900 or 1,000
C) 83	C) <u>83</u> 100	I) 999	I) 9<u>99</u> 1,000
D) 945	D) 9<u>45</u> 900	J) 1,034	J) 1,0<u>34</u> 1,000
E) 376	E) 3<u>76</u> 400	K) 8,267	K) 8,2<u>67</u> 8,300
F) 550	F) 5<u>50</u> 500 or 600	L) 4,444	L) 4,4<u>44</u> 4,400

Round to the nearest 1,000

Ex. Round each number to the **nearest 1,000**.

If the number on the **right** of the 1,000's column is ...

below 500 -round down.
500 -round up or down.
above 500 -round up.

1,000's
⇩

a) 8,<u>099</u> → 8,000   This is nearer to 8,000 so **round down to 8,000**.

b) 8,<u>500</u> → 8,000 or 9,000   This is exactly half way between 8,000 and 9,000 so **round down to 8,000 or round up to 9,000**.

c) 8,<u>888</u> → 9,000   This is nearer to 9,000 so **round up to 9,000**.

Q	A	Q	A
1) Round to the nearest 1,000.	1)		
A) 309	A) <u>309</u> → 0	G) 2,809	G) 2,<u>809</u> → 3,000
B) 500	B) <u>500</u> 0 or 1,000	H) 5,550	H) 5,<u>550</u> 6,000
C) 767	C) <u>767</u> 1,000	I) 67,291	I) 67,<u>291</u> 67,000
D) 1,499·99...	D) 1,<u>499·99...</u> 1,000	J) 39,500	J) 39,<u>500</u> 39,000 or 40,000
E) 6,500	E) 6,<u>500</u> 6,000 or 7,000	K) 23,678	K) 23,<u>678</u> 24,000
F) 3,501	F) 3,<u>501</u> 4,000	L) 99,999	L) 99,<u>999</u> 100,000

Rounding Numbers 3

Round Decimal to the nearest Whole Number.

Rounding Decimals

As with whole numbers, we may increase or decrease a decimal to become a whole number or a decimal with less figures.

We usually '**round**' up or down ...

       '**to the nearest whole number**' - the answer is a whole number only.
   or '**to 1 decimal place**'   - the answer has 1 decimal place (1 d.p.).
   or '**to 2 decimal places**' ...   - the answer has 2 decimal places (2 d.p.) ...

We look at the figure in the column that we are rounding to -
(the units whole number column or 1st. decimal place column or 2nd. decimal place column ...)

If we **round down** - the figure does **not change**.
If we **round up**   - the figure is **increased by 1**. (figures to the **left** may have to change)
Whether we round up or down depends on the number to the **right** of this figure.

Round to the nearest Whole Number

Ex. Round each number to the **nearest whole number**.

If the decimal on the **right** of the whole number column is ...

     below 0·5 -round down.
           0·5 -round up or down.
    above 0·5 -round up.

whole number

a) 3·1   a) 3·**1** ⟶ 3   This is nearer to 3 so **round down to 3**.

b) 3·5   b) 3·**5** ⟶ 3 or 4   This is exactly half way between 3 and 4
                                                   so **round down to 3 or round up to 4**.

c) 3·7   c) 3·**7** ⟶ 4   This is nearer to 4 so **round up to 4**.

Q	A	Q	A
1) Round to the **nearest whole number**.	1)		
A) 0·2	A) 0·**2** 0	G) 28·5001	G) 28·**5001** 29
B) 0·84	B) 0·**84** 1	H) 99·5	H) 99·**5** 99 or 100
C) 3·08	C) 3·**08** 3	I) 555·555...	I) 555·**555...** 556
D) 7·777...	D) 7·**777...** 8	J) 104·19	J) 104·**19** 104
E) 9·5	E) 9·**5** 9 or 10	K) 800·62	K) 800·**62** 801
F) 46·456	F) 46·**456** 46	L) 202·4999...	L) 202·**4999...** 202

Rounding Numbers 4

Round to 1 / 2 Decimal Places.

Round to 1 Decimal Place

Ex. Round each number to **1 decimal place**.

If the number on the **right** of the 1st. decimal place column is ...

below 0·05 -round down.
0·05 -round up or down.
above 0·05 -round up.

1st. decimal place ⇩

a) 4·2<u>3</u>
b) 4·7<u>5</u>
c) 4·86

a) 4·2<u>3</u> ⟶ 4·2   This is nearer to 4·2 so **round down to 4·2**.
b) 4·7<u>5</u> ⟶ 4·7 or 4·8   This is exactly half way between 4·7 and 4·8 so **round down to 4·7 or round up to 4·8**.
c) 4·8<u>61</u> ⟶ 4·9   This is nearer to 4·9 so **round up to 4·9**.

Q	A	Q	A
1) Round to **1 decimal place**.	1)		
A) 0·03	A) 0·0<u>3</u> 0·0	G) 8·55	G) 8·5<u>5</u> 8·5 or 8·6
B) 0·31	B) 0·3<u>1</u> 0·3	H) 34·434	H) 34·4<u>34</u> 34·4
C) 2·09	C) 2·0<u>9</u> 2·1	I) 51·97	I) 51·9<u>7</u> 52·0
D) 5·678	D) 5·6<u>78</u> 5·7	J) 93·05	J) 93·0<u>5</u> 93·0 or 93·1
E) 4·25	E) 4·2<u>5</u> 4·2 or 4·3	K) 246·8499...	K) 246·8<u>499...</u> 246·8
F) 10·555	F) 10·5<u>55</u> 10·6	L) 781·187	L) 781·1<u>87</u> 781·2

Round to 2 Decimal Places

Ex. Round each number to **2 decimal places**.

If the number on the **right** of the 2nd. decimal place column is ...

below 0·005 -round down.
0·005 -round up or down.
above 0·005 -round up.

2nd. decimal place ⇩

a) 8·79<u>22</u>...
b) 8·33<u>5</u>
c) 8·15<u>9</u>

a) 8·79<u>22...</u> ⟶ 8·79   This is nearer to 8·79 so **round down to 8·79**.
b) 8·33<u>5</u> ⟶ 8·33 or 8·34   This is exactly half way between 8·33 and 8·34 so **round down to 8·33 or round up to 8·34**.
c) 8·15<u>9</u> ⟶ 8·16   This is nearer to 8·16 so **round up to 8·16**.

Q	A	Q	A
1) Round to **2 decimal places**.	1)		
A) 0·534	A) 0·53<u>4</u> 0·53	G) 25·007	G) 25·00<u>7</u> 25·01
B) 6·016	B) 6·01<u>6</u> 6·02	H) 18·555	H) 18·55<u>5</u> 18·55 or 18·56
C) 3·7272	C) 3·72<u>72</u> 3·73	I) 2·4456	I) 2·44<u>56</u> 2·45
D) 9·2018	D) 9·20<u>18</u> 9·20	J) 70·0888...	J) 70·08<u>88...</u> 70·09
E) 44·004	E) 44·00<u>4</u> 44·00	K) 333·333...	K) 333·33<u>3...</u> 333·33
F) 1·999	F) 1·99<u>9</u> 2·00	L) 959·595	L) 959·59<u>5</u> 959·59 or 959·60

Rounding Numbers 5

Round to 1,2,3, ... Significant Figures.

Round to 1, 2, 3, ... Significant Figures

Any value has a given number of **significant figures (s.f.)** - '**significant**' means '**important**'.
We may increase or decrease a value to become a value with a given number of significant figures.
This is done because we do not need such accuracy or so we can approximate calculations more easily.

This is also called 'rounding' and involves the work already explained - (**Rounding Numbers 1-4**).

In the following Exs. and Q/A, the **significant** figures are underlined.

In general,

ALL NON-ZERO figures COUNT as **significant**.   Ex. 42·869 has **5 s.f.**

0's at the FRONT of a number do NOT COUNT as **significant**.   Ex. 0·0002 has **1 s.f.**

0's in the MIDDLE of a number COUNT as **significant**.   Ex. 7005 has **4 s.f.**

0's at the END of a number MAY or MAY NOT COUNT as **significant** -
this depends on the original value and how it is rounded.

      Ex. 50·03 has **4 s.f.**

This becomes   50·0   if rounded to **3 s.f.**   (the 0's COUNT   as **significant**)
                         50   if rounded to **2 s.f.**   (the 0   COUNTS as **significant**)
                         50   if rounded to **1 s.f.**   (the 0 does NOT COUNT as **significant** - but ...
                                                                 the 0 **must be written** to show the *place value* of the 5)

ALL 0's at the END of a WHOLE NUMBER **must be written**.

Q	A	Q	A
1) State the number of **significant** figures in each value.	1) The **significant** figures are underlined.	I) 700	700 1 or 700 2 or 700 3 depending on the original value and if / how it was rounded.
A) 456	A) 456 3		
B) 17·92	B) 17·92 4		
C) 0·003	C) 0·003 1	J) 40·0	40·0 3 the value has been 'rounded' to 1 d.p. so both 0's count.
D) 0·81	D) 0·81 2		
E) 205	E) 205 3		
F) 90·009	F) 90·009 5	K) 0·0200	0·0200 3 the value has been 'rounded' to 4 d.p. so the END 0's count.
G) 0·03004	G) 0·03004 4		
H) 0·777...	H) 0·777... infinite		

Q	A	Q	A
1) Round to **1 s.f.**	1) The **s.f.** are underlined.	E) 3,990	E) 4,000
A) 73	A) 70	F) 0·404	F) 0·4
B) 246	B) 200	G) 0·965	G) 1
C) 0·051	C) 0·05	H) 15·08	H) 20
D) 0·0088	D) 0·009		

Q	A	Q	A
1) Round to **2 s.f.**	1) The **s.f.** are underlined.	E) 12·75	E) 13
A) 562	A) 560	F) 30,038	F) 30,000
B) 8795	B) 8800	G) 0·0499	G) 0·050
C) 0·0453	C) 0·045	H) 2·25	H) 2·2 or 2·3
D) 0·606	D) 0·61		

Q	A	Q	A
1) Round to **3 s.f.**	1) The **s.f.** are underlined.	E) 4·0005	E) 4·00
A) 3,434	A) 3,430	F) 0·07068	F) 0·0707
B) 27,068	B) 27,100	G) 896·123	G) 896
C) 5,003	C) 5,000	H) 55·55	H) 55·5 or 55·6
D) 0·1847	D) 0·185		

Rounding Numbers 6   Estimation.

Estimation

It is useful to make estimates of many calculations to compare with the 'actual' answers in order to recognise if any **obvious** errors are made and so correct them by re-calculating.

These can often be carried out **mentally**.

Numbers are 'rounded' for speed - usually working to 1 significant figure is acceptable ...
(see **Rounding Numbers 5**) but we may also 'round' more drastically and not be concerned too much about the 'rules' for rounding.

Estimates are for our **own purposes**.

At first, they may not appear **too close** to actual answers, but with practice the differences become more acceptable and reliable - and we can always make a different estimate.

In the following Exs. and Q/A, any estimate may be different than shown here.

Ex. a) Estimate each answer.
b) Calculate each answer.

1) $339 + 956$
   a) $300 + 900 = 1200$
   or $300 + 1000 = 1300$
   b) 1295

2) $0.051 - 0.0155$
   a) $0.05 - 0.01 = 0.04$
   b) 0.0355

3) 14×27
   a) $10 \times 20 = 200$
   or $10 \times 30 = 300$
   or $14 \times 30 = 420$
   b) 378

4) $58.31 \div 9.48$
   a) $58 \div 9 = 6$ or 7
   or $60 \div 10 = 6$
   b) $6.15...$

5) $\dfrac{427 + 767}{91 - 48}$
   a) $\dfrac{400 + 800}{90 - 50} = \dfrac{1200}{40} = 30$
   b) $\dfrac{1194}{43} = 27.76...$

6) $\sqrt{(7.2)^2 + (3.9)^2}$
   a) $\sqrt{7^2 + 4^2}$
   $\sqrt{49 + 16}$
   $\sqrt{65} \approx \pm 8$
   b) $\sqrt{51.84 + 15.21}$
   $\sqrt{67.05} = \pm 8.18...$

Q	A	Q	A
a) Estimate each answer. b) Calculate each answer.		6) $\dfrac{4.3 \times 9.7}{5.5 \div 1.2}$	6) a) $\dfrac{4 \times 10}{5 \div 1} = \dfrac{40}{5} = 8$ b) $\dfrac{41.71}{4.58...} = 9.10...$
1) $787 - 345$	1) a) $800 - 300 = 500$ b) 442	7) $\dfrac{17.6 + 8.9}{0.4 \times 6.09}$	7) a) $\dfrac{17 + 8}{0.4 \times 6} \approx \dfrac{25}{2} \approx 12$ (it does not help to round to 0) b) $\dfrac{26.5}{2.436} = 10.87...$
2) $0.429 + 0.246$	2) a) $0.4 + 0.2 = 0.6$ b) 0.675		
3) 6.13×5.68	3) a) $6 \times 5 = 30$ b) 34.8184	8) $\sqrt{(3.3)^2 + (5.7)^2}$	8) a) $\sqrt{3^2 + 6^2} = \sqrt{9 + 36} = \sqrt{45} \approx \pm 7$ b) $\sqrt{10.89 + 32.49}$ $\sqrt{43.38} = \pm 6.58...$
4) $891 \div 35$	4) a) $900 \div 30 = 30$ b) $25.45...$		
5) $\dfrac{714 - 176}{59 + 22}$	5) a) $\dfrac{700 - 200}{60 + 20} = \dfrac{500}{80} \approx 6$ b) $\dfrac{538}{81} = 6.64...$	9) $77^2 - 24^2$	9) a) $80^2 - 20^2$ $6400 - 400 = 6000$ b) $5929 - 576 = 5353$

Negative Numbers 1

Negative Numbers. Number Line.

Negative Numbers

Negative numbers are less than 0.   Exs. -0·0001, -1, -2·5, -4$\frac{1}{2}$, -10, -987, -3333·3..., -1,000,000.

The number of numbers is **infinite** and each can have the same **negative value** as **any positive value**.
It is useful to consider the numbers positioned on a **number line** - either horizontal or vertical.

Exs. Any number is **more than** any number on its left.

(minus infinity) -∞   -10 -9 -8 -7 -6 -5 -4 -3 -2 -1 0 1 2 3 4 5 6 7 8 9 10   ∞ (infinity)

(infinity) ∞ Any number is **more than** any number below it.

Exs. a) 4 is greater than -4.   e) -3 is less than 1.
b) 0 is greater than -3.   f) -2 is less than 0.
c) -1 is greater than -2.   g) -4 is less than -3.
d) -5 is greater than -9.   h) -7 is less than -1.

*Note 1
The **negative** or **minus** sign may be called either (usually **minus**).
The sign may be 'long' − or 'short' - .
Ex. −1   'negative 1' or 'minus 1'.
    -1

*Note 2
The **negative** and **positive** numbers are also referred to as
Directed Numbers (since their − or + sign 'directs' their position).

(minus infinity) -∞

It is also useful to consider **negative decimals** and **negative fractions** also positioned on a number line.

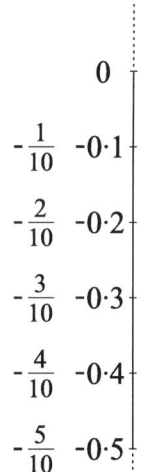

Exs.   -1   -0·9 -0·8 -0·7 -0·6 -0·5 -0·4 -0·3 -0·2 -0·1   0
         -$\frac{9}{10}$ -$\frac{8}{10}$ -$\frac{7}{10}$ -$\frac{6}{10}$ -$\frac{5}{10}$ -$\frac{4}{10}$ -$\frac{3}{10}$ -$\frac{2}{10}$ -$\frac{1}{10}$

Exs. a) 0 is greater than -0·6   e) -0·5 is less than -0·4
b) -0·1 is greater than -0·8   f) -0·9 is less than -0·2
c) -$\frac{2}{10}$ is greater than -$\frac{7}{10}$ g) -$\frac{8}{10}$ is less than -$\frac{1}{10}$
d) -$\frac{4}{10}$ is greater than -$\frac{5}{10}$ h) -$\frac{6}{10}$ is less than -$\frac{3}{10}$

Q	A	Q	A
1) State the largest number.		2) State the smallest number.	
a) -3 or -5	a) -3	a) -8 or -7	a) -8
b) -20 or -16	b) -16	b) -123 or -456	b) -456
c) -0·4 or -0·9	c) -0·4	c) -0·5 or -0·2	c) -0·5
d) -6·83 or -6·07	d) -6·07	d) -99·99 or -33·33	d) -99·99
e) -$\frac{5}{10}$ or -$\frac{6}{10}$	e) -$\frac{5}{10}$	e) -2$\frac{1}{4}$ or -2$\frac{1}{2}$	e) -2$\frac{1}{2}$

Negative Numbers 2   Number Order.

$-\infty$ (minus infinity) -10 -9 -8 -7 -6 -5 -4 -3 -2 -1 0 1 2 3 4 5 6 7 8 9 10 ∞ (infinity)

Number Order
Numbers can be arranged in **ascending** order (from smallest to largest)
or **descending** order (from largest to smallest).

Ex.1
List the numbers in **ascending** order.
-5, -1, -8
⇩
-8, -5, -1

Ex.2
List the numbers in **descending** order.
-7, -4, -6
⇩
-4, -6, -7

Q	A	Q	A
1) List in ascending order.		2) List in descending order.	
a) -3,-7,-2	a) -7,-3,-2	a) -5,-9,-1	a) -1,-5,-9
b) -4,-6,-9	b) -9,-6,-4	b) -8,-2,-10	b) -2,-8,-10
c) -1, 1, -10	c) -10,-1, 1	c) -3,-6, 3	c) 3,-3,-6
d) -3,-8, 4,-5	d) -8,-5,-3, 4	d) -7, 0,-9,-2	d) 0,-2,-7,-9
e) -88,-60,-92,-73	e) -92,-88,-73,-60	e) -32,-18,-50,-45	e) -18,-32,-45,-50

Decimals and Fractions
Decimals **and** Fractions can also be arranged **together**. It is useful to arrange them in a **column**.
It may be possible to arrange the numbers in their given form - but if we are not sure, it is usual to
change the necessary values to **decimals**, arrange the numbers, **then write them in their given form**.
The decimal point is on the **right** of an integer - it is useful to show it.

Ex.1
List the following numbers in **ascending** order.

-3·35 $-\frac{1}{2}$ -2·3 $-3\frac{2}{5}$
⇩   ⇩
-0·5   -3·4

ascending
-0·5   ⇨ $-\frac{1}{2}$
-2·3   -2·3
-3·35   -3·35
-3·4 $-3\frac{2}{5}$

Ex.2
List the following numbers in **descending** order.

$-1\frac{3}{4}$ -1·71   -1·8 $-1\frac{5}{7}$
⇩   ⇩
-1·75   -1·714...

descending
-1·8   ⇨   -1·8
-1·75 $-1\frac{3}{4}$
-1·714... $-1\frac{5}{7}$
-1·71   -1·71

Q	A	Q	A
1) List in **ascending** order -0·4 $-\frac{3}{5}$ -0·3 $-\frac{1}{3}$	1) **ascending** -0·3 ⇨ -0·3 -0·333... $-\frac{1}{3}$ -0·4 -0·4 -0·6 $-\frac{3}{5}$	3) List in **descending** order -5·6 -6·5 $-5\frac{5}{9}$ $-6\frac{2}{3}$	3) **descending** -5·555... ⇨ $-5\frac{5}{9}$ -5·6 -5·6 -6·5 -6·5 -6·666... $-6\frac{2}{3}$
2) List in **ascending** order $-2\frac{9}{10}$ $-1\frac{1}{2}$ $\sqrt[3]{-8}$ -1·7	2) **ascending** -1·5 ⇨ $-1\frac{1}{2}$ -1·7 -1·7 -2· $\sqrt[3]{-8}$ -2·9 $-2\frac{9}{10}$	4) List in **descending** order $-3\frac{3}{10}$ $-\pi$ -3·2 -3	4) **descending** -3· ⇨ -3 -3·141... $-\pi$ -3·2 -3·2 -3·3 $-3\frac{1}{10}$

Negative Numbers 3

Increase / Decrease.

Increase / Decrease

An increase in a number is first considered in the form of 'more than'.
A decrease in a number is first considered in the form of 'less than'.

Ex. Use the number line to find
- A) 3 more than -10
- B) 3 less than -1

A) From -10 move 3 units to the **right** = -7
B) From -1 move 3 units to the **left** = -4

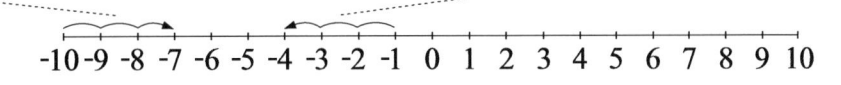

-10 -9 -8 -7 -6 -5 -4 -3 -2 -1 0 1 2 3 4 5 6 7 8 9 10

Q	A	Q	A
1) Find		2) Find	
a) 3 more than -5	a) -2	a) 3 less than -5	a) -8
b) 2 more than -6	b) -4	b) 4 less than -1	b) -5
c) 7 more than -7	c) 0	c) 8 less than 0	c) -8
d) 9 more than -4	d) 5	d) 5 less than 2	d) -3
e) 8 more than -1	e) 7	e) 9 less than 3	e) -6

Ex. Use the number line to find
- A) 0·4 more than -0·9
- B) $\frac{2}{10}$ less than $-\frac{1}{10}$

A) From -0·9 move 0·4 to the **right** = -0·5
B) From $-\frac{1}{10}$ move $\frac{2}{10}$ the **left** = $-\frac{3}{10}$

-1 -0·9 -0·8 -0·7 -0·6 -0·5 -0·4 -0·3 -0·2 -0·1 0 0·1 0·2 0·3 0·4 0·5 0·6 0·7 0·8 0·9 1

$-\frac{9}{10}$ $-\frac{8}{10}$ $-\frac{7}{10}$ $-\frac{6}{10}$ $-\frac{5}{10}$ $-\frac{4}{10}$ $-\frac{3}{10}$ $-\frac{2}{10}$ $-\frac{1}{10}$ $\frac{1}{10}$ $\frac{2}{10}$ $\frac{3}{10}$ $\frac{4}{10}$ $\frac{5}{10}$ $\frac{6}{10}$ $\frac{7}{10}$ $\frac{8}{10}$ $\frac{9}{10}$

Q	A	Q	A
1) Find		2) Find	
a) 0·3 more than -0·7	a) -0·4	a) 0·9 less than -0·1	a) -1
b) 0·6 more than -0·6	b) 0	b) 0·5 less than -0·2	b) -0·7
c) 0·8 more than -0·3	c) 0·5	c) 0·4 less than 0·3	c) -0·1
d) $\frac{4}{10}$ more than $-\frac{8}{10}$	d) $-\frac{4}{10}$	d) $\frac{1}{10}$ less than $-\frac{8}{10}$	d) $-\frac{9}{10}$
e) $\frac{7}{10}$ more than -1	e) $-\frac{3}{10}$	e) $\frac{6}{10}$ less than 0	e) $-\frac{6}{10}$
f) $\frac{5}{10}$ more than $-\frac{4}{10}$	f) $\frac{1}{10}$	f) $\frac{9}{10}$ less than $\frac{4}{10}$	f) $-\frac{5}{10}$

Negative Numbers 4 — Addition. Subtraction.

Operations
Addition and Subtraction
Consider the following examples.

Ex.1 a) $-2 + 4 = 2$ 'Start at -2 and add 4'

b) $4 - 2 = 2$ 'Start at 4 and subtract 2'

The answers are equal - so we can **think** to carry out **similar calculations** in either order.

Ex.2 a) $-13 + 57 = 44$
b) $57 - 13 = 44$

Ex.3 a) $-3 + 2 = -1$ 'Start at -3 and add 2'

b) $2 - 3 = -1$ 'Start at 2 and subtract 3'

The answers are equal - so we can **think** to carry out **similar calculations** in either order.

Also, it is useful to consider, especially for larger numbers, that the subtraction $3 - 2 = 1$ is 'easier' and then just change the answer 1 to -1 (since there are '**more negatives than positives**').

Ex.4 a) $-89 + 64 = -25$
b) $64 - 89 = -25$ (**think** $89 - 64 = 25$ change to -25)

Q	A	Q	A
Calculate			
1) $-1 + 6$	1) 5	11) $-4 + 1$	11) -3
2) $6 - 1$	2) 5	12) $1 - 4$	12) -3
3) $-4 + 7$	3) 3	13) $-7 + 2$	13) -5
4) $7 - 4$	4) 3	14) $2 - 7$	14) -5
5) $-2 + 10$	5) 8	15) $-8 + 0$	15) -8
6) $-9 + 18$	6) 9	16) $0 - 39$	16) -39
7) $-15 + 90$	7) 75	17) $-20 + 12$	17) -8
8) $-32 + 57$	8) 25	18) $10 - 50$	18) -40
9) $-23 + 84$	9) 61	19) $-65 + 41$	19) -24
10) $-49 + 61$	10) 12	20) $76 - 93$	20) -17

Q	A	Q	A
Calculate			
1) $-0.2 + 0.5$	1) 0.3	5) $-0.9 + 0.4$	5) -0.5
2) $-4.6 + 8.7$	2) 4.1	6) $4.6 - 8.7$	6) -4.1
3) $-\frac{3}{10} + \frac{9}{10}$	3) $\frac{6}{10}$	7) $-\frac{5}{10} + \frac{3}{10}$	7) $-\frac{2}{10}$
4) $-2\frac{1}{4} + 7\frac{1}{2}$	4) $5\frac{1}{4}$	8) $8\frac{1}{7} - 9\frac{6}{7}$	8) $-1\frac{5}{7}$

Negative Numbers 5

Addition. Subtraction.

Operations
Addition and Subtraction
Consider the following examples.

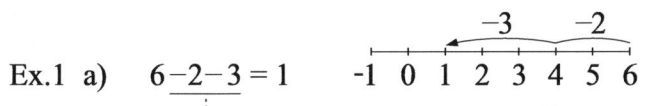

Ex.1 a) $6-2-3 = 1$ -1 0 1 2 3 4 5 6   'Start at 6 and subtract 2 and then subtract 3'

  b) $6 -5 = 1$ -1 0 1 2 3 4 5 6   'Start at 6 and subtract 5'

The answers are equal - so we can see that $-2-3 = -5$
This can also be thought of as **adding negative numbers**.
Calculations may or may not involve positive numbers.

Ex.2 Calculate
  a) $-4-1$ b) $-3-1-6$ c) $2-3-4$ d) $-6+1-2$ e) $-7+8-5+8$

⇩

Method 1 - add the negative numbers together.

  a) $-4-1 = -5$ b) $-3-1-6 = -10$ c) $2-3-4 =$ d) $-6+1-2 =$ e) $-7+8-5+8 =$
  $\quad 2\ -7\ = -5$ $\quad 1-8\ = -7$ $\quad 16-12\ = 4$

(It is useful to underline the negative numbers as 'like terms'.)

Method 2 - add the numbers in turn.

  c) $2-3-4 =$ d) $-6+1-2 =$ e) $-7+8-5+8 =$
  $-1\ -4 = -5$ $\ -5\ -2 = -7$ $\ 1\ -5+8 = 4$
  $\qquad\qquad\qquad\qquad\qquad\qquad\qquad\qquad\quad -4\ +8 = 4$

Q	A	Q	A
Calculate			
1) $-3-2$	1) -5	5) $-4-4-1$	5) -9
2) $-7-1$	2) -8	6) $-9-2-5$	6) -16
3) $-10-59$	3) -69	7) $-20-13-37$	7) -70
4) $-18-82$	4) -100	8) $-56-60-8$	8) -124

Q	A	
Calculate	Method 1	Method 2
1) $1-4-5$	1) $1-4-5 =$ $1\ -9\ = -8$	$1-4-5 =$ $-3\ -5 = -8$
2) $8-9-3$	2) $8-9-3 =$ $8\ -12\ = -4$	$8-9-3 =$ $-1\ -3 = -4$
3) $-2+6-7$	3) $-2+6-7 =$ $6-9\ = -3$	$-2+6-7 =$ $4\ -7 = -3$
4) $-3+3-2$	4) $-3+3-2 =$ $3-5\ = -2$	$-3+3-2 =$ $0\ -2 = -2$
5) $-4-1+8$	5) $-4-1+8 =$ $-5\ +8 = 3$	$-4-1+8 =$ $-5\ +8 = 3$
6) $-5+2+1-4$	6) $-5+2+1-4 =$ $3-9\ = -6$	$-5+2+1-4 =$ $-3\ +1-4 =$ $-2\ -4 = -6$
7) $3-8+7-9$	7) $3-8+7-9 =$ $10-17\ = -7$	$3-8+7-9 =$ $-5\ +7-9 =$ $2\ -9 = -7$

Negative Numbers 6

Addition. Subtraction.

Q	A	Q	A
Calculate			
1) $-0.7 - 0.2$	1) -0.9	5) $-\frac{3}{10} - \frac{2}{10}$	5) $-\frac{5}{10}$
2) $-4.1 - 3.6$	2) -7.7	6) $-4\frac{1}{7} - 4\frac{1}{7}$	6) $-8\frac{2}{7}$
3) $-11.3 - 21.8$	3) -33.1	7) $-1\frac{3}{4} - 5\frac{1}{2}$	7) $-7\frac{1}{4}$
4) $-0.5 - 1.5 - 2.5$	4) -4.5	8) $-3\frac{1}{9} - 2\frac{2}{9} - 1\frac{5}{9}$	8) $-6\frac{8}{9}$

Q	A		
Calculate	Method 1		Method 2
1) $7.5 - 1.5 - 2.5$	1) $7.5 - 1.5 - 2.5 =$ $7.5 \quad -4 \quad = 3.5$		$7.5 - 1.5 - 2.5 =$ $6 \quad - 2.5 = 3.5$
2) $0.2 - 0.3 - 0.4$	2) $0.2 - 0.3 - 0.4 =$ $0.2 \quad -0.7 \quad = -0.5$		$0.2 - 0.3 - 0.4 =$ $-0.1 \quad - 0.4 = -0.5$
3) $-1.9 + 2.1 - 5.8$	3) $-1.9 + 2.1 - 5.8 =$ $2.1 \quad -7.7 \quad = -5.6$		$-1.9 + 2.1 - 5.8 =$ $0.2 \quad - 5.8 = -5.6$
4) $-8.4 + 4.2 - 0.6$	4) $-8.4 + 4.2 - 0.6 =$ $4.2 \quad -9.0 \quad = -4.8$		$-8.4 + 4.2 - 0.6 =$ $-4.2 \quad - 0.6 = -4.8$
5) $\frac{9}{10} - \frac{5}{10} - \frac{1}{10}$	5) $\frac{9}{10} - \frac{5}{10} - \frac{1}{10} =$ $\frac{9}{10} \quad - \frac{6}{10} = \frac{3}{10}$		$\frac{9}{10} - \frac{5}{10} - \frac{1}{10} =$ $\frac{4}{10} \quad - \frac{1}{10} = \frac{3}{10}$
6) $\frac{1}{10} - \frac{6}{10} - \frac{2}{10}$	6) $\frac{1}{10} - \frac{6}{10} - \frac{2}{10} =$ $\frac{1}{10} \quad - \frac{8}{10} = -\frac{7}{10}$		$\frac{1}{10} - \frac{6}{10} - \frac{2}{10} =$ $-\frac{5}{10} \quad - \frac{2}{10} = -\frac{7}{10}$
7) $-3\frac{3}{8} + 2\frac{2}{8} - 5\frac{4}{8}$	7) $-3\frac{3}{8} + 2\frac{2}{8} - 5\frac{4}{8} =$ $2\frac{2}{8} \quad -8\frac{7}{8} = -6\frac{5}{8}$		$-3\frac{3}{8} + 2\frac{2}{8} - 5\frac{4}{8} =$ $-1\frac{1}{8} \quad - 5\frac{4}{8} = -6\frac{5}{8}$
8) $-4\frac{3}{4} + 1\frac{1}{2} - 2\frac{1}{4}$	8) $-4\frac{3}{4} + 1\frac{1}{2} - 2\frac{1}{4} =$ $1\frac{1}{2} \quad -7 \quad = -5\frac{1}{2}$		$-4\frac{3}{4} + 1\frac{1}{2} - 2\frac{1}{4} =$ $-3\frac{1}{4} \quad - 2\frac{1}{4} = -5\frac{1}{2}$

Negative Numbers 7

Addition. Subtraction.

Addition and Subtraction

Addition and Subtraction may involve 2 signs next to each other.
We 'cancel them out' to give 1 sign only - so it is easier to calculate.

A 'simple explanation' is that a $+$ sign does not change a value but a $-$ sign does (it gives the 'opposite').

   Ex.1 $+(1) = +1 = 1$ but $-(1) = -1$

Rules

There are 4 possible combinations of 2 signs.   (**not** usually written)

1) $++ = +$ 2) $+- = -$ 3) $-+ = -$ 4) $-- = +$

   $+$ does not change $+$ $+$ does not change $-$ $-$ changes $+$ $-$ changes $-$

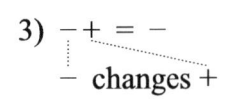

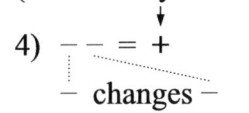

It is useful to cross out the 2 signs and replace them with 1 sign when calculating.

1) $\cancel{++}^{+}$ 2) $\cancel{+-}^{-}$ 3) $\cancel{-+}^{-}$ 4) $\cancel{--}^{+}$

Ex.2 Simplify   a) $++5$ b) $+-3$ c) $-+8$ d) $--1$

   a) $\cancel{++}^{+} 5 = 5$ b) $\cancel{+-}^{-} 3 = -3$ c) $\cancel{-+}^{-} 8 = -8$ d) $\cancel{--}^{+} 1 = 1$

Q	A	Q	A
Simplify		5) $++\frac{1}{2}$	5) $\cancel{++}^{+} \frac{1}{2} = \frac{1}{2}$
1) $++2$	1) $\cancel{++}^{+} 2 = 2$	6) $+-5\frac{3}{8}$	6) $\cancel{+-}^{-} 5\frac{3}{8} = -5\frac{3}{8}$
2) $+-4$	2) $\cancel{+-}^{-} 4 = -4$	7) $-+0\cdot 6$	7) $\cancel{-+}^{-} 0\cdot 6 = -0\cdot 6$
3) $-+9$	3) $\cancel{-+}^{-} 9 = -9$	8) $--3\cdot 4$	8) $\cancel{--}^{+} 3\cdot 4 = 3\cdot 4$
4) $--7$	4) $\cancel{--}^{+} 7 = 7$		

Calculate

Ex.3 a) $3++2$ b) $-7++1$ Ex.4 a) $9+-5$ b) $-1+-8$

   a) $3 \cancel{++}^{+} 2 = 5$ b) $-7 \cancel{++}^{+} 1 = -6$ a) $9 \cancel{+-}^{-} 5 = 4$ b) $-1 \cancel{+-}^{-} 8 = -9$

Ex.5 a) $6-+9$ b) $-4-+4$ Ex.6 a) $2--7$ b) $-3--3$

   a) $6 \cancel{-+}^{-} 9 = -3$ b) $-4 \cancel{-+}^{-} 4 = -8$ a) $2 \cancel{--}^{+} 7 = 9$ b) $-3 \cancel{--}^{+} 3 = 0$

Q	A	Q	A
Calculate		12) $\frac{1}{7}++\frac{5}{7}$	12) $\frac{1}{7} \cancel{++}^{+} \frac{5}{7} = \frac{6}{7}$
1) $1++5$	1) $1 \cancel{++}^{+} 5 = 6$	13) $\frac{4}{8}+-\frac{6}{8}$	13) $\frac{4}{8} \cancel{+-}^{-} \frac{6}{8} = -\frac{2}{8}$
2) $4+-7$	2) $4 \cancel{+-}^{-} 7 = -3$	14) $-\frac{2}{5}-+\frac{2}{5}$	14) $-\frac{2}{5} \cancel{-+}^{-} \frac{2}{5} = -\frac{4}{5}$
3) $8-+6$	3) $8 \cancel{-+}^{-} 6 = 2$	15) $-\frac{1}{2}--\frac{3}{4}$	15) $-\frac{1}{2} \cancel{--}^{+} \frac{3}{4} = \frac{1}{4}$
4) $9--9$	4) $9 \cancel{--}^{+} 9 = 18$	16) $0\cdot 5++0\cdot 1$	16) $0\cdot 5 \cancel{++}^{+} 0\cdot 1 = 0\cdot 6$
5) $-3++2$	5) $-3 \cancel{++}^{+} 2 = -1$	17) $3\cdot 7+-2\cdot 4$	17) $3\cdot 7 \cancel{+-}^{-} 2\cdot 4 = 1\cdot 3$
6) $-5+-4$	6) $-5 \cancel{+-}^{-} 4 = -9$	18) $-1\cdot 2-+1\cdot 6$	18) $-1\cdot 2 \cancel{-+}^{-} 1\cdot 6 = -2\cdot 8$
7) $-7-+8$	7) $-7 \cancel{-+}^{-} 8 = -15$	19) $-9\cdot 3--0\cdot 2$	19) $-9\cdot 3 \cancel{--}^{+} 0\cdot 2 = -9\cdot 1$
8) $-1--3$	8) $-1 \cancel{--}^{+} 3 = 2$	20) $4-+3+-8$	20) $4 \cancel{-+}^{-} 3 \cancel{+-}^{-} 8 = -7$
9) $-2+-6$	9) $-2 \cancel{+-}^{-} 6 = -8$	21) $-2--7-+1$	21) $-2 \cancel{--}^{+} 7 \cancel{-+}^{-} 1 = 4$
10) $-8-+4$	10) $-8 \cancel{-+}^{-} 4 = -12$		
11) $-5--5$	11) $-5 \cancel{--}^{+} 5 = 0$		

Negative Numbers 8   Multiplication.

Multiplication
Multiplication is considered to be **repeated addition** - both calculations are therefore related.

Ex. 1
Consider **adding** -1's which, in short, becomes **multiplying** -1's.

                           and 'in reverse'

$$0 \times -1 = 0 \qquad\qquad -1 \times 0 = 0$$
$$-1 = 1 \times -1 = -1 \qquad -1 \times 1 = -1$$
$$-1-1 = 2 \times -1 = -2 \qquad -1 \times 2 = -2$$
$$-1-1-1 = 3 \times -1 = -3 \qquad -1 \times 3 = -3$$
$$-1-1-1-1 = 4 \times -1 = -4 \qquad -1 \times 4 = -4$$

Ex. 2
Consider **adding** -2's which, in short, becomes **multiplying** -2's.

                           and 'in reverse'

$$0 \times -2 = 0 \qquad\qquad -2 \times 0 = 0$$
$$-2 = 1 \times -2 = -2 \qquad -2 \times 1 = -2$$
$$-2-2 = 2 \times -2 = -4 \qquad -2 \times 2 = -4$$
$$-2-2-2 = 3 \times -2 = -6 \qquad -2 \times 3 = -6$$
$$-2-2-2-2 = 4 \times -2 = -8 \qquad -2 \times 4 = -8$$

As with **addition**, a $+$ sign does not change a value but a $-$ sign does (it gives the 'opposite').
So, if each **positive number** is made **negative**, each **negative answer** is changed and made **positive**.

$$0 \times -1 = 0 \qquad\qquad\qquad\qquad 0 \times -2 = 0$$
$$-1 \times -1 = 1 \qquad\qquad\qquad\qquad -1 \times -2 = 2$$
$$-2 \times -1 = 2 \qquad\qquad\qquad\qquad -2 \times -2 = 4$$
$$-3 \times -1 = 3 \qquad\qquad\qquad\qquad -3 \times -2 = 6$$
$$-4 \times -1 = 4 \qquad\qquad\qquad\qquad -4 \times -2 = 8$$

Combining the above provides a continuous 'pattern' in the tables.

$$4 \times -1 = -4 \qquad\qquad\qquad\qquad 4 \times -2 = -8$$
$$3 \times -1 = -3 \qquad\qquad\qquad\qquad 3 \times -2 = -6$$
$$2 \times -1 = -2 \qquad\qquad\qquad\qquad 2 \times -2 = -4$$
$$1 \times -1 = -1 \qquad\qquad\qquad\qquad 1 \times -2 = -2$$
$$0 \times -1 = 0 \qquad\qquad\qquad\qquad 0 \times -2 = 0$$
$$-1 \times -1 = 1 \qquad\qquad\qquad\qquad -1 \times -2 = 2$$
$$-2 \times -1 = 2 \qquad\qquad\qquad\qquad -2 \times -2 = 4$$
$$-3 \times -1 = 3 \qquad\qquad\qquad\qquad -3 \times -2 = 6$$
$$-4 \times -1 = 4 \qquad\qquad\qquad\qquad -4 \times -2 = 8$$

From the above we can see that when multiplying there are 4 possible combinations of 2 signs.

     1) $+ \times + = +$ 2) $+ \times - = -$ 3) $- \times + = -$ 4) $- \times - = +$

The $+$ sign is usually omitted.

Ex. Calculate
1) $+5 \times +3 \Rightarrow +5 \times +3 = 15$
2) $+4 \times -8 \Rightarrow +4 \times -8 = -32$
3) $-6 \times +9 \Rightarrow -6 \times +9 = -54$
4) $-7 \times -11 \Rightarrow -7 \times -11 = +77$

Q	A	Q	A
Calculate			
1) $+4 \times +2$	1) $+8$ ($=8$)	9) -6×40	9) -240
2) 5×-6	2) -30	10) -10×-15	10) 150
3) -3×7	3) -21	11) 2×-222	11) -444
4) -2×-8	4) $+16$ ($=16$)	12) -30×30	12) -900
5) 9×-9	5) -81	13) -54×-7	13) 378
6) -1×5	6) -5	14) 9×-0.5	14) -4.5
7) -4×-3	7) 12	15) $-\frac{1}{2} \times -\frac{3}{4}$	15) $\frac{3}{8}$
8) 0×-10	8) 0		

Negative Numbers 9 — Division.

Division
The calculations for multiplication are used to explain the calculations for division - since division is the reverse of multiplication.

Ex. 1: $+4 \times +2 = +8 \;\Rightarrow\; +2 = \dfrac{+8}{+4}$

Ex. 2: $-4 \times -2 = +8 \;\Rightarrow\; -2 = \dfrac{+8}{-4}$

Ex. 3: $+4 \times -2 = -8 \;\Rightarrow\; -2 = \dfrac{-8}{+4}$

Ex. 4: $+4 \times -2 = -8 \;\Rightarrow\; +4 = \dfrac{-8}{-2}$

From the above we can see that when dividing there are 4 possible combinations of 2 signs.

$$\dfrac{+}{+} = + \qquad \dfrac{+}{-} = - \qquad \dfrac{-}{+} = - \qquad \dfrac{-}{-} = +$$

The + sign is usually omitted.

Ex. Calculate

1) $\dfrac{+7}{+1} \;\Rightarrow\; \dfrac{+7}{+1} = +7$

2) $\dfrac{+6}{-2} \;\Rightarrow\; \dfrac{+6}{-2} = -3$

3) $\dfrac{-25}{+5} \;\Rightarrow\; \dfrac{-25}{+5} = -5$

4) $\dfrac{-90}{-9} \;\Rightarrow\; \dfrac{-90}{-9} = +10$

Q Calculate	A	Q	A
1) $\dfrac{+6}{+2}$	1) $+3$ ($=3$)	9) $\dfrac{-700}{+100}$	9) -7
2) $\dfrac{-5}{+1}$	2) -5	10) $\dfrac{-5000}{-50}$	10) 100
3) $\dfrac{-6}{-3}$	3) $+2$ ($=2$)	11) $\dfrac{+1}{-2}$	11) $-\dfrac{1}{2}$ ($=-0.5$)
4) $\dfrac{+9}{-3}$	4) -3	12) $\dfrac{-3}{+4}$	12) $-\dfrac{3}{4}$ ($=-0.75$)
5) $\dfrac{+45}{+5}$	5) 9	13) $\dfrac{-9}{-2}$	13) $4\dfrac{1}{2}$ ($=4.5$)
6) $\dfrac{-22}{+2}$	6) -11	14) $\dfrac{+0.8}{-2}$	14) -0.4
7) $\dfrac{-80}{-10}$	7) 8	15) $\dfrac{-0.5}{+0.1}$	15) -5
8) $\dfrac{0}{-9}$	8) $-0 = 0$	16) $\dfrac{-6.3}{-3}$	16) 2.1

*Note Ex. $\dfrac{+1}{-2} = -\dfrac{1}{2}$ and $\dfrac{-1}{+2} = -\dfrac{1}{2}$

Multiplication and Division - can be combined.

Calculate Ex.1 $\dfrac{-2 \times 4}{5} = \dfrac{-8}{5} = -\dfrac{8}{5} = -1\dfrac{3}{5}$

Ex.2 $\dfrac{-7}{-3 \times 9} = \dfrac{-7}{-27} = \dfrac{7}{27}$

Q Calculate	A	Q	A
1) $\dfrac{-3 \times 2}{5}$	1) $= \dfrac{-6}{5} = -\dfrac{6}{5} = -1\dfrac{1}{5}$	5) $\dfrac{-11}{-4 \times 4}$	5) $= \dfrac{-11}{-16} = \dfrac{11}{16}$
2) $\dfrac{4 \times 1}{-9}$	2) $= \dfrac{4}{-9} = -\dfrac{4}{9}$	6) $\dfrac{5}{2 \times -6}$	6) $= \dfrac{5}{-12} = -\dfrac{5}{12}$
3) $\dfrac{-5 \times -7}{8}$	3) $= \dfrac{+35}{8} = 4\dfrac{3}{8}$	7) $\dfrac{-9}{-5 \times -1}$	7) $= \dfrac{-9}{+5} = -1\dfrac{4}{5}$
4) $\dfrac{2 \times -1}{-9}$	4) $= \dfrac{-2}{-9} = \dfrac{2}{9}$	8) $\dfrac{2}{-7 \times -3}$	8) $= \dfrac{2}{+21} = \dfrac{2}{21}$

Negative Numbers 10 — 'Problems'.

Addition / Subtraction

Q	A
1) In relation to Sea Level - Point A is 30m above, Point B is 10m below, Point C is 40m below. Find the difference in heights between Points A and B, A and C, B and C. Height (metres): 30 A, 20, 10, Sea Level 0, -10 B, -20, -30, -40 C	1) The 'number line' may be used to just count in 10's. or Difference in height between A and B = A − B $\quad$ 30 −(−10) = 40m A and C = A − C $\quad$ 30 −(−40) = 70m B and C = B − C $\quad$ −10 −(−40) = 30m
2) Calculate the change in temperature, °C, from 6 a.m. to 6 p.m. for each Town - X, Y, Z. 	Town	°C 6 p.m.	°C 6 a.m.
---	---	---
X	+7	−1
Y	−2	+5
Z	−9	−3		2) The calculations may be done mentally / using a number line. or $\quad$ °C $\quad$ °C $\quad$ 6 p.m. 6 a.m. X $\quad$ +7 −(−1) = +8 °C Y $\quad$ −2 −(+5) = −7 °C Z $\quad$ −9 −(−3) = −6 °C
3) 2 darts are thrown and land in the board - the 2 scores are added to give a Total Score. Calculate all the possible Total Scores. Board: −1, 4, −8	3) There are 6 possible Total Scores. (Calculations may be in a different order.) −1 + −1 = −2 $\quad\quad$ −1 + 4 = 3 4 + 4 = 8 $\quad\quad$ −1 + −8 = −9 −8 + −8 = −16 $\quad\quad$ 4 + −8 = −4

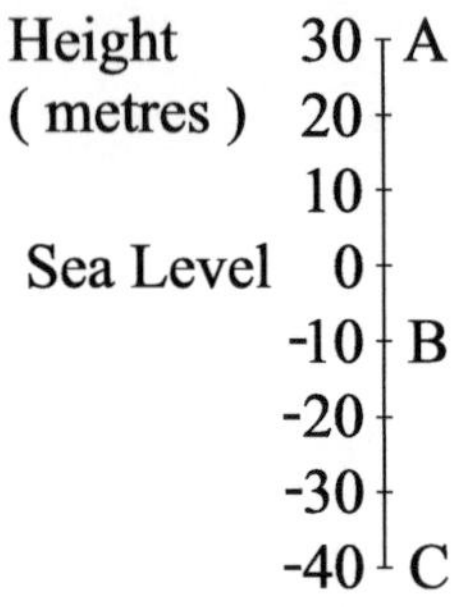

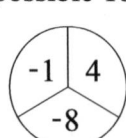

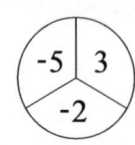

Multiplication / Division

Q	A
1) A machine generates 2 random numbers and multiplies them together to give a 'Winner'. Find the 'Winner' when the 2 random numbers are a) +7 and −7 $\quad\quad$ b) −9 and −20	1) a) +7 × −7 = −49 b) −9 × −20 = +180
2) 2 darts are thrown and land in the board - the 2 scores are multiplied to give a Total Score. Calculate all the possible Total Scores. Board: −5, 3, −2	2) There are 6 possible Total Scores. (Calculations may be in a different order.) −5 × −5 = 25 $\quad$ −5 × 3 = −15 3 × 3 = 9 $\quad$ −5 × −2 = 10 −2 × −2 = 4 $\quad$ 3 × −2 = −6
3) The value of each card is divided in turn by the value of each of the other 2 cards. Calculate all the possible solutions. Cards: −6, 3, −2	3) There are 6 possible solutions. (Calculations may be in a different order.) $\frac{-6}{3} = -2$ $\quad$ $\frac{3}{-6} = -\frac{3}{6}$ $(= -\frac{1}{2})$ $\quad$ $\frac{-2}{-6} = \frac{2}{6}$ $(= \frac{1}{3})$ $\frac{-6}{-2} = 3$ $\quad$ $\frac{3}{-2} = -\frac{3}{2}$ $(= -1\frac{1}{2})$ $\quad$ $\frac{-2}{3} = -\frac{2}{3}$

Negative Numbers 11   Multiplication.

Multiplication
When multiplying more than 2 numbers, the 'Rules' still apply to **pairs** of + or − signs.
The result from 1 pair of signs is then 'paired' with the next sign or result from another pair.
The signs can be 'paired' in any order.

Ex.1 $+2 \times +3 \times +4$

$+2 \times +3 \times +4 = +24$

or

$+2 \times +3 \times +4 =$
$+6 \times +4 = +24$

Ex.2 $-2 \times +3 \times +4$

$-2 \times +3 \times +4 = -24$

or

$-2 \times +3 \times +4 =$
$-6 \times +4 = -24$

Ex.3 $-2 \times -3 \times -4$

$-2 \times -3 \times -4 = -24$

or

$-2 \times -3 \times -4 =$
$+6 \times -4 = -24$

Signs
The following is a summary of the results of the signs when multiplying A) 3 numbers B) 4 numbers - the + or − signs may be in a different order (numbers are not written).

A) 3 numbers. $+++ = +$ $++- = -$ $+-- = +$ $--- = -$

B) 4 numbers. $++++ = +$ $+++- = -$ $++-- = +$ $+--- = -$ $---- = +$

In General when multiplying,

an **EVEN** number of − signs results in a + sign
an **ODD** number of − signs results in a − sign.

*Note A positive number is usually written **without** its + sign.

Q	A	Q	A
Calculate			
1) $+5 \times +2 \times +3$	1) $+30$ $(= 30)$	7) $+1 \times +2 \times +3 \times +4$	7) $+24$
2) $4 \times -1 \times +7$	2) -28	8) $4 \times -3 \times +2 \times +1$	8) -24
3) $-2 \times 6 \times 6$	3) -72	9) $-6 \times 2 \times -5 \times 3$	9) $+180$ $(= 180)$
4) $-3 \times -8 \times 10$	4) $+240$ $(= 240)$	10) $-8 \times -1 \times 1 \times -7$	10) -56
5) $-9 \times +5 \times -2$	5) $+90$ $(= 90)$	11) $-5 \times -5 \times -4 \times -9$	11) $+900$ $(= 900)$
6) $-7 \times -3 \times -4$	6) -84	12) $-1 \times -2 \times -3 \times -2 \times -1$	12) -12

Negative Numbers 12 — Powers (Indices). Roots.

Powers (Indices)

A positive or negative number may be 'raised to a power'.

Ex.1 $(+4)^2 = +4 \times +4 = +16$ Ex.2 $(-4)^2 = -4 \times -4 = +16$ ($= 16$)

Q	A	Q	A
Calculate			
1) $(+3)^2$	1) $+3 \times +3 = 9$	6) $(-8)^2$	6) $-8 \times -8 = 64$
2) $(-3)^2$	2) $-3 \times -3 = 9$	7) $(-1)^2$	7) $-1 \times -1 = 1$
3) $(-3)^3$	3) $-3 \times -3 \times -3 = -27$	8) $(-9)^3$	8) $-9 \times -9 \times -9 = -729$
4) $(-3)^4$	4) $-3 \times -3 \times -3 \times -3 = 81$	9) $(-5)^4$	9) $-5 \times -5 \times -5 \times -5 = 625$
5) $(-3)^5$	5) $-3 \times -3 \times -3 \times -3 \times -3 = -243$	10) $(-2)^5$	10) $-2 \times -2 \times -2 \times -2 \times -2 = -32$

Roots

The root of a number is the opposite of raising a number to a power - it may be positive or negative. Calculating powers helps to explain the roots.

Powers

$(+2)^2 = +2 \times +2 = 4$
$(-2)^2 = -2 \times -2 = 4$
$(+2)^3 = +2 \times +2 \times +2 = 8$
$(-2)^3 = -2 \times -2 \times -2 = -8$
$(+2)^4 = +2 \times +2 \times +2 \times +2 = 16$
$(-2)^4 = -2 \times -2 \times -2 \times -2 = 16$
$(+2)^5 = +2 \times +2 \times +2 \times +2 \times +2 = 32$
$(-2)^5 = -2 \times -2 \times -2 \times -2 \times -2 = -32$

Roots

2 is usually omitted - but **not** other roots.

$\sqrt[2]{4} = +2$ $\sqrt{4} = +2$ In short $\sqrt{4} = \pm 2$
$\sqrt[2]{4} = -2$ $\sqrt{4} = -2$ 'The square root of 4 equals plus or minus 2'

$\sqrt[3]{8} = +2$
$\sqrt[3]{-8} = -2$

$\sqrt[4]{16} = +2$ In short $\sqrt[4]{16} = \pm 2$ 'The fourth root of 16 equals plus or minus 2'
$\sqrt[4]{16} = -2$

$\sqrt[5]{32} = +2$
$\sqrt[5]{-32} = -2$

In General,

an **EVEN** root results in a $+$number OR a $-$number
an **ODD** root for a $+$number results in a $+$number
an **ODD** root for a $-$number results in a $-$number

*Note

From the above it can be seen that an **EVEN root of a NEGATIVE number is NOT possible** with **REAL NUMBERS**. For example, we cannot square a number to give -4.

Exs. $\sqrt[2]{-4}$ ($\sqrt{-4}$) $\sqrt[4]{-16}$ $\sqrt{-8}$ $\sqrt{-2.5}$ $\sqrt{-0.333...}$ $\sqrt[6]{-1}$ $\sqrt{-\pi}$ are **not** possible.

*Note Imaginary Numbers where $-1 = i^2$ so $\sqrt{-1} = i = 1i$ Ex. $\sqrt{-4} = \sqrt{4} \times \sqrt{-1}$
$= \pm 2 \times i = \pm 2i$

Q	A	Q	A
Calculate with **Real Numbers**		7) $\sqrt[3]{-1}$	7) -1
1) $\sqrt[2]{9}$	1) $+3$ OR -3 (± 3)	8) $\sqrt[3]{27}$	8) 3
2) $\sqrt{9}$	2) $+3$ OR -3 (± 3)	9) $\sqrt[3]{-27}$	9) -3
3) $\sqrt{-9}$	3) **Not** possible.	10) $\sqrt[3]{64}$	10) 4
4) $\sqrt{16}$	4) $+4$ OR -4 (± 4)	11) $\sqrt[3]{-64}$	11) -4
5) $\sqrt{1}$	5) $+1$ OR -1 (± 1)	12) $\sqrt[4]{81}$	12) 3 OR -3 (± 3)
6) $\sqrt[3]{1}$	6) $+1$	13) $\sqrt[4]{-81}$	13) **Not** possible.

Negative Numbers 13 — Combined Operations.

Combined Operations
The following Exs. and Q/A combine operations.
*Note   Calculate the value of each bracket ( ) separately.
        Brackets ( ) are ignored on the last line of work.

Calculate

Ex. 1 $(-4 \times 2) + 1$
$\Downarrow$
$(-4 \times 2) + 1 =$
$(\ -8\ ) + 1 = -7$

Ex. 2 $\dfrac{6+3}{-9}$
$\Downarrow$
$\dfrac{6+3}{-9} = \dfrac{9}{-9} = -1$

Ex. 3 $(-2)^2 - (-5)^2$
$\Downarrow$
$(-2)^2 - (-5)^2$
$(-2 \times -2)\ - (-5 \times -5) =$
$(\ 4\ ) - (\ 25\ ) = -21$

Q	A	Q	A
Calculate			
1) $(-3 \times 4) + 2$	1) $(-3 \times 4) + 2 =$ $(\ -12\ ) + 2 = -10$	7) $-3 \times (4+2)$	7) $-3 \times (4+2) =$ $-3 \times (\ 6\ ) = -18$
2) $\dfrac{2+4}{-2}$	2) $\dfrac{2+4}{-2} = \dfrac{6}{-2} = -3$	8) $\dfrac{-1-4}{-1+8}$	8) $\dfrac{-1-4}{-1+8} = \dfrac{-5}{7} = -\dfrac{5}{7}$
3) $(6)^2 - (-7)^2$	3) $(6)^2 - (-7)^2$ $(6 \times 6) - (-7 \times -7) =$ $(\ 36\ ) - (\ 49\ ) = -13$	9) $(8)^2 + (-1)^3$	9) $(8)^2 + (-1)^3$ $(8 \times 8) + (-1 \times -1 \times -1) =$ $(\ 64\ ) \pm (\ -1\ ) = 63$
4) $8 - (5 \times -9)$	4) $8 - (5 \times -9) =$ $8 - (-45) = 53$	10) $(-7 \times -6) - 50$	10) $(-7 \times -6) - 50 =$ $(\ 42\ ) - 50 = -8$
5) $\dfrac{-9+1}{-4}$	5) $\dfrac{-9+1}{-4} = \dfrac{-8}{-4} = 2$	11) $\dfrac{0-1}{1}$	11) $\dfrac{0-1}{1} = \dfrac{-1}{1} = -1$
6) $(-5)^3 - (5)^2$	6) $(-5)^3 - (5)^2$ $(-5 \times -5 \times -5) - (5 \times 5)$ $(\ -125\ ) - (\ 25\ ) = -150$	12) $-2 \times (-9)^2$	12) $-2 \times (-9)^2$ $-2 \times (-9 \times -9)$ $-2 \times (\ 81\ ) = -162$

*Note   In the following Exs. and Q/A, there are **2 answers** when an **EVEN root** is given.

Ex. 1 $\sqrt[2]{4} + 1$
$\Downarrow$
$\sqrt[2]{4} + 1 =$
$+2 + 1 = 3$
or
$-2 + 1 = -1$

Ex. 2 $7 - \sqrt[3]{-8}$
$\Downarrow$
$7 - \sqrt[3]{-8} =$
$7 - -2 = 9$

Q	A	Q	A
Calculate			
1) $\sqrt[2]{9} + 7$	1) $\sqrt[2]{9} + 7 =$ $+3 + 7 = 10$ or $-3 + 7 = 4$	4) $\dfrac{\sqrt{25}}{-2}$	4) $\dfrac{\sqrt{25}}{-2} = \dfrac{5}{-2} = -\dfrac{5}{2} = -2\dfrac{1}{2}$ or $= \dfrac{-5}{-2} = \dfrac{5}{2} = 2\dfrac{1}{2}$
2) $\sqrt[3]{8} - 22$	2) $\sqrt[3]{8} - 22 =$ $2 - 22 = -20$	5) $3 + \sqrt[4]{81}$	5) $3 + \sqrt[4]{81} =$ $3 + 3 = 6$ or $3 + -3 = 0$
3) $6 \times \sqrt[3]{-27}$	3) $6 \times \sqrt[3]{-27} =$ $6 \times -3 = -18$	6) $\sqrt[3]{125} - \sqrt[5]{-1}$	6) $\sqrt[3]{125} - \sqrt[5]{-1} =$ $5 - -1 = 6$

Negative Numbers 14　　　　　　　　　　　　　　　　　　　　　Absolute Value.

Absolute Value
The Absolute Value of a number can be considered to be **how far it is from 0**.
It is always a positive number.

Ex. 3 is 3 away from 0. The absolute value of 3 is 3.　　$|3| = 3$

　　-3 is 3 away from 0. The absolute value of -3 is 3 also.　$|-3| = 3$

　　　　　　The absolute value is shown by 2 vertical bars.

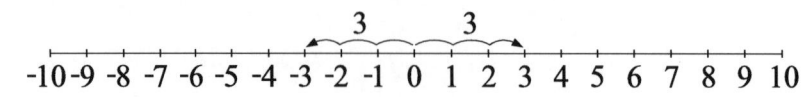

-10 -9 -8 -7 -6 -5 -4 -3 -2 -1 0 1 2 3 4 5 6 7 8 9 10

Q	A	Q	A
1) State the absolute value of each number.	1)	e) $-\dfrac{7}{10}$	e) $\dfrac{7}{10}$
a) 5	a) 5	f) -61·924	f) 61·924
b) -5	b) 5	g) 123	g) 123
c) -2	c) 2	h) $-\sqrt{11}$	h) $\sqrt{11}$
d) 2	d) 2	i) 8,000	i) 8,000

Q	A	Q	A
2) State the value of each number.	2)	e) $\left\|-\dfrac{2}{19}\right\|$	e) $\dfrac{2}{19}$
a) $\|7\|$	a) 7	f) $\|-4·444...\|$	f) 4·444...
b) $\|-7\|$	b) 7	g) $\|111\|$	g) 111
c) $\|1\|$	c) 1	h) $\|-\pi\|$	h) π
d) $\|-1\|$	d) 1	i) $\|-\sqrt{5}\|$	i) $+\sqrt{5}$

Calculations

Ex.　Find the value of:

　　a) $|5-2|$　　　b) $|2-5|$　　　c) $-|9-7|$　　　d) $-|-4 \times 10|$

　　a) $|\,3\,| = 3$　　b) $|-3\,| = 3$　　c) $-|\,2\,| = -2$　　d) $-|-40\,| = -40$

Q	A	Q	A
1) Find the value of:	1)	k) $\left\|-\dfrac{5}{11}+\dfrac{3}{11}\right\|$	k) $\left\|-\dfrac{2}{11}\right\| = \dfrac{2}{11}$
a) $\|8-4\|$	a) 4	l) $-\left\|-\dfrac{1}{3} \times -\dfrac{1}{2}\right\|$	l) $-\left\|\dfrac{1}{6}\right\| = -\dfrac{1}{6}$
b) $\|1-7\|$	b) $\|-6\| = 6$		
c) $-\|3-0\|$	c) $-\|3\| = -3$		
d) $\|-5-5\|$	d) $\|-10\| = 10$	m) $\|2·1 \times 4\|$	m) $\|8·4\| = 8·4$
e) $-\|-2+9\|$	e) $-\|7\| = -7$	n) $-\|7·5 \times -2\|$	n) $-\|-15\| = -15$
f) $-\|6 \times 6\|$	f) $-\|36\| = -36$	o) $\|-\pi \times 9\pi\|$	o) $\|-9\pi^2\| = 9\pi^2$
g) $\|-5 \times 5\|$	g) $\|-25\| = 25$	p) $\|-\sqrt{4} \times \sqrt{1}\|$	p) $\|-2 \times 1\| = \|-2\| = 2$
h) $-\|7 \times -4\|$	h) $-\|-28\| = -28$	q) $\dfrac{\|-8\|}{\|2\|}$	q) $\dfrac{8}{2} = 4$
i) $\|-3 \times -2\|$	i) $\|6\| = 6$	r) $-\left\|\dfrac{6}{-3}\right\|$	r) $-\|-2\| = -2$
j) $-\|-8 \times -10\|$	j) $-\|80\| = -80$		

Money 1 — Units. Addition. 'Problems'.

*Note **Decimals 1-13** explains the work required to understand **Money**.

Money
Units
The basic **unit of currency** used in the United Kingdom / Great Britain (England, Scotland, Wales and Northern Ireland) is the **pound** for which the symbol is **£**.
It is referred to as the Great Britain Pound or Pound Sterling.
The smaller units used are **pence**, usually written **p**, and the system values   1 pound = 100 pence
$$£1 = 100p.$$
The number 100 allows calculations based on the **decimal system**.

Conversion
£ and p can be converted to the different units.

Ex.1
Convert to pence:
A) £3   A) 3·00 × 100 =   300p
B) £18   B) 18·00 × 100 = 1800p
C) £2·57   C) 2·57 × 100 =   257p
D) £46·09   D) 46·09 × 100 = 4609p

Ex.2
Convert to £:
A) 91p   A) 91 ÷ 100 = £0·91
B) 543p   B) 543 ÷ 100 = £5·43
C) 700p   C) 700 ÷ 100 = £7·00 = £7
D) 6280p   D) 6280 ÷ 100 = £62·80

Q	A	Q	A
1) Convert to pence: A) £0·78 B) £54 C) £3·01 D) £90·26	1) A) 0·78 × 100 = 78p B) 54·00 × 100 = 5400p C) 3·01 × 100 = 301p D) 90·26 × 100 = 9026p	2) Convert to £: A) 4p B) 810p C) 2679p D) 3000p	2) A) 04 ÷ 100 = £0·04 B) 810 ÷ 100 = £8·10 C) 2679 ÷ 100 = £26·79 D) 3000 ÷ 100 = £30·00 = £30

Addition
We make sure that the **decimal points are under each other** and the digits are in the correct columns.

Ex.1   £1·23 + £0·45
    £1·23
  + £0·45
    £1·68

Ex.2   £56·79 + £32·85
  £56·79
+ £32·85
  £89·64

Q	A

1) Add £0·61 and £3·38	1) £0·61
	   + £3·38
	      £3·99

2) Calculate £15·21 + £7·54	2) £15·21
	   + £ 7·54
	      £22·75

3) Find the sum of £36·48 and £40·66	3) £36·48
	   + £40·66
	      £77·14

4) Find the total of £1·34, £0·97 and £55·12	4) £ 1·34
	      £ 0·97
	   + £55·12
	      £57·43

5) Work out, in £'s: 86p + £4 + £27·50 + 5p	5) £ 0·86
	      £ 4·00
	      £27·50
	   + £ 0·05
	      £32·41

The following 'problems' involve addition.

Q	A

1) A boy spends £12, £8·09 and 30p.
How much does he spend altogether   a) in £'s ?
b) in p's ?
	1) a) £12·00
	         £ 8·09
	      + £ 0·30
	         £20·39 b) 20·39 × 100 = 2039p

2) A girl saves 75p, 264p and £10·81.
How much does she save altogether   a) in p's ?
b) in £'s ?
	2) a) 75p
	         264p
	      + 1081p
	       1420p b) 14·20 ÷ 100 = £14·20

Money 2

Subtraction. 'Problems'.

Subtraction

We make sure that the **decimal points are under each other** and the digits are in the correct columns.

Ex.1 £5·67 − £3·21

$$\begin{array}{r} £5·67 \\ -£3·21 \\ \hline £2·46 \end{array}$$

Ex.2 £27·92 − £6·58

Method 1
$$\begin{array}{r} £27·9^{8}\!\!\!\!/\,^{1}2 \\ -£6·58 \\ \hline £21·34 \end{array}$$

Method 2
$$\begin{array}{r} £27·9^{1}2 \\ -£6·5_{1}8 \\ \hline £21·34 \end{array}$$

Ex.3 £8 − £2·37

Method 1
$$\begin{array}{r} £8·{}^{7}\!\!\!\!/\,^{9}\!\!\!\!/\,^{1}0 \\ -£2·37 \\ \hline £5·63 \end{array}$$

Method 2
$$\begin{array}{r} £8·{}^{1}0\,{}^{1}0 \\ -£2_{1}·3_{1}\,7 \\ \hline £5·63 \end{array}$$

Always write **00** here

Q	A	Q	A
1) Calculate £9·75 − £4·02	1) £9·75 − £4·02 = £5·73	5) How much is £3·45 less than £5?	5) Method 1: £⁴5̶·⁹0̶¹0 − £3·45 = £1·55 Method 2: £5·¹0¹0 − £3₁·4₁5 = £1·55 write **0 0** here
2) Take £1·61 from £3·86	2) £3·86 − £1·61 = £2·25	6) How much is added to £6·51 to make £9?	6) Method 1: £⁸9̶·⁹0̶¹0 − £6·51 = £2·49 Method 2: £9·¹0¹0 − £6₁·5₁1 = £2·49 write **0 0** here
3) Subtract £8·50 − £2·13	3) Method 1: £8·⁴5̶¹0 − £2·1 3 = £6·37 Method 2: £8·5¹0 − £2·1₁3 = £6·37		
4) Find the difference, in £'s, between £5·74 and 78p.	4) Method 1: £⁴5̶·⁷6̶¹4 − £0·7 8 = £4·96 Method 2: £5·⁷1⁴ − £0₁·7₁8 = £4·96		

The following 'problems' involve subtraction.

Q	A
1) A lady has £8·79 and spends £2·55. How much does she have left?	1) £8·79 − £2·55 = £6·24
2) The cost of a bicycle was £88·18. It is reduced by £10·90 in a sale. What is the new price of the bicycle?	2) Method 1: £8⁷8̶·¹1 8 − £10·90 = £77·28 Method 2: £88·¹18 − £10₁·90 = £77·28
3) A boy saves £1·05 to bring his total savings to £6. How much did he have saved before he added the £1·05?	3) Method 1: £⁵6̶·⁹0̶¹0 − £1·05 = £4·95 Method 2: £6·¹0¹0 − £1₁·0₁5 = £4·95
4) The price of a book is increased by 44p to £3·22. Find the price of the book before the increase, in £'s.	4) Method 1: £²3̶·¹¹2̶ 2 − £0·44 = £2·78 Method 2: £3·¹2¹2 − £0₁·4₁4 = £2·78

Money 3   Subtraction. 'Problems'.

<u>Subtraction</u>

When **each number** in each column can be subtracted from a larger number, then it is usually faster to just subtract.

The following method is more useful when a number in a column is subtracted from a smaller number - but it is shown here for both possibilities.

<u>Method 3</u>

Starting from the **lowest number**, find the **differences** (by subtracting or 'adding on') to
the **next 10p** (10p, 20p, 30p, ... £1)
the **next £1** (£1, £2, £3, ... £10)
the **next £10** (£10, £20, £30, ... £100) ...
The **last difference** should include any number '**left over**' - i.e., smaller than the highest place value.

<u>ADD</u> **the differences** to find the total difference between the given numbers.

The **differences** are usually calculated **mentally** - and only these need be written down.

Ex.1	Ex.2	Ex.3
£5·67 − £3·21	£27·92 − £6·58	£8 − £2·37
£3·21 to £3·30 £0·09	£ 6·58 to £ 6·60 £ 0·02	£2·37 to £2·40 £0·03
£3·30 to £4·00 £0·70	£ 6·60 to £ 7·00 £ 0·40	£2·40 to £3·00 £0·60
£4·00 to £5·67 +£1·67	£ 7·00 to £10·00 £ 3·00	£3·00 to £8·00 +£5·00
£2·46	£10·00 to £27·92 +£17·92	£5·63
1 1	£21·34	
	1	

Q	A	Q	A

1) Calculate £9·75 − £4·02	1) £4·02 to £4·10 £0·08
	£4·10 to £5·00 £0·90
	£5·00 to £9·75 +£4·75
	              £5·73
	             1 1
		4) How much is £3·45 less than £5 ?	4) £3·45 to £3·50 £0·05
			£3·50 to £4·00 £0·50
			£4·00 to £5·00 +£1·00
			             £1·55

2) Subtract £8·50 − £2·13	2) £2·13 to £2·20 £0·07
	£2·20 to £3·00 £0·80
	£3·00 to £8·50 +£5·50
	             £6·37
	             1
		5) Find £12·26 − £6·60	5) £6·60 to £ 7·00 £0·40
			£7·00 to £12·26 +£5·26
			              £5·66

3) Find the difference, in £'s, between £5·74 and 78p.	3) £0·78 to £0·80 £0·02
	£0·80 to £1·00 £0·20
	£1·00 to £5·74 +£4·74
	             £4·96
		6) Calculate £71·31 − £7·89	6) £ 7·89 to £ 7·90 £ 0·01
			£ 7·90 to £ 8·00 £ 0·10
			£ 8·00 to £10·00 £ 2·00
			£10·00 to £71·31 +£61·31
			                £63·42

The following 'problems' involve subtraction.

Q	A

1) A lady has £8·79 and spends £2·55. How much does she have left?	1) Find £2·55 to £2·60 £0·05
	£8·79 £2·60 to £3·00 £0·40
	−£2·55 £3·00 to £8·79 +£5·79
	                                   £6·24
	                                 1 1

2) The cost of a bicycle was £88·18. It is reduced by £10·90 in a sale. What is the new price of the bicycle?	2) Find £10·90 to £11·00 £ 0·10
	£88·18 £11·00 to £20·00 £ 9·00
	−£10·90 £20·00 to £88·18 £68·18
	                                   +£77·28
	                                        1

3) A boy saves £1·05 to bring his total savings to £6. How much did he have saved before he added the £1·05 ?	3) Find £1·05 to £1·10 £0·05 £6·00 £1·10 to £2·00 £0·90 −£1·05 £2·00 to £6·00 +£4·00 £4·95

Money 4

Multiplication

We make sure that the **decimal points are under each other** and the digits are in the correct columns.

Ex.1 £1·34 × 2	Ex.2 £2·87 × 3	Ex.3 £3·25 × 15
£1·34 × 2 £2·68	£2·87 × 3 £8·61 2 2	£ 3·2 5 × 1 5 $1\,6_1 2_2 5$ + 3 2 5 **0** £4 8·7 5

Q	A	Q	A
1) Calculate £1·23 × 3	1) £1·23 × 3 £3·69	5) Evaluate £806·15 × 4	5) £ 806·15 × 4 £3224·60 2 2
2) Multiply £4·98 × 2	2) £4·98 × 2 £9·96 1 1	6) How much is £8·54 times 13 ?	6) £ 6·5 9 × 1 3 $1\,9_1 7_2 7$ + 6 5 9 **0** £8 5·6 7 1 1
3) Find £9·35 × 9	3) £ 9·35 × 9 £84·15 3 4		
4) Find the value of £21·04 × 5	4) £ 21·04 × 5 £105·20 2	7) Calculate 74 × £7·08.	7) £ 7·0 8 × 7 4 $2\,8\,3_3 2$ + 4 9 $5_5$6 **0** £5 2 3·9 2 1 1

The following 'problems' involve multiplication.

Q	A
1) What is the total cost of 4 pens at £3·21 each?	1) £ 3·21 × 4 £12·84
2) A man earns £8·56 per hour. Calculate his pay for 8 hours work.	2) £ 8·56 × 8 £68·48 4 4
3) A girl saves £2·75 each week for 17 weeks. How much does she save in total?	3) £ 2·7 5 × 1 7 $1\,9_5 2_3 5$ + 2 7 5 **0** £4 6·7 5 1
4) Find the total value of 35 chocolate bars at £0·49 each.	4) £ 0·4 9 × 3 5 $2\,4_4 5$ + 1 $4_2$7 **0** £1 7·1 5 1
5) Find the total cost of 10 mobile phones at £50·68 each.	5) £50·68 × 10 = £506·**80**

Money 5 Division. 'Problems'.

Division
We make sure that the **decimal points are under each other** and the digits are in the correct columns.

Ex.1 £2·46 ÷ 2

$$\begin{array}{r}£1·23\\2\overline{)£2·46}\end{array}$$

Ex.2 £7·83 ÷ 3

$$\begin{array}{r}£2·6\,1\\3\overline{)£7·^18\,3}\end{array}$$

Ex.3 £7 ÷ 4

$$\begin{array}{r}£1·7\,5\\4\overline{)£7·^30^20}\end{array}$$

Ex.4 £5·59 ÷ 5

$$\begin{array}{r}£1·1\,1\,\mathbf{8}\\5\overline{)£5·5\,9^40}\end{array} = £1·12 \text{ (nearest 1p)}$$

or

$$\begin{array}{r}£1·1\,1 \text{ r £0·04}\\5\overline{)£5·5\,9}\end{array}$$

*Note The remainder is from this column - it is **4p**.

Q	A	Q	A
1) Calculate £6·93 ÷ 3	1) $3\overline{)£6·93}$ giving £2·31	5) Calculate £0·84 ÷ 4	5) $4\overline{)£0·84}$ giving £0·21
2) Divide £5·48 ÷ 2	2) $2\overline{)£5·^14\,8}$ giving £2·74	6) Divide £32·76 ÷ 6	6) $6\overline{)£32·^27^36}$ giving £ 5·46
3) Find £9 ÷ 4	3) $4\overline{)£9·^10^20}$ giving £2·25	7) Find £41 ÷ 2	7) $2\overline{)£41·^10\,0}$ giving £20·50
4) Evaluate £7·61 ÷ 5	4) $5\overline{)£7·^26^11^10}$ giving £1·5 2 **2** = £1·52 (nearest 1p) or $5\overline{)£7·^26^11}$ giving £1·52 r £0·01	8) Evaluate £8·04 ÷ 8	8) $8\overline{)£8·0\,4^40}$ giving £1·0 0 **5** = £1·00 / £1·01 (nearest 1p) or $8\overline{)£8·0\,4}$ giving £1·00 r £0·04

The following 'problems' involve division.

Q	A
1) Share £8·64 between 2 people.	1) $2\overline{)£8·64}$ giving £4·32
2) 3 equally priced loaves cost a total of £3·48. Find the cost of 1 loaf.	2) $3\overline{)£3·4^18}$ giving £1·16
3) A £5 prize is divided equally between 4 people. Find how much each person receives.	3) $4\overline{)£5·^10^20}$ giving £1·25
4) A pack of 5 batteries costs £1·17. Calculate the cost of 1 battery.	4) $5\overline{)£1·1^17^20}$ giving £0·2 3 **4** = £0·23 (nearest 1p) or $5\overline{)£1·1^17}$ giving £0·23 r £0·02
5) A £94 bicycle is paid for in 8 equal weekly payments. Calculate each weekly payment.	5) $8\overline{)£94·^60^40}$ giving £11·75
6) A £6,999 car is paid for in 10 equal monthly payments. How much is each monthly payment?	6) £699⌐9⌐ ÷ 10 = £699·90
7) 100 dice cost £26. Work out the cost of 1 dice.	7) £⌐26⌐ ÷ 100 = £0·26

Money 6

Bank Account.

Bank Account

A bank offers its customers many different kinds of accounts.
Most customers have at least a **Current Account** - this is briefly explained here.

The primary function of a Current Account is to allow the customer transactions on a day-to-day basis.
These are either **Credit** or **Debit** transactions.

Credit (Pay in) transactions put money into the account - for example, pay in cash, pay in cheques, transfer money from another account.
Regular automatic payments may be set up - for example, to pay in wages / salary, to receive other funds.
These are usually received by a 'Standing Order' - SO or a 'Direct Debit' - DD.

Debit (Pay out) transactions take money out of the account - for example, withdraw cash, pay for goods, pay bills, transfer money to another account.
Regular automatic payments may be set up - for example, to pay a mortgage or car insurance.
These are usually paid by a 'Standing Order' or a 'Direct Debit'.

A **Bank Statement** records all the transactions for the account and provides a '**balance**' - the amount in the account - after each.
It is made available to the customer either on paper (usually monthly) or on computer / phone by means of 'Internet Banking' (more immediately).

*Note The following Bank Statements are simplified versions.

Ex. The Bank Statement shows the following details.

April 1 Bank Account balance £200.
April 1 £80 cash paid in. £50 paid out to K.S.J.
April 4 £20 cash withdrawn. £95 gas bill paid-DD.

Bank Statement

Date	Details	Pay in	Pay out	Balance
April 1				£200.00
April 1	Cash	80.00		280.00
April 1	K.S.J.		50.00	230.00
April 4	Cash		20.00	210.00
April 4	Gas DD		95.00	115.00

Q

Complete each Bank Statement.

1) May 1 Bank Account balance £115.
   May 3 £40 paid out to T.H. £75 cash paid in.
   May 5 £36 paid to X Garage. £60 cheque paid in.

A

1) **Bank Statement**

Date	Details	Pay in	Pay out	Balance
May 1				£115.00
May 3	T.H.		40.00	75.00
May 3	Cash	75.00		150.00
May 4	X Garage		36.00	114.00
May 4	Cheque	60.00		174.00

2) June 1 Bank Account balance £174.
   June 7 £50 Transfer in. £43 phone bill paid-SO.
   June 9 £410 wages from IOU. £60 cash withdrawn.

2) **Bank Statement**

Date	Details	Pay in	Pay out	Balance
June 1				£174.00
June 7	Transfer	50.00		224.00
June 7	Phone SO		43.00	181.00
June 9	IOU	410.00		591.00
June 9	Cash		60.00	531.00

Ready Reckoner 1

Ready Reckoner

A 'Ready Reckoner' is a **table** (or book of tables) of the **results of calculations**.
It is usually used when a task involves the **same number, multiplied many times separately** and therefore saves time since the 'answers' have been written down.
Here, the tables are limited in size.
The work also involves **addition** of the results to obtain results which are not recorded - this may be carried out in more than 1 way.

Ex. The 'Ready Reckoner' gives the results of **multiplying** the number **8**.
Use it to find the **result** -

a) 6×8 a) 48

b) 30×8 b) 240

c) 18×8 c) 8 ---- 64
                      +10 ---- + 80
                       18   144

d) 92×8 d) 2 ---- 16
                       40 ---- 320
                      +50 ---- +400
                       92   736

Ready Reckoner (×8)

Number	Result
1	8 ----($1 \times 8 = 8$)
d) 2	16 ----($2 \times 8 = 16$)
3	24 ($3 \times 8 = 24$)
4	32 : : :
5	40
a) 6	48
7	56
c) 8	64
9	72
c) 10	80
20	160
b) 30	240
d) 40	320
d) 50	400

Q

1) The 'Ready Reckoner' gives the results of **multiplying** the number **17**.
Use it to find the **result** -

a) 6×17

b) 14×17

c) 29×17

d) 75×17

Number	Result
1	17
2	34
3	51
4	68
5	85
6	102
7	119
8	136
9	153
10	170
20	340
30	510
40	680

A

1) a) 102

b)   4 ---- 68
   +10 ---- +170
    14   238

c)   9 ---- 153
   +20 ---- +340
    17   493

d)   5 ----   85
    30 ---- 510
   +40 ---- + 680
    75   1275

2) The 'Ready Reckoner' gives the results of **multiplying** the amount **£0.23**.
Use it to find the **result** -

a) $9 \times £0.23$

b) $22 \times £0.23$

c) $36 \times £0.23$

d) $98 \times £0.23$

Number	Result £
1	0.23
2	0.46
3	0.69
4	0.92
5	1.15
6	1.38
7	1.61
8	1.84
9	2.07
10	2.30
20	4.60
30	6.90
40	9.20
50	11.50

2) a) £2.07

b)   2 ---- £0.46
    +20 ---- +£4.60
     22   £5.06

c)   6 ---- £1.38
    +30 ---- +£6.90
     36   £8.28

d)   8 ---- £ 1.84
     40 ---- £ 9.20
    +50 ---- +£11.50
     98   £22.54

Magic Squares 1

Magic Squares

A Magic Square is a grid of numbers where the values in each of the rows, columns and diagonals add up to the same sum, known as the 'magic number'.

Here, just 3×3 and 4×4 squares are presented.

To complete a magic square, the missing values are calculated (usually mentally).

The 'magic number' may be given - or -

has to be found from the given values by adding the numbers in a row, column or diagonal.

Ex. Complete the magic square.

3		5
	6	
		9

⇨

3		5
	6	
		9

'magic number' = 3 + 6 + 9 = 18

⇨

3	10	5
8	6	4
7	2	9

Q / A

Complete each magic square.

1)

8		
3		7
4		

A 1)

8	1	6
3	5	7
4	9	2

'magic number' = 15

2)

	4	2
1		

'magic number' = 12

A 2)

5	0	7
6	4	2
1	8	3

3)

6		
5	7	
		8

A 3)

6	11	4
5	7	9
10	3	8

'magic number' = 21

4)

		6
0		2

'magic number' = 9

A 4)

4	-1	6
5	3	1
0	7	2

5)

0	2	4
5		

A 5)

1	6	-1
0	2	4
5	-2	3

'magic number' = 6

6)

16			
5		11	8
	6		12
	15		1

'magic number' = 34

A 6)

16	3	2	13
5	10	11	8
9	6	7	12
4	15	14	1

7)

12		14	
13	8		
		5	16
6		4	9

A 7)

12	1	14	7
13	8	11	2
3	10	5	16
6	15	4	9

'magic number' = 34

8)

	4	11	16
10	17		
8			
	14		2

'magic number' = 38

A 8)

7	4	11	16
10	17	6	5
8	3	12	15
13	14	9	2

9)

		7	6
9		2	-5
	-1		8
	10		1

A 9)

0	-3	7	6
9	4	2	-5
-2	-1	5	8
3	10	-4	1

'magic number' = 10

Percentages 1

Change a Percentage to a Fraction. 100% Grid.

Percentage
'**per cent**' means '**per 100**' ... or '**out of 100**' ... or '**for every 100**'.
'**per cent**' is represented by the symbol **%** (which is 100 in a different form !).

Ex. **1%** is read as '1 **per cent**' - it means '1 **per 100**'.

Ex. **5%** is read as '5 **per cent**'.
 5% of the population means '5 people **for every 100** people'.

Change a Percentage to a Fraction.
To change / convert a percentage to a fraction we **divide by 100** (write the percentage '**over 100**').

Ex.1 Change 1% to a fraction. $1\% = \frac{1}{100}$

Ex.2 Express 75% as a fraction. $75\% = \frac{75}{100}$

Ex.3 Change 100% to a fraction. $100\% = \frac{100}{100} = 1$ (This is '1 whole' or 'the whole of an amount'.)

Q	A	Q	A
1. Change each percentage to a fraction.			
1) 3%	1) $\frac{3}{100}$	4) 50%	4) $\frac{50}{100}$
2) 10%	2) $\frac{10}{100}$	5) 87%	5) $\frac{87}{100}$
3) 25%	3) $\frac{25}{100}$	6) 99%	6) $\frac{99}{100}$

100% Grid
A percentage may be represented in a diagram.
In the following Q/A, each grid is 10×10 = 100 squares.
The grid represents 100% and each small square is 1%.

Q	A	Q	A
1. State the percentage of the large square which is shaded. 1)	1) 50% (50 small squares are shaded)	2. Shade each percentage of the large square. 1) 15%	1) Shade **any** 15 small squares
2)	2) 21% (21 small squares are shaded)	2) 77%	2) Shade **any** 77 small squares

Percentages 2 — Change a Percentage to a Fraction - Cancel.

Change a Percentage to a Fraction - Cancel

Given a percentage, it <u>may</u> be possible to **cancel** its fraction form.
(If so, it is **not** always necessary or useful.)
Since the **denominator** is 100, we can only cancel using the **factors of 100** ⇨

1	100
2	50
4	25
5	20
10	

*Note reminder ... '**cancel**' means **divide** numerator and denominator
by the **same** number / **common** factor (other than 1).

Ex. Write each percentage as a fraction **in its lowest terms** (so it can not **cancel** further).

1) $10\% = \dfrac{10}{100} = \dfrac{1}{10}$ ($\div 10$)
'**cancel**' by 10

2) $44\% = \dfrac{44}{100} = \dfrac{11}{25}$ ($\div 4$)

3) $67\% = \dfrac{67}{100}$ no common factors - factors of $\dfrac{67}{1\ \ 67}$ (prime)

Q	A	Q	A
1. Write each percentage as a fraction **in its lowest terms**.			
1) 2%	1) $= \dfrac{2}{100} = \dfrac{1}{50}$ ($\div 2$)	7) 50%	7) $= \dfrac{50}{100} = \dfrac{1}{2}$ ($\div 50$)
2) 8%	2) $= \dfrac{8}{100} = \dfrac{2}{25}$ ($\div 4$)	8) 35%	8) $= \dfrac{35}{100} = \dfrac{7}{20}$ ($\div 5$)
3) 15%	3) $= \dfrac{15}{100} = \dfrac{3}{20}$ ($\div 5$)	9) 60%	9) $= \dfrac{60}{100} = \dfrac{3}{10}$ ($\div 20$)
4) 70%	4) $= \dfrac{70}{100} = \dfrac{7}{10}$ ($\div 10$)	10) 24%	10) $= \dfrac{24}{100} = \dfrac{6}{25}$ ($\div 4$)
5) 40%	5) $= \dfrac{40}{100} = \dfrac{2}{5}$ ($\div 20$)	11) 33%	11) $= \dfrac{33}{100}$ no common factors - factors of $\dfrac{33}{1\ \ 33}$ $\dfrac{\ }{3\ \ 11}$
6) 75%	6) $= \dfrac{75}{100} = \dfrac{3}{4}$ ($\div 25$)	12) 58%	12) $= \dfrac{58}{100} = \dfrac{29}{50}$ ($\div 2$)

Percentages 3   Greater than 100% - Whole Number / Mixed Number.

Percentage greater than 100%
If a percentage is **greater than 100%**, then it will become a **whole number** or a **mixed number** when it is written as a fraction with 100 as its denominator.

Ex. Write each percentage as a fraction **in its lowest terms**.

A) 400%

$$A)\ 400\% = \frac{400}{100} = 4$$

The answer means '4 times the whole amount'

B) 120%

Method 1

$$B)\ 120\% = \frac{120}{100} = \frac{6}{5} = 1\frac{1}{5} \quad (\div 20)$$

Method 2

$$B)\ 120\% = \frac{120}{100} = 1\frac{20}{100} = 1\frac{1}{5} \quad (\div 20)$$

The answer means '$1\frac{1}{5}$ times the whole amount'

*Note
Method 1 will be used in the following Q/A.

Q	A	Q	A
1. Write each percentage as a fraction **in its lowest terms**.			
1) 200%	1) $= \frac{200}{100} = 2$	6) 140%	6) $= \frac{140}{100} = \frac{7}{5} = 1\frac{2}{5}$ ($\div 20$)
2) 500%	2) $= \frac{500}{100} = 5$	7) 170%	7) $= \frac{170}{100} = \frac{17}{10} = 1\frac{7}{10}$ ($\div 10$)
3) 900%	3) $= \frac{900}{100} = 9$	8) 250%	8) $= \frac{250}{100} = \frac{5}{2} = 2\frac{1}{2}$ ($\div 50$)
4) 1800%	4) $= \frac{1800}{100} = 18$	9) 444%	9) $= \frac{444}{100} = \frac{111}{25} = 4\frac{11}{25}$ ($\div 4$)
5) 3000%	5) $= \frac{3000}{100} = 30$	10) 375%	10) $= \frac{375}{100} = \frac{15}{4} = 3\frac{3}{4}$ ($\div 25$)

Percentages 4

Change a Fraction Percentage to a Fraction.

Change a Fraction Percentage to a Fraction

A percentage may be a fraction less than 1% ... Ex. $\frac{1}{2}$% ... or a mixed number ... Ex. $4\frac{8}{15}$%

The percentage can be written as a fraction with a denominator of 100.

The numerator of the fraction is divided by its **denominator and 100**.

Ex. Write each percentage as a fraction.

1) $\frac{1}{2}$%

$$\frac{1}{2}\% = \frac{\frac{1}{2}}{100} = \frac{1}{2 \times 100} = \frac{1}{200}$$

⇧ this step can be omitted

2) $4\frac{8}{15}$%

$(4 \times 15) + 8 = 60$

change to an improper fraction

$$4\frac{8}{15}\% = \frac{4\frac{8}{15}}{100} = \frac{\frac{68}{15}}{100} = \frac{68}{15 \times 100} = \frac{68}{1500}$$

⇧ this step can be omitted

Q	A	Q	A
Write each percentage as a fraction. 1) $\frac{1}{5}$%	1) $= \frac{\frac{1}{5}}{100} = \frac{1}{500}$	7) $4\frac{1}{2}$%	7) $= \frac{4\frac{1}{2}}{100} = \frac{\frac{9}{2}}{100} = \frac{9}{200}$
2) $\frac{3}{4}$%	2) $= \frac{\frac{3}{4}}{100} = \frac{3}{400}$	8) $8\frac{7}{10}$%	8) $= \frac{8\frac{7}{10}}{100} = \frac{\frac{87}{10}}{100} = \frac{87}{1000}$
3) $\frac{9}{10}$%	3) $= \frac{\frac{9}{10}}{100} = \frac{9}{1000}$	9) $3\frac{2}{11}$%	9) $= \frac{3\frac{2}{11}}{100} = \frac{\frac{35}{11}}{100} = \frac{35}{1100}$
4) $\frac{6}{17}$%	4) $= \frac{\frac{6}{17}}{100} = \frac{6}{1700}$	10) $25\frac{5}{9}$%	10) $= \frac{25\frac{5}{9}}{100} = \frac{\frac{230}{9}}{100} = \frac{230}{900}$
5) $\frac{21}{50}$%	5) $= \frac{\frac{21}{50}}{100} = \frac{21}{5000}$	11) $40\frac{11}{20}$%	11) $= \frac{40\frac{11}{20}}{100} = \frac{\frac{811}{20}}{100} = \frac{811}{2000}$
6) $\frac{79}{80}$%	6) $= \frac{\frac{79}{80}}{100} = \frac{79}{8000}$	12) $150\frac{1}{4}$%	12) $= \frac{150\frac{1}{4}}{100} = \frac{\frac{601}{4}}{100} = \frac{601}{400}$

Percentages 5 — Change a Fraction Percentage to a Fraction - Cancel.

Change a Fraction Percentage to a Fraction - Cancel

Given a fraction percentage, it **may** be possible to **cancel** its <u>final</u> fraction form.
(If so, it is **not** always necessary or useful.)
Since the **denominator** is multiplied by 100, we can only cancel using the **factors of 100**.

Ex. Write each percentage as a fraction **in its lowest terms**.

1) $\dfrac{2}{5}\%$

$$\dfrac{2}{5}\% = \dfrac{\frac{2}{5}}{100} = \dfrac{2}{500} \xrightarrow{\div 2} \dfrac{1}{250}$$

2) $3\dfrac{3}{7}\%$

$$3\dfrac{3}{7}\% = \dfrac{3\frac{3}{7}}{100} = \dfrac{\frac{24}{7}}{100} = \dfrac{24}{700} \xrightarrow{\div 4} \dfrac{6}{175}$$

Q	A
Write each percentage as a fraction **in its lowest terms**. 1) $\dfrac{5}{8}\%$	1) $\dfrac{5}{8}\% = \dfrac{\frac{5}{8}}{100} = \dfrac{5}{800} \xrightarrow{\div 5} \dfrac{1}{160}$
2) $\dfrac{8}{9}\%$	2) $\dfrac{8}{9}\% = \dfrac{\frac{8}{9}}{100} = \dfrac{8}{900} \xrightarrow{\div 4} \dfrac{2}{225}$
3) $\dfrac{3}{4}\%$	3) $\dfrac{3}{4}\% = \dfrac{\frac{3}{4}}{100} = \dfrac{3}{400}$
4) $\dfrac{70}{89}\%$	4) $\dfrac{70}{89}\% = \dfrac{\frac{70}{89}}{100} = \dfrac{70}{8900} \xrightarrow{\div 10} \dfrac{7}{890}$
5) $5\dfrac{1}{5}\%$	5) $5\dfrac{1}{5}\% = \dfrac{5\frac{1}{5}}{100} = \dfrac{\frac{26}{5}}{100} = \dfrac{26}{500} \xrightarrow{\div 2} \dfrac{13}{250}$
6) $12\dfrac{1}{2}\%$	6) $12\dfrac{1}{2}\% = \dfrac{12\frac{1}{2}}{100} = \dfrac{\frac{25}{2}}{100} = \dfrac{25}{200} \xrightarrow{\div 25} \dfrac{1}{8}$
7) $33\dfrac{1}{3}\%$	7) $33\dfrac{1}{3}\% = \dfrac{33\frac{1}{3}}{100} = \dfrac{\frac{100}{3}}{100} = \dfrac{100}{300} \xrightarrow{\div 100} \dfrac{1}{3}$
8) $135\dfrac{5}{7}\%$	8) $135\dfrac{5}{7}\% = \dfrac{135\frac{5}{7}}{100} = \dfrac{\frac{950}{7}}{100} = \dfrac{950}{700} \xrightarrow{\div 50} \dfrac{19}{14} = 1\dfrac{5}{14}$

Percentages 6

Change a Fraction to a Percentage.

Change a Fraction to a Percentage.
(Sections A and B are very useful **but** only allow for **some** fractions to be changed to a percentage -
  Sections C allows for **ALL** fractions to be changed to a percentage.)

Section A
If a fraction has a **denominator of 100** then it can be changed / converted to a percentage by writing ... the **numerator** and the **%** sign.

Ex. Change each fraction to a percentage.

1) $\dfrac{3}{100}$ $\dfrac{3}{100} = 3\%$

2) $\dfrac{75}{100}$ $\dfrac{75}{100} = 75\%$

Q	A	Q	A
Change each fraction to a percentage.		4) $\dfrac{100}{100}$	4) 100%
1) $\dfrac{5}{100}$	1) 5%	5) $\dfrac{117}{100}$	5) 117%
2) $\dfrac{42}{100}$	2) 42%	6) $\dfrac{290}{100}$	6) 290%
3) $\dfrac{99}{100}$	3) 99%	7) $\dfrac{800}{100}$	7) 800%

Section B
If a fraction has a **denominator** which is a **factor of 100**, then both numerator and denominator can be multiplied by the same number to give an equal fraction with a **denominator of 100**.
Then it can be changed to a percentage by writing ... the **numerator** and the **%** sign.

Ex. Change each fraction to a percentage.

1) $\dfrac{3}{10}$ $\dfrac{3}{10} \xrightarrow{\times 10} \dfrac{30}{100} = 30\%$

2) $\dfrac{18}{25}$ $\dfrac{18}{25} \xrightarrow{\times 4} \dfrac{72}{100} = 72\%$

Q	A	Q	A
Change each fraction to a percentage.			
1) $\dfrac{9}{10}$	1) $\dfrac{9}{10} \xrightarrow{\times 10} \dfrac{90}{100} = 90\%$	5) $\dfrac{1}{2}$	5) $\dfrac{1}{2} \xrightarrow{\times 50} \dfrac{50}{100} = 50\%$
2) $\dfrac{8}{25}$	2) $\dfrac{8}{25} \xrightarrow{\times 4} \dfrac{32}{100} = 32\%$	6) $\dfrac{1}{4}$	6) $\dfrac{1}{4} \xrightarrow{\times 25} \dfrac{25}{100} = 25\%$
3) $\dfrac{13}{20}$	3) $\dfrac{13}{20} \xrightarrow{\times 5} \dfrac{65}{100} = 65\%$	7) $\dfrac{4}{5}$	7) $\dfrac{4}{5} \xrightarrow{\times 20} \dfrac{80}{100} = 80\%$
4) $\dfrac{13}{50}$	4) $\dfrac{13}{50} \xrightarrow{\times 2} \dfrac{26}{100} = 26\%$	8) $\dfrac{59}{25}$	8) $\dfrac{59}{25} \xrightarrow{\times 4} \dfrac{236}{100} = 236\%$

Percentages 7 — Change a Fraction to a Percentage.

Section C
A **fraction** can be changed to a **percentage** by **multiplying by 100%**.
(Since a **percentage** can be changed to a **fraction** by **dividing by 100** ... we <u>reverse</u> the calculation.)

Ex. Change each fraction to a percentage.

A) $\dfrac{3}{100}$ $\dfrac{3}{\cancel{100}} \times \dfrac{\cancel{100}\%^{1}}{1} = \dfrac{3\%}{1} = 3\%$
cancel by 100

B) $\dfrac{4}{5}$ $\dfrac{4}{\cancel{5}} \times \dfrac{\cancel{100}\%^{20}}{1} = \dfrac{80\%}{1} = 80\%$
cancel by 5

C) $\dfrac{7}{40}$ $\dfrac{7}{\cancel{40}_{2}} \times \dfrac{\cancel{100}\%^{5}}{1} = \dfrac{35\%}{2} = 17\dfrac{1}{2}\%$
cancel by 20

D) $\dfrac{1}{3}$ $\dfrac{1}{3} \times \dfrac{100\%}{1} = \dfrac{100\%}{3} = 33\dfrac{1}{3}\%$
can **not** cancel

Q	A	Q	A
Change each fraction to a percentage.		9) $\dfrac{1}{8}$	9) $\dfrac{1}{\cancel{8}_{2}} \times \dfrac{\cancel{100}\%^{25}}{1} = \dfrac{25\%}{2} = 12\dfrac{1}{2}\%$
1) $\dfrac{29}{100}$	1) $\dfrac{29}{\cancel{100}} \times \dfrac{\cancel{100}\%^{1}}{1} = \dfrac{29\%}{1} = 29\%$	10) $\dfrac{3}{40}$	10) $\dfrac{3}{\cancel{40}_{2}} \times \dfrac{\cancel{100}\%^{5}}{1} = \dfrac{15\%}{2} = 7\dfrac{1}{2}\%$
2) $\dfrac{1}{2}$	2) $\dfrac{1}{\cancel{2}} \times \dfrac{\cancel{100}\%^{50}}{1} = \dfrac{50\%}{1} = 50\%$	11) $\dfrac{7}{80}$	11) $\dfrac{7}{\cancel{80}_{4}} \times \dfrac{\cancel{100}\%^{5}}{1} = \dfrac{35\%}{4} = 8\dfrac{3}{4}\%$
3) $\dfrac{1}{4}$	3) $\dfrac{1}{\cancel{4}} \times \dfrac{\cancel{100}\%^{25}}{1} = \dfrac{25\%}{1} = 25\%$	12) $\dfrac{99}{250}$	12) $\dfrac{99}{\cancel{250}_{5}} \times \dfrac{\cancel{100}\%^{2}}{1} = \dfrac{198\%}{5} = 39\dfrac{3}{5}\%$
4) $\dfrac{4}{5}$	4) $\dfrac{4}{\cancel{5}} \times \dfrac{\cancel{100}\%^{20}}{1} = \dfrac{80\%}{1} = 80\%$	13) $\dfrac{3}{404}$	13) $\dfrac{3}{\cancel{404}_{101}} \times \dfrac{\cancel{100}\%^{25}}{1} = \dfrac{75}{101}\%$
5) $\dfrac{7}{10}$	5) $\dfrac{7}{\cancel{10}} \times \dfrac{\cancel{100}\%^{10}}{1} = \dfrac{70\%}{1} = 70\%$	14) $\dfrac{2}{3}$	14) $\dfrac{2}{3} \times \dfrac{100\%}{1} = \dfrac{200\%}{3} = 66\dfrac{2}{3}\%$
6) $\dfrac{11}{20}$	6) $\dfrac{11}{\cancel{20}} \times \dfrac{\cancel{100}\%^{5}}{1} = \dfrac{55\%}{1} = 55\%$	15) $\dfrac{4}{7}$	15) $\dfrac{4}{7} \times \dfrac{100\%}{1} = \dfrac{400\%}{7} = 57\dfrac{1}{7}\%$
7) $\dfrac{22}{25}$	7) $\dfrac{22}{\cancel{25}} \times \dfrac{\cancel{100}\%^{4}}{1} = \dfrac{88\%}{1} = 88\%$	16) $\dfrac{39}{10}$	16) $\dfrac{39}{\cancel{10}} \times \dfrac{\cancel{100}\%^{10}}{1} = \dfrac{390\%}{1} = 390\%$
8) $\dfrac{39}{50}$	8) $\dfrac{39}{\cancel{50}} \times \dfrac{\cancel{100}\%^{2}}{1} = \dfrac{78\%}{1} = 78\%$	17) $\dfrac{11}{9}$	17) $\dfrac{11}{9} \times \dfrac{100\%}{1} = \dfrac{1100\%}{9} = 122\dfrac{2}{9}\%$

Percentages 8

Change a Whole / Mixed Number to a Percentage.

Section C
Change a Whole Number to a Percentage
Multiply the whole number by 100%.

Ex. Change 4 to a percentage.

$$4 \times 100\% = 400\%$$

Q	A	Q	A
Change each whole number to a percentage.		3) 17	3) $17 \times 100\% = 1700\%$
1) 2	1) $2 \times 100\% = 200\%$	4) 100	4) $100 \times 100\% = 10{,}000\%$
2) 5	2) $5 \times 100\% = 500\%$	5) 345	5) $345 \times 100\% = 34{,}500\%$

Change a Mixed Number to a Percentage

Ex. Change $4\frac{1}{2}$ to a percentage.

Method 1
Multiply the **whole number** by 100%.
Multiply the **fraction** part by 100%.
Add the answers.

$$4 \times 100\% = 400\%$$

$$\frac{1}{\cancel{2}_1} \times \frac{\cancel{100}^{50}\%}{1} = \frac{50\%}{1} = 50\%$$

$$+ \underline{}$$
$$450\%$$

Method 2
Change the Mixed Number to an Improper Fraction - multiply by 100%.

$$4\frac{1}{2} = \frac{9}{2}$$

$$\frac{9}{\cancel{2}_1} \times \frac{\cancel{100}^{50}\%}{1} = \frac{450\%}{1} = 450\%$$

Q	A
Change each mixed number to a percentage. 1) $1\frac{3}{4}$	1) **Method 1** $1 \times 100\% = 100\%$ $\frac{3}{\cancel{4}_1} \times \frac{\cancel{100}^{25}\%}{1} = \frac{75\%}{1} = 75\%$ $+\underline{}$ 175% **Method 2** $1\frac{3}{4} = \frac{7}{4}$ $\frac{7}{\cancel{4}_1} \times \frac{\cancel{100}^{25}\%}{1} = \frac{175\%}{1} = 175\%$
2) $5\frac{9}{25}$	2) **Method 1** $5 \times 100\% = 500\%$ $\frac{9}{\cancel{25}_1} \times \frac{\cancel{100}^{4}\%}{1} = \frac{36\%}{1} = 36\%$ $+\underline{}$ 536% **Method 2** $5\frac{9}{25} = \frac{134}{25}$ $\frac{134}{\cancel{25}_1} \times \frac{\cancel{100}^{4}\%}{1} = \frac{536\%}{1} = 536\%$
3) $2\frac{1}{8}$	3) **Method 1** $2 \times 100\% = 200\%$ $\frac{1}{\cancel{8}_2} \times \frac{\cancel{100}^{25}\%}{1} = \frac{25\%}{2} = 12\frac{1}{2}\%$ $+\underline{}$ $212\frac{1}{2}\%$ **Method 2** $2\frac{1}{8} = \frac{17}{8}$ $\frac{17}{\cancel{8}_2} \times \frac{\cancel{100}^{25}\%}{1} = \frac{425\%}{2} = 212\frac{1}{2}\%$

Percentages 9   Change - Percentage to Decimal / Decimal to Percentage.

Percentages and Decimals
Change a Percentage to a Decimal
A percentage can be changed to a decimal by **dividing by 100** - the decimal point is moved **2 places to the left**.

Ex. Change each percentage to a decimal.

A) 1%	B) 23%	C) 4·5%	D) 67·8%	E) 123%
= 0̂01· ÷100	= 0̂23· ÷100	= 0̂04·5 ÷100	= 0̂67·8 ÷100	= 1̂23· ÷100
= 0·01	= 0·23	= 0·045	= 0·678	= 1·23

*Note Only the answer is required.

Q	A	Q	A
Change each percentage to a decimal.		5) 80%	5) = 0̂80· ÷100 = 0·8
1) 6%	1) = 0̂06· ÷100 = 0·06	6) 2·65%	6) = 0̂02·65 ÷100 = 0·0265
2) 49%	2) = 0̂49· ÷100 = 0·49	7) 0·9%	7) = 0̂00·9 ÷100 = 0·009
3) 7·1%	3) = 0̂07·1 ÷100 = 0·071	8) 384%	8) = 3̂84· ÷100 = 3·84
4) 15·2%	4) = 0̂15·2 ÷100 = 0·152	9) 500%	9) = 5̂00· ÷100 = 5

Change a Decimal to a Percentage.
A decimal can be changed to a percentage by **multiplying by 100** - the decimal point is moved **2 places to the right**.

Ex. Change each decimal to a percentage.

A) 0·05	B) 0·96	C) 0·027	D) 0·804	E) 3·19
0·05̂ ×100	0·96̂ ×100	0·02̂7 ×100	0·80̂4 ×100	3·19̂ ×100
= 5%	= 96%	= 2·7%	= 80·4%	= 319%

*Note Only the answer is required.

Q	A	Q	A
Change each decimal to a percentage.		5) 4·95	5) 4·95̂ ×100 = 495%
1) 0·02	1) 0·02̂ ×100 = 2%	6) 0·5	6) 0·50̂ ×100 = 50%
2) 0·77	2) 0·77̂ ×100 = 77%	7) 0·0308	7) 0·03̂08 ×100 = 3·08%
3) 0·068	3) 0·06̂8 ×100 = 6·8%	8) 9	8) 9·00̂ ×100 = 900%
4) 0·531	4) 0·53̂1 ×100 = 53·1%	9) 0·0014	9) 0·00̂14 ×100 = 0·14%

Percentages 10 — Change a Fraction Percentage to a Decimal.

Change a Fraction Percentage to a Decimal

Given a fraction percentage, Ex. $\frac{1}{2}\%$, Ex. $4\frac{3}{5}\%$, it may be useful to change it to a decimal.

A We first change the **fraction percentage** to a **decimal percentage** -
B   then change the **decimal percentage** to a **decimal**.

Ex. Change each percentage to a decimal.

1) $\frac{5}{8}\%$

A $\frac{5}{8}\% = \dfrac{0{\cdot}6\,2\,5\,\%}{8\overline{)5{\cdot}0^20^40}}$

B $= 0{\cdot}00{\cdot}625 \div 100$
      $= 0{\cdot}00625$

2) $7\frac{1}{3}\%$

A $\frac{1}{3}\% = \dfrac{0{\cdot}3\,3\,3\,...\%}{3\overline{)1{\cdot}0^10^10...}}$

B $7\frac{1}{3}\% = 0{\cdot}07{\cdot}333... \div 100$
              $= 0{\cdot}07333...$

Q	A	Q	A
Change each percentage to a decimal. 1) $\frac{3}{4}\%$	1) $\frac{3}{4}\% = \dfrac{0{\cdot}7\,5\,\%}{4\overline{)3{\cdot}0^20}}$ $= 0{\cdot}00{\cdot}75 \div 100$ $= 0{\cdot}0075$	5) $4\frac{1}{2}\%$	5) $\frac{1}{2}\% = \dfrac{0{\cdot}5\,\%}{2\overline{)1{\cdot}0}}$ $4\frac{1}{2}\% = 0{\cdot}04{\cdot}5 \div 100$ $= 0{\cdot}045$
2) $\frac{1}{8}\%$	2) $\frac{1}{8}\% = \dfrac{0{\cdot}1\,2\,5\,\%}{8\overline{)1{\cdot}0^20^40}}$ $= 0{\cdot}00{\cdot}125 \div 100$ $= 0{\cdot}00125$	6) $9\frac{7}{8}\%$	6) $\frac{7}{8}\% = \dfrac{0{\cdot}8\,7\,5\,\%}{8\overline{)7{\cdot}0^60^40}}$ $9\frac{7}{8}\% = 0{\cdot}09{\cdot}875 \div 100$ $= 0{\cdot}09875$
3) $\frac{5}{7}\%$	3) $\frac{5}{7}\% = \dfrac{0{\cdot}7\,1\,4\,...\%}{7\overline{)5{\cdot}0^10^30...}}$ $= 0{\cdot}00{\cdot}714... \div 100$ $= 0{\cdot}00714...$	7) $1\frac{4}{9}\%$	7) $\frac{4}{9}\% = \dfrac{0{\cdot}4\,4\,4\,\%}{9\overline{)4{\cdot}0^40^40...}}$ $1\frac{4}{9}\% = 0{\cdot}01{\cdot}444... \div 100$ $= 0{\cdot}01444...$
4) $\frac{9}{10}\%$	4) $\frac{9}{10}\% = \dfrac{0{\cdot}9\,\%}{10\overline{)9{\cdot}0}}$ $= 0{\cdot}00{\cdot}9 \div 100$ $= 0{\cdot}009$	8) $63\frac{11}{20}\%$	8) $\frac{11}{20}\% = \dfrac{0{\cdot}5\,5\,\%}{20\overline{)11{\cdot}0^{10}0}}$ $63\frac{11}{20}\% = 0{\cdot}63{\cdot}55 \div 100$ $= 0{\cdot}6355$

Percentages 11   Whole Amount / Parts. Compare Percentage / Fraction / Decimal.

Whole Amount / Parts

Since a <u>Whole Amount</u> is <u>100%</u>, given '**a part (as a %)**' then the '**remaining part (as a %)**' can be calculated.

The values of the 'Whole Amount' or 'parts' do not need to be given.

Ex. In a class of pupils, 40% are boys. What percentage of the class are girls?

$$\begin{array}{ll} \text{Class} & 100\% \\ -\text{Boys} & 40\% \\ \hline \text{Girls} & 60\% \end{array}$$

60% of the class are girls.

We do not need to be given **how many** pupils or boys or girls are in the class — (**this is known** in order to calculate the percentage in the first place !)

Q	A	Q	A
1) In a class, 55% are girls. What percentage are boys?	1) Class 100% − Girls 55% Boys 45%	5) A family had an annual income of £20,000. It spent 88·5%, saving the rest. What percentage is saved?	5) Income 100% − Spent 88·5% Saved 11·5%
2) A box holds red and blue pens. 70% of the pens are red. What percentage are blue?	2) Pens 100% − Red 70% Blue 30%	6) Teams A, B and C compete. Team A wins 34% of the points. Team B wins 27% of the points. What percentage of the points does Team C win?	6) Team A 34% + Team B 27% Team A/B 61% Points 100% − Team A/B 61% Team C 39%
3) 26% of a set of numbers are odd. What percentage are even?	3) Set 100% − Odd 26% Even 74%		
4) If $7\frac{1}{2}\%$ of a company's products have at least 1 fault, what percentage have no fault?	4) Products 100% − Fault $7\frac{1}{2}\%$ No Fault $92\frac{1}{2}\%$	7) 5 girls share a cake equally. What percentage does each get?	7) $\dfrac{\text{Cake}}{\text{Girls}} = \dfrac{100\%}{5}$ $= 20\%$

Compare Percentage / Fraction / Decimal

To compare a Percentage / Fraction / Decimal it is usually easiest to convert a Percentage / Fraction to a Decimal and compare all numbers in the decimal form. (Here, just 1 of each kind is given.)
It may be possible to compare **mentally**.

Given several numbers to arrange in order, it is useful to list them in a **column** in **descending** order from top to bottom (even if **ascending** order is required) and then indicate the order or list them on a line.

*Note see **Percentages 9** (Percentage to Decimal) / **Fractions 29** (Fraction to Decimal).

Ex. Write in ascending order

$\dfrac{1}{2}$, 40% , 0·6 ⇨

ascending ↑
$$\begin{array}{rl} 0·6 &= 0·6 \\ \dfrac{1}{2} &= 0·5 \\ 40\% &= 0·4 \end{array}$$

Q	A	Q	A
1) Write in **ascending** order $\dfrac{3}{4}$, 0·77 , 73%	1) ascending ↑ $0·77 = 0·77$ $\dfrac{3}{4} = 0·75$ $73\% = 0·73$	3) State the smallest value. $0·08$, 9% , $\dfrac{3}{50}$	3) $9\% = 0·09$ $0·08 = 0·08$ smallest $\dfrac{3}{50} = 0·06$
2) Write in **descending** order 15% , $\dfrac{1}{5}$, 0·1	2) descending $\dfrac{1}{5} = 0·2$ $15\% = 0·15$ ↓ $0·1 = 0·1$	4) State the largest value. $\dfrac{2}{3}$, 67% , 0·654	4) largest $67\% = 0·67$ $\dfrac{2}{3} = 0·666...$ $0·654 = 0·654$

Percentages 12　　　　　　　　　　　　　　Percentage of a Value - using Fractions.

Percentage of a Value

A **Percentage of a Value** can be calculated using either fractions or decimals -
(we often have a choice which to use).

Fractions

Method 1　　　　　　　　　　　　　　　　(Method 2 - is explained in **Percentages 14**.)

The Q/A will allow us to **cancel** - once - **before** multiplying .
(It may be possible to cancel **more than once** - but this will not be shown in the following Q/A.)

*Note

Fractions 19-21 explain how to multiply / cancel.
Since we can cancel in **many** ways, the 'work' for each Q/A may be different than shown here.

Ex.1　Find　10% of 50　　　　　　　　　　　Ex.2　Find　70% of 9 km

'of' becomes 'times'

$$\frac{10}{100} \times \frac{50}{1} = \frac{10}{2} = 5$$

cancel by 50

$$\frac{70}{100} \times \frac{9}{1} = \frac{63}{10} = 6\frac{3}{10} \text{ km}$$

cancel by 10

Q	A	Q	A
Find 1) 10% of 30	1) $\frac{10}{100} \times \frac{30}{1} = \frac{30}{10} = 3$ cancel by 10	6) 15% of 19 km	6) $\frac{15}{100} \times \frac{19}{1} = \frac{57}{20} = 2\frac{17}{20}$ km cancel by 5
2) 40% of 40 m	2) $\frac{40}{100} \times \frac{40}{1} = \frac{80}{5} = 16$ m cancel by 20	7) 66% of 12 g	7) $\frac{66}{100} \times \frac{12}{1} = \frac{396}{50} = 7\frac{46}{50} = 7\frac{23}{25}$ g cancel by 2　　cancel by 2
3) 9% of 800 eggs	3) $\frac{9}{100} \times \frac{800}{1} = 72$ eggs cancel by 100	8) $3\frac{1}{2}$% of 28 °C	8) $3\frac{1}{2}\% = \frac{3\frac{1}{2}}{100} = \frac{\frac{7}{2}}{100} = \frac{7}{200}$ $\frac{7}{200} \times \frac{28}{1} = \frac{49}{50}$ °C cancel by 4
4) 75% of £28	4) $\frac{75}{100} \times \frac{28}{1} = \frac{84}{4} = £21$ cancel by 25	9) $80\frac{3}{4}$% of £360	9) $80\frac{3}{4}\% = \frac{80\frac{3}{4}}{100} = \frac{\frac{323}{4}}{100} = \frac{323}{400}$ $\frac{323}{400} \times \frac{360}{1} = \frac{2907}{10} = £290\frac{7}{10}$ 　　　　　　　　　　＝ £290·70 cancel by 40 (it can make sense to change the final 　fraction to a decimal - as here)
5) 99% of 50 miles	5) $\frac{99}{100} \times \frac{50}{1} = \frac{99}{2} = 49\frac{1}{2}$ miles cancel by 50		

Percentages 13

Percentage of a Value - using Decimals.

Percentage of a Value
Decimals
Method 1 (Method 2 - is explained in **Percentages 14**.)

We change a percentage to a decimal **then** multiply by the amount.

*Note 1
Decimals 8-9 explain how to multiply decimals - if a calculator is not used !
(... 'reminder' ...
'Count the number of <u>decimal places</u> in the '**question**' ... then ...
count the same number of <u>decimal places</u> in the '**answer**' so as to position the decimal point.
Here, these values are <u>underlined</u>.)

*Note 2
Percentages 9 explains how to change a percentage to a decimal.
Here, the percentage is first shown as a fraction with denominator 100 as a 'reminder' to divide it by 100.

Ex.1 Find 10% of 50

'**of**' becomes '**times**'

$\frac{10}{100} \times 50 =$ 50 × 0·1 = 5·0

0·1 × 50 = 5

Ex.2 Find 31% of 57·1 kg

$\frac{31}{100} \times 57·1 =$

0·31 × 57·1 = 17·701 kg

```
   57·1
 × 0·31
   571
+17130
 17·701
```

Q	A	Q	A
Find 1) 20% of 90	1) $\frac{20}{100} \times 90 =$ 0·2 × 90 = 18 90 × 0·2 = 18·0	6) 4·4% of 2 litres	6) $\frac{4·4}{100} \times 2 =$ 0·044 × 2 = 0·088 litres 0·044 × 2 = 0·088
2) 60% of 60 kg	2) $\frac{60}{100} \times 60 =$ 0·6 × 60 = 36 kg 60 × 0·6 = 36·0	7) 98·7% of 1·1 km	7) $\frac{98·7}{100} \times 1·1 =$ 0·987 × 1·1 = 1·0857 km 0·987 × 1·1 = 1·0857
3) 25% of 700 m	3) $\frac{25}{100} \times 700 =$ 0·25 × 700 = 175 m 0·25 × 700 = 175·00	8) 0·5% of 45·9 mph	8) $\frac{0·5}{100} \times 45·9 =$ 0·005 × 45·9 = 0·2295 mph 45·9 × 0·005 = 0·2295
4) 53% of £94	4) $\frac{53}{100} \times 94 =$ 0·53 × 94 = £49·82 0·53 × 94 = 49·82	9) $3\frac{4}{5}$% of 8·2 hrs	9) $\left(\frac{4}{5} = 0·8\right)$ $\frac{3·8}{100} \times 8·2 =$ 0·038 × 8·2 = 0·3116 hrs
5) 140% of 80 g	5) $\frac{140}{100} \times 80 =$ 1·4 × 80 = 112 g 1·4 × 80 = 112·0		

Percentages 14

Percentage of a Value - using Fractions / Decimals.

Percentage of a Value
Fractions - Method 2
Decimals - Method 2

Since we usually write a decimal percentage as a fraction with denominator 100 ...
in both Methods - we can decide to **multiply the numerators** first.

Ex.1 Calculate 10% of 90

$$\frac{10}{100} \times \frac{90}{1} = \frac{900}{100} = 9$$

Ex.2 Calculate 45% of £70

$$\left.\frac{45}{100} \times \frac{70}{1} = \frac{3150}{100}\right\} = 31\frac{50}{100} = £31\frac{1}{2}$$
$$= £31 \cdot 50$$

cancel by 50

We can easily **choose here** to have a fraction / decimal answer.

Ex.3 Calculate 8·2% of 3 km

$$\frac{8 \cdot 2}{100} \times \frac{3}{1} = \frac{24 \cdot 6}{100} = 0 \cdot 246 \text{ km}$$

Q	A	Q	A
Calculate		7) $33\frac{1}{3}$% of 84p	7) $\left(33\frac{1}{3} = \frac{100}{3}\right)$ cancel by 100 $\frac{100}{300} \times \frac{84}{1} = \frac{8400}{300} = \frac{84}{3} = 28$p
1) 20% of 50	1) $\frac{20}{100} \times \frac{50}{1} = \frac{1000}{100} = 10$		
2) 4% of 600 kg	2) $\frac{4}{100} \times \frac{600}{1} = \frac{2400}{100} = 24$ kg	8) $2\frac{1}{2}$% of £17	8) $\left(2\frac{1}{2} = \frac{5}{2}\right)$ cancel by 5 $\left.\frac{5}{200} \times \frac{17}{1}\right\} = £\frac{85}{200} = £\frac{17}{40}$ $= £0 \cdot 425$
3) 650% of 30 g	3) $\frac{650}{100} \times \frac{30}{1} = \frac{19500}{100} = 195$ g		
4) 11% of £9	4) $\left.\frac{11}{100} \times \frac{9}{1}\right\} = £\frac{99}{100}$ $= £0 \cdot 99$	9) $22\frac{3}{4}$% of 500 litres	9) $\left(22\frac{3}{4} = \frac{91}{4}\right)$ $\frac{91}{400} \times \frac{500}{1} =$ cancel by 100 $\left.\frac{45500}{400} = \frac{455}{4}\right\} = 113\frac{3}{4}$ $= 113 \cdot 75$ litres
5) 8% of 80 km	5) $\frac{8}{100} \times \frac{80}{1} =$ cancel by 20 $\left.\frac{640}{100}\right\} = 6\frac{40}{100} = 6\frac{2}{5}$ km $= 6 \cdot 4$ km	10) 1·5% of 40 hrs	10) $\frac{1 \cdot 5}{100} \times \frac{40}{1} = \frac{60}{100} = \frac{3}{5}$ hr $= 0 \cdot 6$ hr
		11) 77·7% of 7 °F	11) $\frac{77 \cdot 7}{100} \times \frac{7}{1} = \frac{543 \cdot 9}{100} = 5 \cdot 439$ °F
6) 96% of 202 m	6) $\frac{96}{100} \times \frac{202}{1} =$ cancel by 4 $\left.\frac{19392}{100}\right\} = 193\frac{92}{100} = 193\frac{23}{25}$ m $= 193 \cdot 92$ m	12) 56·8% of 32·9 mph	12) $\frac{56 \cdot 8}{100} \times \frac{32 \cdot 9}{1} = \frac{1868 \cdot 72}{100}$ $= 18 \cdot 6872$ mph

Percentages 15 — Percentage Increase / Decrease.

Increase / Decrease of a Value by a Percentage
Method 1
The **given percentage** increase / decrease is used to calculate the **actual** increase / decrease - this **amount** is then **added** / **subtracted** to give the value required.

Ex.1 Increase 40 by 10%
(this means 'Increase 40 by **10% of 40**')

calculate 10% of 40

$\dfrac{10}{100} \times \dfrac{40}{1} = \dfrac{400}{100} = 4$ Increase

$\begin{array}{r} 40 \\ +\text{add } 4 \\ \hline 44 \end{array}$

Ex.2 Decrease 40 by 10%
(this means 'Decrease 40 by **10% of 40**')

calculate 10% of 40

$\dfrac{10}{100} \times \dfrac{40}{1} = \dfrac{400}{100} = 4$ Decrease

$\begin{array}{r} 40 \\ -\text{subtract } 4 \\ \hline 36 \end{array}$

*Note This method of calculation will be used.

Q	A	Q	A
1) Increase 6 cars by 50%	1) 50% of 6 $\dfrac{50}{100} \times \dfrac{6}{1} = \dfrac{300}{100} = 3$ $6 + 3 = 9$ cars	5) Decrease 8 mph by 25%	5) 25% of 8 $\dfrac{25}{100} \times \dfrac{8}{1} = \dfrac{200}{100} = 2$ $8 - 2 = 6$ mph
2) Increase 70 km by 20%	2) 20% of 70 $\dfrac{20}{100} \times \dfrac{70}{1} = \dfrac{1400}{100} = 14$ $70 + 14 = 84$ km	6) Decrease £50 by 90%	6) 90% of 50 $\dfrac{90}{100} \times \dfrac{50}{1} = \dfrac{4500}{100} = 45$ $50 - 45 = £5$
3) Increase 11 by 89%	3) 89% of 11 $\dfrac{89}{100} \times \dfrac{11}{1} = \dfrac{979}{100} = 9\dfrac{79}{100}$ $11 + 9\dfrac{79}{100} = 20\dfrac{79}{100}$ ----- or ----- 89% of 11 $\dfrac{89}{100} \times \dfrac{11}{1} = \dfrac{979}{100} = 9.79$ $11 + 9.79 = 20.79$	7) Decrease 2·5 g by 41·5%	7) 41·5% of 2·5 $\dfrac{41 \cdot 5}{100} \times \dfrac{2 \cdot 5}{1} = \dfrac{103 \cdot 75}{100} = 1 \cdot 0375$ $2.5 - 1.0375 = 1 \cdot 4625$ g
4) Increase £440 by $6\dfrac{2}{5}$%	4) $6\dfrac{2}{5}$% of 440 $\dfrac{32}{500} \times \dfrac{440}{1} = \dfrac{14080}{500}$ cancel by 20 $= 28\dfrac{80}{500} = 28\dfrac{4}{25}$ $440 + 28\dfrac{4}{25} = £468\dfrac{4}{25}$ ----- or ----- $6\dfrac{2}{5}$% of 440 $\dfrac{32}{500} \times \dfrac{440}{1} = \dfrac{14080}{500} = 28 \cdot 16$ $440 + 28.16 = £468 \cdot 16$	8) Decrease 343 hrs by 3%	8) 3% of 343 $\dfrac{3}{100} \times \dfrac{343}{1} = \dfrac{1029}{100} = 10\dfrac{29}{100}$ $343 - 10\dfrac{29}{100} = 332\dfrac{71}{100}$ hrs ----- or ----- 3% of 343 $\dfrac{3}{100} \times \dfrac{343}{1} = \dfrac{1029}{100} = 10 \cdot 29$ $343 - 10.29 = 332 \cdot 71$ hrs

Percentages 16

Percentage Increase / Decrease. 'Multiplier'.

Increase / Decrease of a Value by a Percentage
Method 2

The **given percentage** increase / decrease is added / subtracted from **100%**.
The **new percentage** - which can be referred to as the **'multiplier'** -
is then used to calculate the value required.

Ex.1   Increase 40 by 10%
(this means 'Increase 40 by **10% of 40**')

original	100%
add increase	+ 10%
calculate	110% of 40
	'multiplier'

$$\frac{110}{100} \times \frac{40}{1} = \frac{4400}{100} = 44$$

Ex.2   Decrease 40 by 10%
(this means 'Decrease 40 by **10% of 40**')

original	100%
subtract decrease	− 10%
calculate	90% of 40
	'multiplier'

$$\frac{90}{100} \times \frac{40}{1} = \frac{360}{100} = 36$$

*Note This method of calculation will be used.

Q	A	Q	A
1) Increase 6 cars by 50%	1) 100% + 50% 150% of 6 'multiplier' $\frac{150}{100} \times \frac{6}{1} = \frac{900}{100} = 9$ cars	5) Decrease 8 mph by 25%	5) 100% − 25% 75% of 8 'multiplier' $\frac{75}{100} \times \frac{8}{1} = \frac{600}{100} = 6$ mph
2) Increase 70 km by 20%	2) 100% + 20% 120% of 70 'multiplier' $\frac{120}{100} \times \frac{70}{1} = \frac{8400}{100} = 84$ km	6) Decrease £50 by 90%	6) 100% − 90% 10% of 50 'multiplier' $\frac{10}{100} \times \frac{50}{1} = \frac{500}{100} = £5$
3) Increase 11 by 89%	3) 100% + 89% 189% of 11 'multiplier' $\frac{189}{100} \times \frac{11}{1} = \frac{2079}{100} \Big\} = 20\frac{79}{100}$ $ \Big\} = 20 \cdot 79$	7) Decrease 2·5 g by 41·5%	7) 100% − 41·5% 58·5% of 2·5 'multiplier' $\frac{58 \cdot 5}{100} \times \frac{2 \cdot 5}{1} = \frac{146 \cdot 25}{100}$ $ = 1 \cdot 4625$ g
4) Increase £440 by $6\frac{2}{5}$%	4) 100% + $6\frac{2}{5}$% $106\frac{2}{5}$% of 440 'multiplier' $\frac{532}{500} \times \frac{440}{1} = \frac{234080}{500}$ $= 468\frac{80}{500} = £468\frac{16}{100} = £468 \cdot 16$	8) Decrease 343 hrs by 3%	8) 100% − 3% 97% of 343 'multiplier' $\frac{97}{100} \times \frac{343}{1} = \frac{33271}{100}$ $ \Big\} = 332\frac{71}{100}$ hrs $ \Big\} = 332 \cdot 71$ hrs

Percentages 17 — Original Value of Increased / Decreased Value.

The Original Value of an Increased / Decreased Value

The **original value** of a value that has been **increased** / **decreased** by a **given percentage** can be calculated.

The **original value** is considered to be **100%**.

(This can be considered as the **reverse** of the work explained in **Percentage 15** and **16**)

Method 1 (using an equation)

The **given percentage** increase / decrease is added to / subtracted from **100%** - the **new percentage** is then used to calculate the value required.

Ex.1 A value is **increased** by 10% to 44. Calculate the original value.

Let ... original value   100% = x
add increase   + 10%
increased value   110% = 44

We now 'set up' and solve an **equation** since we can think of this as ...

110% of x is 44. Calculate x.

$$\frac{110}{100} \times x = 44$$

$$x = 44 \times \frac{100}{110} = \frac{4400}{110} = 40$$

'Check' 10% of 40 = 4 and add on = 44 ✓

Ex.2 A number is **decreased** by 10% to 36. Calculate the original number.

Let ... original number   100% = x
subtract decrease   − 10%
decreased number   90% = 36

We now 'set up' and solve an **equation** since we can think of this as ...

90% of x is 36. Calculate x.

$$\frac{90}{100} \times x = 36$$

$$x = 36 \times \frac{100}{90} = \frac{3600}{90} = 40$$

'Check' 10% of 40 = 4 and take off = 36 ✓

Q	A
1) A value is increased by 20% to 72. Calculate the original value.	1) 100% = x + 20% 120% = 72 $\frac{120}{100} \times x = 72$ $x = 72 \times \frac{100}{120} = \frac{7200}{120} = 60$
2) A number is decreased by 30% to 56. Calculate the original number.	2) 100% = x − 30% 70% = 56 $\frac{70}{100} \times x = 56$ $x = 56 \times \frac{100}{70} = \frac{5600}{70} = 80$
3) The price of a cake is increased by 5% to £9·45. State its price before the increase.	3) 100% = x + 5% 105% = 9·45 $\frac{105}{100} \times x = 9\cdot45$ $x = 9\cdot45 \times \frac{100}{105} = \frac{945}{105} = £9$
4) In a sale, the original price of a guitar was reduced by 50% to £180. Calculate its price before the sale.	4) 100% = x − 50% 50% = 180 $\frac{50}{100} \times x = 180$ $x = 180 \times \frac{100}{50} = \frac{18000}{50} = £360$
5) A company increases the number of workers by 77% to 2124. Calculate the number of workers before the increase.	5) 100% = x + 77% 177% = 2124 $\frac{177}{100} \times x = 2124$ $x = 2124 \times \frac{100}{177} = \frac{212400}{177}$ = 1200 workers
6) 96·7% of a substance is burnt off in a process leaving 0·693kg residue. Find the weight of the substance at the start of the process.	6) 100% = x − 96·7% 3·3% = 0·693 $\frac{3\cdot3}{100} \times x = 0\cdot693$ $x = 0\cdot693 \times \frac{100}{3\cdot3} = \frac{69\cdot3}{3\cdot3} = 21$kg

Percentages 18

Original Value of Increased / Decreased Value.

The Original Value of an Increased / Decreased Value
Method 2 (using ratio / direct proportion)
The **given percentage** increase / decrease is added to / subtracted from **100%** - the **new percentage** is then used to calculate the value required.

Ex.1 A value is **increased** by 10% to 44. Calculate the original value.

original value   100%
add increase   + 10%
increased value 110% = **44**

We now use **ratio / direct proportion** to ...

find   1% = $\frac{44}{110}$

find   100% = $\frac{44}{110} \times 100$

           = $\frac{4400}{110}$ = 40

'Check' 10% of 40 = 4 and add on = 44 ✓

Ex.2 A number is **decreased** by 10% to 36. Calculate the original number.

original value   100%
subtract decrease − 10%
decreased value 90% = **36**

We now use **ratio / direct proportion** to ...

find   1% = $\frac{36}{90}$

find   100% = $\frac{36}{90} \times 100$

           = $\frac{3600}{90}$ = 40

'Check' 10% of 40 = 4 and take off = 36 ✓

Q	A	
1) A value is increased by 20% to 72. Calculate the original value.	1) 100% + 20% **120% = 72**	1% = $\frac{72}{120}$ 100% = $\frac{72}{120} \times 100 = \frac{7200}{120} = 60$
2) A number is decreased by 30% to 56. Calculate the original number.	2) 100% − 30% **70% = 56**	1% = $\frac{56}{70}$ 100% = $\frac{56}{70} \times 100 = \frac{5600}{70} = 80m$
3) The price of a cake is increased by 5% to £9.45. State its price before the increase.	3) 100% + 5% **105% = 9.45**	1% = $\frac{9.45}{105}$ 100% = $\frac{9.45}{105} \times 100 = \frac{945}{105} = £9$
4) In a sale, the original price of a guitar was reduced by 50% to £180. Calculate its price before the sale.	4) 100% − 50% **50% = 180**	1% = $\frac{180}{50}$ 100% = $\frac{180}{50} \times 100 = \frac{18000}{50} = £360$
5) A company increases the number of workers by 77% to 2124. Calculate the number of workers before the increase.	5) 100% + 77% **177% = 2124**	1% = $\frac{2124}{177}$ 100% = $\frac{2124}{177} \times 100 = \frac{212400}{177}$ = 1200 workers
6) 96.7% of a substance is burnt off in a process leaving 0.693kg residue. Find the weight of the substance at the start of the process.	6) 100 % − 96.7% **3.3% = 0.693**	1% = $\frac{0.693}{3.3}$ 100% = $\frac{0.693}{3.3} \times 100 = \frac{69.3}{3.3} = 21kg$

Percentages 19　　　　　　　　　　Express One Value as a Percentage of Another Value.

Express One Value as a Percentage of Another Value.
To express one value **as a percentage** of another value,
we first express one value **as a fraction of** the other value -
then change the fraction to a percentage by **multiplying by 100%**.

Both values must have the **same units of measure** - (changing larger units to smaller units is easier).

Ex.1　Express 3 as a percentage of 6.
(first - Express 3 **as a fraction of** 6)

$$\frac{3}{6} \times \frac{100\%}{1} = \frac{300\%}{6} = 50\%$$

*Note This method of calculation will be used.

Ex.2　Express 19m as a percentage of 21m.
(first - Express 19m **as a fraction of** 21m)

$$\frac{19}{21} \times \frac{100\%}{1} = \frac{1900\%}{21} = 90\frac{10}{21}\%$$
$$= 90 \cdot 476...\%$$

Q	A
1) Express 2 as a percentage of 8.	1) $\frac{2}{8} \times \frac{100\%}{1} = \frac{200\%}{8} = 25\%$
2) In a group of 60 pupils, 48 are girls. What percentage of the group are girls?	2) $\frac{48}{60} \times \frac{100\%}{1} = \frac{4800\%}{60} = 80\%$
3) Express 3 °C as a percentage of 7 °C.	3) $\frac{3}{7} \times \frac{100\%}{1} = \frac{300\%}{7} = 42\frac{6}{7}\% = 42 \cdot 85...\%$
4) It rains on 9 days in May. State the number of days on which it rains as a percentage of the days in May.	4) $\frac{9}{31} \times \frac{100\%}{1} = \frac{900\%}{31} = 29\frac{1}{31}\% = 29 \cdot 03...\%$ (May = 31 days)
5) Express 1·5**mm** as a percentage of 5**cm**.	5) 5cm = 50 mm　$\frac{1 \cdot 5}{50} \times \frac{100\%}{1} = \frac{150\%}{50} = 3\%$
6) Express 2 **mins** as a percentage of 50 **secs**.	6) 2 mins = 2× 60 secs = 120 secs　$\frac{120}{50} \times \frac{100\%}{1} = \frac{12000\%}{50} = 240\%$

Either or **both** of the **fraction values** may have to be calculated before the percentage is calculated.

Ex.1　In a box of 50 beads, 15 are red and the rest are blue. What percentage of the beads are blue.

Blue beads = 50 − Red beads
　　　　　= 50 − 15
　　　　　= 35

$$\frac{35}{50} \times \frac{100\%}{1} = \frac{3500\%}{50} = 70\%$$

Q	A
1) There are 3 faulty drills in a batch of 200. What percentage of drills are not faulty?	1) Not faulty = 200 − faulty = 200 − 3 = 197　$\frac{197}{200} \times \frac{100\%}{1} = \frac{19700\%}{200} = 98\frac{100}{200}\% = 98\frac{1}{2}\%$ = 98·5%
2) 700 litres of white paint are mixed with 23 litres of red paint. What percentage of the total is red paint?	2) Total = 700 + 23 = 723 litres　$\frac{23}{723} \times \frac{100\%}{1} = \frac{2300\%}{723} = 3\frac{131}{723}\% = 3 \cdot 18...\%$
3) Team A scores 10 points. Team B scores 4 times as many points. Express the points scored by Team B as a percentage of the total points scored.	3) Team B = 4 × 10 = 40 points　Total = 10 + 40 = 50 points　$\frac{40}{50} \times \frac{100\%}{1} = \frac{4000\%}{50} = 80\%$

Percentages 20 — Percentage Change.

Percentage Change
To express the **change** (increase / decrease) in a value **as a percentage** of the **original value**, we first express the **change** as a **fraction** of the **original value** - then convert the fraction to a percentage by **multiplying by 100%**.

$$\text{percentage increase} = \frac{\text{actual increase}}{\text{original value}} \times 100\% \qquad \text{percentage decrease} = \frac{\text{actual decrease}}{\text{original value}} \times 100\%$$

Ex.1 The price of a pen is increased by 14p from 70p. By what percentage has the price increased?

$$\frac{14}{70} \times \frac{100\%}{1} = \frac{1400\%}{70} = 20\%$$

*Note This method of calculation will be used.

Q	A
1) The price of a book is decreased by £4 from £9. What is the percentage decrease in price?	1) $\frac{4}{9} \times \frac{100\%}{1} = \frac{400\%}{9} = 44\frac{4}{9}\% = 44.44...\%$
2) The number of pupils in a school rises by 16 from 800. By what percentage has the number of pupils risen?	2) $\frac{16}{800} \times \frac{100\%}{1} = \frac{1600\%}{800} = 2\%$
3) Calculate the percentage increase in water when 38 litres are added to a tank which already contains 40 litres.	3) $\frac{38}{40} \times \frac{100\%}{1} = \frac{3800\%}{40} = 95\%$

The **actual change** or **original value** may have to be calculated before the percentage change is calculated.

Ex.1 The price of a pen increases from 70p to 84p. By what percentage has the price increased?

actual increase = 84 - 70 = 14p

$$\frac{14}{70} \times \frac{100\%}{1} = \frac{1400\%}{70} = 20\%$$

Q	A
1) The price of a lamp is reduced from £25 to £10. Calculate the percentage decrease in price?	1) actual decrease = 25 - 10 = £15 $\frac{15}{25} \times \frac{100\%}{1} = \frac{1500\%}{25} = 60\%$
2) By using lighter materials in making a machine, its weight is reduced by 3·6 kg to 8·4 kg. What is the percentage reduction in weight?	2) original value = 3·6 + 8·4 = 12 kg $\frac{3.6}{12} \times \frac{100\%}{1} = \frac{360\%}{12} = 30\%$
3) It is planned to increase the length of a 99m path to 103·95m. By what percentage will the length of the path be increased?	3) actual increase = 103·95 - 99 = 4·95m $\frac{4.95}{99} \times \frac{100\%}{1} = \frac{495\%}{99} = 5\%$
4) What is the percentage increase in temperature when it rises 18 °F to 20 °F.	4) original value = 20 - 18 = 2 °F $\frac{18}{2} \times \frac{100\%}{1} = \frac{1800\%}{2} = 900\%$
5) The crowd capacity of a stadium is reduced by 160 to 39,840. Find the percentage reduction in the crowd capacity.	5) original value = 160 + 39,840 = 40,000 Cancel by 8,000 $\frac{160}{40,000} \times \frac{100\%}{1} = \frac{16,000\%}{40,000} = \frac{2}{5}\% = 0.4\%$
6) A car's speed changes to 76 kmph from 67 kmph. State the percentage change in the speed of the car.	6) actual change = 76 - 67 = 9 kmph $\frac{9}{67} \times \frac{100\%}{1} = \frac{900\%}{67} = 13\frac{29}{67}\% = 13.43...\%$ increase

Percentages 21 — Inflation - Deflation.

Inflation - Deflation

Inflation is the **rate of increase in prices** of goods and services - expressed **as a percentage**.
Deflation is the **rate of decrease in prices** of goods and services - expressed **as a percentage**.

The terms are used to relate to the general state of the country's economy.

The Consumer Prices Index (CPI) and the Retail Prices Index (RPI) are the 2 main measures in the UK. Each records the prices of 1,000's of things that people commonly spend money on - food, clothes, water, gas, electricity, rent, appliances, bus/train fares, cars, petrol, sweets, etc. - and tracks any changes in the prices over a given period, typically 1 year.

An 'average' figure is then produced which states in general whether there is **inflation** or **deflation** in the economy - this is usually 'around' 5%.

The more that is spent on an item, the more importance or 'weight' is given to it in the calculation - for example, more of a person's income is spent on heating than postage stamps.

Percentage calculations are similar to those already explained.

Inflation   percentage increase = $\dfrac{\text{price increase}}{\text{last year's price}} \times 100\%$

Deflation   percentage decrease = $\dfrac{\text{price decrease}}{\text{last year's price}} \times 100\%$

Ex.1 The price of a loaf is increased by 4p from 80p. Based on this item alone, what is the rate of **inflation**?

$\dfrac{4}{80} \times \dfrac{100\%}{1} = \dfrac{400\%}{80} = 5\%$

Q	A
1) A train fare is increased by £2 from £50. Based on this item, what is the rate of **inflation**?	1) $\dfrac{2}{50} \times \dfrac{100\%}{1} = \dfrac{200\%}{50} = 4\%$
2) Petrol is reduced from £1·20 per litre by 3p. Based on this item, what is the rate of **deflation**?	2) $\dfrac{3p}{120p} \times \dfrac{100\%}{1} = \dfrac{300\%}{120} = 2\dfrac{60}{120} = 2\dfrac{1}{2}\%$
3) A packet of sweets is raised by 7p from £1. Based on this item, what is the rate of **inflation**?	3) $\dfrac{7p}{100p} \times \dfrac{100\%}{1} = \dfrac{700\%}{100} = 7\%$
4) Water rates are reduced from £600 by £20. Based on this item, what is the rate of **deflation**?	4) $\dfrac{20}{600} \times \dfrac{100\%}{1} = \dfrac{2000\%}{600} = 3\dfrac{200}{600} = 3\dfrac{1}{3}\%$

Ex.2 If the rate of **inflation** is 5%, find the price of a toy based on last year's price of £9.

5% of £9
$\dfrac{5}{100} \times \dfrac{9}{1} = \dfrac{45}{100} = 0.45$

£9
+ add
0·45   Increase
———
£9·45

Q	A
1) If the rate of **inflation** is 6%, find the price of a can of beans based on last year's price of 50p.	1) 6% of 50p $\dfrac{6}{100} \times \dfrac{50}{1} = \dfrac{300}{100} = 3$ 50 + add 3 Increase —— 53p
2) If the rate of **deflation** is 4%, find the price of a computer based on last year's price of £325.	2) 4% of £325 $\dfrac{4}{100} \times \dfrac{325}{1} = \dfrac{1300}{100} = 13$ £325 − subtract 13 Decrease —— £312
3) If the rate of **inflation** is 2%, find the price of house rent based on last year's price of £120 p/wk.	3) 2% of £120 $\dfrac{2}{100} \times \dfrac{120}{1} = \dfrac{240}{100} = 2.40$ £120 + add 2·40 Increase ——— £122·40

Percentages 22 — Profit / Loss.

Profit / Loss
The **cost price** of an item is the cost of making or first buying it.
The **selling price** of an item is the price it is then sold for.
A **profit** is made if the **selling price** is **more than** the **cost price** ... **Profit = Selling Price - Cost Price**
A **loss** is made if the **cost price** is **more than** the **selling price** ... **Loss = Cost Price - Selling Price**

*Note In the following Exs. and Q/A, s.p. = selling price   c.p. = cost price

Ex.1
The selling price of a toy is £8.
The cost price is £5.
What is the profit?

s.p.   = £8
−c.p.   = £5
profit = £3

Ex.2
The cost price of a plant is £8.
The selling price is £5.
What is the loss?

c.p.   = £8
−s.p.   = £5
loss   = £3

Q	A	Q	A
1) The selling price of a pen is £6. The cost price is £4. What is the profit?	1) s.p. = £6 −c.p. = £4 profit = £2	4) A car is bought for £9,999 and sold for £5,678. What is the loss?	4) c.p. = £9,999 −s.p. = £5,678 loss = £4,321
2) The cost price of a tie is £5. The selling price is £2. What is the loss?	2) c.p. = £5 −s.p. = £2 loss = £3	5) If a magazine which cost 61p is sold for 88p, what is the profit?	5) s.p. = 88p −c.p. = 61p profit = 27p
3) A hat is bought for £7·50 and sold for £14·90. Find the profit.	3) s.p. = £14·90 −c.p. = £ 7·50 profit = £ 7·40	6) A diary is bought for £1·70 and sold for £0·85. Calculate the loss.	6) c.p. = £1·70 −s.p. = £0·85 loss = £0·85

The **cost price** or **selling price** may have to be calculated.

Ex.1
The selling price of a game is £9.
The profit is £1.
What is the cost price?

s.p.   = £9
−profit = £1
c.p.   = £8

Ex.2
The cost price of a carpet is £56.
The profit is £43.
What is the selling price?

c.p.   = £56
+profit = £43
s.p.   = £99

Q	A	Q	A
1) The selling price of a pan is £7. The profit is £3. What is the cost price?	1) s.p. = £7 −profit = £3 c.p. = £4	4) An iron is bought for £5·90 and sold at a loss of £2·85. What is the selling price?	4) c.p. = £5·90 −loss = £2·85 s.p. = £3·05
2) The cost price of a rug is £15. The profit is £75. Calculate the selling price.	2) c.p. = £15 +profit = £75 s.p. = £90	5) What is the cost price of a van sold for £8,490 at a loss of £490 ?	5) s.p. = £8,490 +loss = £ 490 c.p. = £8,980
3) A bicycle is sold for £62. The loss is £39. What is the cost price?	3) s.p. = £ 62 +loss = £ 39 c.p. = £101	6) Find the selling price of a comic costing 33p and sold at a loss of 33p.	6) c.p. = 33p −loss = 33p s.p. = 0p free!

131

Percentages 23

Percentage Profit / Loss.

Percentage Profit / Loss

To express the **profit / loss** as a **percentage** of the **cost price**, we first express the **profit / loss** as a **fraction** of the **cost price** - then change the fraction to a percentage by **multiplying by 100%**.

$$\text{percentage profit} = \frac{\text{profit}}{\text{cost price}} \times 100\% \qquad \text{percentage loss} = \frac{\text{loss}}{\text{cost price}} \times 100\%$$

*Note Here, the profit / loss is expressed as a percentage of the **cost price** - rather than the **selling price**.

Ex.1 The cost price of a bag is £5. It is sold at a profit of £1. What is the percentage profit?
$\frac{1}{5} \times \frac{100\%}{1} = \frac{100\%}{5} = 20\%$

Ex.2 A vase is bought for £8 and sold at a loss of £6. What is the percentage loss?
$\frac{6}{8} \times \frac{100\%}{1} = \frac{600\%}{8} = 75\%$

Q	A
1) The cost price of a C.D. is £4. It is sold at a profit of £2. What is the percentage profit?	1) $\frac{2}{4} \times \frac{100\%}{1} = \frac{200\%}{4} = 50\%$
2) Tables are bought for £60 and sold at a loss of £36. Calculate the percentage loss.	2) $\frac{36}{60} \times \frac{100\%}{1} = \frac{3600\%}{60} = 60\%$
3) £560 loss is made when a painting is sold which cost £7,000 to buy. What is the percentage loss?	3) $\frac{560}{7,000} \times \frac{100\%}{1} = \frac{56,000\%}{7,000} = 8\%$

The **actual profit / loss** or **cost** may have to be calculated before the percentage profit / loss is calculated.

Ex.1 A box which cost 10p to make is sold for 17p. What is the percentage profit?
actual profit = 17 - 10 = 7p
$\frac{7}{10} \times \frac{100\%}{1} = \frac{700\%}{10} = 70\%$

Ex.2 A box is sold for 17p giving a loss of 3p. What is the percentage loss?
cost = 17 + 3 = 20p
$\frac{3}{20} \times \frac{100\%}{1} = \frac{300\%}{20} = 15\%$

Q	A
1) The selling price of a bracelet is £100 making a profit of £20. Calculate the percentage profit.	1) cost price = 100 - 20 = £80 $\frac{20}{80} \times \frac{100\%}{1} = \frac{2000\%}{80} = 25\%$
2) A watch is bought for £40 but sold for only £28. What is the percentage loss?	2) actual loss = 40 - 28 = £12 $\frac{12}{40} \times \frac{100\%}{1} = \frac{1200\%}{40} = 30\%$
3) Chocolate bars are sold for 81p making a loss of 9p. Calculate the percentage loss.	3) cost price = 81 + 9 = 90p $\frac{9}{90} \times \frac{100\%}{1} = \frac{900\%}{90} = 10\%$
4) The cost of making a ball is £3·20. If it is sold for £6 what is the percentage profit?	4) actual profit = 6 - 3·20 = £2·80 $\frac{2·80}{3·20} \times \frac{100\%}{1} = \frac{280\%}{3·20} = 87·5\%$
5) What is the percentage loss on a car with a selling price of £3,960 second hand but which cost £9,000 when new?	5) actual loss = 9,000 - 3,960 = £5,040 $\frac{5040}{9000} \times \frac{100\%}{1} = \frac{504000\%}{9000} = 56\%$
6) Dice are made with a selling price of 8p giving a profit of 1p. State the percentage profit.	6) cost price = 8 - 1 = 7p $\frac{1}{7} \times \frac{100\%}{1} = \frac{100\%}{7} = 14\frac{2}{7}\% = 14·28...\%$

Percentages 24   Credit / Higher Purchase.

Credit / Higher Purchase

A business may allow a customer to purchase an item by making several regular 'part' payments / 'repayments' over a period of time rather than making just one 'full' payment for the whole amount.
A **deposit** - an initial payment - may be required immediately and may be a percentage of its price.

This method of payment is called '**buying on Credit**' / '**buying on Hire Purchase**' - often called '**H.P.**'
(**Credit** is providing the funds to allow the customer to make the purchase.
  Think of **Hire Purchase** as '**Hire** the item but **Purchase** the item - eventually - **as well !**')

It allows for the use of the item while it is still being payed for and is usually offered for 'more expensive' items which a customer may not be able to afford immediately.
It may also be offered for 'less expensive' items ...
for example, if several are bought or if a business uses a catalogue to sell its products.

In general, a business will **add a charge** for such a purchase since it is lending the customer the funds to make the purchase immediately.
The charge is added to the cost of the item and forms part of the repayments.

Difference between '**cash price**' and '**credit price**'.
It is useful to know what the **difference** is between paying the full amount immediately
- usually called the '**cash price**', since it can be paid for with cash
- and paying over a period of time (weeks / months / years) - usually called the '**credit price**'.
Some types of calculations are considered here - others are possible.

Ex. The **cash price** of a television is £200.
The **credit price** is 6 monthly payments of £40.

Calculate the difference between the cash price and the credit price.

```
              £ 40
           ×     6
credit price = £240
cash price   = −£200
difference   = £ 40
```

Q	A
1) The **cash price** of a bicycle is £150. The **credit price** is 9 weekly payments of £20. Calculate the difference between the cash price and the credit price.	1) £ 20 × 9 credit price = £180 cash price = −£150 difference = £ 30
2) The **cash price** of a computer is £487. The **credit price** is a **deposit** of £100 followed by 12 monthly payments of £38·30. Calculate the difference between the cash price and the credit price.	2) deposit £100·00 £ 38·30 × 12 payments = £459·60 ⇨ +£459·60 credit price = £559·60 cash price = −£487·00 difference = £ 72·60
3) The cash price of a tent is £160. The tent can also be bought on Hire Purchase with a **deposit of 10% of the cash price** followed by 8 weekly payments of £23. Calculate the difference between the cash price and the H.P. price.	3) deposit 10% of £160 $\frac{10}{100} \times \frac{160}{1} = \frac{1600}{100} = £16$ £ 23 × 8 payments = £184 ⇨ +£184 H.P. price = £200 cash price = −£160 difference = £ 40

Percentages 25 — Credit / Higher Purchase.

Q	A
4) The cash price of a washing machine is £450. The credit price is a **deposit of 20% of the cash price** followed by 12 monthly payments of £45. Calculate the difference between the cash price and the credit price.	4) deposit 20% of £450 $\frac{20}{100} \times \frac{450}{1} = \frac{9000}{100} = £90$ £45 × 12 payments = £540 ⇒ +£540 credit price = £630 cash price = −£450 difference = £180
5) The **cash price** of a car is £6,400. The credit price is a **deposit of 30% of the cash price** followed by 36 monthly payments of £200. Calculate the difference between the cash price and the credit price.	5) deposit 30% of £6400 $\frac{30}{100} \times \frac{6400}{1} = \frac{192000}{100} = £1920$ £200 × 36 payments = £7200 ⇒ +£7200 credit price = £9120 cash price = −£6400 difference = £2720
6) The cash price of exercise equipment is £800. The H.P. price is a **deposit of 25% of the cash price** followed by 16 weekly payments of £50. a) Calculate the difference between the cash price and the H.P. price. b) Express the **difference** as a percentage of the **cash price**.	6) a) deposit 25% of £800 $\frac{25}{100} \times \frac{800}{1} = \frac{20000}{100} = £200$ £50 × 16 payments = £800 ⇒ +£800 H.P. price = £1000 cash price = −£800 difference = £200 b) $\frac{\text{difference}}{\text{cash price}} \times 100\%$ $\frac{200}{800} \times \frac{100\%}{1} = \frac{20000}{800} = 25\%$
7) The cash price of a boat is £3,000. The credit price is a **deposit of 50% of the cash price** followed by 30 monthly payments of £65. a) Calculate the difference between the cash price and the credit price. b) Express the **difference** as a percentage of the **cash price**.	7) a) deposit 50% of £3,000 $\frac{50}{100} \times \frac{3,000}{1} = \frac{150000}{100} = £1500$ £65 × 30 payments = £1950 ⇒ +£1950 credit price = £3450 cash price = −£3000 difference = £450 b) $\frac{\text{difference}}{\text{cash price}} \times 100\%$ $\frac{450}{3000} \times \frac{100\%}{1} = \frac{45000}{3000} = 15\%$

Percentages 26   Wage / Salary.

Wage and Salary

An employer usually pays an employee either ...

a **wage** - usually an amount per hour of work, usually paid weekly.
  (employees - for ex., nurse, driver, shop assistant, factory worker.)

a **salary** - usually a yearly amount, paid monthly (the exact number of hours of work are not specified).
  (employees - for ex., doctor, teacher, office manager, architect, M.P.)

The following terms are used in relation to Wage and Salary:

Gross Pay	The original pay - the employee **earns** the amount as **income**.
Deductions	Amounts deducted from the Gross Pay. [ex. Income Tax, National Insurance (N.I.) - both are 'Government payments'.]
Tax Allowance	The pay allowed to be earned before **Income Tax** is deducted. [ex. different allowance if person is single, married, has child.]
Taxable Pay	The pay which is **taxed** ... (Gross Pay − Tax Allowance).
Tax Rate	The tax deducted from each £1 of Taxable Pay (usually stated as a percentage).
Nett Pay	The amount left **after deductions** which the employee **actually receives**.

Q	A

1) Miss A. works a 40 hour week and earns £8 per hour.
 Calculate her pay for the week.
	1) £ 8
	    × 40hrs
	    £320

2) Mr. B. earns £517 per week.
 Calculate his pay for a 52-week year.
	2) £ 517
	    × 52weeks
	    1 034
	 + 25 850
	 £26,884

3) Mrs. C. earns a salary of £2,900 per month. Calculate her salary for the year.	3) £ 2,900 × 12months £34,800
4) Miss. D. earns an annual salary of £21,600. Calculate her salary for a month.	4) £ 1,8 0 0 per month 12 ⟌ £2 1⁹6 0 0 months

5) Mr. E. works a 35 hour week and earns £9 per hour.
 He pays £42 Income Tax and £18 N.I.
 	
 Calculate a) the gross pay for the week.
            b) the total deductions.
            c) the nett pay.
	5)a) £ 9 b) Income Tax £42
	      × 35hrs + N.I. + 18
	      £315 deductions £60
	
	 c) gross pay £315
	     − deductions − £ 60
	     nett pay £255

6) Mrs. F. earns a monthly salary of £4k.
 She pays £956 Income Tax and £240 N.I.
 	
 Calculate a) the total deductions.
            b) the monthly nett pay.
	6) (£1k = £1,000 £4k = **£4,000**)
	
	 a) Income Tax £ 956
	     + N.I. + £ 240
	     deductions £1,196
	
	 b) gross pay **£4,000**
	     − deductions − £1,196
	     nett pay £2,804

Percentages 27 Wage / Salary.

Q	A

7) Miss G. earns £300 per week.
Her Income Tax allowance is £100.
The tax rate is 10% on her taxable pay.
She also pays £25 N.I.
 	
Calculate a) the taxable pay.
b) the tax paid.
c) the total deductions.
d) the nett pay.
	7) a) gross pay £300
	   − tax allowance − £100
	   taxable pay £200
	
	b) tax = 10% of £200
	   = $\frac{10}{100} \times \frac{200}{1} = \frac{2000}{100}$ = £20
	
	c) Income Tax £20
	   + N.I. + £25
	   deductions £45
	
	d) gross pay £300
	   − deductions − £ 45
	   nett pay £255

8) Mr. H. earns £670 per week.
His Income Tax allowance is £270.
The tax rate is 20% on the taxable pay.
He pays £59 N.I.
 	
Calculate a) the taxable pay.
b) the tax paid.
c) the total deductions.
d) the nett pay.
	8) a) gross pay £670
	   − tax allowance − £270
	   taxable pay £400
	
	b) tax = 20% of £400
	   = $\frac{20}{100} \times \frac{400}{1} = \frac{8000}{100}$ = £80
	
	c) Income Tax £ 80
	   + N.I. + £ 59
	   deductions £139
	
	d) gross pay £670
	   − deductions − £139
	   nett pay £531

9) Mrs. I. earns a salary of £3,000 per month.
Her Income Tax allowance is £1,000.
The tax rate is 28% on the taxable pay.
She pays £210 N.I.
 	
Calculate a) the taxable pay.
b) the tax paid.
c) the total deductions.
d) the nett pay.
	9) a) gross pay £3,000
	   − tax allowance − £1,000
	   taxable pay £2,000
	
	b) tax = 28% of £2000
	   = $\frac{28}{100} \times \frac{2000}{1} = \frac{56000}{100}$ = £560
	
	c) Income Tax £560
	   + N.I. + £210
	   deductions £770
	
	d) gross pay £3,000
	   − deductions − £ 770
	   nett pay £2,230

10) Miss. J. earns a basic wage of £150 per week and 10% commission on the sales she makes.
 	
If her sales for the week are £4,000 what is her gross pay?
	10) Commission
	   10% of £4000
	   $\frac{10}{100} \times \frac{4000}{1} = \frac{40000}{100}$ = £400
	
	   basic pay £150
	   + commission + £400
	   gross pay £550

Percentages 28   Value Added Tax (V.A.T.)

V.A.T.
V.A.T. (or VAT) stands for Value Added Tax and is a tax added to most goods and services.
The tax is set by the government and was introduced in the U.K. in 1973.

A business with a given minimum sales must register and charge V.A.T.

Goods and services fit into the following 5 V.A.T. categories - (examples in each category are given)

Standard Rate	(Exs. clothes, cars, petrol, sweets, bicycles, toys)
Reduced Rate	(Exs. domestic electricity / gas, carry cots, alter an empty home)
Zero Rate	(Exs. most food, child's clothing, household water, books)
Exempt from V.A.T.	(Exs. school education, health care, postage stamps, insurance)
Outside the scope of V.A.T.	(Exs. charity donations, welfare services, public toll bridge)

*Note
In the following Exs. and Q/A the VAT rates may be different than the actual rates set at the present time.

Ex.
A T.V. is sold for £200 + VAT.
Calculate
a) the VAT at 20%
b) the total price.

a) VAT
   20% of £200
   $= \dfrac{20}{100} \times \dfrac{£200}{1} = \dfrac{4000}{100} = £40$
   or
   $= 0.2 \times £200 = £40$

b) price   £200
   + VAT   + £ 40
   total   £240

Q / A

1) A bicycle's price is £300 + VAT.
Calculate
a) the VAT at 20%
b) the total price.

1) a) VAT
   20% of £300
   $= \dfrac{20}{100} \times \dfrac{£300}{1} = \dfrac{6000}{100} = £60$
   or
   $= 0.2 \times £300 = £60$

   b) price   £300
      + VAT   + £ 60
      total   £360

2) A phone bill is £124 + VAT.
Calculate
a) the VAT at 15%
b) the total bill.

2) a) VAT
   15% of £124
   $= \dfrac{15}{100} \times \dfrac{£124}{1} = \dfrac{1860}{100} = £18.60$
   or
   $= 0.15 \times £124 = £18.60$

   b) cost   £124.00
      + VAT   + £ 18.60
      total   £142.60

3) A carry cot sells for £89 + VAT.
Calculate
a) the VAT at 10%
b) the total price.

3) a) VAT
   10% of £89
   $= \dfrac{10}{100} \times \dfrac{£89}{1} = \dfrac{890}{100} = £8.90$
   or
   $= 0.1 \times £89 = £8.90$

   b) price   £89.00
      + VAT   + £ 8.90
      total   £97.90

4) A gas bill is for £61·01 + VAT.
Calculate
a) the VAT at 5%
b) the total cost.

4) a) VAT
   5% of £61·01
   $= \dfrac{5}{100} \times \dfrac{£61.01}{1} = \dfrac{305.05}{100} = £3.0505$
   or
   $= 0.05 \times £61.01 = £3.0505$
   $= £3.05$ (to the nearest 1p)

   b) cost   £61·01
      + VAT   + £ 3·05
      total   £64·06

Percentages 29

Original Value - now includes V.A.T.

<u>The Original Value of an item that now includes V.A.T.</u>
*Note **Percentages 17-18** explain similar calculations.

The **original value** of an item that now includes V.A.T. can be calculated.
The **original value** is considered to be **100%**.

The **VAT percentage** increase is added to **100%** -
the **new percentage** is then used to calculate the value required.

The <u>method</u> of **ratio / direct proportion** is shown.

Ex. The price of a pen is £9 and includes 20% VAT.

Calculate
a) the price before VAT is added.

b) the VAT.

a) original value 100%
    add VAT   + 20%
    total value   120% = £9

   $1\% = \dfrac{9}{120}$

   $100\% = \dfrac{9}{120} \times 100$

   $= \dfrac{900}{120}$

   $= £7{\cdot}50$

b) VAT

   $1\% = \dfrac{9}{120}$

   $20\% = \dfrac{9}{120} \times 20$

   $= \dfrac{180}{120}$

   $= £1{\cdot}50$

or

b) VAT

   price including VAT   £9·00
   − price excluding VAT   − £7·50
                     VAT   £1·50

Q	A

1) The price of a computer is £600 and includes VAT at 20%.
 	
Calculate
a) the price before VAT is added.
 	
b) the VAT.
	1) a) 100%
	       + 20%
	       120% = £600
	
	$1\% = \dfrac{600}{120}$
	
	$100\% = \dfrac{600}{120} \times 100$
	
	$= \dfrac{60000}{120}$
	
	$= £500$
	
	b) price including VAT £600
	− price excluding VAT − £500
	                  VAT £100
	
	or
	
	$1\% = \dfrac{600}{120}$
	
	$20\% = \dfrac{600}{120} \times 20$
	
	$= \dfrac{12000}{120}$
	
	$= £100$

Percentages 30

Original Value - now includes V.A.T.

Q	A	
2) The price of a bicycle is £96 and includes VAT at 20%. Calculate a) the price before VAT is added. b) the VAT.	2) a) $\quad$ 100% $\quad +\ 20\%$ $\quad \overline{120\%} = £\,96$ $\quad 1\% = \dfrac{96}{120}$ $\quad 100\% = \dfrac{96}{120} \times 100$ $\quad\quad\quad = \dfrac{9600}{120}$ $\quad\quad\quad = £80$	b) $\quad$ price including VAT $\quad$ £96 $\quad -$ price excluding VAT $\quad -£80$ $\quad\quad\quad\quad\quad\quad$ VAT $\quad\quad$ £16 or $\quad 1\% = \dfrac{96}{120}$ $\quad 20\% = \dfrac{96}{120} \times 20$ $\quad\quad\quad = \dfrac{1920}{120}$ $\quad\quad\quad = £16$
3) A car is sold for £9,350 including VAT at 10%. Calculate a) the cost before VAT is added. b) the VAT.	3) a) $\quad$ 100% $\quad +\ 10\%$ $\quad \overline{110\%} = £\,9350$ $\quad 1\% = \dfrac{9350}{110}$ $\quad 100\% = \dfrac{9350}{110} \times 100$ $\quad\quad\quad = \dfrac{935000}{110}$ $\quad\quad\quad = £8,500$	b) $\quad$ cost including VAT $\quad$ £9,350 $\quad -$ cost excluding VAT $\quad -£8,500$ $\quad\quad\quad\quad\quad\quad$ VAT $\quad\quad$ £ 850 or $\quad 1\% = \dfrac{9350}{110}$ $\quad 10\% = \dfrac{9350}{110} \times 10$ $\quad\quad\quad = \dfrac{93500}{110}$ $\quad\quad\quad = £850$
4) A gas bill for £129·57 includes VAT at 5%. Calculate a) the cost before VAT is added. b) the VAT.	4) a) $\quad$ 100% $\quad +\ 5\%$ $\quad \overline{105\%} = £\,129\cdot57$ $\quad 1\% = \dfrac{129\cdot57}{105}$ $\quad 100\% = \dfrac{129\cdot57}{105} \times 100$ $\quad\quad\quad = \dfrac{12957}{105}$ $\quad\quad\quad = £123\cdot40$	b) $\quad$ cost including VAT $\quad$ £129·57 $\quad -$ cost excluding VAT $\quad -£123\cdot40$ $\quad\quad\quad\quad\quad\quad$ VAT $\quad\quad$ £ 6·17 or $\quad 1\% = \dfrac{129\cdot57}{105}$ $\quad 5\% = \dfrac{129\cdot57}{105} \times 5$ $\quad\quad\quad = \dfrac{647\cdot85}{105}$ $\quad\quad\quad = £6\cdot17$
5) The price of a chess game is £5 and includes VAT at 17·5%. Calculate a) the price before VAT is added. b) the VAT.	5) a) $\quad$ 100% $\quad +\ 17\cdot5\%$ $\quad \overline{117\cdot5\%} = £\,5$ $\quad 1\% = \dfrac{5}{117\cdot5}$ $\quad 100\% = \dfrac{5}{117\cdot5} \times 100$ $\quad\quad\quad = \dfrac{500}{117\cdot5}$ $\quad\quad\quad = £4\cdot255\ldots$ $\quad\quad\quad = £4\cdot26$ $\quad\quad\quad$ (nearest 1p)	b) $\quad$ price including VAT $\quad$ £5·00 $\quad -$ price excluding VAT $\quad -£4\cdot26$ $\quad\quad\quad\quad\quad\quad$ VAT $\quad\quad$ £0·74 or $\quad 1\% = \dfrac{5}{117\cdot5}$ $\quad 17\cdot5\% = \dfrac{5}{117\cdot5} \times 17\cdot5\%$ $\quad\quad\quad = \dfrac{87\cdot5}{117\cdot5}$ $\quad\quad\quad = £0\cdot744\ldots$ $\quad\quad\quad = £0\cdot74$ $\quad\quad\quad$ (nearest 1p)

Percentages 31 — Simple Interest - Formula.

Interest on Money
A bank / building society will a) **pay interest** to a customer who **saves / invests** money in it
b) **charge interest** to a customer who **loans / borrows** money from it.

The amount of money that a customer saves / invests or loans / borrows is called the '**principal**'.
The **interest** is a **sum of money** which is calculated as a **percentage** of the **principal**
and is paid / charged for each full **year** (or part year) of an agreed period.

Simple Interest and Compound Interest
There are 2 types of interest -
Simple Interest - the interest is the **same** for each year - (this is **simple !**)
Compound Interest - the interest is **different** for each year - (this is **compound / not simple !**)

Simple Interest - Formula
The interest is the **same** for each year.

Ex.1 Calculate the Simple Interest charged to loan £50 for **1 year** at 10% per year.

The calculation is
10% of £50

$$\text{Interest} = \frac{10}{100} \times \frac{£50}{1} = \frac{£500}{100} = £5$$

*Note
If the loan was for **2 years,** the Interest of £5 would be **charged again** ...

Year 1 Interest = £ 5
+ Year 2 Interest = £ 5
Total Interest = £10

The calculation can then be set out as in Ex.2

Ex.2 Calculate the Simple Interest charged to loan £50 for **2 years** at 10% per year.

The calculation is
10% of £50 for **2** years

$$\text{Interest} = \frac{10}{100} \times \frac{£50}{1} \times \frac{2}{1} = \frac{£1000}{100} = £10$$

However, it is usual to **rearrange the calculation** and present it as follows ...

$$\text{Interest} = \frac{£50 \times 10 \times 2}{100} = \frac{£1000}{100} = £10$$

... **and in general** the **formula** is written ...

$$I = \frac{P \times R \times T}{100}$$

The letters represent ...
I = Simple Interest (the **total** amount of money paid / charged)
P = Principal (the amount of money invested / borrowed)
R = Rate of interest (the percentage of the Principal to be calculated each year -
           this may be written in Latin as '**per annum**' or just '**p.a**')
T = Time of investment / loan (in years)

Q	A
1) Calculate the Simple Interest charged to borrow £80 for 3 years at 10% per annum.	1) $I = \frac{£80 \times 10 \times 3}{100} = \frac{£2400}{100} = £24$
2) Calculate the Simple Interest paid on a loan of £7,500 for 4 years at 6% per year.	2) $I = \frac{£7500 \times 6 \times 4}{100} = \frac{£180000}{100} = £1,800$
3) £961 is invested for 6 years at 7% p.a. Calculate the Simple Interest received.	3) $I = \frac{£961 \times 6 \times 7}{100} = \frac{£40362}{100} = £403 \cdot 62$
4) Calculate the Simple Interest received on savings of £333 for $5\frac{1}{2}$ years at 2·9% p.a.	4) $I = \frac{£333 \times 2 \cdot 9 \times 5 \cdot 5}{100} = \frac{£5311 \cdot 35}{100} = £53 \cdot 1135$ $= £53 \cdot 11$ (nearest 1p)

Percentages 32

Simple Interest - Formula.

Simple Interest

The Simple Interest **formula** can also be used to find P (the Principal), R (the Rate) or T (the Time).
It is useful to **rearrange the formula** to keep the 'unknown' value (P / R / T) on the left of the = sign.

$$I = \frac{P \times R \times T}{100} \quad \Rightarrow \quad \frac{P \times R \times T}{100} = I$$

We substitute the given values into the formula and **solve the equation**.

Ex. Calculate the sum invested that would receive £12 Simple Interest in 4 years at 5% per annum.

(calculate P)

$$\frac{P \times R \times T}{100} = I$$

$$\frac{P \times 5 \times 4}{100} = £12$$

... reminder ... 'Rules for Equations' ...

IF a value moves across the equal sign we reverse the operation involving that value.

$$\frac{20P}{100} = £12$$

$$P = \frac{£12 \times 100}{20}$$

$$P = £60$$

The sum invested is £60.

Q	A
1) Calculate the sum borrowed that would be charged £96 Simple Interest in 3 years at 8% per annum.	1) (calculate P) $$\frac{P \times R \times T}{100} = I$$ $$\frac{P \times 8 \times 3}{100} = £96$$ $$\frac{24P}{100} = £96$$ $$P = \frac{£96 \times 100}{24}$$ $$P = £400$$ The sum borrowed is £400.
2) Calculate the rate of Simple Interest needed to give interest of £198 on savings of £550 in 6 years.	2) (calculate R) $$\frac{P \times R \times T}{100} = I$$ $$\frac{£550 \times R \times 6}{100} = £198$$ $$\frac{£3300R}{100} = £198$$ $$R = \frac{198 \times 100}{3300}$$ $$R = 6$$ The rate of interest is 6%.

Percentages 33 — Simple Interest - Formula.

Q	A
3) Calculate the time required for an investment of £7,000 to produce £1,400 interest if 5% p.a. Simple Interest is offered by a bank.	3) (calculate T) $$\frac{P \times R \times T}{100} = I$$ $$\frac{£7000 \times 5 \times T}{100} = £1400$$ $$\frac{£35000T}{100} = £1400$$ $$T = \frac{1400 \times 100}{35000}$$ $$T = 4$$ The time required is 4 years.
4) Calculate the sum borrowed that would be charged £495 Simple Interest in 9 years at 11% per annum.	4) (calculate P) $$\frac{P \times R \times T}{100} = I$$ $$\frac{P \times 11 \times 9}{100} = £495$$ $$\frac{99P}{100} = £495$$ $$P = \frac{£495 \times 100}{99}$$ $$P = £500$$ The sum borrowed is £500.
5) Savings of 6 **grand** produced Simple Interest of £870 in 2 years. Calculate the rate of interest.	5) (1 **grand** = £1,000) (calculate R) $$\frac{P \times R \times T}{100} = I$$ $$\frac{£6000 \times R \times 2}{100} = £870$$ $$\frac{£12000R}{100} = £870$$ $$R = \frac{870 \times 100}{12000}$$ $$R = 7·25$$ The rate of interest is 7·25%.

Percentages 34 — Compound Interest

Compound Interest

The Rate of Interest is the **same** for each year - the interest is **different** for each year.
The **interest** at the end of **each year** is **added to the principal**.
As the **principal increases each year** - **the interest** on the principal **increases each year**.

Ex. Calculate the Compound Interest charged to borrow £50 for **3 years** at 10% per year.

Year 1 The calculation is 10% of **£50**

$$\text{Interest} = \frac{10}{100} \times \frac{£50}{1} = \frac{£500}{100} = £5$$

Principal £50
+ Interest +£ 5
New Principal £55

Year 2 The calculation is 10% of **£55**

$$\text{Interest} = \frac{10}{100} \times \frac{£55}{1} = \frac{£550}{100} = £5.50$$

+ Interest +£ 5·50
New Principal £60·50

Year 3 The calculation is 10% of **£60·50**

$$\text{Interest} = \frac{10}{100} \times \frac{£60.50}{1} = \frac{£605}{100} = £6.05$$

+ Interest +£ 6·05
New Principal £66·55

Total Interest = **£16·55**

Q

1) Calculate the Compound Interest received on savings of £70 for 3 years at 10% per annum.

A

1)
Year 1
$$I = \frac{10}{100} \times \frac{£70}{1} = \frac{£700}{100} = £7$$
P £70
+ I +£ 7
New P £77

Year 2
$$I = \frac{10}{100} \times \frac{£77}{1} = \frac{£770}{100} = £7.70$$
+ I +£ 7·70
New P £84·70

Year 3
$$I = \frac{10}{100} \times \frac{£84.70}{1} = \frac{£847}{100} = £8.47$$
+ I +£ 8·47
New P £93·17

Total Interest = **£23·17**

2) Calculate the Compound Interest charged on a loan of £400 for 3 years at 10% p.a.

2)
Year 1
$$I = \frac{10}{100} \times \frac{£400}{1} = \frac{£4000}{100} = £40$$
P £400
+ I +£ 40
New P £440

Year 2
$$I = \frac{10}{100} \times \frac{£440}{1} = \frac{£4400}{100} = £44$$
+ I +£ 44
New P £484

Year 3
$$I = \frac{10}{100} \times \frac{£484}{1} = \frac{£4840}{100} = £48.40$$
+ I +£ 48·40
New P £532·40

Total Interest = **£132·40**

Percentages 35　　　　　　　　　　　　　　　　　　　　　　　　　　　　　　　　Compound Interest.

Q	A
3) A bank offers a loan of £3,000 at 6% per year Compound Interest for 2 years. a) What is the interest charged? b) What is the total sum repaid? c) The total sum repaid is arranged in 24 monthly payments. State the amount of each payment.	**3)** Year 1 $I = \frac{6}{100} \times \frac{£3000}{1} = £180$　　　　P　£3000 　　　　　　　　　　　　　　　　　　　　+　I　+£ 180 　　　　　　　　　　　　　　　　　　　New P　£3180 Year 2 $I = \frac{6}{100} \times \frac{£3180}{1} = £190.80$　　+　I　+£ 190.80 　　　　　　　　　　　　　　　　　　　New P　£3370.80 Total Interest = £370.80　　　　　b) 　　　　a) a) £ 370.80 b) £3370.80 c) $\frac{£3370.80}{24} = £140.45$
4) A sum of £700 is invested for 3 years at $7\frac{1}{2}$ % per annum Compound Interest. How much interest is received?	**4)** Year 1 $I = \frac{7.5}{100} \times \frac{£700}{1} = £52.50$　　　　P　£700 　　　　　　　　　　　　　　　　　　　+　I　+£ 52.50 　　　　　　　　　　　　　　　　　New P　£752.50 Year 2 　　　　　　　　　　　(£ 56.4375) $I = \frac{7.5}{100} \times \frac{£752.50}{1} = £ 56.44$　+　I　+£ 56.44 　　　　　　　　　　　　　　　　　New P　£808.94 Year 3 　　　　　　　　　　　(£ 60.6705) $I = \frac{7.5}{100} \times \frac{£808.94}{1} = £ 60.67$　+　I　+£ 60.67 　　　　　　　　　　　　　　　　　New P　£869.61 Total Interest = **£169.61**
5) Calculate the Compound Interest received on an investment of £200 for $1\frac{1}{2}$ years at 8% p.a. **Interest is paid every $\frac{1}{2}$ year**.	**5)** 　　　　　　　$T = \frac{1}{2}$ year - **half** the interest is received. $\frac{1}{2}$ year $I = \frac{8}{100} \times \frac{£200}{1} \times \frac{0.5}{1} = £ 8$　　　P　£200 　　　　　　　　　　　　　　　　+　I　+£ 8 　　　　　　　　　　　　　　　New P　£208 1 year $I = \frac{8}{100} \times \frac{£208}{1} \times \frac{0.5}{1} = £ 8.32$　+　I　+£ 8.32 　　　　　　　　　　　　　　　New P　£216.32 $1\frac{1}{2}$ years 　　　　　　　　　　　　(£ 8.6528) $I = \frac{8}{100} \times \frac{£216.32}{1} \times \frac{0.5}{1} = £ 8.65$　+　I　+£ 8.65 　　　　　　　　　　　　　　　New P　£224.97 Total Interest = **£24.97**

Percentages *36 — Compound Interest - Formula.

Compound Interest - Formula

Percentages 34-35 show Method 1 of calculating Compound Interest.
The work involved is lengthy and time consuming - even for just 2 / 3 years.

Method 2 uses a **formula** to calculate the '**Amount**' (= Principal + Total Compound Interest) ...
... so **Total Compound Interest = 'Amount' - Principal**.
This is shorter and quicker - and is especially appropriate for a larger number of years.

The following helps explain **how the formula is arrived at**.

*Note Ex.	⇨	(in general)	Method 3 -
$110\% = 100\% + 10\%$		$100\% + R\%$	**only calculates in Year 3** (the year required)
$= \frac{100}{100} + \frac{10}{100}$		$= \frac{100}{100} + \frac{R}{100}$	Method 4 -
$= 1 + 0.1$		$= 1 + \frac{R}{100}$	(as shown in **Percentages 15**) **calculates each year** to show what the answers to Method 3 would be !
$\mathbf{110\% = 1.1}$			

Ex. Calculate the Compound Interest charged to borrow £5,000 for **3 years** at 10% per year.

Method 3

Year 1

$A = $ 110% of £5000   (Principal) P

Method 4

Year 1

Amount   100%
Increase + 10%
New Amount   110% of £5000

$\frac{110}{100} \times \frac{£5000}{1} = £5500$

Year 2

$A = $ 110% of (Year 1)
$A = $ = 110% of (110% of £5000)

Year 2

Amount   100%
Increase + 10%
New Amount   110% of £5500

$\frac{110}{100} \times \frac{£5500}{1} = £6050$

Year 3

$A = $ 110% of [Year 2]
$A = $ 110% of [110% of (110% of £5000)]
$A = 1.1 \times [1.1 \times (1.1 \times £5000)]$
$A = (1.1)^3 \times £5000 = £6655$

Year 3

Amount   100%
Increase + 10%
New Amount   110% of £6050

$\frac{110}{100} \times \frac{£6050}{1} = £6655$

in general

$A = \left(1 + \frac{R}{100}\right)^n \times P$

Amount repaid	£6655
−Amount borrowed	−£5000
Compound Interest	**£1655**

The **formula** is usually written

$$\boxed{A = P\left(1 + \frac{R}{100}\right)^n}$$

The letters represent ...

$A = $ 'Amount' (Principal + total Compound Interest after **n** years)
$P = $ Principal (the amount of money invested / borrowed)
$R = $ Rate of interest (the percentage of the Principal for each year)
$n = $ number of years (of investment / loan)

Percentages *37 — Compound Interest - Formula.

Compound Interest - Formula
Method 2

Ex. Calculate the Compound Interest charged to borrow £50 for 3 years at 10% per year.

$$A = P\left(1 + \frac{R}{100}\right)^n \qquad P = £50,\ R = 10,\ n = 3$$

$$A = £50\left(1 + \frac{10}{100}\right)^3$$

$$A = £50(1 + 0.1)^3$$

$$A = £50(1.1)^3$$

$$A = £50(1.331)$$

$$A = £66.55$$

Amount	£66.55
− Principal	− £50
Compound Interest	**£16.55**

Q 1) Calculate the Compound Interest received on savings of £70 for 3 years at 10% per annum.

A 1)

$$A = P\left(1 + \frac{R}{100}\right)^n \qquad P = £70,\ R = 10,\ n = 3$$

$$A = £70\left(1 + \frac{10}{100}\right)^3$$

$$A = £70(1 + 0.1)^3$$

$$A = £70(1.1)^3$$

$$A = £70(1.331)$$

$$A = £93.17$$

Amount	£93.17
− Principal	− £70
Compound Interest	**£23.17**

Q 2) Calculate the Compound Interest charged on a loan of £800 for 5 years at 9% p.a.

A 2)

$$A = P\left(1 + \frac{R}{100}\right)^n \qquad P = £800,\ R = 9,\ n = 5$$

$$A = £800\left(1 + \frac{9}{100}\right)^5$$

$$A = £800(1 + 0.09)^5$$

$$A = £800(1.09)^5$$

$$A = £800(1.538...)$$

$$A = £1230.899...$$

$$= £1230.90 \text{ (nearest 1p)}$$

Amount	£1230.90
− Principal	− £800
Compound Interest	**£430.90**

Percentages *38

Compound Interest - Formula.

Q	A
3) Calculate the Compound Interest received on an investment of £6,000 for 9 years at 7% per annum.	3) $A = P\left(1+\dfrac{R}{100}\right)^n$ $P = £6000$, $R = 7$, $n = 9$ $A = £6000\left(1+\dfrac{7}{100}\right)^9$ $A = £6000(1+0.07)^9$ $A = £6000(1.07)^9$ $A = £6000(1.838...)$ $A = £11030.755...$ $= £11030.76$ (nearest 1p) Amount £11030.76 −Principal −£ 6000 Compound Interest £ 5030.76
4) Calculate the Compound Interest received on savings of £255 for 16 years at 14% per annum.	4) $A = P\left(1+\dfrac{R}{100}\right)^n$ $P = £255$, $R = 14$, $n = 16$ $A = £255\left(1+\dfrac{14}{100}\right)^{16}$ $A = £255(1+0.14)^{16}$ $A = £255(1.14)^{16}$ $A = £255(8.137...)$ $A = £2074.998...$ $= £2075$ (nearest 1p) Amount £2075 −Principal −£ 255 Compound Interest £1820
5) Calculate the Compound Interest received on an investment of £1,000 for 8 years at 4·3% per annum.	5) $A = P\left(1+\dfrac{R}{100}\right)^n$ $P = £1000$, $R = 4.3$, $n = 8$ $A = £1000\left(1+\dfrac{4.3}{100}\right)^8$ $A = £1000(1+0.043)^8$ $A = £1000(1.043)^8$ $A = £1000(1.400...)$ $A = £1400.472...$ $= £1400.47$ (nearest 1p) Amount £1400.47 −Principal −£1000 Compound Interest £ 400.47

Percentages 39 — Appreciation / Depreciation.

Appreciation / Depreciation.

Appreciation - the **rise in value** of an **item** ⎫
Depreciation - the **fall in value** of an **item** ⎬ (often called an **asset**) - over a period of time.

Here, the methods of calculation to express a rise / fall in value are **similar to** those used when calculating **compound interest**.

Method 1

Given a **value**, a **percentage of that value** is calculated and either **added to it** (appreciation) or **subtracted from it** (depreciation) at the **end of a year** to give a **new value**.
The **same percentage** of this **new value** is then calculated and again **added / subtracted** at the **end of the next year**, and so on ...
(... a **different percentage** may be calculated - this is not shown here.)

Q	A
1) The value of a violin is £800. It is expected to **appreciate** each year by 12% of its value at the start of the year. Find, after 2 years a) the total appreciation b) its expected new value.	**1)** Year 1 $\text{App} = \dfrac{12}{100} \times \dfrac{£800}{1} = \dfrac{£9600}{100} = £96$ Year 2 $\text{App} = \dfrac{12}{100} \times \dfrac{£896}{1} = \dfrac{£10752}{100} = £107.52$ a) Total Appreciation = £203.52 £ 800 +Appreciation +£ 96 Value £ 896 +Appreciation +£ 107.52 b) New Value £1003.52
2) A painting is valued at £5,000. It is expected to **appreciate** each year by 20% of its value at the start of the year. Calculate, at the end of 3 years a) the total appreciation b) its expected new value.	**2)** Year 1 $\text{App} = \dfrac{20}{100} \times \dfrac{£5000}{1} = £1000$ Year 2 $\text{App} = \dfrac{20}{100} \times \dfrac{£6000}{1} = £1200$ Year 3 $\text{App} = \dfrac{20}{100} \times \dfrac{£7200}{1} = £1440$ a) Total Appreciation = £3640 £ 5000 +Appreciation +£ 1000 Value £ 6000 +Appreciation +£ 1200 New Value £ 7200 +Appreciation +£ 1440 b) New Value £ 8640
3) A camera is bought for £300. It is expected to **depreciate** each year by 9% of its value at the start of the year. Find, at the end of 2 years a) the total depreciation b) its expected new value.	**3)** Year 1 $\text{Dep} = \dfrac{9}{100} \times \dfrac{£300}{1} = £27$ Year 2 $\text{Dep} = \dfrac{9}{100} \times \dfrac{£273}{1} = £24.57$ a) Total Depreciation = £51.57 £300 −Depreciation −£ 27 Value £273 −Depreciation −£ 24.57 b) New Value £248.43
4) The value of a car is £10,000. It is expected to **depreciate** each year by 10% of its value at the start of the year. Calculate, at the end of 3 years a) the total depreciation b) its expected value.	**4)** Year 1 $\text{Dep} = \dfrac{10}{100} \times \dfrac{£10000}{1} = £1000$ Year 2 $\text{Dep} = \dfrac{10}{100} \times \dfrac{£9000}{1} = £900$ Year 3 $\text{Dep} = \dfrac{10}{100} \times \dfrac{£8100}{1} = £810$ a) Total Depreciation = £2710 £10000 −Depreciation −£ 1000 Value £ 9000 −Depreciation −£ 900 New Value £ 8100 −Depreciation −£ 810 b) New Value £ 7290

Percentages *40 — Appreciation / Depreciation - Formula.

Appreciation / Depreciation - Formula
The **Formula for Compound Interest** (Percentages *36) can be **adapted** -
(but it is easier to keep the **same letters**) - to calculate both **Appreciation / Depreciation**.
Method 2 (using **Formula**)

Q	A
1) The value of a gold ring is £300. It is expected to **appreciate** each year by 6% of its value at the start of the year. Find a) its expected new value at the end of 7 years b) the total appreciation.	1) a) $A = P\left(1 + \dfrac{R}{100}\right)^n$ $\quad P = £300,\ R = 6,\ n = 7$ $A = £300\left(1 + \dfrac{6}{100}\right)^7$ $A = £300(1 + 0\cdot06)^7$ $A = £300(1\cdot06)^7$ $A = £300(1\cdot503...)$ $A = £451\cdot089...$ $\quad = £451\cdot09$ (nearest 1p) b) A (new value) £451·09 −P (old value) −£300 **appreciation £151·09**
2) The value of a camera is £500. It is expected to **depreciate** each year by 10% of its value at the start of the year. Find a) its expected new value at the end of 8 years b) the total depreciation.	2) a) $A = P\left(1 - \dfrac{R}{100}\right)^n$ *Note $P = £500,\ R = 10,\ n = 8$ $A = £500\left(1 - \dfrac{10}{100}\right)^8$ $A = £500(1 - 0\cdot1)^8$ $A = £500(0\cdot9)^8$ $A = £500(0\cdot430...)$ $A = £215\cdot233...$ $\quad = £215\cdot23$ (nearest 1p) b) P (old value) £500 −A (new value) −£215·23 **depreciation £284·77**
3) The value of a van is £9,000. It is expected to **depreciate** each year by 8·5% of its value at the start of the year. Find a) its expected new value at the end of 6 years b) the total depreciation.	3) a) $A = P\left(1 - \dfrac{R}{100}\right)^n$ $P = £9000,\ R = 8\cdot5,\ n = 6$ $A = £9000\left(1 - \dfrac{8\cdot5}{100}\right)^6$ $A = £9000(1 - 0\cdot085)^6$ $A = £9000(0\cdot915)^6$ $A = £9000(0\cdot586...)$ $A = £5281\cdot643...$ $\quad = £5281\cdot64$ (nearest 1p) b) P (old value) £9000 −A (new value) −£5281·64 **depreciation £3718·36**

Percentages 41

Percentage Increase / Decrease. Combine 'Multipliers'.

Increase / Decrease of a Value by a Percentage
(*Note The work of **Percentages 16** is now continued.)

Combine 'Multipliers'
A value can be increased / decreased by a **percentage** of itself -
the **original value** is considered to be **100%**.

The **given percentage** increase / decrease is added / subtracted from **100%**.
The **new percentage** - which can be referred to as the '**multiplier**' -
is then used to calculate the value required.
The 'multiplier' is often expressed as a decimal.

We may combine 'multipliers' in order to continue a calculation and reduce the amount of work.
If required, the resulting **change** in a value is found by subtracting the lower value from the higher one.

Ex.1
a) Increase 40 by 10% then
  increase the result by 10%.
b) State the value of the
  combined 'multipliers'.
c) State the value of the total increase.

Method 1

a) original   100%
   add increase + 10%
   calculate   110% of 40
                   'multiplier'

   $\dfrac{110}{100} \times \dfrac{40}{1} = \dfrac{4400}{100} = 44$

   calculate   110% of 44
                   'multiplier'

   $\dfrac{110}{100} \times \dfrac{44}{1} = \dfrac{4840}{100} = 48 \cdot 4$

b) 110% × 110%
   'multiplier' 'multiplier'

   $\dfrac{110}{100} \times \dfrac{110}{100} = 121\%$

c) 48·4
   −40·0
   ─────
    8·4

Method 2

a) original   100%
   add increase + 10%
   calculate   110% of (110% of 40)
                   'multiplier'   'multiplier'

   1·1   ×   1·1   × 40 =
         1·21   × 40 = 48·4

b) 121% = 1·21

c) 48·4
   −40·0
   ─────
    8·4

*Note **Method 2** only is used in the following Q/A.

Q	A

1) a) Increase 60 by 20% then increase the result by 20%.
b) State the value of the combined 'multipliers'.
c) State the value of the total increase.
	1) a) 100%
	   + 20%
	   ─────
	   120% of (120% of 60)
	   'multiplier' 'multiplier'
	
	   1·2 × 1·2 × 60 =
	        1·44 × 60 = 86·4
	
	b) 144% = 1·44
	
	c) 86·4
	   −60·0
	   ─────
	    26·4

Percentages 42 — Percentage Increase / Decrease. Combine 'Multipliers'.

Q	A
2) a) Decrease 80 by 10% then decrease the result by 10%. b) State the value of the combined 'multipliers' of the calculation. c) State the value of the total decrease.	2) a) $\quad$ 100% $\quad\;\;-\;$ 10% $\quad\;\;\;$ 90% of (90% of 80) $\quad$ 'multiplier' 'multiplier' $\quad\;\;\;$ 0·9 $\;\times\;$ 0·9 $\;\times\;$ 80 = $\quad\quad\quad\;\;$ 0·81 $\;\times\;$ 80 = 64·8 b) 81% = 0·81 c) $\;\;$ 80·0 $\;\;-64·8$ $\;\;\;\;15·2$
3) a) Increase 500 by 30% then decrease the result by 30%. b) State the value of the combined 'multipliers' of the calculation. c) State the value of the total change.	3) a) $\;$ decrease $\quad\;\;$ increase $\quad\;\;$ 100% $\quad\quad\quad$ 100% $\;-$ 30% $\quad\quad\;\;+$ 30% $\quad\;\;$ 70% of (130% of 500) $\;$ 'multiplier' 'multiplier' $\quad\;$ 0·7 $\;\times\;$ 1·3 $\;\times\;$ 500 = $\quad\quad\quad$ 0·91 $\;\times\;$ 500 = 455 b) 91% = 0·91 c) $\;\;$ 500 $\;\;-455$ $\quad\;\;$ 45 (decrease)
4) £900 is invested for 4 years at 2·5% Compound Interest per annum. Calculate a) the total amount in the account after 4 years. b) the total Interest received.	4) a) $\;$ 100 % $\;+\;$ 2·5% $\;\;\;$ 102·5% $\;$ 'multiplier' Year $\;$ 1 $\quad\;$ 2 $\quad\;$ 3 $\quad\;$ 4 $\;\;$ 1·025 × 1·025 × 1·025 × 1·025 × £900 = £993·43... or $\;(1·025)^4 \times £900 = \;\;1·103... \times £900 = £993·43...$ b) $\;\;£993·43$ $\;\;-£900·00$ $\;\;£\;\;93·43$
5) The amount of liquid in a container is increased by 10% every hour. State the amount of liquid 8 hours after it was measured to be 80 litres.	5) a) $\;$ 100% $\;+\;$ 10% $\;\;\;$ 110% $\;$ 'multiplier' $\;\;$ 1·1 $\quad\quad\quad\quad$ 8 hours $\quad\quad\quad\quad\;\;\downarrow$ $\quad\quad\;(1·1)^8 \;\times 80 =$ $\quad\quad\;$ 2·143... × 80 = 171·487... litres
6) An employee earns £25,000 per year. An increase of 3% per year is received followed by annual increases of 4% per year and 5% per year. Calculate a) the amount earned per year after 3 years. b) the total increase in pay per year received after 3 years.	6) a) $\;$ 100% $\quad\;\;$ 100% $\quad\;\;$ 100% $\;+\;$ 3% $\quad\;+\;$ 4% $\quad\;+\;$ 5% $\;\;\;$ 103% $\quad\;\;$ 104% $\quad\;\;$ 105% 'multiplier' 'multiplier' 'multiplier' $\;$ 1·03 $\;\times\;$ 1·04 $\;\times\;$ 1·05 $\;\times £25000 =$ $\quad\quad\quad\;\;$ 1·12476 $\quad\quad\;\;\times £25000 = £28119$ b) $\;\;£28119$ $\;\;-£25000$ $\;\;£\;\;3119$
7) The value of a silver ring is £600. It is expected to **appreciate** each year by 15% of its value at the start of the year. Find, at the end of 2 years a) its expected new value b) the total appreciation.	7) a) $\;$ 100% $\;+\;$ 15% $\;\;\;$ 115% $\;$ 'multiplier' Year $\;$ 1 $\quad\;$ 2 $\;$ 1·15 × 1·15 × £600 = $\quad$ 1·3225 × £600 = £793·50 b) $\;\;£793·50$ $\;\;-£600·00$ $\;\;£193·50$
8) A motorbike is valued at £7000. It is expected to **depreciate** each year by 20% of its value at the start of the year. Find, at the end of 7 years a) its expected new value b) the total depreciation.	8) a) $\;$ 100% $\;-\;$ 20% $\;\;\;$ 80% $\;$ 'multiplier' $\quad\quad\quad\quad$ 7 years $\;(0·8)^7 \;\times £7000 =$ $\;$ 0·209... × £7000 = £1468·00(64) b) $\;\;£7000$ $\;\;-£1468$ $\;\;£5532$

Percentages 43

Percentages - Fractions - Decimals.

<u>Percentages - Fractions - Decimals</u>
It is useful to be able to **mentally** interchange the following Percentages / Fractions / Decimals.
Equivalent fractions are included.
To help practice, each of the columns can be the Q/A in turn.

Q/A	Q/A					Q/A	Q/A	Q/A					Q/A
1%	$\frac{1}{100}$					0·01	26%	$\frac{26}{100}$	$\frac{13}{50}$				0·26
2%	$\frac{2}{100}$	$\frac{1}{50}$				0·02	27%	$\frac{27}{100}$					0·27
3%	$\frac{3}{100}$					0·03	28%	$\frac{28}{100}$	$\frac{14}{50}$	$\frac{7}{25}$			0·28
4%	$\frac{4}{100}$	$\frac{2}{50}$	$\frac{1}{25}$			0·04	29%	$\frac{29}{100}$					0·29
5%	$\frac{5}{100}$		$\frac{1}{20}$			0·05	30%	$\frac{30}{100}$	$\frac{15}{50}$	$\frac{6}{20}$	$\frac{3}{10}$		0·3
6%	$\frac{6}{100}$	$\frac{3}{50}$				0·06	31%	$\frac{31}{100}$					0·31
7%	$\frac{7}{100}$					0·07	32%	$\frac{32}{100}$	$\frac{16}{50}$	$\frac{8}{25}$			0·32
8%	$\frac{8}{100}$	$\frac{4}{50}$	$\frac{2}{25}$			0·08	33%	$\frac{33}{100}$					0·33
9%	$\frac{9}{100}$					0·09	34%	$\frac{34}{100}$	$\frac{17}{50}$				0·34
10%	$\frac{10}{100}$	$\frac{5}{50}$	$\frac{2}{20}$	$\frac{1}{10}$		0·1	35%	$\frac{35}{100}$		$\frac{7}{20}$			0·35
11%	$\frac{11}{100}$					0·11	36%	$\frac{36}{100}$	$\frac{18}{50}$	$\frac{9}{25}$			0·36
12%	$\frac{12}{100}$	$\frac{6}{50}$	$\frac{3}{25}$			0·12	37%	$\frac{37}{100}$					0·37
13%	$\frac{13}{100}$					0·13	38%	$\frac{38}{100}$	$\frac{19}{50}$				0·38
14%	$\frac{14}{100}$	$\frac{7}{50}$				0·14	39%	$\frac{39}{100}$					0·39
15%	$\frac{15}{100}$		$\frac{3}{20}$			0·15	40%	$\frac{40}{100}$	$\frac{20}{50}$	$\frac{10}{25}$	$\frac{8}{20}$	$\frac{4}{10}$ $\frac{2}{5}$	0·4
16%	$\frac{16}{100}$	$\frac{8}{50}$	$\frac{4}{25}$			0·16	41%	$\frac{41}{100}$					0·41
17%	$\frac{17}{100}$					0·17	42%	$\frac{42}{100}$	$\frac{21}{50}$				0·42
18%	$\frac{18}{100}$	$\frac{9}{50}$				0·18	43%	$\frac{43}{100}$					0·43
19%	$\frac{19}{100}$					0·19	44%	$\frac{44}{100}$	$\frac{22}{50}$	$\frac{11}{25}$			0·44
20%	$\frac{20}{100}$	$\frac{10}{50}$	$\frac{5}{25}$	$\frac{4}{20}$	$\frac{2}{10}$ $\frac{1}{5}$	0·2	45%	$\frac{45}{100}$		$\frac{9}{20}$			0·45
21%	$\frac{21}{100}$					0·21	46%	$\frac{46}{100}$	$\frac{23}{50}$				0·46
22%	$\frac{22}{100}$	$\frac{11}{50}$				0·22	47%	$\frac{47}{100}$					0·47
23%	$\frac{23}{100}$					0·23	48%	$\frac{48}{100}$	$\frac{24}{50}$	$\frac{12}{25}$			0·48
24%	$\frac{24}{100}$	$\frac{12}{50}$	$\frac{6}{25}$			0·24	49%	$\frac{49}{100}$					0·49
25%	$\frac{25}{100}$		$\frac{5}{20}$		$\frac{1}{4}$	0·25	50%	$\frac{50}{100}$	$\frac{25}{50}$	$\frac{10}{20}$	$\frac{5}{10}$	$\frac{1}{2}$	0·5

Percentages 44

Percentages - Fractions - Decimals

Q/A	Q/A	Q/A
51%	$\frac{51}{100}$	0·51
52%	$\frac{52}{100}$ $\frac{26}{50}$ $\frac{13}{25}$	0·52
53%	$\frac{53}{100}$	0·53
54%	$\frac{54}{100}$ $\frac{27}{50}$	0·54
55%	$\frac{55}{100}$ $\frac{11}{20}$	0·55
56%	$\frac{56}{100}$ $\frac{28}{50}$ $\frac{14}{25}$	0·56
57%	$\frac{57}{100}$	0·57
58%	$\frac{58}{100}$ $\frac{29}{50}$	0·58
59%	$\frac{59}{100}$	0·59
60%	$\frac{60}{100}$ $\frac{30}{50}$ $\frac{15}{25}$ $\frac{12}{20}$ $\frac{6}{10}$ $\frac{3}{5}$	0·6
61%	$\frac{61}{100}$	0·61
62%	$\frac{62}{100}$ $\frac{31}{50}$	0·62
63%	$\frac{63}{100}$	0·63
64%	$\frac{64}{100}$ $\frac{32}{50}$ $\frac{16}{25}$	0·64
65%	$\frac{65}{100}$ $\frac{13}{20}$	0·65
66%	$\frac{66}{100}$ $\frac{33}{50}$	0·66
67%	$\frac{67}{100}$	0·67
68%	$\frac{68}{100}$ $\frac{34}{50}$ $\frac{17}{25}$	0·68
69%	$\frac{69}{100}$	0·69
70%	$\frac{70}{100}$ $\frac{35}{50}$ $\frac{14}{20}$ $\frac{7}{10}$	0·7
71%	$\frac{71}{100}$	0·71
72%	$\frac{72}{100}$ $\frac{36}{50}$ $\frac{18}{25}$	0·72
73%	$\frac{73}{100}$	0·73
74%	$\frac{74}{100}$ $\frac{37}{50}$	0·74
75%	$\frac{75}{100}$ $\frac{15}{20}$ $\frac{3}{4}$	0·75

Q/A	Q/A	Q/A
76%	$\frac{76}{100}$ $\frac{38}{50}$ $\frac{19}{25}$	0·76
77%	$\frac{77}{100}$	0·77
78%	$\frac{78}{100}$ $\frac{39}{50}$	0·78
79%	$\frac{79}{100}$	0·79
80%	$\frac{80}{100}$ $\frac{40}{50}$ $\frac{20}{25}$ $\frac{16}{20}$ $\frac{8}{10}$ $\frac{4}{5}$	0·8
81%	$\frac{81}{100}$	0·81
82%	$\frac{82}{100}$ $\frac{41}{50}$	0·82
83%	$\frac{83}{100}$	0·83
84%	$\frac{84}{100}$ $\frac{42}{50}$ $\frac{21}{25}$	0·84
85%	$\frac{85}{100}$ $\frac{17}{20}$	0·85
86%	$\frac{86}{100}$ $\frac{43}{50}$	0·86
87%	$\frac{87}{100}$	0·87
88%	$\frac{88}{100}$ $\frac{44}{50}$ $\frac{22}{25}$	0·88
89%	$\frac{89}{100}$	0·89
90%	$\frac{90}{100}$ $\frac{45}{50}$ $\frac{18}{20}$ $\frac{9}{10}$	0·9
91%	$\frac{91}{100}$	0·91
92%	$\frac{92}{100}$ $\frac{46}{50}$ $\frac{23}{25}$	0·92
93%	$\frac{93}{100}$	0·93
94%	$\frac{94}{100}$ $\frac{47}{50}$	0·94
95%	$\frac{95}{100}$ $\frac{19}{20}$	0·95
96%	$\frac{96}{100}$ $\frac{48}{50}$ $\frac{24}{25}$	0·96
97%	$\frac{97}{100}$	0·97
98%	$\frac{98}{100}$ $\frac{49}{50}$	0·98
99%	$\frac{99}{100}$	0·99
100%	$\frac{100}{100}$ $\frac{50}{50}$ $\frac{25}{25}$ $\frac{20}{20}$ $\frac{10}{10}$ $\frac{5}{5}$ $\frac{4}{4}$ $\frac{2}{2}$ $\frac{1}{1}$	1

Percentages 45

Percentages - Fractions - Decimals.

Percentages - Fractions - Decimals

Q/A		Q/A	Q/A
$33\frac{1}{3}$ %	33·333... %	$\frac{1}{3}$	0·333...
$66\frac{2}{3}$ %	66·666... %	$\frac{2}{3}$	0·666...

Q/A		Q/A			Q/A
$12\frac{1}{2}$ %	12·5 %	$\frac{1}{8}$			0·125
25 %		$\frac{2}{8}$	$\frac{1}{4}$		0·25
$37\frac{1}{2}$ %	37·5 %	$\frac{3}{8}$			0·375
50 %		$\frac{4}{8}$	$\frac{2}{4}$	$\frac{1}{2}$	0·5
$62\frac{1}{2}$ %	62·5 %	$\frac{5}{8}$			0·625
75 %		$\frac{6}{8}$	$\frac{3}{4}$		0·75
$87\frac{1}{2}$ %	87·5 %	$\frac{7}{8}$			0·875
100 %		$\frac{8}{8}$	$\frac{4}{4}$	$\frac{2}{2}$ $\frac{1}{1}$	1

Number Order 1   Ascending / Descending.

Number Order
Numbers can be arranged in **ascending** order (from smallest to largest)
                         or **descending** order (from largest to smallest).

Given several numbers to arrange in order, it is useful to list them in a **column**.
It is easier to compare numbers if the digits are in their correct columns ... Th. H T U.
It is easy to then list the numbers in a **row**, in **either** order, if required.

Ex.1
List the following numbers in **ascending** order.
200   9   855   87   846

ascending
9
87
200
846
855

Ex.2
List the following numbers in **descending** order.
1,621   333   2,080   44   533

descending
2,080
1,621
533
333
44

Q	A	Q	A
1) List in **ascending** order 271 56 400 384 65	1) **ascending** 56 65 271 384 400	4) List in **descending** order 555 5,555 55 55,555	4) **descending** 55,555 5,555 555 55
2) List in **descending** order 77 109 95 112 81	2) **descending** 112 109 95 81 77	5) List in **ascending** order 480 300 720 190 610	5) **ascending** 190 300 480 610 720
3) List in **ascending** order 49 42 243 149 342 43	3) **ascending** 42 43 49 149 243 342	6) List in **descending** order 1,234 2,341 3,124 2,431 1,432 2,143 3,214	6) **descending** 3,214 3,124 2,431 2,341 2,143 1,432 1,234

Number Order 2   Decimals / Fractions.

<u>Number Order</u>
Numbers - including decimals **and** fractions - can be arranged <u>**together**</u> ...
     ... in **ascending** order (from smallest to largest)
     or **descending** order (from largest to smallest).

It may be possible to arrange the numbers in their given form - but if we are not sure, it is usual to change any necessary values to **decimals**, arrange the numbers, **then write them in their given form**. The decimal point is on the **right** of a whole number - it is useful to show it.

Ex.1
List the following numbers in **ascending** order.

$3.35 \quad \frac{1}{2} \quad 2.3 \quad 0.6 \quad 3\frac{2}{5}$

⇩   ⇩
0.5 3.4

ascending
0.5 ⇨	$\frac{1}{2}$
0.6	0.6
2.3	2.3
3.35	3.35
3.4	$3\frac{2}{5}$

Ex.2
List the following numbers in **descending** order.

$1\frac{3}{4} \quad 1.71 \quad 17 \quad 1.8 \quad 1\frac{5}{7}$

⇩   ⇩
1.75 $1.714...$

descending
17. ⇨	17
1.8	1.8
1.75	$1\frac{3}{4}$
1.714...	$1\frac{5}{7}$
1.71	1.71

Q | **A**

1) List in **ascending** order
$4.4 \quad 4.38 \quad 4\frac{4}{5} \quad 4.44 \quad 4\frac{3}{8}$

1) **ascending**
4.375 ⇨	$4\frac{3}{8}$
4.38	4.38
4.4	4.4
4.44	4.44
4.8	$4\frac{4}{5}$

2) List in **descending** order
$5.6 \quad 0.111 \quad 5\frac{5}{6} \quad 9 \quad \frac{1}{9}$

2) **descending**
9. ⇨	9
5.83...	$5\frac{5}{6}$
5.6	5.6
0.111...	$\frac{1}{9}$
0.111	0.111

3) List in **ascending** order
$\frac{1}{4} \quad 2.01 \quad \frac{2}{7} \quad 0.201 \quad 1$

3) **ascending**
0.201 ⇨	0.201
0.25	$\frac{1}{4}$
0.285...	$\frac{2}{7}$
1.	1
2.01	2.01

4) List in **descending** order
$7.49 \quad 8\frac{5}{9} \quad 74.5 \quad 10\frac{9}{10} \quad 50$

4) **descending**
74.5
50
$10\frac{9}{10}$
$8\frac{5}{9}$
7.49

*Note
It is possible to arrange these numbers in their given form.

5) List in **ascending** order
$3.11 \quad 1.8^2 \quad \pi \quad 3\frac{1}{8} \quad \sqrt{9}$ (positive value)

5) **ascending**
3. ⇨	$\sqrt{9}$ ($\sqrt{9} = \pm 3$)
3.11	3.11
3.125	$3\frac{1}{8}$
3.141...	π
3.24	1.8^2

Number Grids 1

Number Grids.

Number Grids

Number Grids are used to summarise the results of several similar calculations for a range of numbers - usually in order - sometimes to help consider patterns of numbers.

Here, just a sample of numbers for each operation $+, -, \times, \div$ is presented.

It is usually expected to carry out the calculations **mentally** (but work may be written as necessary).

The Grid consists of columns and rows of 'boxes' or 'cells'.
The 1st. Number is written in a **column** and the 2nd. Number in a **row** (or vice versa).
Each 1st. Number is combined with each 2nd. Number - the result of each calculation is written in the 'box' of the intersection of the row and column of each pair of numbers.

Ex. Complete the Number Grid.

operation ↘ 2nd. Number ⇨

+	3	4
1		
2		

1st. Number

⇨

+	3	4
1	1+3=4	1+4=5
2	2+3=5	2+4=6

1st. Number

+	3	4
1	4	5
2	5	6

1st. Number

Q — Complete each Number Grid.

1) 2nd. Number

+	4	5
3		
5		

A 1)

+	4	5
3	7	8
5	9	10

5) 2nd. Number

×	3	9
2		
7		

A 5)

×	3	9
2	6	18
7	21	63

2) 2nd. Number

+	3	-1	-2
2			
0			
-2			
-4			

A 2)

+	3	-1	-2
2	5	1	0
0	3	-1	-2
-2	1	-3	-4
-4	-1	-5	-6

6) 2nd. Number

×	4	-1	-2
2			
0			
-2			

A 6)

×	4	-1	-2
2	8	-2	-4
0	0	0	0
-2	-8	2	4

3) 2nd. Number

−	1	7
9		
8		

A 3)

−	1	7
9	8	2
8	7	1

7) 2nd. Number

÷	3	9
18		
27		

A 7)

÷	3	9
18	6	2
27	9	3

4) 2nd. Number

−	4	-1	-6
6			
0			
-1			
-3			

A 4)

−	4	-1	-6
6	2	7	12
0	-4	1	6
-1	-5	0	5
-3	-7	-2	3

8) 2nd. Number

÷	2	-1	-4
4			
0			
-8			

A 8)

÷	2	-1	-4
4	2	-4	-1
0	0	0	0
-8	-4	8	2

Indices / Powers 1   Index Form.

Indices / Powers
(The plural of <u>index</u> is <u>indices</u>)
The **index** or **power** of a number indicates **how the number is <u>multiplied by itself</u>**.
It is written at the 'top right' of a number and usually slightly smaller in size.
The number (or **base**) and its index / power can be calculated as a number without an index / power.

Indices / Powers 1, 2, 3, ...

index / power

Exs. $4^1, 4^2, 4^3$ The values are in '**index form**'.

base   The values can be calculated ...

$4^1 = 4 = 4$ *Note **index 1 / power 1** is often omitted.

$4^2 = 4 \times 4 = 16$

$4^3 = 4 \times 4 \times 4 = 64$

$4^4 = 4 \times 4 \times 4 \times 4 = 256$

$4^5 = 4 \times 4 \times 4 \times 4 \times 4 = 1024$

*Note $(-4)^2 = -4 \times -4 = 16$
       $-4^2 = -4 \times 4 = -16$

We refer to the **number** being ' ... **raised to the power of ...** ' and then state the **index / power**.

More usually ...

Exs. 4^1 is read as '**4 raised to the power of 1**' or '**4 to the power 1**'

4^2 is read as '**4 raised to the power of 2**' or '**4 to the power 2**' or '**4 squared**'

4^3 is read as '**4 raised to the power of 3**' or '**4 to the power 3**' or '**4 cubed**'

4^4 is read as '**4 raised to the power of 4**' or '**4 to the power 4**' or '**4 to the 4th**'

4^5 is read as '**4 raised to the power of 5**' or '**4 to the power 5**' or '**4 to the 5th**'

Ex.1 Calculate		Ex.2 Write in **index form**	
a) 3^5	a) $3 \times 3 \times 3 \times 3 \times 3 = 243$	a) $9 \times 9 \times 9$	a) 9^3
b) $(-2)^3$	b) $-2 \times -2 \times -2 = -8$	b) $-7 \times -7 \times -7 \times -7$	b) $(-7)^4$

Q	A	Q	A
Calculate		Write in **index form**	
1) 9^2	1) $9 \times 9 = 81$	1) $2 \times 2 \times 2 \times 2$	1) 2^4
2) $(-3)^3$	2) $-3 \times -3 \times -3 = -27$	2) 3×3	2) 3^2
3) 2^5	3) $2 \times 2 \times 2 \times 2 \times 2 = 32$	3) $-8 \times -8 \times -8 \times -8 \times -8 \times -8$	3) $(-8)^6$
4) 0^7	4) $0 \times 0 \times 0 \times 0 \times 0 \times 0 \times 0 = 0$	4) $5 \times 5 \times 5$	4) 5^3
5) $(-5)^4$	5) $-5 \times -5 \times -5 \times -5 = 625$	5) $1 \times 1 \times 1 \times 1 \times 1 \times 1 \times 1$	5) 1^7
6) 1^8	6) $1 \times 1 \times 1 \times 1 \times 1 \times 1 \times 1 \times 1 = 1$	6) $10 \times 10 \times 10 \times 10 \times 10$	6) 10^5
7) 6^1	7) 6	7) -9	7) $(-9)^1$
8) 10^6	8) $10 \times 10 \times 10 \times 10 \times 10 \times 10 = 1,000,000$	8) $4 \times 4 \times 4 \times 4 \times 4 \times 4 \times 4 \times 4$	8) 4^8
9) -8^2	9) $-8 \times 8 = -64$	9) $-7 \times 7 \times 7$	9) $-7^3 = -(7)^3$

Indices / Powers 2

Decimals / Fractions. Index Form - Operations.

Decimals and Fractions

Decimals and Fractions - usually written in **brackets** ( ) for clarity - can also have indices / powers.

Ex.1 Calculate

a) $(0.1)^2$ a) $0.1 \times 0.1 = 0.01$

b) $\left(\dfrac{3}{5}\right)^2$ b) $\dfrac{3}{5} \times \dfrac{3}{5} = \dfrac{9}{25}$

Q	A	Q	A
Calculate		6) $\left(-\dfrac{1}{2}\right)^2$	6) $-\dfrac{1}{2} \times -\dfrac{1}{2} = \dfrac{1}{4}$
1) $(0.3)^2$	1) $0.3 \times 0.3 = 0.09$		
2) $(-0.4)^3$	2) $-0.4 \times -0.4 \times -0.4 = -0.064$	7) $\left(\dfrac{3}{4}\right)^3$	7) $\dfrac{3}{4} \times \dfrac{3}{4} \times \dfrac{3}{4} = \dfrac{27}{64}$
3) $(1.25)^2$	3) $1.25 \times 1.25 = 1.5625$	8) $\left(\dfrac{2}{5}\right)^4$	8) $\dfrac{2}{5} \times \dfrac{2}{5} \times \dfrac{2}{5} \times \dfrac{2}{5} = \dfrac{16}{625}$
4) $(6.7)^4$	4) $6.7 \times 6.7 \times 6.7 \times 6.7 = 2015.1121$	9) $\left(3\dfrac{1}{2}\right)^2$	9) $3\dfrac{1}{2} \times 3\dfrac{1}{2} = \dfrac{7}{2} \times \dfrac{7}{2} = \dfrac{49}{4} = 12\dfrac{1}{4}$
5) $(0.8)^5$	5) $0.8 \times 0.8 \times 0.8 \times 0.8 \times 0.8 = 0.32768$		

Operations

Operations with numbers and their indices are possible - some are shown here.

Q	A	Q	A
Calculate		10) $\dfrac{\left(\dfrac{1}{3}\right)^2}{\left(\dfrac{3}{4}\right)^3}$	10) $\dfrac{\dfrac{1}{3} \times \dfrac{1}{3}}{\dfrac{3}{4} \times \dfrac{3}{4} \times \dfrac{3}{4}} = \dfrac{\dfrac{1}{9}}{\dfrac{27}{64}}$ $= \dfrac{1}{9} \times \dfrac{64}{27} = \dfrac{64}{243}$
1) $2^5 + 9^2$	1) $2 \times 2 \times 2 \times 2 \times 2 + 9 \times 9$ $= 32 + 81 = 113$		
2) $4^3 - 4^1$	2) $4 \times 4 \times 4 - 4$ $= 64 - 4 = 60$		Find the value of the bracket first.
		11) $(4^2)^3$	11) $(4 \times 4)^3 = (16)^3$ $= 16 \times 16 \times 16 = 4,096$
3) $(-6)^3 - 5^3$	3) $-6 \times -6 \times -6 - 5 \times 5 \times 5$ $= -216 - 125 = -341$	12) $(4^3)^2$	12) $(4 \times 4 \times 4)^2 = (64)^2$ $= 64 \times 64 = 4,096$
4) $8^2 \times (-3)^2$	4) $8 \times 8 \times (-3 \times -3)$ $= 64 \times 9 = 576$	13) $(1^2 + 2^2)^3$	13) $(1 \times 1 + 2 \times 2)^3$ $= (1 + 4)^3$ $= (5)^3$ $= 5 \times 5 \times 5 = 125$
5) $5^4 \times 10^3$	5) $5 \times 5 \times 5 \times 5 \times 10 \times 10 \times 10$ $= 625 \times 1,000 = 625,000$		
6) $\dfrac{-2^3}{1^5}$	6) $= \dfrac{-2 \times 2 \times 2}{1 \times 1 \times 1 \times 1 \times 1} = \dfrac{-8}{1} = -8$	14) $\sqrt{3^2 + 4^2}$	14) $\sqrt{3 \times 3 + 4 \times 4}$ $= \sqrt{9 + 16}$ $= \sqrt{25}$ $= \pm 5$ ($= +5$ or -5)
7) $\dfrac{5^2}{10^2}$	7) $= \dfrac{5 \times 5}{10 \times 10} = \dfrac{25}{100} = \dfrac{1}{4} = 0.25$		
8) $\dfrac{(0.1)^2}{(0.7)^3}$	8) $\dfrac{0.1 \times 0.1}{0.7 \times 0.7 \times 0.7} = \dfrac{0.01}{0.343}$ $= 0.029...$	15) $\sqrt{9^2 - 8^2}$	15) $\sqrt{9 \times 9 - 8 \times 8}$ $= \sqrt{81 - 64}$ $= \sqrt{17}$ $= \pm 4.123...$ ($= +4.123...$ or $-4.123...$)
9) $\dfrac{(3.6)^4}{(1.5)^2}$	9) $\dfrac{3.6 \times 3.6 \times 3.6 \times 3.6}{1.5 \times 1.5}$ $= \dfrac{167.9616}{2.25} = 74.6496$		

Indices / Powers 3 — Multiplication - Addition of Indices.

Multiplication of Numbers with the Same Base

Multiplication of the **same number (base)** with indices / powers can be **simplified**.

Ex. Write as a **single power** of the given number:

1) $3^2 \times 3^4$

$3^2 \times 3^4 =$

$\underline{3 \times 3} \times \underline{3 \times 3 \times 3 \times 3} = 3^6$

or

$3^2 \times 3^4 = 3^6$ the given powers are **added**.

In general, $x^a \times x^b = x^{a+b}$

2) $4^4 \times 4$

$4^4 \times 4 =$

$\underline{4 \times 4 \times 4 \times 4} \times \underline{4} = 4^5$

or

$4^4 \times 4^1 = 4^5$

*Note power 1

---

*Note In the following Q/A, the answer shows both methods, as in the Exs.

Q	A	
Write as a single power of the given number:		
1) $9^5 \times 9^2$	1) $9^5 \times 9^2 = 9 \times 9 \times 9 \times 9 \times 9 \times 9 \times 9 = 9^7$	or $9^5 \times 9^2 = 9^7$
2) 5×5^7	2) $5 \times 5^7 = 5 \times 5 \times 5 \times 5 \times 5 \times 5 \times 5 \times 5 = 5^8$	or $5^1 \times 5^7 = 5^8$
3) $(-3)^2 \times (-3)^4$	3) $(-3)^2 \times (-3)^4 = -3 \times -3 \times -3 \times -3 \times -3 \times -3 = (-3)^6$	or $(-3)^2 \times (-3)^4 = (-3)^6$
4) $(0.1)^3 \times (0.1)^2$	4) $(0.1)^3 \times (0.1)^2 = 0.1 \times 0.1 \times 0.1 \times 0.1 \times 0.1 = (0.1)^5$	or $(0.1)^3 \times (0.1)^2 = (0.1)^5$
5) $\left(\frac{1}{3}\right)^2 \times \left(\frac{1}{3}\right)^2$	5) $\left(\frac{1}{3}\right)^2 \times \left(\frac{1}{3}\right)^2 = \frac{1}{3} \times \frac{1}{3} \times \frac{1}{3} \times \frac{1}{3} = \left(\frac{1}{3}\right)^4$	or $\left(\frac{1}{3}\right)^2 \times \left(\frac{1}{3}\right)^2 = \left(\frac{1}{3}\right)^4$

*Note In the following Q/A, the answer only shows the given powers being **added**.

Q	A	Q	A
Write as a single power of the given number:			
1) $2^7 \times 2$	1) $2^7 \times 2^1 = 2^8$	5) $(4.4)^3 \times (4.4)^2$	5) $(4.4)^3 \times (4.4)^2 = (4.4)^5$
2) $5^2 \times 5^4$	2) $5^2 \times 5^4 = 5^6$	6) $-3^2 \times 3^3 \times 3^4$	6) $-3^2 \times 3^3 \times 3^4 = -3^9$
3) $(-8)^5 \times (-8)^3$	3) $(-8)^5 \times (-8)^3 = (-8)^8$	7) $7^4 \times 7^4 \times 7^4$	7) $7^4 \times 7^4 \times 7^4 = 7^{12}$
4) $\left(\frac{2}{5}\right)^3 \times \left(\frac{2}{5}\right)^4$	4) $\left(\frac{2}{5}\right)^3 \times \left(\frac{2}{5}\right)^4 = \left(\frac{2}{5}\right)^7$	8) $9 \times 9^2 \times 9 \times 9^3$	8) $9^1 \times 9^2 \times 9^1 \times 9^3 = 9^7$

160

Indices / Powers 4 — Division - Subtraction of Indices.

Division of the Same Number

Division of the **same number** with indices / powers can be **simplified**.

Ex. Write as a **single power** of the given number:

a) $\dfrac{5^3}{5^2}$ cancel here as with fractions

$$\dfrac{5^3}{5^2} = \dfrac{\cancel{5} \times \cancel{5} \times 5}{\cancel{5} \times \cancel{5}} = \dfrac{5}{1} = 5^1$$

or

$\dfrac{5^3}{5^2} = 5^1$ The powers are **subtracted**.

b) $\dfrac{4^5}{4^3}$

$$\dfrac{4^5}{4^3} = \dfrac{\cancel{4} \times \cancel{4} \times \cancel{4} \times 4 \times 4}{\cancel{4} \times \cancel{4} \times \cancel{4}} = \dfrac{4^2}{1} = 4^2$$

or

$\dfrac{4^5}{4^3} = 4^2$

*Note c) $\dfrac{0^2}{0^1} = \dfrac{0 \times 0}{0} = \dfrac{0}{0}$ ⇐ Division by 0 is **not** allowed / defined.

So, $\dfrac{0^2}{0^1} = 0^1$ this calculation is **not** allowed.

In general, $\dfrac{x^a}{x^b} = x^{a-b}$ $\begin{cases} x \neq 0 \\ \text{'}x\text{ is not equal to 0'} \end{cases}$

*Note In the following Q/A, the answer shows both methods, as in the Exs.

Q	A
Write as a **single power** of the given number:	
1) $\dfrac{2^5}{2^4}$	1) $\dfrac{2^5}{2^4} = \dfrac{\cancel{2} \times \cancel{2} \times \cancel{2} \times \cancel{2} \times 2}{2 \times 2 \times 2 \times 2} = \dfrac{2}{1} = 2^1$ or $\dfrac{2^5}{2^4} = 2^1$
2) $\dfrac{7^6}{7^3}$	2) $\dfrac{7^6}{7^3} = \dfrac{\cancel{7} \times \cancel{7} \times \cancel{7} \times 7 \times 7 \times 7}{7 \times 7 \times 7} = \dfrac{7^3}{1} = 7^3$ or $\dfrac{7^6}{7^3} = 7^3$

*Note In the following Q/A, the answer only shows the given powers being **subtracted**.

Q	A	Q	A
Write as a **single power** of the given number:		5) $\dfrac{10^{10}}{10^5}$	5) $\dfrac{10^{10}}{10^5} = 10^5$
1) $\dfrac{5^6}{5^2}$	1) $\dfrac{5^6}{5^2} = 5^4$	6) $\dfrac{89^6}{89^4}$	6) $\dfrac{89^6}{89^4} = 89^2$
2) $\dfrac{2^4}{2}$	2) $\dfrac{2^4}{2^1} = 2^3$ ⇐ power 1	7) $\dfrac{\left(\tfrac{1}{2}\right)^5}{\left(\tfrac{1}{2}\right)^2}$	7) $\dfrac{\left(\tfrac{1}{2}\right)^5}{\left(\tfrac{1}{2}\right)^2} = \left(\tfrac{1}{2}\right)^3$
3) $\dfrac{(-3)^5}{(-3)^3}$	3) $\dfrac{(-3)^5}{(-3)^3} = (-3)^2$	8) $\dfrac{(0 \cdot 8)^4}{(0 \cdot 8)^2}$	8) $\dfrac{(0 \cdot 8)^4}{(0 \cdot 8)^2} = (0 \cdot 8)^2$
4) $\dfrac{1^7}{1^6}$	4) $\dfrac{1^7}{1^6} = 1^1$		

*Note Division of the same number with indices / powers may be presented with the ÷ sign.

Ex. Write as a **single power** of the given number:

$5^3 \div 5^2$ ⇒ $5^3 \div 5^2 = 5^1$

This is usually changed to the fraction form $\dfrac{5^3}{5^2}$ and work will be shown here in this form - it is more 'mathematical' for explanations and calculations.

Indices / Powers 5 — Multiplication / Division - Addition / Subtraction of Indices.

Multiplication and Division of the Same Number
Multiplication and **Division** of the **same number** with indices / powers can be **simplified**. We combine the methods in **Indices / Powers 3-4**.

Ex. Write as a **single power** of the given number:

a) $\dfrac{6^2 \times 6^3}{6^4}$

$\dfrac{6^2 \times 6^3}{6^4} = \dfrac{\cancel{6}\times\cancel{6}\times\cancel{6}\times\cancel{6}\times 6}{\cancel{6}\times\cancel{6}\times\cancel{6}\times\cancel{6}} = \dfrac{6^1}{1} = 6^1$

or

$\dfrac{6^2 \times 6^3}{6^4} = \dfrac{6^5}{6^4} = 6^1$

b) $\dfrac{7^4 \times 7^2}{7^2 \times 7^1}$

$\dfrac{7^4 \times 7^2}{7^2 \times 7^1} = \dfrac{\cancel{7}\times\cancel{7}\times\cancel{7}\times 7\times 7\times 7}{\cancel{7}\times\cancel{7}\times\cancel{7}} = \dfrac{7^3}{1} = 7^3$

or

$\dfrac{7^4 \times 7^2}{7^2 \times 7^1} = \dfrac{7^6}{7^3} = 7^3$

*Note In the following Q/A, the answer shows both methods, as in the Exs.

Q	A
Write as a **single power** of the given number:	
1) $\dfrac{5^4 \times 5^1}{5^3}$	1) $= \dfrac{\cancel{5}\times\cancel{5}\times\cancel{5}\times 5\times 5}{\cancel{5}\times\cancel{5}\times\cancel{5}} = \dfrac{5^2}{1} = 5^2$ or $\dfrac{5^4 \times 5^1}{5^3} = \dfrac{5^5}{5^3} = 5^2$
2) $\dfrac{3^3 \times 3^3}{3^2 \times 3^2}$	2) $= \dfrac{\cancel{3}\times\cancel{3}\times\cancel{3}\times\cancel{3}\times 3\times 3}{\cancel{3}\times\cancel{3}\times\cancel{3}\times\cancel{3}} = \dfrac{3^2}{1} = 3^2$ or $\dfrac{3^3 \times 3^3}{3^2 \times 3^2} = \dfrac{3^6}{3^4} = 3^2$

*Note In the following Q/A, the answer only shows the given powers being **calculated**.

Q	A	Q	A
Write as a **single power** of the given number:			
1) $\dfrac{9^3 \times 9^2}{9^1}$	1) $\dfrac{9^3 \times 9^2}{9^1} = \dfrac{9^5}{9^1} = 9^4$	5) $\dfrac{8^6 \times 8^2}{8^3 \times 8^3}$	5) $\dfrac{8^6 \times 8^2}{8^3 \times 8^3} = \dfrac{8^8}{8^6} = 8^2$
2) $\dfrac{3^4 \times 3^5}{3^2}$	2) $\dfrac{3^4 \times 3^5}{3^2} = \dfrac{3^9}{3^2} = 3^7$	6) $\dfrac{2^4 \times 2^4}{2^1 \times 2^2}$	6) $\dfrac{2^4 \times 2^4}{2^1 \times 2^2} = \dfrac{2^8}{2^3} = 2^5$
3) $\dfrac{4^6}{4^2 \times 4^3}$	3) $\dfrac{4^6}{4^2 \times 4^3} = \dfrac{4^6}{4^5} = 4^1$	7) $\dfrac{(-5)^7}{(-5)^2 \times (-5)^1}$	7) $\dfrac{(-5)^7}{(-5)^2 \times (-5)^1} = \dfrac{(-5)^7}{(-5)^3} = (-5)^4$
4) $\dfrac{7^9}{7^5 \times 7}$	4) $\dfrac{7^9}{7^5 \times 7^1} = \dfrac{7^9}{7^6} = 7^3$	8) $\dfrac{6^6 \times 6^6}{6^3 \times 6^4 \times 6^3}$	8) $\dfrac{6^6 \times 6^6}{6^3 \times 6^4 \times 6^3} = \dfrac{6^{12}}{6^{10}} = 6^2$

Indices / Powers 6 — Raise Power of Number to a Power - Multiplication of Indices.

Raising a Power of a Number to a Power

Raising a power of a number **to a power** can be **simplified**.
We use the idea of **multiplication** in **Indices / Powers 3** to help explain.

Ex. Write as a **single power** of the given number:

A. $(4^3)^2$ — The value **inside the bracket** is raised to the **power 2**.

$$(4^3)^2 = 4^3 \times 4^3 = 4^6$$

or

$$(4^3)^2 = 4^6 \quad \text{The powers are multiplied.}$$

B. $(5^2)^4$ — The value **inside the bracket** is raised to the **power 4**.

$$(5^2)^4 = 5^2 \times 5^2 \times 5^2 \times 5^2 = 5^8$$

or

$$(5^2)^4 = 5^8$$

In general, $(x^a)^b = x^{ab}$

*Note: In the following Q/A, the answer shows both methods, as in the Exs.

Write as a single power of the given number:

1. $(7^4)^2$

   1. $= 7^4 \times 7^4 = 7^8$
   or
   $(7^4)^2 = 7^8$

2. $(6^2)^3$

   2. $= 6^2 \times 6^2 \times 6^2 = 6^6$
   or
   $(6^2)^3 = 6^6$

3. $(8^3)^4$

   3. $= 8^3 \times 8^3 \times 8^3 \times 8^3 = 8^{12}$
   or
   $(8^3)^4 = 8^{12}$

4. $(2^2)^2$

   4. $= 2^2 \times 2^2 = 2^4$
   or
   $(2^2)^2 = 2^4$

5. $(4^1)^1$

   5. $= 4^1$
   or
   $(4^1)^1 = 4^1$

6. $((-3)^3)^3$

   6. $= (-3)^3 \times (-3)^3 \times (-3)^3 = (-3)^9$
   or
   $((-3)^3)^3 = (-3)^9$

7. $(1^6)^5$

   7. $= 1^6 \times 1^6 \times 1^6 \times 1^6 \times 1^6 = 1^{30}$
   or
   $(1^6)^5 = 1^{30}$

8. $(9^1)^2$

   8. $= 9^1 \times 9^1 = 9^2$
   or
   $(9^1)^2 = 9^2$

9. $(10^6)^1$

   9. $= 10^6$
   or
   $(10^6)^1 = 10^6$

10. $(0^1)^1$

    10. $= 0^1$
    or
    $(0^1)^1 = 0^1$

11. $((0{\cdot}5)^4)^2$

    11. $= (0{\cdot}5)^4 \times (0{\cdot}5)^4 = (0{\cdot}5)^8$
    or
    $((0{\cdot}5)^4)^2 = (0{\cdot}5)^8$

12. $\left(\left(\frac{1}{4}\right)^2\right)^3$

    12. $= \left(\frac{1}{4}\right)^2 \times \left(\frac{1}{4}\right)^2 \times \left(\frac{1}{4}\right)^2 = \left(\frac{1}{4}\right)^6$
    or
    $\left(\left(\frac{1}{4}\right)^2\right)^3 = \left(\frac{1}{4}\right)^6$

Indices / Powers 7

Index 0 / Power 0

We use the idea of **division** in **Indices / Powers 4** to help explain.

Consider the examples of division -

a) $\dfrac{5^3}{5^3}$

$\dfrac{5^3}{5^3} = \dfrac{\cancel{5}\times\cancel{5}\times\cancel{5}}{\cancel{5}\times\cancel{5}\times\cancel{5}} = \dfrac{1}{1} = 1$

or $\dfrac{5^3}{5^3} = 5^0$ $\Big\}\; 5^0 = 1$

b) $\dfrac{9^2}{9^2}$

$\dfrac{9^2}{9^2} = \dfrac{\cancel{9}\times\cancel{9}}{\cancel{9}\times\cancel{9}} = \dfrac{1}{1} = 1$

or $\dfrac{9^2}{9^2} = 9^0$ $\Big\}\; 9^0 = 1$

---

*Note c) $\dfrac{0^2}{0^2} = \dfrac{0\times 0}{0\times 0} = \dfrac{0}{0}$ ⇦ Division by 0 is **not** allowed / defined.

So, $\dfrac{0^2}{0^2} = 0^0$ this calculation is **not** allowed and therefore 0^0 is **not** allowed.

Excluding 0 the value of **any number raised to the power 0 = 1**.

In general, $x^0 = 1 \quad (x \neq 0)$

Ex.1 State the value of 4^0

$4^0 = 1$

Ex.2 Calculate $3^2 \times 7^0$

$3^2 \times 7^0$
$= 3 \times 3 \times 1 = 9$

Q	A	Q	A
1) State the value of 2^0	1) $2^0 = 1$	9) Calculate $(3^0)^4$	9) $(3^0)^4 = (1)^4$ $= 1\times 1\times 1\times 1 = 1$
2) Calculate $2^3 \times 8^0$	2) $2^3 \times 8^0$ $= 2\times 2\times 2 \times 1 = 8$	10) Calculate $(3^4)^0$	10) $(3^4)^0 = 1$
3) Calculate $5^0 + 4^2$	3) $5^0 + 4^2$ $= 1 + 4\times 4$ $= 1 + 16 = 17$	11) Calculate $(-6^9)^0$	11) $(-6^9)^0 = 1$
4) Calculate $1^0 - 6^0$	4) $1^0 - 6^0$ $= 1 - 1 = 0$	12) Write as a **single power** of the given number: a. $2^4 \times 2^0$	12) a. $2^4 \times 2^0 = 2^4$ or $2^4 \times 1 = 2^4$
5) Calculate $\dfrac{2^2}{4^0}$	5) $\dfrac{2^2}{4^0} = \dfrac{2\times 2}{1} = \dfrac{4}{1} = 4$	b. $\dfrac{9^7}{9^0}$	b. $\dfrac{9^7}{9^0} = 9^7$ or $\dfrac{9^7}{9^0} = \dfrac{9^7}{1} = 9^7$
6) Calculate $\dfrac{-(5)^0}{5^1}$	6) $\dfrac{-(5)^0}{5^1} = \dfrac{-1}{5}$		
7) State the value of $(-7\cdot 7)^0$	7) $(-7\cdot 7)^0 = 1$	c. $(8^0)^5$	c. $(8^0)^5 = 8^0$
8) State the value of $\left(\dfrac{2}{3}\right)^0$	8) $\left(\dfrac{2}{3}\right)^0 = 1$	d. $(8^5)^0$	d. $(8^5)^0 = 8^0$

Indices / Powers 8   Negative Indices / Powers.

Negative Indices / Powers -1, -2, -3, ...

We use the idea of **division** in **Indices / Powers 4** to help explain.

Consider the examples of division -

a) $\dfrac{8^2}{8^3}$ ⇩

$\dfrac{8^2}{8^3} = \dfrac{\overset{1}{\cancel{8}} \times \overset{1}{\cancel{8}}}{\underset{1}{\cancel{8}} \times \underset{1}{\cancel{8}} \times 8} = \dfrac{1}{8^1}$

or

$\dfrac{8^2}{8^3} = 8^{-1}$

$\Biggr\} \; 8^{-1} = \dfrac{1}{8^1}$

b) $\dfrac{8^2}{8^4}$ ⇩

$\dfrac{8^2}{8^4} = \dfrac{\overset{1}{\cancel{8}} \times \overset{1}{\cancel{8}}}{\underset{1}{\cancel{8}} \times \underset{1}{\cancel{8}} \times 8 \times 8} = \dfrac{1}{8^2}$

or

$\dfrac{8^2}{8^4} = 8^{-2}$

$\Biggr\} \; 8^{-2} = \dfrac{1}{8^2}$

*Note

c) $\dfrac{0^2}{0^3} = \dfrac{0 \times 0}{0 \times 0 \times 0} = \dfrac{0}{0}$ ⇦ Division by 0 is **not** allowed / defined.

So, $\dfrac{0^2}{0^3} = 0^{-1} = \dfrac{1}{0^1}$ this calculation is **not** allowed and therefore $0^{\text{-ve number}}$ is **not** allowed.

Excluding 0 the value of **any number raised to a negative power** is equal to

$$\dfrac{1}{\text{the number raised to the positive power}}$$

In general, $x^{-a} = \dfrac{1}{x^a}$ ($x \neq 0$)

Ex.1 State the value of 6^{-1}

⇩

$6^{-1} = \dfrac{1}{6^1} = \dfrac{1}{6}$

Ex.2 Write as a **single power** of the given number:

$\dfrac{3^1}{3^5}$ ⇨ $\dfrac{3^1}{3^5} = 3^{-4}$

Ex.3 Calculate

$2^2 \times 4^{-2}$ ⇨ $2^2 \times 4^{-2}$

$= 2^2 \times \dfrac{1}{4^2} = \dfrac{2 \times 2}{4 \times 4} = \dfrac{4}{16} = \dfrac{1}{4}$

Ex.4 Calculate

$5^4 \times 5^{-3}$ ⇨ $5^4 \times 5^{-3}$

$= 5^4 \times \dfrac{1}{5^3} = \dfrac{\overset{1}{\cancel{5}} \times \overset{1}{\cancel{5}} \times \overset{1}{\cancel{5}} \times 5}{\underset{1}{\cancel{5}} \times \underset{1}{\cancel{5}} \times \underset{1}{\cancel{5}}} = \dfrac{5}{1} = 5$

or

$5^4 \times 5^{-3} = 5^1 = 5$

*Note In the following Q/A, the answer shows both methods, as in the above Exs. a) and b).

Q	A
Write as a **single power** of the given number: 1) $\dfrac{5^3}{5^4}$	1) $\dfrac{5^3}{5^4} = \dfrac{\overset{1}{\cancel{5}} \times \overset{1}{\cancel{5}} \times \overset{1}{\cancel{5}}}{\underset{1}{\cancel{5}} \times \underset{1}{\cancel{5}} \times \underset{1}{\cancel{5}} \times 5} = \dfrac{1}{5^1} = 5^{-1}$ or $\dfrac{5^3}{5^4} = 5^{-1}$
2) $\dfrac{7^2}{7^5}$	2) $\dfrac{7^2}{7^5} = \dfrac{\overset{1}{\cancel{7}} \times \overset{1}{\cancel{7}}}{\underset{1}{\cancel{7}} \times \underset{1}{\cancel{7}} \times 7 \times 7 \times 7} = \dfrac{1}{7^3} = 7^{-3}$ or $\dfrac{7^2}{7^5} = 7^{-3}$

Indices / Powers 9

Negative Indices / Powers.

Negative Indices / Powers $-1, -2, -3, \ldots$

Q	A	Q	A
Calculate 1) 4^{-1}	1) $= \dfrac{1}{4^1} = \dfrac{1}{4}$	14) Write as a **single power** of the given number: a. $4^3 \times 4^{-1}$	14) a. $4^3 \times 4^{-1} = 4^2$
2) -4^{-1}	2) $= -\dfrac{1}{4^1} = -\dfrac{1}{4}$	b. $2^{-8} \times 2^2$	b. $2^{-8} \times 2^2 = 2^{-6}$
3) 3^{-2}	3) $= \dfrac{1}{3^2} = \dfrac{1}{3\times 3} = \dfrac{1}{9}$	c. $6^{-4} \times 6^{-3}$	c. $6^{-4} \times 6^{-3} = 6^{-7}$
4) $8^2 \times 2^{-1}$	4) $= 8^2 \times \dfrac{1}{2^1} = \dfrac{8 \times 8}{2}$ $= \dfrac{64}{2} = 32$	d. $3^{-5} \times 3^0$	d. $3^{-5} \times 3^0 = 3^{-5}$
5) $9^{-4} \times 9^3$	5) $= \dfrac{1}{9^4} \times 9^3 = \dfrac{9 \times 9 \times 9}{9 \times 9 \times 9 \times 9} = \dfrac{1}{9}$	e. $\dfrac{7^1}{7^3}$	e. $\dfrac{7^1}{7^3} = 7^{-2}$
		f. $\dfrac{6^{-3}}{6^0}$	f. $\dfrac{6^{-3}}{6^0} = 6^{-3}$
6) $1^{-2} \times 2^{-1}$	6) $= \dfrac{1}{1^2} \times \dfrac{1}{2^1} =$ $= \dfrac{1}{1 \times 1} \times \dfrac{1}{2} = \dfrac{1}{2}$	g. $\dfrac{8^{-2}}{8^4}$	g. $\dfrac{8^{-2}}{8^4} = 8^{-6}$
7) $5^{-1} \times (-5)^{-2}$	7) $= \dfrac{1}{5^1} \times \dfrac{1}{(-5)^2} =$ $= \dfrac{1}{5 \times -5 \times -5} = \dfrac{1}{125}$	h. $\dfrac{1}{9^7}$	h. $\dfrac{1}{9^7} = 9^{-7}$ or $\dfrac{1}{9^7} = \dfrac{9^0}{9^7} = 9^{-7}$
8) 1^{-1}	8) $= \dfrac{1}{1^1} = \dfrac{1}{1} = 1$	i. $\dfrac{6^1 \times 6^2}{6^4}$	i. $\dfrac{6^1 \times 6^2}{6^4} = \dfrac{6^3}{6^4} = 6^{-1}$
9) $\dfrac{7^{-1}}{2^2}$	9) $= \dfrac{1}{7^1} \times \dfrac{1}{2^2}$ $= \dfrac{1}{7 \times 2 \times 2} = \dfrac{1}{28}$	j. $\dfrac{4^5}{4^2 \times 4^7}$	j. $\dfrac{4^5}{4^2 \times 4^7} = \dfrac{4^5}{4^9} = 4^{-4}$
10) $\dfrac{3^{-2}}{3^2}$	10) $= \dfrac{1}{3^2} \times \dfrac{1}{3^2}$ $= \dfrac{1}{3 \times 3 \times 3 \times 3} = \dfrac{1}{81}$	k. $\dfrac{2^1 \times 2^3}{2^2 \times 2^5}$	k. $\dfrac{2^1 \times 2^3}{2^2 \times 2^5} = \dfrac{2^4}{2^7} = 2^{-3}$
11) $6^0 + 6^{-1}$	11) $= 1 + \dfrac{1}{6} = 1\dfrac{1}{6}$	l. $(5^2)^{-4}$	l. $(5^2)^{-4} = 5^{-8}$
12) $(3^{-1})^5$	12) $= \left(\dfrac{1}{3}\right)^5$ $= \dfrac{1}{3} \times \dfrac{1}{3} \times \dfrac{1}{3} \times \dfrac{1}{3} \times \dfrac{1}{3} = \dfrac{1}{243}$	m. $(5^{-2})^4$	m. $(5^{-2})^4 = 5^{-8}$
		n. $(7^{-5})^0$	n. $(7^{-5})^0 = 7^0$
13) $(3^5)^{-1}$	13) $= (3 \times 3 \times 3 \times 3 \times 3)^{-1}$ $= (243)^{-1} = \dfrac{1}{243}$	o. $(7^0)^{-5}$	o. $(7^0)^{-5} = 7^0$

Indices / Powers 10 — Division - Cancel Powers.

Cancel Powers of the Same Number

We can **cancel powers of the same number** by **mentally calculating** the division - when we are familiar with the work already seen.
(The Q/A may be presented as already seen.)

Ex. Write as a **single power** of the given number and **calculate** the value of :

a) $\dfrac{5^2}{5^1} \Rightarrow$ cancel by 5^1: $\dfrac{5^2}{5^1} = 5^1 = 5$

b) $\dfrac{4^2}{4^5} \Rightarrow$ cancel by 4^2: $\dfrac{4^2}{4^5} = \dfrac{1}{4^3} = 4^{-3} = \dfrac{1}{64}$

c) $\dfrac{6^1 \times 6^5}{6^3} \Rightarrow$ cancel by 6^1: $\dfrac{6^1 \times 6^5}{6^3} \Rightarrow$ cancel by 6^2: $\dfrac{6^1 \times 6^5}{6^3} = 6^3 = 216$

or

c) $\dfrac{6^1 \times 6^5}{6^3} \Rightarrow$ cancel by 6^3: $\dfrac{6^1 \times 6^5}{6^3} = 6^3 = 216$

(This way is quicker and is used in the following Q/A)

Q	A	Q	A
Write as a **single power** of the given number and **calculate** the value of : 1) $\dfrac{7^2}{7^1}$	1) cancel by 7^1: $\dfrac{7^2}{7^1} = 7^1 = 7$	5) $\dfrac{3^6 \times 3^1}{3^2}$	5) cancel by 3^2: $\dfrac{3^6 \times 3^1}{3^2} = 3^5 = 243$
2) $\dfrac{8^1}{8^3}$	2) cancel by 8^1: $\dfrac{8^1}{8^3} = \dfrac{1}{8^2} = 8^{-2} = \dfrac{1}{64}$	6) $\dfrac{5^3}{5^2 \times 5^5}$	6) cancel by 5^3: $\dfrac{5^3}{5^2 \times 5^5} = \dfrac{1}{5^4} = 5^{-4} = \dfrac{1}{625}$
3) $\dfrac{9^5}{9^2}$	3) cancel by 9^2: $\dfrac{9^5}{9^2} = 9^3 = 729$	7) $\dfrac{4^2 \times 4^8}{4^5 \times 4^2}$	7) cancel by 4^2: $\dfrac{4^2 \times 4^8}{4^5 \times 4^2}$ ⇩ cancel by 4^5: $\dfrac{4^2 \times 4^8}{4^5 \times 4^2} = 4^3 = 64$
4) $\dfrac{2^3}{2^7}$	4) cancel by 2^3: $\dfrac{2^3}{2^7} = \dfrac{1}{2^4} = 2^{-4} = \dfrac{1}{16}$		

Indices / Powers *11

Positive Fraction Indices.

*Note Calculator
The Exs. & Q/A in **Indices / Powers *11-*14** can / should **mainly** be answered without using a calculator.
The positive / negative fraction index may be input - (see **Calculator**).

Fraction Indices - Positive Fraction (numerator = 1)
A number with a fraction index can be written as a number without an index.
  Consider the following examples.
  When multiplying the same numbers (base) the given powers are **added**. (see **Indices / Powers 3**)

Ex.1
1) a) $9^{\frac{1}{2}} \times 9^{\frac{1}{2}} = 9^1$
   c) $\sqrt[2]{9} \times \sqrt[2]{9} = 9^1$
   b) $3 \times 3 = 9$

2) a) $8^{\frac{1}{3}} \times 8^{\frac{1}{3}} \times 8^{\frac{1}{3}} = 8^1$
   c) $\sqrt[3]{8} \times \sqrt[3]{8} \times \sqrt[3]{8} = 8$
   b) $2 \times 2 \times 2 = 8$

3) a) $81^{\frac{1}{4}} \times 81^{\frac{1}{4}} \times 81^{\frac{1}{4}} \times 81^{\frac{1}{4}} = 81^1$
   c) $\sqrt[4]{81} \times \sqrt[4]{81} \times \sqrt[4]{81} \times \sqrt[4]{81} = 81$
   b) $3 \times 3 \times 3 \times 3 = 81$

The index / power 1 is usually omitted.

From line a) we can 'calculate' line b) must be true and therefore obtain line c).

so $9^{\frac{1}{2}} = \sqrt[2]{9} = 3$ so $8^{\frac{1}{3}} = \sqrt[3]{8} = 2$ so $81^{\frac{1}{4}} = \sqrt[4]{81} = 3$

$$\boxed{\text{In general, } x^{\frac{1}{n}} = \sqrt[n]{x}}$$ (the **denominator** of the fraction = the **root** of the number)

*Note also $-3 \times -3 = 9$
Here, only the **positive root answer** is considered for clarity.

Ex.2   Find the value of $16^{\frac{1}{2}}$ ⇒ $= \sqrt[2]{16} = 4$ or   ⇒ $= \sqrt{16} = 4$
       *Note 'root 2' may be omitted.

Q	A	Q	A
State the value of			
1) $1^{\frac{1}{2}}$	1) $= \sqrt[2]{1} = 1$	10) $16^{\frac{1}{4}}$	10) $= \sqrt[4]{16} = 2$
2) $4^{\frac{1}{2}}$	2) $= \sqrt[2]{4} = 2$	11) $256^{\frac{1}{4}}$	11) $= \sqrt[4]{256} = 4$
3) $16^{\frac{1}{2}}$	3) $= \sqrt[2]{16} = 4$	12) $625^{\frac{1}{4}}$	12) $= \sqrt[4]{625} = 5$
4) $25^{\frac{1}{2}}$	4) $= \sqrt[2]{25} = 5$	13) $1^{\frac{1}{5}}$	13) $= \sqrt[5]{1} = 1$
5) $27^{\frac{1}{3}}$	5) $= \sqrt[3]{27} = 3$	14) $32^{\frac{1}{5}}$	14) $= \sqrt[5]{32} = 2$
6) $64^{\frac{1}{3}}$	6) $= \sqrt[3]{64} = 4$	15) $243^{\frac{1}{5}}$	15) $= \sqrt[5]{243} = 3$
7) $125^{\frac{1}{3}}$	7) $= \sqrt[3]{125} = 5$	16) $100{,}000^{\frac{1}{5}}$	16) $= \sqrt[5]{100{,}000} = 10$
8) $-8^{\frac{1}{3}}$	8) $= \sqrt[3]{-8} = -2$	17) $-1^{\frac{1}{5}}$	17) $= \sqrt[5]{-1} = -1$
9) $-27^{\frac{1}{3}}$	9) $= \sqrt[3]{-27} = -3$	18) $-32^{\frac{1}{5}}$	18) $= \sqrt[5]{-32} = -2$

Q	A	Q	A
State the value of			
1) $2^{\frac{1}{2}}$	1) $= \sqrt[2]{2} = 1 \cdot 414...$	3) $7^{\frac{1}{4}}$	3) $= \sqrt[4]{7} = 1 \cdot 626...$
2) $50^{\frac{1}{3}}$	2) $= \sqrt[3]{50} = 3 \cdot 684...$	4) $100^{\frac{1}{5}}$	4) $= \sqrt[5]{100} = 2 \cdot 511...$

Indices / Powers *12

Negative Fraction Indices.

Fraction Indices - Negative Fraction (numerator = 1)
(Indices / Powers 8-9 explains negative powers.)
Excluding 0 the value of **any number raised to a negative power is equal to**

$$\frac{1}{\text{the number raised to the positive power}}$$

Ex.1 State the value of $9^{-\frac{1}{2}}$

$9^{-\frac{1}{2}} = \frac{1}{9^{\frac{1}{2}}} = \frac{1}{\sqrt[2]{9}} = \frac{1}{3}$

In general, $x^{-\frac{1}{n}} = \frac{1}{x^{\frac{1}{n}}} = \frac{1}{\sqrt[n]{x}}$

Q	A	Q	A
State the value of			
1) $1^{-\frac{1}{2}}$	1) $= \frac{1}{1^{\frac{1}{2}}} = \frac{1}{\sqrt[2]{1}} = \frac{1}{1} = 1$	11) $16^{-\frac{1}{4}}$	11) $= \frac{1}{16^{\frac{1}{4}}} = \frac{1}{\sqrt[4]{16}} = \frac{1}{2}$
2) $4^{-\frac{1}{2}}$	2) $= \frac{1}{4^{\frac{1}{2}}} = \frac{1}{\sqrt[2]{4}} = \frac{1}{2}$	12) $81^{-\frac{1}{4}}$	12) $= \frac{1}{81^{\frac{1}{4}}} = \frac{1}{\sqrt[4]{81}} = \frac{1}{3}$
3) $36^{-\frac{1}{2}}$	3) $= \frac{1}{36^{\frac{1}{2}}} = \frac{1}{\sqrt[2]{36}} = \frac{1}{6}$	13) $625^{-\frac{1}{4}}$	13) $= \frac{1}{625^{\frac{1}{4}}} = \frac{1}{\sqrt[4]{625}} = \frac{1}{5}$
4) $81^{-\frac{1}{2}}$	4) $= \frac{1}{81^{\frac{1}{2}}} = \frac{1}{\sqrt[2]{81}} = \frac{1}{9}$	14) $10,000^{-\frac{1}{4}}$	14) $= \frac{1}{10,000^{\frac{1}{4}}} = \frac{1}{\sqrt[4]{10,000}} = \frac{1}{10}$
5) $8^{-\frac{1}{3}}$	5) $= \frac{1}{8^{\frac{1}{3}}} = \frac{1}{\sqrt[3]{8}} = \frac{1}{2}$	15) $32^{-\frac{1}{5}}$	15) $= \frac{1}{32^{\frac{1}{5}}} = \frac{1}{\sqrt[5]{32}} = \frac{1}{2}$
6) $27^{-\frac{1}{3}}$	6) $= \frac{1}{27^{\frac{1}{3}}} = \frac{1}{\sqrt[3]{27}} = \frac{1}{3}$	16) $243^{-\frac{1}{5}}$	16) $= \frac{1}{243^{\frac{1}{5}}} = \frac{1}{\sqrt[5]{243}} = \frac{1}{3}$
7) $64^{-\frac{1}{3}}$	7) $= \frac{1}{64^{\frac{1}{3}}} = \frac{1}{\sqrt[3]{64}} = \frac{1}{4}$	17) $1024^{-\frac{1}{5}}$	17) $= \frac{1}{1024^{\frac{1}{5}}} = \frac{1}{\sqrt[5]{1024}} = \frac{1}{4}$
8) $125^{-\frac{1}{3}}$	8) $= \frac{1}{125^{\frac{1}{3}}} = \frac{1}{\sqrt[3]{125}} = \frac{1}{5}$	18) $100,000^{-\frac{1}{5}}$	18) $= \frac{1}{100,000^{\frac{1}{5}}} = \frac{1}{\sqrt[5]{100,000}} = \frac{1}{10}$
9) $-8^{-\frac{1}{3}}$	9) $= \frac{1}{-8^{\frac{1}{3}}} = \frac{1}{\sqrt[3]{-8}} = \frac{1}{-2} = -\frac{1}{2}$	19) $-1^{-\frac{1}{5}}$	19) $= \frac{1}{-1^{\frac{1}{5}}} = \frac{1}{\sqrt[5]{-1}} = \frac{1}{-1} = -1$
10) $-64^{-\frac{1}{3}}$	10) $= \frac{1}{-64^{\frac{1}{3}}} = \frac{1}{\sqrt[3]{-64}} = \frac{1}{-4} = -\frac{1}{4}$	20) $-32^{-\frac{1}{5}}$	20) $= \frac{1}{-32^{\frac{1}{5}}} = \frac{1}{\sqrt[5]{-32}} = \frac{1}{-2} = -\frac{1}{2}$

Q	A	Q	A
Calculate			
1) $8^{-\frac{1}{2}}$	1) $= \frac{1}{8^{\frac{1}{2}}} = \frac{1}{\sqrt[2]{8}} = \frac{1}{2.828...} = 0.353...$	3) $20^{-\frac{1}{4}}$	3) $= \frac{1}{20^{\frac{1}{4}}} = \frac{1}{\sqrt[4]{20}} = \frac{1}{2.114...} = 0.472...$
2) $3^{-\frac{1}{3}}$	2) $= \frac{1}{3^{\frac{1}{3}}} = \frac{1}{\sqrt[3]{3}} = \frac{1}{1.442...} = 0.693...$	4) $90^{-\frac{1}{5}}$	4) $= \frac{1}{90^{\frac{1}{5}}} = \frac{1}{\sqrt[5]{90}} = \frac{1}{2.459...} = 0.406...$

Indices / Powers *13

Positive Fraction Indices.

Fraction Indices - Positive Fraction (numerator = 2,3,4, ...)

To help explain, **raising a power** of a number **to a power** is used - see **Indices / Powers 6**.

Ex.1 a) $(8^{\frac{1}{3}})^2$ The value **inside the bracket** is raised to the **power 2**.   b) $(8^2)^{\frac{1}{3}}$

$$(8^{\frac{1}{3}})^2 = 8^{\frac{1}{3}} \times 8^{\frac{1}{3}} = 8^{\frac{2}{3}}$$

or

$$(8^{\frac{1}{3}})^2 = 8^{\frac{2}{3}} \quad \text{The powers are \textbf{multiplied}.} \qquad (8^2)^{\frac{1}{3}} = 8^{\frac{2}{3}} \quad \text{The powers are \textbf{multiplied}.}$$

From the above, we can calculate using ...

Method 1	Method 2
$8^{\frac{2}{3}} = (8^{\frac{1}{3}})^2$	$8^{\frac{2}{3}} = (8^2)^{\frac{1}{3}}$

It is useful to see $8^{\frac{2}{3}}$ — the numerator as the '**power**' — the denominator is the **root**

In general, $x^{\frac{m}{n}} = (x^{\frac{1}{n}})^m$ and $x^{\frac{m}{n}} = (x^m)^{\frac{1}{n}}$

Ex.2 Find the value of $8^{\frac{2}{3}}$

Method 1
$$8^{\frac{2}{3}} = (8^{\frac{1}{3}})^2 = (\sqrt[3]{8})^2 = (2)^2 = 4$$

Method 2
$$8^{\frac{2}{3}} = (8^2)^{\frac{1}{3}} = (64)^{\frac{1}{3}} = (\sqrt[3]{64}) = 4$$

Method 1 is preferred since it is easier to find the root and 'power' of a smaller number.
(... as in **Indices / Powers *14** also.)
Method 2 is only shown in the following Q/A 1-3.

Q Calculate	A Method 1	Method 2
1) $27^{\frac{2}{3}}$	1) $= (27^{\frac{1}{3}})^2 = (\sqrt[3]{27})^2 = (3)^2 = 9$	$= (27^2)^{\frac{1}{3}} = (729)^{\frac{1}{3}} = (\sqrt[3]{729}) = 9$
2) $64^{\frac{2}{3}}$	2) $= (64^{\frac{1}{3}})^2 = (\sqrt[3]{64})^2 = (4)^2 = 16$	$= (64^2)^{\frac{1}{3}} = (4096)^{\frac{1}{3}} = (\sqrt[3]{4096}) = 16$
3) $125^{\frac{2}{3}}$	3) $= (125^{\frac{1}{3}})^2 = (\sqrt[3]{125})^2 = (5)^2 = 25$	$= (125^2)^{\frac{1}{3}} = (15625)^{\frac{1}{3}} = (\sqrt[3]{15625}) = 25$
4) $1{,}000^{\frac{2}{3}}$	4) $= (1{,}000^{\frac{1}{3}})^2 = (\sqrt[3]{1{,}000})^2 = (10)^2 = 100$	
5) $16^{\frac{3}{4}}$	5) $= (16^{\frac{1}{4}})^3 = (\sqrt[4]{16})^3 = (2)^3 = 8$	
6) $81^{\frac{3}{4}}$	6) $= (81^{\frac{1}{4}})^3 = (\sqrt[4]{81})^3 = (3)^3 = 27$	
7) $625^{\frac{3}{4}}$	7) $= (625^{\frac{1}{4}})^3 = (\sqrt[4]{625})^3 = (5)^3 = 125$	
8) $32^{\frac{3}{5}}$	8) $= (32^{\frac{1}{5}})^3 = (\sqrt[5]{32})^3 = (2)^3 = 8$	
9) $9^{\frac{3}{2}}$	9) $= (9^{\frac{1}{2}})^3 = (\sqrt[2]{9})^3 = (3)^3 = 27$	
10) $8^{\frac{5}{3}}$	10) $= (8^{\frac{1}{3}})^5 = (\sqrt[3]{8})^5 = (2)^5 = 32$	
11) $81^{\frac{5}{4}}$	11) $= (81^{\frac{1}{4}})^5 = (\sqrt[4]{81})^5 = (3)^5 = 243$	
12) $7^{\frac{5}{3}}$	12) $= (7^{\frac{1}{3}})^5 = (\sqrt[3]{7})^5 = (1{\cdot}912...)^5 = 25{\cdot}615...$	
13) $-8^{\frac{2}{3}}$	13) $= (-8^{\frac{1}{3}})^2 = (\sqrt[3]{-8})^2 = (-2)^2 = 4$	
14) $-32^{\frac{3}{5}}$	14) $= (-32^{\frac{1}{5}})^3 = (\sqrt[5]{-32})^3 = (-2)^3 = -8$	

Indices / Powers *14   Negative Fraction Indices.

Fraction Indices - Negative Fraction (numerator = 2,3,4,...)

(Indices / Powers 8-9 explains negative powers.)

Excluding 0 the value of **any number raised to a negative power** is equal to

$$\frac{1}{\text{the number raised to the positive power}}$$

Ex.1 State the value of $8^{-\frac{2}{3}}$ (The Ex. and Q/A relate to **Method 1** ... as in **Indices / Powers *13**.)

$$8^{-\frac{2}{3}} = \frac{1}{8^{\frac{2}{3}}} = \frac{1}{(8^{\frac{1}{3}})^2} = \frac{1}{(\sqrt[3]{8})^2} = \frac{1}{(2)^2} = \frac{1}{4}$$

In general, $x^{-\frac{m}{n}} = \frac{1}{x^{\frac{m}{n}}} = \frac{1}{(x^{\frac{1}{n}})^m} = \frac{1}{(\sqrt[n]{x})^m}$ and $x^{-\frac{m}{n}} = \frac{1}{x^{\frac{m}{n}}} = \frac{1}{(x^m)^{\frac{1}{n}}} = \frac{1}{\sqrt[n]{x^m}}$

Q	A
State the value of	
1) $1^{-\frac{2}{3}}$	1) $\frac{1}{1^{\frac{2}{3}}} = \frac{1}{(1^{\frac{1}{3}})^2} = \frac{1}{(\sqrt[3]{1})^2} = \frac{1}{(1)^2} = \frac{1}{1} = 1$
2) $27^{-\frac{2}{3}}$	2) $\frac{1}{27^{\frac{2}{3}}} = \frac{1}{(27^{\frac{1}{3}})^2} = \frac{1}{(\sqrt[3]{27})^2} = \frac{1}{(3)^2} = \frac{1}{9}$
3) $-64^{-\frac{2}{3}}$	3) $\frac{1}{-64^{\frac{2}{3}}} = \frac{1}{(-64^{\frac{1}{3}})^2} = \frac{1}{(\sqrt[3]{-64})^2} = \frac{1}{(-4)^2} = \frac{1}{16}$
4) $125^{-\frac{2}{3}}$	4) $\frac{1}{125^{\frac{2}{3}}} = \frac{1}{(125^{\frac{1}{3}})^2} = \frac{1}{(\sqrt[3]{125})^2} = \frac{1}{(5)^2} = \frac{1}{25}$
5) $1{,}000^{-\frac{2}{3}}$	5) $\frac{1}{1{,}000^{\frac{2}{3}}} = \frac{1}{(1{,}000^{\frac{1}{3}})^2} = \frac{1}{(\sqrt[3]{1{,}000})^2} = \frac{1}{(10)^2} = \frac{1}{100}$
6) $16^{-\frac{3}{4}}$	6) $\frac{1}{16^{\frac{3}{4}}} = \frac{1}{(16^{\frac{1}{4}})^3} = \frac{1}{(\sqrt[4]{16})^3} = \frac{1}{(2)^3} = \frac{1}{8}$
7) $81^{-\frac{3}{4}}$	7) $\frac{1}{81^{\frac{3}{4}}} = \frac{1}{(81^{\frac{1}{4}})^3} = \frac{1}{(\sqrt[4]{81})^3} = \frac{1}{(3)^3} = \frac{1}{27}$
8) $256^{-\frac{3}{4}}$	8) $\frac{1}{256^{\frac{3}{4}}} = \frac{1}{(256^{\frac{1}{4}})^3} = \frac{1}{(\sqrt[4]{256})^3} = \frac{1}{(4)^3} = \frac{1}{64}$
9) $4^{-\frac{5}{2}}$	9) $\frac{1}{4^{\frac{5}{2}}} = \frac{1}{(4^{\frac{1}{2}})^5} = \frac{1}{(\sqrt{4})^5} = \frac{1}{(2)^5} = \frac{1}{64}$
10) $9^{-\frac{3}{2}}$	10) $\frac{1}{9^{\frac{3}{2}}} = \frac{1}{(9^{\frac{1}{2}})^3} = \frac{1}{(\sqrt{9})^3} = \frac{1}{(3)^3} = \frac{1}{27}$
11) $-8^{-\frac{7}{3}}$	11) $\frac{1}{-8^{\frac{7}{3}}} = \frac{1}{(-8^{\frac{1}{3}})^7} = \frac{1}{(\sqrt[3]{-8})^7} = \frac{1}{(-2)^7} = \frac{1}{-128} = -\frac{1}{128}$
12) $16^{-\frac{5}{4}}$	12) $\frac{1}{16^{\frac{5}{4}}} = \frac{1}{(16^{\frac{1}{4}})^5} = \frac{1}{(\sqrt[4]{16})^5} = \frac{1}{(2)^5} = \frac{1}{32}$
13) $100^{-\frac{5}{2}}$	13) $\frac{1}{100^{\frac{5}{2}}} = \frac{1}{(100^{\frac{1}{2}})^5} = \frac{1}{(\sqrt{100})^5} = \frac{1}{(10)^5} = \frac{1}{100{,}000}$
14) $5^{-\frac{2}{3}}$	14) $\frac{1}{5^{\frac{2}{3}}} = \frac{1}{(5^{\frac{1}{3}})^2} = \frac{1}{(\sqrt[3]{5})^2} = \frac{1}{(1 \cdot 709...)^2} = \frac{1}{2 \cdot 924...} = 0 \cdot 341...$

Indices / Powers *15

Fraction Indices - Write Number in Index Form

Fraction Indices - Write Number in Index Form

The work in **Indices / Powers *11-*14** can be **reversed** - a number to a given **root** or **root and 'power'** can be written as a number with its **index in fraction form**.

Ex.1 Write each number in index form.

1) $\sqrt{16} \Rightarrow = \sqrt[2]{16} = 16^{\frac{1}{2}}$
   *Note 'root 2' may be omitted.

2) $\dfrac{1}{\sqrt[3]{27}} \Rightarrow = \dfrac{1}{27^{\frac{1}{3}}} = 27^{-\frac{1}{3}}$

Q	A	Q	A
Write in Index form.		8) $\dfrac{1}{\sqrt{9}}$	8) $= \dfrac{1}{\sqrt[2]{9}} = \dfrac{1}{9^{\frac{1}{2}}} = 9^{-\frac{1}{2}}$
1) $\sqrt[2]{9}$	1) $= 9^{\frac{1}{2}}$	9) $\dfrac{1}{\sqrt[2]{34}}$	9) $= \dfrac{1}{34^{\frac{1}{2}}} = 34^{-\frac{1}{2}}$
2) $\sqrt{43}$	2) $= \sqrt[2]{43} = 43^{\frac{1}{2}}$	10) $\dfrac{1}{\sqrt[3]{7}}$	10) $= \dfrac{1}{7^{\frac{1}{3}}} = 7^{-\frac{1}{3}}$
3) $\sqrt[3]{8}$	3) $= 8^{\frac{1}{3}}$	11) $\dfrac{1}{\sqrt[4]{81}}$	11) $= \dfrac{1}{81^{\frac{1}{4}}} = 81^{-\frac{1}{4}}$
4) $\sqrt[3]{3}$	4) $= 3^{\frac{1}{3}}$	12) $\dfrac{1}{\sqrt[5]{55}}$	12) $= \dfrac{1}{55^{\frac{1}{5}}} = 55^{-\frac{1}{5}}$
5) $\sqrt[4]{16}$	5) $= 16^{\frac{1}{4}}$	13) $\dfrac{1}{\sqrt[5]{-32}}$	13) $= \dfrac{1}{-32^{\frac{1}{5}}} = -32^{-\frac{1}{5}}$
6) $\sqrt[5]{100}$	6) $= 100^{\frac{1}{5}}$		
7) $\sqrt[3]{-64}$	7) $= -64^{\frac{1}{3}}$		

Ex.2 Write each number in power form.

1) $(\sqrt[3]{64})^2 \Rightarrow = (64^{\frac{1}{3}})^2 = 64^{\frac{2}{3}}$

2) $\dfrac{1}{(\sqrt[4]{81})^3} \Rightarrow = \dfrac{1}{(81^{\frac{1}{4}})^3} = \dfrac{1}{81^{\frac{3}{4}}} = 81^{-\frac{3}{4}}$

Q	A	Q	A
Write in power form.		10) $\dfrac{1}{(\sqrt{6})^3}$	10) $= \dfrac{1}{(\sqrt[2]{6})^3} = \dfrac{1}{(6^{\frac{1}{2}})^3} = \dfrac{1}{6^{\frac{3}{2}}} = 6^{-\frac{3}{2}}$
1) $(\sqrt[3]{8})^2$	1) $= (8^{\frac{1}{3}})^2 = 8^{\frac{2}{3}}$	11) $\dfrac{1}{(\sqrt[2]{49})^3}$	11) $= \dfrac{1}{(49^{\frac{1}{2}})^3} = \dfrac{1}{49^{\frac{3}{2}}} = 49^{-\frac{3}{2}}$
2) $(\sqrt[5]{32})^3$	2) $= (32^{\frac{1}{5}})^3 = 32^{\frac{3}{5}}$	12) $\dfrac{1}{(\sqrt[4]{2^3})}$	12) $= \dfrac{1}{(2^3)^{\frac{1}{4}}} = \dfrac{1}{2^{\frac{3}{4}}} = 2^{-\frac{3}{4}}$
3) $(\sqrt[2]{9})^5$	3) $= (9^{\frac{1}{2}})^5 = 9^{\frac{5}{2}}$	13) $\dfrac{1}{(\sqrt{5^7})}$	13) $= \dfrac{1}{(\sqrt[2]{5^7})} = \dfrac{1}{(5^7)^{\frac{1}{2}}} = \dfrac{1}{5^{\frac{7}{2}}} = 5^{-\frac{7}{2}}$
4) $(\sqrt{4})^3$	4) $= (\sqrt[2]{4})^3 = (4^{\frac{1}{2}})^3 = 4^{\frac{3}{2}}$	14) $\dfrac{1}{(\sqrt[4]{2^3})}$	14) $= \dfrac{1}{(2^3)^{\frac{1}{4}}} = \dfrac{1}{2^{\frac{3}{4}}} = 2^{-\frac{3}{4}}$
5) $(\sqrt[6]{2^5})$	5) $= (2^5)^{\frac{1}{6}} = 2^{\frac{5}{6}}$	15) $\dfrac{1}{(\sqrt[3]{-8})^7}$	15) $= \dfrac{1}{(-8^{\frac{1}{3}})^7} = \dfrac{1}{-8^{\frac{7}{3}}} = -8^{-\frac{7}{3}}$
6) $(\sqrt[5]{7^4})$	6) $= (7^4)^{\frac{1}{5}} = 7^{\frac{4}{5}}$		
7) $(\sqrt{5^3})$	7) $= (\sqrt[2]{5^3}) = (5^3)^{\frac{1}{2}} = 5^{\frac{3}{2}}$		
8) $(\sqrt[5]{-32})^7$	8) $= (-32^{\frac{1}{5}})^7 = -32^{\frac{7}{5}}$		
9) $\dfrac{1}{(\sqrt[5]{32})^2}$	9) $= \dfrac{1}{(32^{\frac{1}{5}})^2} = \dfrac{1}{32^{\frac{2}{5}}} = 32^{-\frac{2}{5}}$		

BIDMAS 1 — Order of Operations.

*Note For clarity, only whole numbers are considered to explain BIDMAS.

BIDMAS

It is accepted for calculations that the **priority / importance** of operations is **generally** - in order - **B**rackets **I**ndices **D**ivision **M**ultiplication **A**ddition **S**ubtraction - in short ... BIDMAS.

BIDMAS is the abbreviation formed by the initial letter of the words of the operations and pronounced as a word. Such a word is called an **acronym**. It is an easily remembered help.

If we do **not** perform calculations with the correct **priority / importance** then errors occur.

It is usually clear in most topics of Mathematics what calculations are needed and in which order - we are not constantly referring to BIDMAS - it becomes 'instinctive'.

For a given number of operations we can calculate in any suitable order.
For example, we may add 2 numbers before having to multiply.

> **A**ddition and **S**ubtraction are **inverses** of each other and have **equal priority**.
>
> **D**ivision and **M**ultiplication are **inverses** of each other and have **equal priority**.
>
> **I**ndices - '**powers**' and '**roots**' are **inverses** of each other and have **equal priority**.
>
> (**I**ndices includes '**roots**' - since a root can be expressed as a **fraction index / power**.)

The example shows that $\sqrt[2]{9^1} = 9^{\frac{1}{2}}$ $9^{\frac{1}{2}} \times 9^{\frac{1}{2}} = 9^1$ We add the powers when multiplying
$\sqrt[2]{9^1} \times \sqrt[2]{9^1} = 9^1$ numbers with the same base.
$3 \times 3 = 9$ The power 1 is usually omitted.

The following Examples and Q/A do not refer to any particular topic.
The calculations follow BIDMAS as explained above.

With $+$ and $-$ and no brackets we calculate Left to Right.

Ex.1	Ex.2	Ex.3	Ex.4	Ex.5	Ex.6
$3+2 = 5$	$3+2-1$	$(3+2)-1$	$3+(2-1)$	$9-4-2$	$9-(4-2)$
	$5\ -1 = 4$	$(\ 5\ )-1 = 4$	$3+(\ 1\ ) = 4$	$5\ -2 = 3$	$9-(\ 2\ ) = 7$
		calculate ( ) first	calculate ( ) first		calculate ( ) first

---

With $\times$ and $\div$ and no brackets we calculate Left to Right.

Ex.1	Ex.2	Ex.3	Ex.4	Ex.5	Ex.6
$2\times 4 = 8$	$3\times 8 \div 4$	$(3\times 8)\div 4$	$3\times(8\div 4)$	$8\div 4\div 2$	$8\div(4\div 2)$
	$24\ \div 4 = 6$	$(\ 24\ )\div 4 = 6$	$3\times(\ 2\ ) = 6$	$2\ \div 2 = 1$	$8\div(\ 2\ ) = 4$
		calculate ( ) first	calculate ( ) first		calculate ( ) first

Q	A	Q	A
1. Calculate		2. Calculate	
1) $7+4$	1) 11	1) 5×7	1) 35
2) $5+4-8$	2) $5+4-8$ $\ \ 9\ -8 = 1$	2) $4\times 6\div 3$	2) $4\times 6\div 3$ $\ \ 24\ \div 3 = 3$
3) $(7+6)-3$	3) $(7+6)-3$ $(\ 13\ )-3 = 10$	3) $(2\times 10)\div 5$	3) $(2\times 10)\div 5$ $(\ 20\ )\div 5 = 4$
4) $7+(6-3)$	4) $7+(6-3)$ $7+(\ 3\ ) = 10$	4) $2\times(10\div 5)$	4) $2\times(10\div 5)$ $2\times(\ 2\ ) = 4$
5) $16-9-2$	5) $16-9-2$ $\ \ 7\ -2 = 5$	5) $36\div 6\div 2$	5) $36\div 6\div 2$ $\ \ 6\ \div 2 = 3$
6) $16-(9-2)$	6) $16-(9-2)$ $16-(\ 7\ ) = 9$	6) $36\div(6\div 2)$	6) $36\div(6\div 2)$ $36\div\ 3\ = 12$

BIDMAS 2 — Order of Operations.

Calculate
- Ex.1: $8 \div 4 \times 3$ → $2 \times 3 = 6$
- Ex.2: $(8 \div 4) \times 1$ → $(2) \times 1 = 2$
- Ex.3: $8 \div (4 \times 1)$ → $8 \div (4) = 2$
- Ex.4: $(8 \div 4) \times 2$ → $(2) \times 2 = 4$
- Ex.5: $8 \div (4 \times 2)$ → $8 \div (8) = 1$
- Ex.6: $0 \div (1 \times 1)$ → $0 \div (1) = 0$

Q	A	Q	A
1. Calculate			
1) $63 \div 9 \times 7$	1) $63 \div 9 \times 7$ $7 \times 7 = 49$	4) $(30 \div 6) \times 5$	4) $(30 \div 6) \times 5$ $(5) \times 5 = 25$
2) $(55 \div 5) \times 1$	2) $(55 \div 5) \times 1$ $(11) \times 1 = 11$	5) $30 \div (6 \times 5)$	5) $30 \div (6 \times 5)$ $30 \div (30) = 1$
3) $55 \div (5 \times 1)$	3) $55 \div (5 \times 1)$ $55 \div (5) = 11$	6) $9 \div (3 \times 0)$	6) $9 \div (3 \times 0)$ $9 \div (0) =$ undefined

Calculate
- Ex.1: $2 \times 3 + 4$ → $6 + 4 = 10$
- Ex.2: $2 \times (3 + 4)$ → $2 \times (7) = 14$
- Ex.3: $3 + 4 \times 2$ → $3 + 8 = 11$
- Ex.4: $(3 + 4) \times 2$ → $(7) \times 2 = 14$
- Ex.5: $5 \times 8 - 1$ → $40 - 1 = 39$
- Ex.6: $5 \times (8 - 1)$ → $5 \times (7) = 35$
- Ex.7: $8 - 1 \times 5$ → $8 - 5 = 3$
- Ex.8: $(8 - 1) \times 5$ → $(7) \times 5 = 35$

Q	A	Q	A
1. Calculate			
1) $7 \times 9 + 2$	1) $7 \times 9 + 2$ $63 + 2 = 65$	5) $3 \times 20 - 4$	5) $3 \times 20 - 4$ $60 - 4 = 56$
2) $7 \times (9 + 2)$	2) $7 \times (9 + 2)$ $7 \times (11) = 77$	6) $3 \times (20 - 4)$	6) $3 \times (20 - 4)$ $3 \times (16) = 48$
3) $9 + 2 \times 7$	3) $9 + 2 \times 7$ $9 + 14 = 23$	7) $20 - 4 \times 3$	7) $20 - 4 \times 3$ $20 - 12 = 8$
4) $(9 + 2) \times 7$	4) $(9 + 2) \times 7$ $(11) \times 7 = 77$	8) $(20 - 4) \times 3$	8) $(20 - 4) \times 3$ $(16) \times 3 = 48$

Calculate
- Ex.1: $12 \div 4 + 8$ → $3 + 8 = 11$
- Ex.2: $12 \div (4 + 8)$ → $12 \div (12) = 1$
- Ex.3: $10 + 6 \div 2$ → $10 + 3 = 13$
- Ex.4: $(10 + 6) \div 2$ → $(16) \div 2 = 8$
- Ex.5: $36 \div 9 - 3$ → $4 - 3 = 1$
- Ex.6: $36 \div (9 - 3)$ → $36 \div (6) = 6$
- Ex.7: $70 - 20 \div 5$ → $70 - 4 = 66$
- Ex.8: $(70 - 20) \div 5$ → $50 \div 5 = 10$

Q	A	Q	A
1. Calculate			
1) $20 \div 5 + 5$	1) $20 \div 5 + 5$ $4 + 5 = 9$	5) $72 \div 9 - 1$	5) $72 \div 9 - 1$ $8 - 1 = 7$
2) $20 \div (5 + 5)$	2) $20 \div (5 + 5)$ $20 \div (10) = 2$	6) $72 \div (9 - 1)$	6) $72 \div (9 - 1)$ $72 \div (8) = 9$
3) $14 + 35 \div 7$	3) $14 + 35 \div 7$ $14 + 5 = 19$	7) $60 - 30 \div 6$	7) $60 - 30 \div 6$ $60 - 5 = 55$
4) $(14 + 35) \div 7$	4) $(14 + 35) \div 7$ $(49) \div 7 = 7$	8) $(60 - 30) \div 6$	8) $(60 - 30) \div 6$ $30 \div 6 = 5$

BIDMAS 3 — Order of Operations.

Calculate

Ex.1	Ex.2	Ex.3	Ex.4	Ex.5	Ex.6
$1+3^2$	5^2-9	7×2^3	$6^2 \div 3$	4^2+1^5	$(9-4)^2$
$1+9=10$	$25-9=16$	$7\times 8=56$	$36\div 3=12$	$16+1=17$	$(\ 5\ )^2=25$

Q	A	Q	A
1. Calculate			
1) 8^2+2	1) 8^2+2 $64+2=66$	4) $32\div 2^4$	4) $32\div 2^4$ $32\div 16=2$
2) $50-3^3$	2) $50-3^3$ $50-27=23$	5) 5^2+5^3	5) 5^2+5^3 $25+125=150$
3) $7^2\times 10$	3) $7^2\times 10$ $49\times 10=490$	6) $(2\times 3)^2$	6) $(2\times 3)^2$ $(\ 6\ )^2=36$

Calculate

Ex.1	Ex.2	Ex.3	Ex.4	Ex.5	Ex.6
$4+\sqrt{9}$	$\sqrt{4}-1$	$9\times\sqrt{25}$	$\sqrt[3]{8}\div 2$	$\sqrt{49}+\sqrt[3]{64}$	$6^2-\sqrt{36}$
$4+\ 3=7$	$2-1=1$	$9\times 5=45$	$2\div 2=1$	$7+\ 4=11$	$36-\ 6=30$

Q	A	Q	A
1. Calculate			
1) $\sqrt{1}+1$	1) $\sqrt{1}+1$ $1+1=2$	4) $12\div\sqrt{16}$	4) $12\div\sqrt{16}$ $12\div 4=3$
2) $8+\sqrt{81}$	2) $8+\sqrt{81}$ $8+9=17$	5) $\sqrt[3]{1000}-\sqrt{100}$	5) $\sqrt[3]{1000}-\sqrt{100}$ $10-10=0$
3) $\sqrt[3]{125}\times 11$	3) $\sqrt[3]{125}\times 11$ $5\times 11=55$	6) $\sqrt{64}+3^2$	6) $\sqrt{64}+3^2$ $8+9=17$

Calculate

Ex.1	Ex.2	Ex.3	Ex.4
$10\div 2+8\times 4-1$	$50-6\div 3+5\times 2$	$9-(5+3)\div 2+4^2$	$7+\sqrt{9}\times(8\div 4)-2^3$
$5+32-1=36$	$50-2+10=58$	$9-(\ 8\ )\div 2+16$ $9-\ 4\ +16=21$	$7+3\times(\ 2\ )-8$ $7+\ 6\ -8=5$

Q	A	Q	A
1. Calculate		6) $9-5+3\times(8-6)^3$	6) $9-5+3\times(8-6)^3$ $4+3\times(\ 2\ )^3$ $4+3\times\ 8$ $4+\ 24\ =28$
1) $11\times 3-9\div 3+3$	1) $11\times 3-9\div 3+3$ $33-3+3=33$		
2) $20\div 4+7-2\times 6$	2) $20\div 4+7-2\times 6$ $5+7-12=0$	7) $49\div\sqrt{49}+(7-3)\times 4^2$	7) $49\div\sqrt{49}+(7-3)\times 4^2$ $49\div 7+\ 4\times 16$ $7+\ 64$ $=71$
3) $8+3\times(7-2)+8\div 4$	3) $8+3\times(7-2)+8\div 4$ $8+3\times(\ 5\ )+2$ $8+15+2=25$		
		8) $20-6\times 3\div(2+2^2)$	8) $20-6\times 3\div(2+2^2)$ $20-18\div(2+4)$ $20-18\div 6$ $20-3=17$
4) $(5+1)\div 6\times 9-1\times 7$	4) $(5+1)\div 6\times 9-1\times 7$ $6\div 6\times 9-7$ $1\times 9-7$ $9-7=2$		
		9) $30\div 6\times\sqrt{3^2+4^2}$	9) $30\div 6\times\sqrt{3^2+4^2}$ $5\times\sqrt{9+16}$ $5\times\sqrt{25}$ $5\times 5=25$
5) $6^2\div 3^2+4\times 4-4$	5) $6^2\div 3^2+4\times 4-4$ $36\div 9+16-4$ $4+16-4=16$		

Sequences 1   Find the Next Term.

<u>Sequence</u>
A sequence is a set of numbers in ascending or descending order which are **related by a rule**.
Here, the **rule** is concerned with the operation of **addition, subtraction, multiplication** or **division**.
Each value in a sequence is also called a **term**. The numbers form a **pattern**.

<u>Find the Next Term in a Sequence - Addition / Subtraction</u>
To **find the next term** in a sequence, the **differences** between the given values are calculated -
these are related by a **simpler rule** or form a **simpler pattern** which is used to decide the relationship.
The **differences** involve addition or subtraction -
these are usually **calculated mentally** but should be written down.

Ex. Find the next 2 terms in the sequence
1, 3, 5, 7, —, —

1, 3, 5, 7, —, — ⇨ 1, 3, 5, 7, **9, 11**   The next 2 terms are **9** and **11**.
differences +2 +2 +2   differences +2 +2 +2 **+2 +2**
(the + sign can be omitted)

The **difference** between the values is a **constant +2**. Ex. 3−1 = **2**
This is then added to the last term to find the next term, and so on ... Ex. 7 +2 = **9**

Q	A
Find the next 2 terms in each sequence.	
1) 1, 4, 7, 10, —, —	1) 1, 4, 7, 10, **13, 16** differences +3 +3 +3 +3 +3
2) 24, 19, 14, 9, —, —	2) 24, 19, 14, 9, **4, −1** differences −5 −5 −5 −5 −5
3) −17, −13, −9, −5, —, —	3) −17, −13, −9, −5, **−1, 3** differences +4 +4 +4 +4 +4
4) −2, −6, −10, −14, —, —	4) −2, −6, −10, −14, **−18, −22** differences −4 −4 −4 −4 −4
5) 0·7, 1·3, 1·9, 2·5, —, —	5) 0·7, 1·3, 1·9, 2·5, **3·1, 3·7** differences +0·6 +0·6 +0·6 +0·6 +0·6
6) $\frac{1}{2}$, $1\frac{1}{4}$, 2, $2\frac{3}{4}$, —, —	6) $\frac{1}{2}$, $1\frac{1}{4}$, 2, $2\frac{3}{4}$, **$3\frac{1}{2}$, $4\frac{1}{4}$,** differences $\frac{3}{4}$ $\frac{3}{4}$ $\frac{3}{4}$ $\frac{3}{4}$ $\frac{3}{4}$
7) 1, 2, 4, 7, 11, —, —	7) 1, 2, 4, 7, 11, **16, 22** differences +1 +2 +3 +4 +5 +6 differences +1 +1 +1 +1 +1
8) 5, 14, 22, 29, —, —	8) 5, 14, 22, 29, **35, 40** differences +9 +8 +7 +6 +5 differences −1 −1 −1 −1
9) 1, 3, 7, 13, —, —	9) 1, 3, 7, 13, **21, 31** differences +2 +4 +6 +8 +10 differences +2 +2 +2 +2
10) 30, 28, 24, 18, —, —	10) 30, 28, 24, 18, **10, 0** differences −2 −4 −6 −8 −10 differences −2 −2 −2 −2

176

Sequences 2

Find the Next Term.

Q	A
11) 34, 26, 19, 13, __ , __	11) 34, 26, 19, 13, **8**, **4** differences −8 −7 −6 −5 −4 differences +1 +1 +1 +1
12) 18, 15, 11, 6, __ , __	12) 18, 15, 11, 6, **0**, **−7** differences −3 −4 −5 −6 −7 differences −1 −1 −1 −1
13)a) 0, 1, 4, 9, 16, __ , __ b) Continue the dot pattern for the next 2 terms. • :: ::: ::::	13)a) 0, 1, 4, 9, 16, **25**, **36** differences +1 +3 +5 +7 +9 +11 differences +2 +2 +2 +2 +2 b) Number of Dots 1 4 9 16 25 36 +3 +5 +7 +9 +11 The sequence should also be recognised as the Square Numbers.

14)a) Continue the dot / number pattern for the next 2 terms.
 	
• :: :::
 	
1   1+2   1+2+3
= 1   3   6
 	
b) Continue the number pattern in a) for a **further** 5 terms.
	14)a) The sequence is referred to as the 'Triangular Numbers'.
	
	1 1+2 1+2+3 1+2+3+4 1+2+3+4+5
	= 1 3 6 **10** **15**
	     +2 +3 +4 +5
	
	b) = **21** **28** **36** **45** **55**
	    +6 +7 +8 +9 +10

15) Continue the sequence for the next 4 terms.
 	
1, 1, 2, 3, 5, 8, 13, __ , __ , __ , __
	15) The sequence is referred to as the 'Fibonacci Sequence' and often occurs in nature. It is named after the Italian mathematician Fibonacci, (1170 − 1230).
	
	1, 1, 2, 3, 5, 8, 13, **21**, **34**, **55**, **89**
	differences 0 +1 +1 +2 +3 +5 +8 +13 +21 +34
	
	The differences are the same terms as the sequence (other than 0) − which allows the 'next terms' to be found.
	
	It should also be recognised that the **sum of 2 consecutive terms** produces the next term.
	  1+1=2
	    1+2=3
	      2+3=5
	        3+5=8
	          5+8=13 ...

Sequences 3 Find the Next Term in a Sequence.

Q

16)a) State the first 14 terms of the Fibonacci Sequence.
   b) Divide each term of the Fibonacci Sequence by the previous term.
      Consider the **decimal** results.

A

16)a) 1, 1, 2, 3, 5, 8, 13, 21, 34, 55, 89, 144, 233, 377

b) $\frac{1}{1}$ $\frac{2}{1}$ $\frac{3}{2}$ $\frac{5}{3}$ $\frac{8}{5}$ $\frac{13}{8}$ $\frac{21}{13}$ $\frac{34}{21}$ $\frac{55}{34}$ $\frac{89}{55}$ $\frac{144}{89}$ $\frac{233}{144}$ $\frac{377}{233}$

= 1 2 1·5 1·666... 1·6 1·625 1·615... 1·619... 1·617... 1·618... 1·617... 1·6180... 1·6180...

As the terms get larger the ratio is approximately 1·618... and is an irrational number.
The number is referred to as the **Golden Section** or **Golden Ratio**.
(It appears in nature and is used in art and architecture.)

Q | A

17)
a) Continue the sequence for the next 3 rows.

b) Continue the sequence for the sum of the numbers in each row for the next 3 rows.

a)

```
              1
           1     1        c)
        1     2     1
     1     3     3     1
  1     4     6     4     1
```

b) Sum of Row

1
2
4
8
16

17)
The sequence is referred to as 'Pascal's Triangle'.
It has many uses in advanced Mathematics.
It is named after the French mathematician
Blaise Pascal, (1623-1662).

a)
```
                  1
                1   1
              1   2   1
            1   3   3   1
          1   4   6   4   1
        1   5   10  10  5   1
      1   6   15  20  15  6   1
    1   7   21  35  35  21  7   1
```

b) Sum of Row

1
2
4
8
16
32
64
128

The first and last number on each row is 1.
The other numbers are found by
adding the 2 nearest numbers from the row above.

Exs.   1 + 1   1 + 2   3 + 3   6 + 4
       = 2   = 3   = 6   = 10

b)   differences

1
2   + 1
4   + 2
8   + 4
16   + 8
32   +16
64   +32
128   +64

The differences are the same terms as the sequence.

c) List and state the name of the sequence of numbers in the diagonal marked c).

c) 1, 3, 6, 10, 15, 21, ...
The sequence is the 'Triangular Numbers'.
(see Q/A 14)

178

Sequences 4 — Find the Next Term in a Sequence.

Find the Next Term in a Sequence - Multiplication / Division

To **find the next term** in a sequence, we can usually **recognise** that the terms are being multiplied or divided. Finding **differences** - as with addition / subtraction - **does not <u>usually</u> help** find the next term. We check by **dividing** pairs of consecutive terms to find the relation.

Ex. Find the next 2 terms in the sequence
  1, 3, 9, 27, —, —

$\dfrac{3}{1} = 3 \quad \dfrac{9}{3} = 3 \quad \dfrac{27}{9} = 3$

1, 3, 9, 27, **81, 243**
rule ×3 ×3 ×3 ×3 ×3

The next 2 terms are **81** and **243**.

If the **differences** are calculated ...
  1, 3, 9, 27, —, —
differences +2 +6 +18
             ×3 ×3
differences +4 +12
             ×3

The **rule** (operation) between the values is a **constant** ×3. Ex. 1×3 = 3, 3×3 = 9
This is applied to the last term to find the next term, and so on ... Ex. 27×3 = **81**

Q	A
Find the next 2 terms in each sequence.	[see also **Sequences 3** Q/A 16b)]
1) 1, 2, 4, 8, —, —	1) 1, 2, 4, 8, **16, 32** rule ×2 ×2 ×2 ×2 ×2
2) 160, 80, 40, 20, —, —	2) 160, 80, 40, 20, **10, 5** rule ÷2 ÷2 ÷2 ÷2 ÷2 $\dfrac{80}{160} = \dfrac{1}{2} \quad \dfrac{40}{80} = \dfrac{1}{2}$
3) 1, 5, 25, 125, —, —	3) 1, 5, 25, 125, **625, 3125** rule ×5 ×5 ×5 ×5 ×5
4) 96, 48, 24, 12, —, —	4) 96, 48, 24, 12, **6, 3** rule ÷2 ÷2 ÷2 ÷2 ÷2
5) 3, 30, 300, 3000, —, —	5) 3, 30, 300, 3000, **30 000, 300 000** rule ×10 ×10 ×10 ×10 ×10
6) 2, 6, 18, 54, —, —	6) 2, 6, 18, 54, **162, 486** rule ×3 ×3 ×3 ×3 ×3
7) 1024, 256, 64, 16, —, —	7) 1024, 256, 64, 16, **4, 1** rule ÷4 ÷4 ÷4 ÷4 ÷4
8) 1, 1, 2, 6, 24, —, —	8) 1, 1, 2, 6, 24, **120, 720** rule ×1 ×2 ×3 ×4 ×5 ×6
9) 3600, 600, 120, 30, 10, —, —	9) 3600, 600, 120, 30, 10, **5, 5** rule ÷6 ÷5 ÷4 ÷3 ÷2 ÷1
10) -7, -14, -28, -56, —, —	10) -7, -14, -28, -56, **-112, -224** rule ×2 ×2 ×2 ×2 ×2
11) -288, -144, -72, -36, -18, —, —	11) -288, -144, -72, -36, -18, **-9** rule ÷2 ÷2 ÷2 ÷2 ÷2
12) $\dfrac{1}{64}, \dfrac{1}{32}, \dfrac{1}{16}, \dfrac{1}{8}$, —, —	12) $\dfrac{1}{64}, \dfrac{1}{32}, \dfrac{1}{16}, \dfrac{1}{8}, \dfrac{1}{4}, \dfrac{1}{2}$ rule ×2 ×2 ×2 ×2 ×2

Proportion 1

Direct Proportion.

Direct Proportion

2 related values are in **direct proportion** if ...
    as **1 value** <u>increases</u> then **the other value** <u>increases</u> in the <u>**same proportion**</u> (<u>**at the same rate**</u>).
and as **1 value** <u>decreases</u> then **the other value** <u>decreases</u> in the <u>**same proportion**</u> (<u>**at the same rate**</u>).

(... their **ratio** <u>is the same</u> as the values change)

*<u>Note 1</u> We are <u>not</u> usually <u>told</u> that 2 values are in direct proportion - we have to <u>recognise</u> this.
*<u>Note 2</u> We consider that the <u>units for each value are identical</u>.
      (for example, in Ex. 1 (below) ... the <u>pens</u> are all the same kind and each pen costs the same.)

Given 2 values in direct proportion <u>and</u> the change in 1 value, we calculate the change in the other value.

Method 1	Method 2
<u>The Unitary Method</u>	We need to **understand** and **mentally plan** Method 1 and carry out the **same calculations** - usually set out in **fraction form**.
We first express the direct proportion in **unitary form** - one of the values is **1** - in order to calculate the change in the other value.	
	(... but we do not <u>write</u> the **unitary form**.)
(It is often useful to write the **unitary form** - even if it is not asked for directly.)	

   Ex.1   4 pens cost a total of £8.
            What is the cost of 3 pens?

We find the cost of **1** pen in order to find the cost of 3 pens.

Method 1:
$\div 4$ (4 pens cost £8) $\div 4$
      1 pen costs £2
$\times 3$ (3 pens cost £6) $\times 3$

Method 2:
$\times \frac{3}{4}$ (4 pens cost £8) $\times \frac{3}{4}$
      3 pens costs £6

Q	A

1) 5 mugs cost a total of £15. Find the cost of 4 mugs.

1)
Method 1:
$\div 5$ (5 mugs ... £15)
    1 mug ... £ 3
$\times 4$ (4 mugs ... £12)

Method 2:
$\times \frac{4}{5}$ (5 mugs ... £15)
    4 mugs ... £12

2) 5 mugs cost a total of £15. Find the cost of 9 mugs.

2)
Method 1:
$\div 5$ (5 mugs ... £15)
    1 mug ... £ 3
$\times 9$ (9 mugs ... £27)

Method 2:
$\times \frac{9}{5}$ (5 mugs ... £15)
    9 mugs ... £27

3) 6 boxes contain a total of 240 sweets. How many sweets are in 11 boxes?

3)
Method 1:
$\div 6$ (6 boxes ... 240 sweets)
    1 box ... 40 sweets
$\times 11$ (11 boxes ... 440 sweets)

Method 2:
$\times \frac{11}{6}$ (6 boxes ... 240 sweets)
    11 boxes ... 440 sweets

Proportion 2

Direct Proportion.

Q	A

4) 9 cans have a total capacity of 45 *l*. Find the capacity of 50 cans.	4) **Method 1**: ÷9 (9 cans ... 45 *l* ; 1 can ... 5 *l*) ×50 (50 cans ... 250 *l*) **Method 2**: × $\frac{50}{9}$ (9 cans ... 45 *l* ; 50 cans ... 250 *l*)
5) 0·3m of a tin tube weighs 0·66kg. Find the weight of 0·18m of the tube.	5) **Method 1**: ÷0·3 (0·3m ... 0·66kg ; 1m ... 2·2kg) ×0·18 (0·18m ... 0·396kg) **Method 2**: × $\frac{0·18}{0·3}$ (0·3m ... 0·66kg ; 0·18m ... 0·396kg)
6) 4·4cm³ of an alloy weighs 35·2g. Find the weight of 6·7cm³ of the alloy.	6) **Method 1**: ÷4·4 (4·4cm³ ... 35·2g ; 1cm³ ... 8g) ×6·7 (6·7cm³ ... 53·6g) **Method 2**: × $\frac{6·7}{4·4}$ (4·4cm³ ... 35·2g ; 6·7cm³ ... 53·6g)
7) A machine makes 1875 discs in 75mins. How many discs are made in an hour?	7) **Method 1**: ÷75 (75mins ... 1875 discs ; 1min ... 25 discs) ×60 (1hr = 60mins ... 1500 discs) **Method 2**: × $\frac{60}{75}$ (75mins ... 1875 discs ; 1hr = 60mins ... 1500 discs)
8) 20 eggs are used to make 16 cakes. How many eggs are in 36 cakes?	8) **Method 1**: ÷16 (16 cakes ... 20 eggs ; 1 cake ... 1·25 eggs) ×36 (36 cakes ... 45 eggs) **Method 2**: × $\frac{36}{16}$ (16 cakes ... 20 eggs ; 36 cakes ... 45 eggs)
9) 16 cakes are made from 20 eggs. How many cakes can 36 eggs make?	9) **Method 1**: ÷20 (20 eggs ... 16 cakes ; 1 egg ... 0·75 cakes) ×36 (36 eggs ... 27 cakes) **Method 2**: × $\frac{36}{20}$ (20 eggs ... 16 cakes ; 36 eggs ... 27 cakes)
10) 12 pancakes can be made from a mixture which requires the following ingredients: 110g flour 2 eggs 200ml milk 70ml water 50g butter pinch of salt State the amount of each ingredient required to make 18 pancakes.	10) *Note* When given several calculations involving the same numbers - in this case, 12 and 18 - it is quicker and easier to use Method 2 (which is used here). We need to multiply each ingredient by $\frac{18}{12} = \frac{3}{2}$ (easier!) (÷6 top and bottom) flour 110g × $\frac{3}{2}$ = $\frac{330}{2}$ = 165g eggs 2 × $\frac{3}{2}$ = $\frac{6}{2}$ = 3 milk 200ml × $\frac{3}{2}$ = $\frac{600}{2}$ = 300ml water 70ml × $\frac{3}{2}$ = $\frac{210}{2}$ = 105ml butter 50g × $\frac{3}{2}$ = $\frac{150}{2}$ = 75g salt pinch × $\frac{3}{2}$ = $\frac{3 \text{ pinches}}{2}$ = $1\frac{1}{2}$ pinches!

Proportion 3

Direct Proportion - Graph.

Q

11) The graph shows the weight (kg) of different lengths (m) of a tube.

Tube Length /Weight

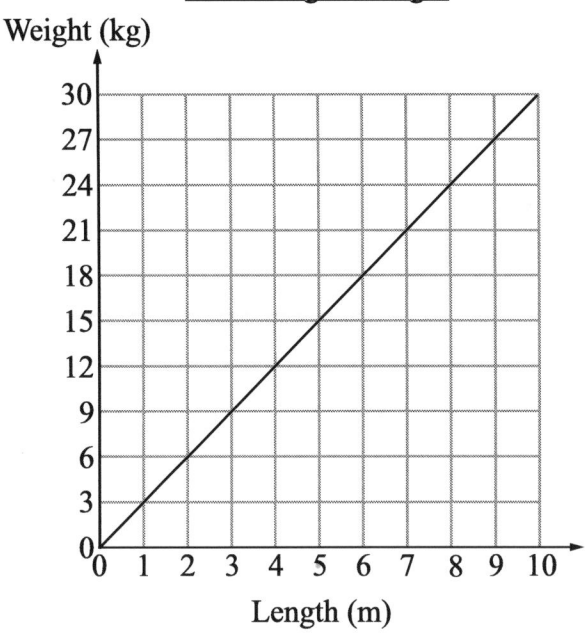

Use the graph to find
a) The weight of tube, length 4m.
b) The length of tube which has a weight of 27kg.

A

11)

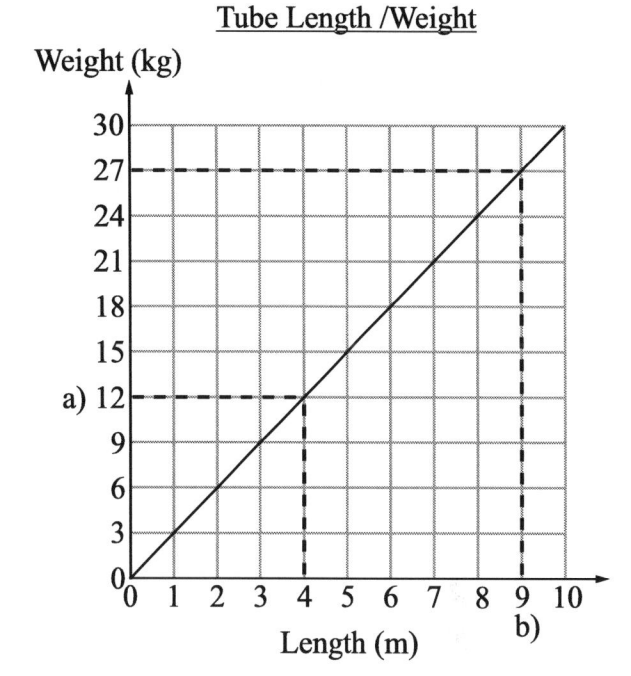

a) 12kg.
b) 9m.

Q

12) Use the table of values to draw the graph showing the volume of liquid (litres) in a tank over a period of time (sec).

Time (sec)	0	1	2	3	4	5
Volume (l)	0	10	20	30	40	50

A

12)

T	0	1	2	3	4	5
V	0	10	20	30	40	50

Plot (T, V)

(0, 0)
(1, 10)
(2, 20)
(3, 30)
(4, 40)
(5, 50)

Liquid in Tank

Proportion 4

Inverse Proportion.

Inverse Proportion
2 related values are in **inverse** proportion if ...
  as **1 value increases** then **the other value decreases** in the **same proportion** (**at the same rate**).
  and as **1 value decreases** then **the other value increases** in the **same proportion** (**at the same rate**).

*Note 1 We are not usually told that 2 values are in inverse proportion - we have to recognise this.
*Note 2 We consider that the units for each value are identical.
      (for example, in Ex. 1 ... the men all do the same amount of work on each day.)

Given 2 values in inverse proportion and the change in 1 value, we calculate the change in the other value.

Method 1
The Unitary Method
We first express the indirect proportion in **unitary form** - one of the values is **1** -
in order to calculate the change in the other value.

(It is often useful to write the **unitary form** - even if it is not asked for directly.)

  Ex.1   4 men build a wall in 6 days.
          How long would it take just 3 men?

We find how long it would take **1** man in order to find how long it would take 3 men.

$$\div 4 \begin{pmatrix} 4 \text{ men take} & 6 \text{ days} \\ 1 \text{ man takes} & 24 \text{ days} \\ 3 \text{ men take} & 8 \text{ days} \end{pmatrix} \times 4 \\ \div 3$$

*Note
We can think that ...
1 man will take 4 times longer than 4 men - so ×4
3 men will take $\frac{1}{3}$ of the time of 1 man - so ÷3

We can also think that ...
4 men × 6 days = 24 days work to be divided
                    between 3 men.

Method 2
We need to **understand** and **mentally plan** Method 1 and carry out the **same / equivalent calculations** - usually set out in **fraction form**.

(... but we do not write the **unitary form**.)

$$\times \frac{3}{4} \begin{pmatrix} 4 \text{ men take} & 6 \text{ days} \\ \\ 3 \text{ men take} & 8 \text{ days} \end{pmatrix} \times \frac{4}{3}$$

*Note
The **inverse** of $\times \frac{3}{4}$ is $\times \frac{4}{3}$

- the numerator and denominator change places.

Q	A

	Method 1	Method 2
1) 8 men build a wall in 6 days. How long would it take just 4 men?	1) $\div 8 \begin{pmatrix} 8 \text{ men ... } & 6 \text{ days} \\ 1 \text{ man ... } & 48 \text{ days} \\ 4 \text{ men ... } & 12 \text{ days} \end{pmatrix} \times 8 \\ \div 4$	$\times \frac{4}{8} \begin{pmatrix} 8 \text{ men ... } & 6 \text{ days} \\ \\ 4 \text{ men ... } & 12 \text{ days} \end{pmatrix} \times \frac{8}{4}$
	We may 'mentally calculate' that if the number of **men** is **halved** - the **time** is **doubled** .	
2) 5 pumps can fill a water tank in 4 hours. How long would it take just 2 pumps?	2) $\div 5 \begin{pmatrix} 5 \text{ pumps ... } & 4 \text{ hrs} \\ 1 \text{ pump ... } & 20 \text{ hrs} \\ 2 \text{ pumps ... } & 10 \text{ hrs} \end{pmatrix} \times 5 \\ \div 2$	$\times \frac{2}{5} \begin{pmatrix} 5 \text{ pumps ... } & 4 \text{ hrs} \\ \\ 2 \text{ pumps ... } & 10 \text{ hrs} \end{pmatrix} \times \frac{5}{2}$

Proportion 5

Inverse Proportion.

Q	A

Method 1   **Method 2**

3) It takes 8 days for 7 women to make a path.
How many women would make the path in 7 days?

3) Method 1:
÷8 (8 days ... 7 women) ×8
×7 (1 day ... 56 women) ÷7
    7 days ... 8 women

Method 2:
×7/8 (8 days ... 7 women) ×8/7
    7 days ... 8 women

4) A quantity of food will last 10 pigs for 9 days.
How long would it last 15 pigs?

4) Method 1:
÷10 (10 pigs ... 9 days) ×10
×15 (1 pig ... 90 days) ÷15
    15 pigs ... 6 days

Method 2:
×15/10 (10 pigs ... 9 days) ×10/15
    15 pigs ... 6 days

5) A group of students were placed in 3 teams of 20 students.
If the students are now placed in 6 teams, how many are in each team?

5) Method 1:
÷3 (3 teams ... 20 students) ×3
×6 (1 team ... 60 students) ÷6
    6 teams ... 10 students

Method 2:
×6/3 (3 teams ... 20 students) ×3/6
    6 teams ... 10 students

We may 'mentally calculate' that if the teams are **doubled** - the number of pupils in a team is **halved**.

6) An order for grain is packed in 72 sacks, each weighing 25kg.
How many sacks are needed if each weighs only 20kg?

6) Method 1:
÷25 (25 kg ... 72 sacks) ×25
×20 (1 kg ... 1800 sacks) ÷20
    20 kg ... 90 sacks

Method 2:
×20/25 (25 kg ... 72 sacks) ×25/20
    20 kg ... 90 sacks

7) On a section of road repair, 440 traffic cones are to be placed approximately 3m apart.
How many cones are needed if they are placed approximately 4m apart?

7) Method 1:
÷3 (3 m ... 440 cones) ×3
×4 (1 m ... 1320 cones) ÷4
    4 m ... 330 cones

Method 2:
×4/3 (3 m ... 440 cones) ×3/4
    4 m ... 330 cones

8) A car travels for 7 hours at an average speed of 40 mph.

a) What speed is needed to complete the journey in 5 hours?

b) How long does the same journey take at 30 mph?

8) a) Method 1:
÷7 (7 hrs ... 40 mph) ×7
×5 (1 hr ... 280 mph) ÷5
    5 hrs ... 56 mph

Method 2:
×5/7 (7 hrs ... 40 mph) ×7/5
    5 hrs ... 56 mph

b) Method 1:
÷40 (40 mph ... 7 hrs) ×40
×30 (1 mph ... 280 hrs) ÷30
    30 mph ... $9\frac{10}{30}$ hrs
    = $9\frac{1}{3}$ hrs
    = 9hrs 20mins

Method 2:
×30/40 (40 mph ... 7 hrs) ×40/30
    30 mph ... $9\frac{10}{30}$ hrs
    = $9\frac{1}{3}$ hrs
    = 9hrs 20mins

Proportion 6

Inverse Proportion - Graph.

Q	A
9) The graph shows the different number of men needed to build a wall and the number of days they would take. **Men / Days for Wall** Use the graph to estimate a) How many men it would take to build the wall in 5 days. b) How many days it would take 8 men to build the wall.	9) **Men / Days for Wall** a) 2 men b) $1\frac{1}{4}$ days
10) a) Use the table of values to draw the graph showing the volume of liquid (litres) in a tank over a period of time (min). | Time (sec)	0	1	2	4	8 | |---|---|---|---|---|---| | Volume (l) | 80 | 50 | 30 | 10 | 0 | **Liquid in Tank** Use the graph to estimate a) the volume of liquid in the tank at 6 min. b) the time when 40 litres are in the tank.	10) | T | 0 | 1 | 2 | 4 | 8 | |---|---|---|---|---|---| | V | 80 | 50 | 30 | 10 | 0 | Plot (T, V) (0, 80) (1, 50) (2, 30) (3, 10) (8, 0) **Liquid in Tank** a) 3·5 litres b) 1·4 min

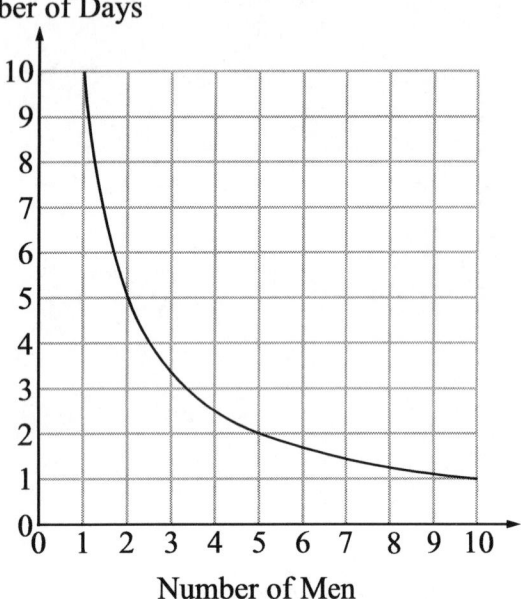

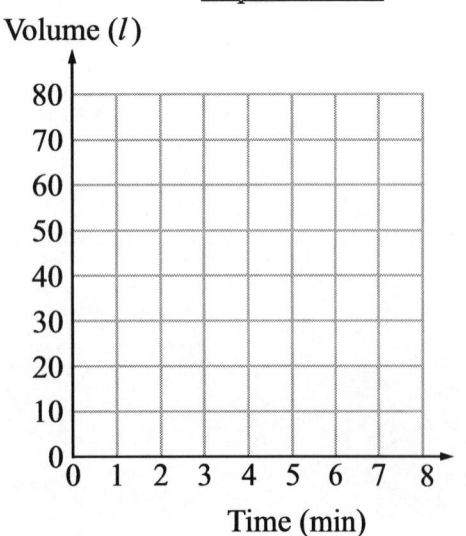

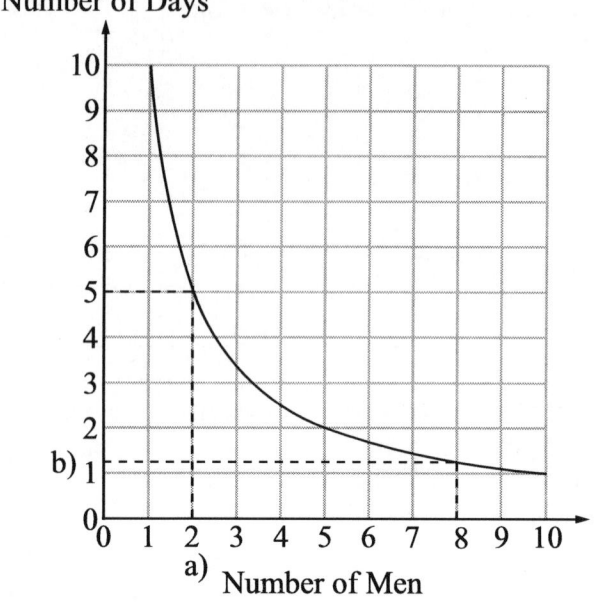

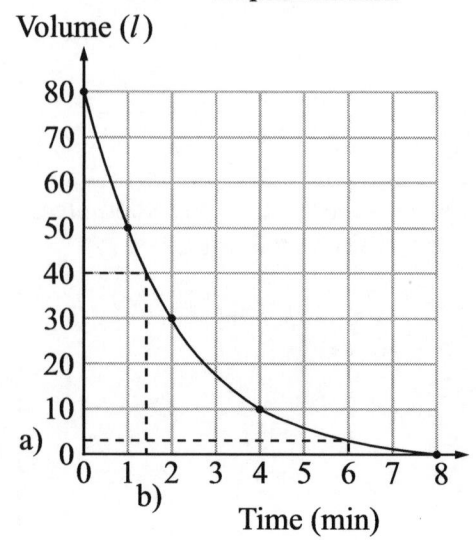

Ratio 1 — Equivalent Ratios. Simplest Form.

Ratio

Ratio is a way of comparing **related values**.
The comparison is based on **multiplication** ... how many **times** one value is more / less than another (rather than how <u>much</u> more / less than another).
The **units** of the values being compared **must be the same** - they should be shown in the work but they **cancel each other out** and so a ratio is **always** presented as **just numbers** (with **no units**).

If A and B are values, then to compare them we write

     ' the **ratio** of A to B '   or   ' the **ratio** of B to A '

In short ...   A : B   B : A
           ↑— note the **:** symbol
This is usually read as A 'to' B   B 'to' A

Consider the following different Examples of ratios of **Girls** to **Boys**.

Ex. a)

 Girls : Boys
  1 : 1 ⇐ **simplest form**
  2 : 2
  3 : 3
  4 : 4
  ⋮   ⋮

For every 1 Girl there is 1 Boy.
As the Girls **increase by 1**
 the Boys **increase by 1** also.

Ex. b)

 Girls : Boys
  1 : 2 ⇐ **simplest form**
  2 : 4
  3 : 6
  4 : 8
  ⋮   ⋮

For every 1 Girl there are 2 Boys.
As the Girls **increase by 1**
 the Boys **increase by 2**.
The Boys can <u>not</u> increase by 1.

Ex. c)

 Girls : Boys
  7 : 5 ⇐ **simplest form**
 14 : 10
 21 : 15
 28 : 20
  ⋮   ⋮

For every 7 Girls there are 5 Boys.
As the Girls **increase by 7**
 the Boys **increase by 5**.
The Girls can <u>not</u> increase by 1,2,3,4,5,6.
The Boys can <u>not</u> increase by 1,2,3,4.

Equivalent Ratios

In each Ex. (Above), the given ratios in the list are all **equivalent** to each other.
(this is the <u>same idea</u> as <u>equivalent fractions</u>.)

Simplest Form

In each Ex. (above), the **simplest form** means the **smallest possible whole numbers** for those ratios.
(this is the <u>same idea</u> as writing a <u>fraction in its lowest terms</u> which can <u>not cancel further</u>.)

A ratio can compare **any number** of values.

Ex.1

  ' the **ratio** of A to B to C '

In short ...   A : B : C

Ex.2

  ' the **ratio** of W to X to Y to Z '

In short ...   W : X : Y : Z

Ex.3
The following ratios compare the number of coloured sweets in a box.

 Red : Yellow : Green
  4 : 2 : 3   ⇐ **simplest form**
  8 : 4 : 6
 12 : 6 : 9
 16 : 8 : 12
 20 : 10 : 15
 ⋮ ⋮ ⋮

For every 4 Red sweets there are 2 Yellow and 3 Green.

As the Red **increase by 4**   Red can <u>not</u> increase by 1,2,3.
the Yellow **increase by 2** and   Yellow can <u>not</u> increase by 1.
the Green **increase by 3**.   Green can <u>not</u> increase by 1,2.

Ratio 2

State Ratios.

Ratio

Ex. There are 15 girls and 10 boys in a class.

a) What is the ratio of girls to boys.

   a) Girls : Boys
      15 : 10 (for every 15 Girls there are 10 Boys)

b) What is the ratio of boys to girls.

   b) Boys : Girls
      10 : 15 (for every 10 Boys there are 15 Girls)

Q	A
1) There are 400 boys and 380 girls in a school. a) What is the ratio of girls to boys. b) What is the ratio of boys to girls.	1) a) Girls : Boys b) Boys : Girls 380 : 400 400 : 380
2) A box contains 25 red pens and 50 blue pens. a) What is the ratio of red pens to blue pens. b) What is the ratio of blue pens to red pens.	2) a) Red pens : Blue pens b) Blue pens : Red pens 25 : 50 50 : 25
3) Team X wins £1,000 and Team Y wins £900. What is the ratio of the amount of money Team Y wins to the amount Team X wins.	3) Team Y : Team X £900 : £1,000 = 900 : 1,000 ← no units (£) ... just numbers
4) In a classroom there are 17 girls, 18 boys and 2 teachers. What is the ratio of girls to boys to teachers.	4) Girls : Boys : Teachers 17 : 18 : 2
5) Line A = 6m, Line B = 7m and Line C = 2m. What is the ratio of the lengths of the Lines A : B : C.	5) Line A : Line B : Line C 6m : 7m : 2m = 6 : 7 : 2 ← no units (m) ... just numbers
6) 4 children weigh 35kg, 32kg, 37kg, 30kg. State the ratio of their weights in the given order.	6) 35kg : 32kg : 37kg : 30kg = 35 : 32 : 37 : 30 ← no units (kg) ... just numbers

7) A store sold 80 cameras, 30 computers and 30 televisions.
What is the ratio of the items sold.
	7) Cameras : Computers : Televisions
	    80 : 30 : 30
	
	*Note Here the order of the items could be changed since a particular order is not asked for ...
	
	Ex. Televisions : Cameras : Computers
	      30 : 80 : 30

Ratio 3   Simplest Form.

Ratio - Simplest Form

We should **divide** the given values by their **highest common factor** (if they have one) in order to present the ratio in its **simplest form**.

*Note 1

If we do not divide by the **highest** common factor (because we do not recognise it / it is easier to divide by smaller common factors) then we should **keep dividing** by **smaller** common factors to obtain the **simplest form**.

*Note 2

a) In the following Exs. and Q/A the minimum detail is given, for simplicity.
b) We **always** keep the **order** of the given values - unless required otherwise .

Ex. Write each ratio in its **simplest form**.

a) 8 : 4

÷4 (8 : 4
= 2 : 1

b) £15 : £35

÷5 (£15 : £35
= 3 : 7

c) 6m : 90cm

6m : 90cm
÷30 (600**cm** : 90**cm** ⇐ (same units **cm**)
= 20 : 3

*Note
It is usual to change larger units to smaller units to avoid fractions / decimals.

Q	A	Q	A
Write each ratio in its **simplest form**. 1) 2 : 6	1) ÷2 (2 : 6 = 1 : 3	9) £5 : 50p	9) £5 : 50p ÷50 (= 500p : 50p = 10 : 1
2) 45 : 25	2) ÷5 (45 : 25 = 9 : 5	10) 75m : 2km	10) 75m : 2km ÷25 (= 75**m** : 2,000**m** = 3 : 80
3) grey : white tiles tiles	3) grey : white ÷2 (4 : 6 = 2 : 3	11) 1kg : 600g	11) 1kg : 600g ÷200 (= 1,000**g** : 600**g** = 5 : 3
4) 15 : 9 : 18	4) ÷3 (15 : 9 : 18 = 5 : 3 : 6	12) 440ml : 8 l	12) 440ml : 8 l ÷40 (= 440**ml** : 8,000**ml** = 11 : 200
5) 14 : 28 : 35 : 14	5) ÷7 (14 : 28 : 35 : 14 = 2 : 4 : 5 : 2	13) 9m : 9cm : 9mm	13) 9m : 9cm : 9mm ÷9 (= 9,000**mm** : 90**mm** : 9**mm** = 1,000 : 10 : 1
6) £70 : £50	6) ÷10 (£70 : £50 = 7 : 5	14) 54mins : 2hrs	14) 54mins : 2hrs ÷6 (= 54**mins** : 120**mins** = 9 : 20
7) 39kg : 13kg	7) ÷13 (39kg : 13kg = 3 : 1		
8) 8m : 12m : 16m	8) ÷4 (8m : 12m : 16m = 2 : 3 : 4		

Ratio 4

Simplest Form - Fraction Values.

Ratio - Simplest Form
Fraction Values

*Note In the following Exs. and Q/A, **units** (ex. £, kg, cm, mins, ...) of the values being compared are omitted for simplicity.

A) When the **denominators** are **equal** (as in the following Exs.) they are regarded as the 'same units' and **cancel each other out** - the **numerators** become the whole numbers.

Ex. Write each ratio in its simplest form.

a) $\frac{1}{4} : \frac{3}{4}$

$\frac{1}{4} : \frac{3}{4}$ 'same units'
$= 1 : 3$

b) $\frac{10}{11} : \frac{4}{11}$

$\frac{10}{11} : \frac{4}{11}$ 'same units'
$= 10 : 4$
$\div 2$
$= 5 : 2$

c) $\frac{2}{5} : 3$

$\frac{2}{5} : 3$
$= \frac{2}{5} : \frac{15}{5}$ 'same units'
$= 2 : 15$

d) $\frac{5}{6} : 7\frac{1}{6}$

$\frac{5}{6} : 7\frac{1}{6}$
$= \frac{5}{6} : \frac{43}{6}$ 'same units'
$= 5 : 43$

A whole number or mixed number is changed to an **improper fraction**.

Q	A	Q	A
Write each ratio in its **simplest form**.		7) $\frac{1}{2} : 2\frac{1}{2}$	7) $\frac{1}{2} : 2\frac{1}{2}$ $= \frac{1}{2} : \frac{5}{2}$ $= 1 : 5$
1) $\frac{2}{3} : \frac{1}{3}$	1) $\frac{2}{3} : \frac{1}{3}$ $= 2 : 1$		
2) $\frac{5}{9} : \frac{7}{9}$	2) $\frac{5}{9} : \frac{7}{9}$ $= 5 : 7$	8) $6\frac{1}{4} : \frac{3}{4}$	8) $6\frac{1}{4} : \frac{3}{4}$ $= \frac{25}{4} : \frac{3}{4}$ $= 25 : 3$
3) $\frac{12}{13} : \frac{8}{13}$	3) $\frac{12}{13} : \frac{8}{13}$ $= 12 : 8$ $\div 4$ $= 3 : 2$	9) $\frac{6}{7} : 8\frac{4}{7}$	9) $\frac{6}{7} : 8\frac{4}{7}$ $= \frac{6}{7} : \frac{60}{7}$ $= 6 : 60$ $\div 6$ $= 1 : 10$
4) $\frac{21}{22} : \frac{9}{22}$	4) $\frac{21}{22} : \frac{9}{22}$ $= 21 : 9$ $\div 3$ $= 7 : 3$	10) $5\frac{5}{8} : 1\frac{7}{8}$	10) $5\frac{5}{8} : 1\frac{7}{8}$ $= \frac{45}{8} : \frac{15}{8}$ $= 45 : 15$ $\div 15$ $= 3 : 1$
5) $\frac{5}{6} : 4$	5) $\frac{5}{6} : 4$ $= \frac{5}{6} : \frac{24}{6}$ $= 5 : 24$		
6) $9 : \frac{2}{9}$	6) $9 : \frac{2}{9}$ $= \frac{81}{9} : \frac{2}{9}$ $= 81 : 2$	11) $3\frac{2}{25} : 3\frac{21}{25}$	11) $3\frac{2}{25} : 2\frac{21}{25}$ $= \frac{77}{25} : \frac{71}{25}$ $= 77 : 71$

Ratio 5

Simplest Form - Fraction Values.

Ratio - Simplest Form
Fraction Values

*Note In the following Exs. and Q/A, **units** (ex. £, kg, cm, mins, ...) of the values being compared are omitted for simplicity.

B) If the **denominators** are **not** equal then we need to **make them equal** - they are then regarded as the 'same units' and **cancel each other out** - the **numerators** become the whole numbers.

We find **equivalent fractions** with a **common denominator**

(This is the same method as finding a common denominator in order to add / subtract fractions - see **Fractions 11**.)

Ex. Write each ratio in its simplest form.

a) $\frac{1}{2} : \frac{3}{4}$

$\times 2 \left(\frac{1}{2} : \frac{3}{4} \right) \times 1$

$= \frac{2}{4} : \frac{3}{4}$

'same units'

$= 2 : 3$

b) $\frac{5}{8} : \frac{6}{7}$

$\times 7 \left(\frac{5}{8} : \frac{6}{7} \right) \times 8$

$= \frac{35}{56} : \frac{48}{56}$

'same units'

$= 35 : 48$

c) $3\frac{3}{4} : \frac{5}{6}$

$= \times 3 \left(\frac{15}{4} : \frac{5}{6} \right) \times 2$

$= \frac{45}{12} : \frac{10}{12}$

'same units'

$= 45 : 10$

$\div 5$

$= 9 : 2$

Q	A	Q	A
Write each ratio in its **simplest form**. 1) $\frac{5}{6} : \frac{1}{3}$	1) $\times 1 \left(\frac{5}{6} : \frac{1}{3} \right) \times 2$ $= \frac{5}{6} : \frac{2}{6}$ $= 5 : 2$	5) $\frac{11}{20} : \frac{11}{12}$	5) $\times 3 \left(\frac{11}{20} : \frac{11}{12} \right) \times 5$ $= \frac{33}{60} : \frac{55}{60}$ $= 33 : 55$ $\div 11$ $= 3 : 5$
2) $\frac{4}{5} : \frac{13}{15}$	2) $\times 3 \left(\frac{4}{5} : \frac{13}{15} \right) \times 1$ $= \frac{12}{15} : \frac{13}{15}$ $= 12 : 13$	6) $\frac{3}{8} : 2\frac{1}{10}$	6) $\frac{3}{8} : 2\frac{1}{10}$ $= \times 5 \left(\frac{3}{8} : \frac{21}{10} \right) \times 4$ $= \frac{15}{40} : \frac{84}{40}$ $= 15 : 84$ $\div 3$ $= 5 : 28$
3) $\frac{2}{7} : \frac{7}{9}$	3) $\times 9 \left(\frac{2}{7} : \frac{7}{9} \right) \times 7$ $= \frac{18}{63} : \frac{49}{63}$ $= 18 : 49$	7) $4\frac{2}{3} : 3\frac{1}{2}$	7) $4\frac{2}{3} : 3\frac{1}{2}$ $= \times 2 \left(\frac{14}{3} : \frac{7}{2} \right) \times 3$ $= \frac{28}{6} : \frac{21}{6}$ $= 28 : 21$ $\div 7$ $= 4 : 3$
4) $\frac{5}{6} : \frac{3}{8}$	4) $\times 4 \left(\frac{5}{6} : \frac{3}{8} \right) \times 3$ $= \frac{20}{24} : \frac{9}{24}$ $= 20 : 9$		

Ratio 6 — Simplest Form - Decimal Values.

Ratio - Simplest Form
Decimal Values

*Note In the following Exs. and Q/A, **units** (ex. £, kg, cm, mins, ...) of the values being compared are omitted for simplicity.

We **multiply** the decimal values ×10, ×100, ×1000 ... to obtain whole numbers.

Ex. Write each ratio in its simplest form.

a) 0·4 : 0·7
   ×10 (0·4 : 0·7
       =   4 : 7

b) 1·2 : 0·8
   ×10 (1·2 : 0·8
       = 12 : 8
   ÷4 (
       =   3 : 2

c) 0·3 : 0·55
   ×100 (0·3 : 0·55
        = 30 : 55
   ÷5   (
        =   6 : 11

d) 0·78 : 6
   ×100 (0·78 : 6
        = 78 : 600
   ÷6   (
        = 13 : 100

Q	A	Q	A
Write each ratio in its **simplest form**. 1) 0·9 : 0·1	1) ×10 (0·9 : 0·1 = 9 : 1	7) 8·88 : 8	7) ×100 (8·88 : 8 = 888 : 800 ÷8 (= 111 : 100
2) 3·2 : 2	2) ×10 (3·2 : 2 = 32 : 20 ÷4 (= 8 : 5	8) 0·02 : 0·028	8) ×1000 (0·02 : 0·028 = 20 : 28 ÷4 (= 5 : 7
3) 5·6 : 5·6	3) ×10 (5·6 : 5·6 = 56 : 56) or ÷56 (= 1 : 1 ÷5·6	9) 9·909 : 6	9) ×1000 (9·909 : 6 = 9909 : 6000 ÷3 (= 3303 : 2000
4) 10 : 4·7	4) ×10 (10 : 4·7 = 100 : 47	10) 0·8 : 0·4 : 0·6	10) ×10 (0·8 : 0·4 : 0·6 = 8 : 4 : 6 ÷2 (= 4 : 2 : 3
5) 0·15 : 0·71	5) ×100 (0·15 : 0·71 = 15 : 71		
6) 6·3 : 0·09	6) ×100 (6·3 : 0·09 = 630 : 9 ÷9 (= 70 : 1	11) 5 : 0·5 : 0·05	11) ×100 (5 : 0·5 : 0·05 = 500 : 50 : 5 ÷5 (= 100 : 10 : 1

Ratio 7 — Unitary Form.

Ratio - Unitary Form
A ratio can be expressed in the **unitary** form (one of the values is 1) ... **1 : n** or **n : 1**
The number **n** is written as a decimal or whole number.

In this form, the ratio states that one value is **n times** the other value.

We **divide** one value **by itself** in order to obtain the number **1** and the other value by the **same number**.

Ex.1 Write each ratio in the form **1 : n**

a) 2 : 6

    1 : n
÷2 (2 : 6
   = 1 : 3

b) 3m : 4m

    1 : n
÷3 (3m : 4m
   = 1 : $\frac{4}{3}$
   = 1 : 1·333...

Ex.2 Write each ratio in the form **n : 1**

a) 5 : 10

    n : 1
    5 : 10) ÷10
  = $\frac{5}{10}$: 1
  = $\frac{1}{2}$: 1
  = 0·5 : 1

b) 9kg : 4g

    n : 1
    9kg : 4g
  = 9000g : 4g) ÷4
  = 2250 : 1

Q	A	Q	A
Write each ratio in the form **1 : n**	**1 : n**	Write each ratio in the form **n : 1**	**n : 1**
1) 4 : 8	1) ÷4 (4 : 8 = 1 : 2	1) 8 : 4	1) 8 : 4) ÷4 = 2 : 1
2) 100 : 600	2) ÷100 (100 : 600 = 1 : 6	2) 90 : 30	2) 90 : 30) ÷30 = 3 : 1
3) 5g : 7g	3) ÷5 (5g : 7g = 1 : $\frac{7}{5}$ (= $1\frac{2}{5}$) = 1 : 1·4	3) 11ml : 2ml	3) 11ml : 2ml) ÷2 = $\frac{11}{2}$ (= $5\frac{1}{2}$) : 1 = 5·5 : 1
4) 6 : 3	4) ÷6 (6 : 3 = 1 : $\frac{3}{6}$ = 1 : 0·5	4) 9 : 12	4) 9 : 12) ÷12 = $\frac{9}{12}$: 1 = 0·75 : 1
5) 9 : 34	5) ÷9 (9 : 34 = 1 : $\frac{34}{9}$ (= $3\frac{7}{9}$) = 1 : 3·777...	5) £56 : 80p	5) £56 : 80p = 5600p : 80p) ÷80 = 70 : 1
6) 2cm : 1m	6) 2cm : 1m = 2cm : 100cm ÷2 (= 1 : 50	6) 41min : 1hr	6) 41min : 1hr = 41min : 60min) ÷60 = $\frac{41}{60}$: 1 = 0·683... : 1
7) £5 : 25p	7) £5 : 25p = 500p : 25p ÷500 (= 1 : 0·05		

192

Ratio 8 — Equivalent Ratios - Find Unknown Value.

Equivalent Ratios - Find Unknown Value
Given 2 **equivalent ratios**, we can calculate 1 unknown value.

Ex.1 If $x : 3$ and $8 : 12$ are equivalent, calculate x.

$$\div 4 \begin{pmatrix} 8 : 12 \\ x : 3 \end{pmatrix} \div 4 \quad \Leftarrow \quad \frac{12}{3} = 4$$

$$x = \frac{8}{4} = 2$$

We **divide** the given values on the 'right-hand side' $\frac{12}{3}$ to obtain the number **4** and then ...

divide the given value on the 'left-hand side' by the same number **4** to obtain the value of x.

Ex.2 If $9 : 5$ and $90 : y$ are equivalent, calculate y.

$$\times 10 \begin{pmatrix} 9 : 5 \\ 90 : y \end{pmatrix} \times 10 \quad \Leftarrow \quad \frac{90}{9} = 10$$

$$y = 5 \times 10 = 50$$

We **divide** the given values on the 'left-hand side' $\frac{90}{9}$ to obtain the number **10** and then ...

multiply the given value on the 'right-hand side' by the same number **10** to obtain the value of y.

Q	A
Find the value of the letter for each pair of **equivalent ratios**.	*Note The minimum work is shown.
1) $x : 7$ and $10 : 14$	1) $\div 2 \begin{pmatrix} 10 : 14 \\ x : 7 \end{pmatrix} \div 2 \quad x = \frac{10}{2} = 5$
2) $6 : y$ and $18 : 33$	2) $\div 3 \begin{pmatrix} 18 : 33 \\ 6 : y \end{pmatrix} \div 3 \quad y = \frac{33}{3} = 11$
3) $15 : 20$ and $k : 4$	3) $\div 5 \begin{pmatrix} 15 : 20 \\ k : 4 \end{pmatrix} \div 5 \quad k = \frac{15}{5} = 3$
4) $72 : 56$ and $9 : n$	4) $\div 8 \begin{pmatrix} 72 : 56 \\ 9 : n \end{pmatrix} \div 8 \quad n = \frac{56}{8} = 7$
5) $1 : 8$ and $c : 80$	5) $\times 10 \begin{pmatrix} 1 : 8 \\ c : 80 \end{pmatrix} \times 10 \quad c = 1 \times 10 = 10$
6) $150 : p$ and $3 : 2$	6) $\times 50 \begin{pmatrix} 3 : 2 \\ 150 : p \end{pmatrix} \times 50 \quad p = 2 \times 50 = 100$
7) $4 : 9$ and $44 : t$	7) $\times 11 \begin{pmatrix} 4 : 9 \\ 44 : t \end{pmatrix} \times 11 \quad t = 9 \times 11 = 99$
8) In a class, the ratio of boys to girls is $5 : 6$. If there are 15 boys, how many girls are there?	8) boys : girls $\times 3 \begin{pmatrix} 5 : 6 \\ 15 : g \end{pmatrix} \times 3 \quad g = 6 \times 3 = 18$ girls
9) Red and white paint is mixed in the ratio $4 : 1$. How much red paint is mixed with $20\,l$ of white?	9) red : white $\times 20 \begin{pmatrix} 4 : 1 \\ r : 20 \end{pmatrix} \times 20 \quad r = 4 \times 20 = 80\,l$ red paint
10) The ratio of the lengths of 2 squares is $9 : 7$. The length of the largest square is $76 \cdot 5$m. Calculate the length of the smallest square.	10) large square : small square $\times 8 \cdot 5 \begin{pmatrix} 9 : 7 \\ 76 \cdot 5 : s \end{pmatrix} \times 8 \cdot 5 \quad s = 7 \times 8 \cdot 5 = 59 \cdot 5$m

Ratio 9 — Divide / Share in a Given Ratio.

Divide / Share in a Given Ratio.
We can divide / share a value in a given ratio.

Method 1 - Divide the value by the total ratio values.

Ex. Divide 20 in the ratio 2 : 3.

We need to find 2 values, in the same ratio as 2 : 3, which total 20.
From the given information, we find the value of '1 share' from the **Total** values.

```
2 : 3    Total
2 + 3 =  5
↓   ↓
↓   ↓
  +   = 20
```

We relate the 2 **Totals** by **dividing** them ... $\frac{20}{5} = 4$ (1 'share')
and **multiply** the ratio values by this 4.

⇨
```
2 : 3    Total
2 + 3 =  5
↓×4 ↓×4  ↓×4
8 + 12 = 20
```
8, 12 are the required values.

Q	A	Q	A
1) Divide 30 in the ratio 4 : 1.	1) 4 : 1 Total 4 + 1 = 5 ↓ ↓ ↓×6 24 + 6 = 30 24, 6	6) Divide 0·99g in the ratio 7 : 4.	6) 7 : 4 Total 7 + 4 = 11 ↓ ↓ ↓×0·09 0·63 + 0·36 = 0·99 0·63g, 0·36g
2) Divide 56 in the ratio 2 : 5.	2) 2 : 5 Total 2 + 5 = 7 ↓ ↓ ↓×8 16 + 40 = 56 16, 40	7) Divide £68·50 in the ratio 19 : 31.	7) 19 : 31 Total 19 + 31 = 50 ↓ ↓ ↓×1·37 26·03 + 42·47 = 68·50 £26·03, £42·47
3) Divide 900g in the ratio 7 : 8.	3) 7 : 8 Total 7 + 8 = 15 ↓ ↓ ↓×60 420 + 480 = 900 420g, 480g	8) In a class of 25 pupils the ratio of girls to boys is 3 : 2. How many girls and boys are there?.	8) girls : boys 3 : 2 Total 3 + 2 = 5 ↓ ↓ ↓×5 15 + 10 = 25 15 girls, 10 boys
4) Divide 39m in the ratio 3 : 10.	4) 3 : 10 Total 3 + 10 = 13 ↓ ↓ ↓×3 9 + 30 = 39 9m, 30m	9) In a survey of 400, 'Yes' and 'No' votes are in the ratio of 9 : 16. How many of each vote are there?	9) Yes : No 9 : 16 Total 9 + 16 = 25 ↓ ↓ ↓×16 144 + 256 = 400 'Yes' 144, 'No' 256
5) Share £18 in the ratio 6 : 1 : 2.	5) 6 : 1 : 2 Total 6 + 1 + 2 = 9 ↓ ↓ ↓ ↓×2 12 + 2 + 4 = 18 £12, £2, £4	10) The angles of a triangle are in the ratio 5 : 7 : 6. Find the size of each angle.	10) 5 : 7 : 6 Total 5 + 7 + 6 = 18 ↓ ↓ ↓ ↓×10 50 + 70 + 60 = 180 50°, 70°, 60° (180° in △)

Q	A
11) 10 Red, 13 Yellow and 21 White cans of paint are bought. 40 cans are mixed in the ratio 2 : 3 : 5. State how many cans of each colour paint are not used.	11) R : Y : W 2 : 3 : 5 Total 2 + 3 + 5 = 10 ↓ ↓ ↓ ↓×4 8 + 12 + 20 = 40 cans used (+) 2 + 1 + 1 = 4 cans not used 10 + 13 + 21 = 44 cans bought

Ratio 10 — Divide / Share in a Given Ratio.

Divide / Share in a Given Ratio
We can divide / share a value in a given ratio.
Method 2 - Find the value of 'each share' as a fraction of the whole.

Ex. Divide 20 in the ratio 2 : 3.

There are 2 + 3 = 5 'shares'.

smaller value $= \frac{2}{5}$ of 20
$= \frac{2}{5} \times \frac{20}{1} = \frac{40}{5} = 8$

larger value $= \frac{3}{5}$ of 20
$= \frac{3}{5} \times \frac{20}{1} = \frac{60}{5} = 12$

*Note
Fractions may 'cancel' - this is not shown here.

Q	A	Q	A
1) Divide 15 in the ratio 4 : 1.	1) 4 + 1 = 5 'shares'. $\frac{4}{5} \times \frac{15}{1} = \frac{60}{5} = 12$ $\frac{1}{5} \times \frac{15}{1} = \frac{15}{5} = 3$	6) Divide 0·55mg in the ratio 9 : 2.	6) 9 + 2 = 11 'shares'. $\frac{9}{11} \times \frac{0·55}{1} = \frac{4·95}{11} = 0·45$mg $\frac{2}{11} \times \frac{0·55}{1} = \frac{1·10}{11} = 0·1$mg
2) Divide 80 in the ratio 3 : 7.	2) 3 + 7 = 10 'shares'. $\frac{3}{10} \times \frac{80}{1} = \frac{240}{10} = 24$ $\frac{7}{10} \times \frac{80}{1} = \frac{560}{10} = 56$	7) Divide £82 in the ratio 17 : 23.	7) 17 + 23 = 40 'shares'. $\frac{17}{40} \times \frac{82}{1} = \frac{1394}{40} = £34·85$ $\frac{23}{40} \times \frac{82}{1} = \frac{1886}{40} = £47·15$
3) Divide 600g in the ratio 11 : 9.	3) 11 + 9 = 20 'shares'. $\frac{11}{20} \times \frac{600}{1} = \frac{6600}{20} = 330$g $\frac{9}{20} \times \frac{600}{1} = \frac{5400}{20} = 270$g	8) In a class of 27 pupils the ratio of girls to boys is 5 : 4. How many girls and boys are there?	8) 5 + 4 = 9 'shares'. $\frac{5}{9} \times \frac{27}{1} = \frac{135}{9} = 15$ girls $\frac{4}{9} \times \frac{27}{1} = \frac{108}{9} = 12$ boys
4) Divide 27m in the ratio 5 : 8.	4) 5 + 8 = 13 'shares'. $\frac{5}{13} \times \frac{27}{1} = \frac{135}{13} = 10·3...$m $\frac{8}{13} \times \frac{27}{1} = \frac{216}{13} = 16·6...$m	9) Divide 2ml in the ratio 3 : 5.	9) 3 + 5 = 8 'shares'. $\frac{3}{8} \times \frac{2}{1} = \frac{6}{8} = 0·75$ml $\frac{5}{8} \times \frac{2}{1} = \frac{10}{8} = 1·25$ml
5) Share £49 in the ratio 1 : 2 : 4.	5) 1 + 2 + 4 = 7 'shares'. $\frac{1}{7} \times \frac{49}{1} = \frac{49}{7} = £7$ $\frac{2}{7} \times \frac{49}{1} = \frac{98}{7} = £14$ $\frac{4}{7} \times \frac{49}{1} = \frac{196}{7} = £28$	10) The angles of a triangle are in the ratio 6 : 2 : 7. Find the size of each angle.	10) 6 + 2 + 7 = 15 'shares'. $\frac{6}{15} \times \frac{180}{1} = \frac{1080}{15} = 72°$ $\frac{2}{15} \times \frac{180}{1} = \frac{360}{15} = 24°$ $\frac{7}{15} \times \frac{180}{1} = \frac{1260}{15} = 84°$ (180° in △)

Q	A
11) A 90g compound is required of chemicals A, B and C in a weight (g) ratio of 2 : 1 : 3. 30g of A, 15g of B and 50g of C are available. State if the compound can be made.	11) 2 + 1 + 3 = 6 'shares'. $\frac{2}{6} \times \frac{90}{1} = \frac{180}{6} = 30$g $\frac{1}{6} \times \frac{90}{1} = \frac{90}{6} = 15$g $\frac{3}{6} \times \frac{90}{1} = \frac{270}{6} = 45$g The compound can be made - 5g of C is not used. 50g − 45g = 5g

Ratio 11

Actual difference between Ratio

Given the **actual difference** between ratio values then the **real values** of the ratio can be calculated.

Ex.1 2 numbers are in the ratio 4 : 3.
The **difference** between the numbers is 5.
Calculate the numbers.

$$\begin{array}{c} 4 : 3 \quad \textbf{difference} \\ 4 - 3 = \dfrac{5}{1} \text{ 'share'} \left. \right\} = 5 \ (1 \text{ 'share'}) \\ \times 5 \downarrow \quad \downarrow \\ 20 - 15 = 5 \text{ 'check'} \end{array}$$

We relate the **actual difference** ...
to the **ratio difference** ... to find the value of 1 'share'.
We **multiply** the ratio values by this **5**
to find the **real values.**

20 , 15 are the numbers required.

Ex.2 2 numbers are in the ratio 5 : 8.
The **difference** between the numbers is 21.
Calculate the numbers.

$$\begin{array}{c} 5 : 8 \quad \textbf{difference} \\ -5 \ 8 = \dfrac{21}{3} \text{ 'shares'} \left. \right\} = 7 \ (1 \text{ 'share'}) \\ \times 7 \downarrow \quad \downarrow \\ -35 \quad 56 = 21 \text{ 'check'} \end{array}$$

35 , 56 are the numbers required.

Q	A	Q	A
1) 2 numbers are in the ratio 7 : 6. The **difference** between the numbers is 4. Calculate the numbers.	1) 7 : 6 **difference** $7 - 6 = \dfrac{4}{1} \} = 4$ ×4 ↓ ↓ 28 − 24 = 4 'check' 28 , 24 required	5) A and B share money in the ratio 8 : 3. A receives £30 more than B. Calculate the amounts A and B receive.	5) A : B 8 : 3 **difference** $8 - 3 = \dfrac{30}{5} \} = 6$ ×6 ↓ ↓ 48 − 18 = 30 'check' A£48 , B£18
2) 2 numbers are in the ratio 9 : 7. The **difference** between the numbers is 6. Calculate the numbers.	2) 9 : 7 **difference** $9 - 7 = \dfrac{6}{2} \} = 3$ ×3 ↓ ↓ 27 − 21 = 6 'check' 27 , 21 required	6) A farmer has sheep and cows in the ratio 15 : 4. There are 55 more sheep than cows. Calculate the number of sheep (S) and cows (C).	6) S : C 15 : 4 **difference** $15 - 4 = \dfrac{55}{11} \} = 5$ ×5 ↓ ↓ 75 − 20 = 55 'check' S 75, C 20
3) 2 numbers are in the ratio 4 : 5. The **difference** between the numbers is 9. Calculate the numbers.	3) 4 : 5 **difference** $-4 \ 5 = \dfrac{9}{1} \} = 9$ ×9 ↓ ↓ −36 45 = 9 'check' 36 , 45 required	7) A mother is 26 years older than her daughter. Their ages are in the ratio 8 : 21. Calculate their ages.	7) D : M 8 : 21 **difference** $-8 \ 21 = \dfrac{26}{13} \} = 2$ ×2 ↓ ↓ −16 42 = 26 'check' Daughter 16yrs Mother 42yrs
4) 2 numbers are in the ratio 6 : 11. The **difference** between the numbers is 40. Calculate the numbers.	4) 6 : 11 **difference** $-6 \ 11 = \dfrac{40}{5} \} = 8$ ×8 ↓ ↓ −48 88 = 40 'check' 48 , 88 required	8) The number of points for Teams X,Y and Z are in the ratio 33 : 37 : 19. Team Y scored 40 points more than Team X. Calculate Team Z points.	8) X : Y 33 : 37 **difference** $-33 \ 37 = \dfrac{40}{4} \} = 10$ ×10 ↓ ↓ −330 370 = 40 'check' Team Z 19 × 10 = 190 points

Ratio 12

Ratio of 3 values - given 2 values

The ratio of 3 values can be found - given the **fraction of the whole** of each of 2 of the values.

Ex.

An amount of money is shared in the ratio A : B : C.

A receives $\frac{1}{4}$ of the amount.

B receives $\frac{5}{8}$ of the amount.

State the ratio A : B : C.

We find the **total** 'share' of A + B and **subtract from 1** (whole) to find C's 'share'.

$$\times 2 \left(\frac{\overset{A}{1}}{4} + \frac{\overset{B}{5}}{8} \right) \times 1$$

$$= \frac{2}{8} + \frac{5}{8} = \frac{7}{8}$$

$$C = 1 - \frac{7}{8} = \frac{1}{8}$$

A : B : C
$\left(\frac{2}{8} : \frac{5}{8} : \frac{1}{8} \right)$
2 : 5 : 1

Q	A	Q	A
1) An amount of money is shared in the ratio A : B : C. A receives $\frac{4}{9}$ of the amount. B receives $\frac{1}{3}$ of the amount. State the ratio A:B:C.	1) $\times 1 \left(\frac{\overset{A}{4}}{9} + \frac{\overset{B}{1}}{3} \right) \times 3$ $= \frac{4}{9} + \frac{3}{9} = \frac{7}{9}$ $C = 1 - \frac{7}{9} = \frac{2}{9}$ A : B : C $\left(\frac{4}{9} : \frac{3}{9} : \frac{2}{9} \right)$ 4 : 3 : 2	4) Voters stated 'Yes', 'No' or 'Undecided'. $\frac{7}{20}$ are 'Undecided'. 40% vote 'Yes'. State the vote ratio Yes:No:Undecided.	4) $40\% = \frac{40}{100} = \frac{8}{20}$ U Y $\frac{7}{20} + \frac{8}{20} = \frac{15}{20}$ $N = 1 - \frac{15}{20} = \frac{5}{20}$ Y : N : U $\left(\frac{8}{20} : \frac{5}{20} : \frac{7}{20} \right)$ 8 : 5 : 7 or $\frac{7}{20} = \frac{35}{100} = 35\%$ (×5) 35% + 40% = 75% $N = 1 - 75\% = 25\%$ Y : N : U $\left(40\% : 25\% : 35\% \right) \div 5\%$ 8 : 5 : 7
2) Paint is mixed in the ratio Red:White:Blue. Red is $\frac{2}{5}$ of the mix. Blue is $\frac{1}{6}$ of the mix. State the paint ratio Red:White:Blue.	2) $\times 6 \left(\frac{\overset{R}{2}}{5} + \frac{\overset{B}{1}}{6} \right) \times 5$ $= \frac{12}{30} + \frac{5}{30} = \frac{17}{30}$ $W = 1 - \frac{17}{30} = \frac{13}{30}$ R : W : B $\left(\frac{12}{30} : \frac{13}{30} : \frac{5}{30} \right)$ 12 : 13 : 5		
3) Gold, Silver, Bronze medals are won by a team over a year. $\frac{5}{12}$ of the medals are Gold. $\frac{5}{18}$ of the medals are Silver. State the medal ratio Gold:Silver:Bronze.	3) $\times 3 \left(\frac{\overset{G}{5}}{12} + \frac{\overset{S}{5}}{18} \right) \times 2$ $= \frac{15}{36} + \frac{10}{36} = \frac{25}{36}$ $B = 1 - \frac{25}{36} = \frac{11}{36}$ G : S : B $\left(\frac{15}{36} : \frac{10}{36} : \frac{11}{36} \right)$ 15 : 10 : 11	5) A,B,C,D form a compound of X kg. $A = \frac{2}{9}$ of X. $B = \frac{1}{5}$ of X. C kg = D kg. State the ratio A : B : C : D.	5) $\times 5 \left(\frac{\overset{A}{2}}{9} + \frac{\overset{B}{1}}{5} \right) \times 9$ $= \frac{10}{45} + \frac{9}{45} = \frac{19}{45}$ $C+D = 1 - \frac{19}{45} = \frac{26}{45} \div 2$ $C = D = \frac{13}{45}$ A : B : C : D $\left(\frac{10}{45} : \frac{9}{45} : \frac{13}{45} : \frac{13}{45} \right)$ 10 : 9 : 13 : 13

Foreign Exchange 1   Currency Units / Exchange Rate.

Foreign Exchange
Foreign Exchange is the exchange of **currencies** (money) used in different countries.

The basic **unit of currency** used in the United Kingdom / Great Britain (England, Scotland, Wales and Northern Ireland) is the **pound** for which the symbol is **£**.
It is referred to as the Great Britain Pound or Pound Sterling.
The smaller units used are **pence**, usually written **p**, and the system values **£**1 = 100**p**.
The number 100 allows calculations based on the decimal system.

Similar **currency systems** are used in other countries - a few are shown in the table (below).
Some countries only use a basic unit and have no smaller units - ex. Japan.

In order to spend money in another country - for example, when on foreign holiday / business - the currency of that country is needed.
The **Exchange Rate** determines the value at which a currency is exchanged - the table (below) lists how much the UK £1 can be exchanged for in the currencies shown.
The Exchange Rate is affected by many factors and is **continually changing** during the day!

The banks, Post Office and Bureau de Change, for example, usually state a rate for the whole day for relatively small amounts for people travelling abroad on holiday / business.
(Travellers Cheques can also be obtained which are then cashed abroad.)

A 3-letter code is used internationally to distinguish each currency - Ex. **GBP** (**G**reat **B**ritain **P**ound).

*Note
The actual Exchange Rates given here (to just 2 decimal places) are not important and may be different if we checked them today ... it is the idea that matters !

Exchange rates

Country	Unit of currency	3-letter code	Unit symbol / system	Exchange Rate
United Kingdom	British Pound	GBP	£1 = 100 pence	
** European Union	Euro	EUR	€1 = 100 cents	€1·16
Australia	Australian Dollar	AUD	$1 = 100 cents	$2·09
China	Yuan	CNY	¥1 = 100 fens	¥10·50
Denmark	Danish Kroner	DKK	kr1 = 100 ores	kr8·73
Egypt	Egyptian Pound	EGP	E£1 = 100 piastres	E£9·25
India	Indian Rupee	INR	₨1 = 100 paises	₨80
Japan	Yen	JPY	¥1	¥150
Mexico	Peso	MXN	$1 = 100 centavos	$21·98
Nigeria	Naira	NGN	₦1 = 100 kobos	₦242
Switzerland	Swiss Franc	CHF	CHF1 = 100 centimes	CHF1·70
South Africa	South African Rand	ZAR	R1 = 100 cents	R13·25
U.S.A.	American Dollar	USD	$1 = 100 cents	$1·44

** European Union (EU) includes France, Germany, Italy, ... and many more countries.

*Note
The table (above) is used for the following Exs. and Q/A.

Foreign Exchange 2

Exchange Pound.

Exchange Pound

To exchange British Pounds for another currency ...

multiply ... the currency **rate of exchange** × the number of **Pounds**

Ex. Exchange £200 for Euros.

£1 = €1·16

Exchange = €1·16 × 200   (' for every £1 we have €1·16 ')

= €232

Q	A	Q	A
1) Exchange £750 for Euros.	1) £1 = €1·16 Exchange = €1·16 × 750 = €870	4) Exchange £900 for Japanese Yens.	4) £1 = ¥150 Exchange = ¥150 × 900 = ¥135000
2) Exchange £80 for Australian Dollars.	2) £1 = $2·09 Exchange = $2·09 × 80 = $167·20	5) Exchange £300 for Nigerian Nairas.	5) £1 = ₦242 Exchange = ₦242 × 300 = ₦72600
3) Exchange £641 for Danish Kroner.	3) £1 = kr8·73 Exchange = kr8·73 × 641 = kr5595·93	6) Exchange £50 for Swiss Francs.	6) £1 = CHF1·70 Exchange = CHF1·70 × 50 = CHF85

To exchange another currency for British Pounds -

divide ... $\dfrac{\text{the number of \textbf{currency units}}}{\text{rate of exchange}}$

Ex. Exchange 580 Euros for British Pounds.

£1 = €1·16

Exchange = $\dfrac{€580}{€1·16}$ (' for every €1·16 we have £1 ')

= £500

Q	A	Q	A
1) Exchange 870 Euros for British Pounds.	1) £1 = €1·16 Exchange = $\dfrac{€870}{€1·16}$ = £750	4) Exchange 160,000 Indian Rupees For British Pounds.	4) £1 = Rs80 Exchange = $\dfrac{Rs160,000}{Rs80}$ = £2,000
2) Exchange 3150 Chinese Yuans for British Pounds.	2) £1 = ¥10·50 Exchange = $\dfrac{¥3150}{¥10·50}$ = £300	5) Exchange 800 Swiss Francs for British Pounds.	5) £1 = CHF1·70 Exchange = $\dfrac{CHF800}{CHF1·70}$ = £470·588... = £470·59 (nearest 1p)
3) Exchange 400 Egyptian Pounds for British Pounds.	3) £1 = E£9·25 Exchange = $\dfrac{E£400}{E£9·25}$ = £43·243... = £43·24 (nearest 1p)	6) Exchange 1,000 American Dollars for British Pounds.	6) £1 = $1·44 Exchange = $\dfrac{\$1,000}{\$1·44}$ = £694·444... = £694·44 (nearest 1p)

Foreign Exchange 3

Compare Price. Exchange 1 Unit.

Compare Price

The price of (identical) items in different countries can be compared by using the Exchange Rate.
In the following Exs. and Q/A, prices are compared in British Pounds.

Ex. A computer in Britain costs £357 and in France costs €406.
Calculate the difference in price.

(Change the Euros to Pounds)

$$£1 = €1 \cdot 16$$

$$\text{Exchange} = \frac{€406}{€1 \cdot 16} = £350$$

	Britain	£357
−	France	−£350
	Difference	£7

Q	A
1) A watch in Britain costs £27 and in Mexico costs $549·50. Calculate the difference in price.	1) £1 = $21·98 Exchange = $\frac{\$549 \cdot 50}{\$21 \cdot 98}$ = £25 Britain £27 − Mexico −£25 Difference £2
2) A camera in Britain costs £100 and in Nigeria costs ₦25,000. Calculate the difference in price.	2) £1 = ₦242 Exchange = $\frac{₦25000}{₦242}$ = £103·305... = £103·31 (nearest 1p) Nigeria £103·31 − Britain −£100·00 Difference £3·31
3) A car in Britain costs £6,000 and in South Africa costs R79,500. Calculate the difference in price.	3) £1 = R13·25 Exchange = $\frac{R79500}{R13 \cdot 25}$ = £6,000 Britain £6,000 − South Africa −£6,000 Difference £0
4) A pen in Britain costs £0·33 and in India costs Rs33·33. Calculate the difference in price.	4) £1 = Rs80 Exchange = $\frac{Rs33 \cdot 33}{Rs80}$ = £0·416... = £0·42 (nearest 1p) India £0·42 − Britain −£0·33 Difference £0·09

Exchange 1 Unit

The Exchange Rate can be used to find how many Pounds can be exchanged for 1 unit of currency.

Ex. State how many Pounds can be exchanged for   a) 1 Euro   b) 1 Nigerian Naira

a) €1·16 = £1

$$€1 = \frac{£1}{1 \cdot 16} = £0 \cdot 862...$$
$$= £0 \cdot 86$$
(nearest 1p)

b) ₦242 = £1

$$₦1 = \frac{£1}{242} = £0 \cdot 0041...$$
(less than 1p)

Q	A
1) State how many Pounds can be exchanged for 1 unit of each currency. a) 1 Danish Kroner	1) a) kr8·73 = £1 kr1 = $\frac{£1}{8 \cdot 73}$ = £0·114... = £0·11 (nearest 1p)
b) 1 Japanese Yen	b) ¥150 = £1 ¥1 = $\frac{£1}{150}$ = £0·006... = £0·01 (nearest 1p)
c) 1 American Dollar	c) $1·44 = £1 $1 = $\frac{£1}{1 \cdot 44}$ = £0·694... = £0·69 (nearest 1p)

Foreign Exchange 4

Exchange 2 Foreign Currencies

Exchange Rates exist directly between countries.
The Exchange Rates here can be used to exchange 2 foreign currencies by relating each to the Pound.

Exchange the given currency for Pounds - then - exchange the Pounds for the required currency.

Ex. Exchange €870 for American Dollars.

Exchange the Euros for Pounds.	Exchange the Pounds for American Dollars.
£1 = €1·16	£1 = $1·44
Exchange = $\frac{€870}{€1·16}$ = £750	Exchange = $1·44 × 750 = $1,080

Q	A
1) Exchange €145 for American Dollars.	1) £1 = €1·16 £1 = $1·44 Exchange = $\frac{€145}{€1·16}$ = £125 Exchange = $1·44 × 125 = $180
2) Exchange (USD)$500 for Euros.	2) £1 = $1·44 £1 = €1·16 Exchange = $\frac{$500}{$1·44}$ = £347·222... Exchange = €1·16 × 347·222... = €402·777... = €402·78 (nearest 1 cent) The number does not have to be 'rounded' here.
3) Exchange ₦242,000 for Danish Kroners.	3) £1 = ₦242 £1 = kr8·73 Exchange = $\frac{₦242,000}{₦242}$ = £1,000 Exchange = kr8·73 × 1,000 = kr8730
4) Exchange R81 for Japanese Yens.	4) £1 = R13·25 £1 = ¥150 Exchange = $\frac{R81}{R13·25}$ = £6·113... Exchange = ¥150 × 6·113... = ¥916·981... = ¥917 (nearest 1 Yen)
5) Exchange CHF10 for Egyptian Pounds.	5) £1 = CHF1·70 £1 = E£9·25 Exchange = $\frac{CHF10}{CHF1·70}$ = £5·882... Exchange = E£9·25 × 5·882... = E£54·411... = E£54·41 (nearest 1 piastre)
6) Exchange MXN65,940 for INR. (The 3-letter code may be used to indicate the currency)	6) GBP1 = MXN21·98 GBP1 = INR80 Exchange = $\frac{MXN65,940}{MXN21·98}$ = GBP3,000 Exchange = INR80 × 3,000 = INR240,000
7) Exchange AUD209 for CNY.	7) GBP1 = AUD2·09 GBP1 = CNY10·50 Exchange = $\frac{AUD209}{AUD2·09}$ = GBP100 Exchange = CNY10·50 × 100 = CNY1,050

Standard Form 1   Write a Number in Standard Form.

Standard Form

Standard Form is a way of expressing numbers in a different manner.
It is particularly useful for expressing quite large / small numbers.

Any number can be written in Standard Form, which is expressed as

$$A \times 10^n \quad (1 \leq A < 10)$$

where   A is a number equal to or greater than 1 but less than 10, usually written

and   n is an integer (... -4, -3, -2, -1, 0, 1, 2, 3, 4, ...)

Write a Number in Standard Form

If the number is **already** in the form of A (equal to or greater than 1 but less than 10)
we multiply by 10^0.   ($10^0 = 1$)

  Ex. Write 5·67 in Standard Form.

    The number is **already** in the form of A, so we do not need to change it ...

$$\begin{aligned} A \quad &\times 10^n \\ 5\cdot 6\,7 \\ = 5\cdot 6\,7 &\times 1 \end{aligned}$$

Standard Form = **5·67 × 10^0**   (only the answer is required)

If the number is **equal to or greater than 10**,
we need to **divide** by a power of 10 (10, 100, 1000, ...) to make the number in the form of A
    and   **multiply** by the **same power** of 10 so the number keeps the same value.

  Ex. Write 56·7 in Standard Form.

    The number is **greater than 10**, so we need to change it ...

$$A \quad \times 10^n$$
$$\div 10 \quad \text{do opposite so number keeps the same value}$$
$$5\,6\cdot 7$$
$$= 5\cdot 6\,7 \times 10$$

Standard Form = **5·67 × 10^1**   (only the answer is required)

If the number is **less than 1**,
we need to **multiply** by a power of 10 (10, 100, 1000, ...) to make the number in the form of A
    and   **divide** by the **same power** of 10 so the number keeps the same value ...

but we change **divide** by 10, 100, 1000, ... to **multiply** by $\frac{1}{10}, \frac{1}{100}, \frac{1}{1000}, \ldots$

$$= 10^{-1}, 10^{-2}, 10^{-3}, \ldots$$

so we can have just 1 rule in Standard Form of **multiplying** by a power of 10.

  Ex. Write 0·567 in Standard Form.

    The number is **less than 10**, so we need to change it ...

$$A \quad \times 10^n$$
$$\times 10 \quad \text{do opposite so number keeps the same value}$$
$$0\cdot 5\,6\,7$$
$$= 0\,5\cdot 6\,7 \times \frac{1}{10}$$

Standard Form = **5·67 × 10^{-1}**   (only the answer is required)

Standard Form 2

Write a number in Standard Form.

*Note
Indices / Powers 1,7,8 explains the following ...
Reminder ...

$$\vdots$$
$$10^5 = 100,000$$
$$10^4 = 10,000$$
$$10^3 = 1,000$$
$$10^2 = 100$$
$$10^1 = 10$$
$$10^0 = 1$$
$$10^{-1} = \frac{1}{10}$$
$$10^{-2} = \frac{1}{100}$$
$$10^{-3} = \frac{1}{1,000}$$
$$10^{-4} = \frac{1}{10,000}$$
$$10^{-5} = \frac{1}{100,000}$$
$$\vdots$$

Q	A	Q	A
Write in Standard Form. 1) 1·234	move point 0 places. 1) $= 1·234 \times 10^0$	6) 0·456	move point 1 place RIGHT. ×10, do opposite 6) 0·4 5 6 $= 0\ 4·5\ 6 \times \frac{1}{10}$ $= 4·5\ 6 \times 10^{-1}$
2) 60·59	move point 1 place LEFT. ÷10, do opposite 2) 6 0·5 9 $= 6·0\ 5\ 9 \times 10$ $= 6·0\ 5\ 9 \times 10^1$	7) 0·038719	move point 2 places RIGHT. ×100, do opposite 7) 0·0 3 8 7 1 9 $= 0\ 0\ 3·8\ 7\ 1\ 9 \times \frac{1}{100}$ $= 3·8\ 7\ 1\ 9 \times 10^{-2}$
3) 584·736	move point 2 places LEFT. ÷100, do opposite 3) 5 8 4·7 3 6 $= 5·8\ 4\ 7\ 3\ 6 \times 100$ $= 5·8\ 4\ 7\ 3\ 6 \times 10^2$	8) 0·005	move point 3 places RIGHT. ×1,000, do opposite 8) 0·0 0 5 $= 0\ 0\ 0\ 5· \times \frac{1}{1,000}$ $= 5 \times 10^{-3}$
4) 9999·99...	move point 3 places LEFT. ÷1000, do opposite 4) 9 9 9 9·9 9 ... $= 9·9\ 9\ 9\ 9\ 9 ... \times 1,000$ $= 9·9\ 9\ 9\ 9\ 9 ... \times 10^3$	9) 0·000201	move point 4 places RIGHT. ×10,000, do opposite 9) 0·0 0 0 2 0 1 $= 0\ 0\ 0\ 0\ 2·0\ 1 \times \frac{1}{10,000}$ $= 2·0\ 1 \times 10^{-4}$
5) 35682	move point 4 places LEFT. ÷10,000, do opposite 5) 3 5 6 8 2· $= 3·5\ 6\ 8\ 2 \times 10,000$ $= 3·5\ 6\ 8\ 2 \times 10^4$		

Standard Form 3

Change Standard Form to a Number.

Change Standard Form to a Number

A number in Standard Form can be written as an 'ordinary number'.

Ex. Write as an ordinary number.

1) $2 \cdot 847 \times 10^0$

    move point 0 places
    $= 2 \cdot 847 \times 1$
    $= 2 \cdot 847$

2) $1 \cdot 3953 \times 10^2$

    move point 2 places RIGHT
    $= 1 \cdot 3953 \times 100$
    $= 139 \cdot 53$

3) $4 \cdot 109 \times 10^{-3}$

    move point 3 places LEFT
    $= 0004 \cdot 109 \times \dfrac{1}{1,000}$
    $= 0 \cdot 004109$

(in each Ex. only the answer is required)

Q	A	Q	A
Write as an ordinary number. 1) $5 \cdot 4321 \times 10^0$	move point 0 places 1) $5 \cdot 4321 \times 10^0$ $= 5 \cdot 4321 \times 1$ $= 5 \cdot 4321$	6) $2 \cdot 27 \times 10^{-1}$	move point 1 place LEFT 6) $2 \cdot 27 \times 10^{-1}$ $= 0 2 \cdot 27 \times \dfrac{1}{10}$ $= 0 \cdot 227$
2) $8 \cdot 967 \times 10^1$	move point 1 place RIGHT 2) $8 \cdot 967 \times 10^1$ $= 8 \cdot 967 \times 10$ $= 89 \cdot 67$	7) $6 \cdot 508 \times 10^{-2}$	move point 2 places LEFT 7) $6 \cdot 508 \times 10^{-2}$ $= 6 \cdot 508 \times \dfrac{1}{100}$ $= 0 \cdot 06508$
3) $4 \cdot 05 \times 10^2$	move point 2 places RIGHT 3) $4 \cdot 05 \times 10^2$ $= 4 \cdot 05 \times 100$ $= 405 \cdot$ $= 405$	8) $3 \cdot 33... \times 10^{-3}$	move point 3 places LEFT 8) $3 \cdot 33... \times 10^{-3}$ $= 3 \cdot 33... \times \dfrac{1}{1,000}$ $= 0 \cdot 00333...$
4) $7 \cdot 777... \times 10^3$	move point 3 places RIGHT 4) $7 \cdot 777... \times 10^3$ $= 7 \cdot 777... \times 1,000$ $= 7777 \cdot 777...$	9) 4×10^{-5}	move point 5 places LEFT 9) 4×10^{-5} $= 4 \cdot \times \dfrac{1}{100,000}$ $= 0 \cdot 00004$
5) 9×10^6	move point 6 places RIGHT 5) 9×10^6 $= 9 \cdot \times 1,000,000$ $= 9,000,000 \cdot$ $= 9,000,000$		

Standard Form 4　　　　　　　　　　　　　　　　　　　　　　　Addition. Subtraction.

Standard Form - Calculations
Calculations can be carried out with numbers presented in Standard Form.

Addition and Subtraction
To add or subtract, numbers in Standard Form have to be changed to ordinary numbers.

Ex. Calculate, giving the answer in Standard Form.

$$(1\cdot234 \times 10^2) + (7\cdot50918 \times 10^3)$$

$$= 1\cdot234 \times 100 \;+\; 7\cdot50918 \times 1{,}000$$
$$= 123\cdot4 \;+\; 7509\cdot18$$

```
   123·40
 + 7509·18
 ---------
   7632·58
```
$$= 7\cdot632\,58 \times 10^3$$

Q	**A**
Calculate, giving the answer in Standard Form.
1) $(9\cdot8765 \times 10^4) + (4\cdot3027 \times 10^3)$	1) $(9\cdot8765 \times 10^4) + (4\cdot3027 \times 10^3)$ $= 9\cdot8765 \times 10{,}000 + 4\cdot3027 \times 1{,}000$ $= 98765\cdot \;+\; 4302\cdot7$ $\;98765\cdot0$ $+4302\cdot7$ $\overline{103067\cdot7}$ $= 1\cdot030677 \times 10^5$
2) $(5\cdot5284 \times 10^2) + (8\cdot13 \times 10^{-2})$	2) $(5\cdot5284 \times 10^2) + (8\cdot13 \times 10^{-2})$ $= 5\cdot5284 \times 100 \;+\; 008\cdot13 \times \dfrac{1}{100}$ $= 552\cdot84 \;+\; 0\cdot0813$ $\;552\cdot8400$ $+\,0\cdot0813$ $\overline{552\cdot9213}$ $= 5\cdot529213 \times 10^2$
3) $(6\cdot767 \times 10^5) - (2\cdot0002 \times 10^3)$	3) $(6\cdot767 \times 10^5) - (2\cdot0002 \times 10^3)$ $= 6\cdot76700 \times 100{,}000 - 2\cdot0002 \times 1{,}000$ $= 676700\cdot0 \;-\; 2000\cdot2$ $\;676700\cdot0$ $-2000\cdot2$ $\overline{674699\cdot8}$ $= 6\cdot746998 \times 10^5$
4) $(7\cdot95 \times 10^{-3}) - (3 \times 10^{-4})$	4) $(7\cdot95 \times 10^{-3}) - (3 \times 10^{-4})$ $= 0007\cdot95 \times \dfrac{1}{1{,}000} - 0003\cdot \times \dfrac{1}{10{,}000}$ $= 0\cdot00795 \;-\; 0\cdot0003$ $\;0\cdot00795$ $-\;0\cdot00030$ $\overline{0\cdot00765}$ $= 7\cdot65 \times 10^{-3}$

Standard Form 5

Multiplication. Division.

Multiplication and Division
To multiply or divide, 2 Methods can be used.

Method 1
The Standard Form numbers are changed to ordinary numbers - then multiply or divide.

Method 2
To multiply - **multiply** the 'A' parts - then the 'powers of 10' parts separately by **adding the indices**.
To divide   -   **divide** the 'A' parts - then the 'powers of 10' parts separately by **subtracting the indices**.

This method is more appropriate with very large / small numbers to avoid writing 'lots of' 0's.

(*Note **Indices / Powers 3-4** explain **adding / subtracting indices**.)

Multiplication

Ex. Calculate, giving the answer in Standard Form.

Method 1	Method 2
$(3.4 \times 10^3) \times (8 \times 10^5)$	$(3.4 \times 10^3) \times (8 \times 10^5)$
$= 3400 \times 800000$	$= 27.2 \times 10^{3+5=8}$
$= 2720000000$	$= 2.72 \times 10^1 \times 10^8$
$= 2.72 \times 10^9$	$= 2.72 \times 10^9$

Q / A

Calculate, giving the answer in Standard Form.

1) $(1.2 \times 10^4) \times (8 \times 10^3)$

A1)

Method 1:
$(1.2 \times 10^4) \times (8 \times 10^3)$
$= 12000 \times 8000$
$= 96000000$
$= 9.6 \times 10^7$

Method 2:
$(1.2 \times 10^4) \times (8 \times 10^3)$
$= 9.6 \times 10^{4+3=7}$
$= 9.6 \times 10^7$

---

2) $(4.97 \times 10^5) \times (3.6 \times 10^6)$

A2)

Method 1:
$(4.97 \times 10^5) \times (3.6 \times 10^6)$
$= 497000 \times 3600000$
$= 1789200000000$
$= 1.7892 \times 10^{12}$

Method 2:
$(4.97 \times 10^5) \times (3.6 \times 10^6)$
$= 17.892 \times 10^{5+6=11}$
$= 1.7892 \times 10^1 \times 10^{11}$
$= 1.7892 \times 10^{12}$

---

3) $(5.025 \times 10^7) \times (9.9 \times 10^{-3})$

A3)

Method 1:
$(5.025 \times 10^7) \times (9.9 \times 10^{-3})$
$= 50250000 \times 0.0099$
$= 497475$
$= 4.97475 \times 10^5$

Method 2:
$(5.025 \times 10^7) \times (9.9 \times 10^{-3})$
$= 49.7475 \times 10^{7-3=4}$
$= 4.97475 \times 10^1 \times 10^4$
$= 4.97475 \times 10^5$

---

4) $(2.7 \times 10^{-4}) \times (2.7 \times 10^{-2})$

A4)

Method 1:
$(2.7 \times 10^{-4}) \times (2.7 \times 10^{-2})$
$= 0.00027 \times 0.027$
$= 0.00000729$
$= 7.29 \times 10^{-6}$

Method 2:
$(2.7 \times 10^{-4}) \times (2.7 \times 10^{-2})$
$= 7.29 \times 10^{-4-2=-6}$
$= 7.29 \times 10^{-6}$

Standard Form 6

Division.

Division

Ex. Calculate, giving the answer in Standard Form.

Method 1	Method 2
$\dfrac{(3\cdot 4 \times 10^3)}{(8 \times 10^5)}$	$\dfrac{(3\cdot 4 \times 10^3)}{(8 \times 10^5)}$
$= \dfrac{3400}{800000}$	$= 0\cdot 425 \times 10^{3-5=-2}$
$= 0\cdot 00425$	$= 0\overset{\frown}{4}\cdot 25 \times 10^{-1} \times 10^{-2}$
$= 4\cdot 25 \times 10^{-3}$	$= 4\cdot 25 \times 10^{-3}$

Q — Calculate, giving the answer in Standard Form.

1) $\dfrac{(7\cdot 7 \times 10^6)}{(2\cdot 5 \times 10^2)}$

A 1)

Method 1:
$\dfrac{(7\cdot 7 \times 10^6)}{(2\cdot 5 \times 10^2)}$
$= \dfrac{7700000}{250}$
$= 30800$
$= 3\cdot 08 \times 10^4$

Method 2:
$\dfrac{(7\cdot 7 \times 10^6)}{(2\cdot 5 \times 10^2)}$
$= 3\cdot 08 \times 10^{6-2=4}$
$= 3\cdot 08 \times 10^4$

2) $\dfrac{(8\cdot 1 \times 10^7)}{(5\cdot 9 \times 10^4)}$

A 2)

Method 1:
$\dfrac{(8\cdot 1 \times 10^7)}{(5\cdot 9 \times 10^4)}$
$= \dfrac{81000000}{59000}$
$= 1372\cdot 881...$
$= 1\cdot 372... \times 10^3$

Method 2:
$\dfrac{(8\cdot 1 \times 10^7)}{(5\cdot 9 \times 10^4)}$
$= 1\cdot 372... \times 10^{7-4=3}$
$= 1\cdot 372... \times 10^3$

3) $\dfrac{(1\cdot 49 \times 10^3)}{(6\cdot 24 \times 10^8)}$

A 3)

Method 1:
$\dfrac{(1\cdot 49 \times 10^3)}{(6\cdot 24 \times 10^8)}$
$= \dfrac{1490}{624000000}$
$= 0\cdot 000002387...$
$= 2\cdot 387... \times 10^{-6}$

Method 2:
$\dfrac{(1\cdot 49 \times 10^3)}{(6\cdot 24 \times 10^8)}$
$= 0\cdot 2387... \times 10^{3-8=-5}$
$= 0\overset{\frown}{2}\cdot 387... \times 10^{-1} \times 10^{-5}$
$= 2\cdot 387... \times 10^{-6}$

*Note 2·387... e-6 (or just -6) may show on a calculator.

4) $\dfrac{(4 \times 10^{-3})}{(9\cdot 7 \times 10^5)}$

A 4)

Method 1:
$\dfrac{(4 \times 10^{-3})}{(9\cdot 7 \times 10^5)}$
$= \dfrac{0\cdot 004}{970000}$
$= 0\cdot 000000004123...$
$= 4\cdot 123... \times 10^{-9}$

Method 2:
$\dfrac{(4 \times 10^{-3})}{(9\cdot 7 \times 10^5)}$
$= 0\cdot 4123... \times 10^{-3-5=-8}$
$= 0\overset{\frown}{4}\cdot 123... \times 10^{-1} \times 10^{-8}$
$= 4\cdot 123... \times 10^{-9}$

*Note 4·123... e-9 (or just -9) may show on a calculator.

Standard Form 7   'Problems'.

The following 'problems' involve Standard Form.

Q	A

1) Calculate the area of a rectangle, length 98·7m, breadth 65m. State the area using Standard Form.
 	
    65m
    98·7m
	1) $A = l \times b$ move point 3 places LEFT.
	    $= 98.7 \times 65$
	    $= 6415.5 \text{ m}^2$ ⇨ $\div 1000$ do opposite
	           6 4 1 5·5
	    $= 6.4155 \times 1,000$
	    $= 6.4155 \times 10^3 \text{ m}^2$

2) Divide the 1st Prime Number by the 4th Prime Number. State the answer in Standard Form.	2) move point 1 place RIGHT.
	    $\frac{2}{7} = 0.285714...$ ⇨ $\times 10$ do opposite
	              0·2857...
	    $= 02.857... \times \frac{1}{10}$
	    $= 2.857... \times 10^{-1}$

3) The weights of 2 items are given as (3.0405×10^2) kg and (1.89×10^{-2}) kg.
A) Find their total weight, in kg.
B) Express their total weight in Standard Form.
	3) $(3.0405 \times 10^2) + (1.89 \times 10^{-2})$
	    $= 3.0405 \times 100 + 001.89 \times \frac{1}{100}$
	    $= 304.05 + 0.0189$
	    $ 304.05$
	    $+ 0.0189$
	A) $\overline{304.0689}$ kg
	B) $= 3.040689 \times 10^2$ kg

4) A plane flies at (1.23×10^3) kmph for (8×10^{-2}) hrs.
A) How far does the plane fly in this time?
B) Express the distance in Standard Form.
	4) <u>Method 1</u>
	    Distance = Speed × Time
	    $= (1.23 \times 10^3) \times (8 \times 10^{-2})$
	    $= 1230 \times 0.08$
	A) $= 98.4$ km
	B) $= 9.84 \times 10^1$ km
	
	    <u>Method 2</u>
	    Distance = Speed × Time
	    $= (1.23 \times 10^3) \times (8 \times 10^{-2})$
	    $= 9.84 \times 10^{3-2=1}$
	    $= 9.84 \times 10^0 \times 10^1$
	B) $= 9.84 \times 10^1$ km
	A) $= 98.4$ km

5) (2.4×10^3) tonnes of earth are moved daily for a road-building project.
A) How many days does it take to remove (4.2×10^5) tonnes of earth?
B) Express the answer in Standard Form.
	5) <u>Method 1</u> <u>Method 2</u>
	    $\dfrac{(4.2 \times 10^5)}{(2.4 \times 10^3)}$ $\dfrac{(4.2 \times 10^5)}{(2.4 \times 10^3)}$
	    $= \dfrac{420000}{2400}$
	A) $= 175$ days. $= 1.75 \times 10^{5-3=2}$
	B) $= 1.75 \times 10^2$ days. B) $= 1.75 \times 10^2$ days.
	                           A) $= 175$ days.

Standard Form 8 — 'Problems'.

Q	A

6) The mass of the moon is ($7 \cdot 343 \times 10^{19}$) tonnes. The mass of the earth is 81 times bigger than that of the moon.
 	
Calculate the mass of the earth, giving the answer in Standard Form.
	6) Earth = 81 × Moon
	         = $81 \times (7 \cdot 343 \times 10^{19})$
	         = $594 \cdot 783 \times 10^{19}$
	         = $5 \cdot 94783 \times 10^2 \times 10^{19}$
	         = $5 \cdot 94783 \times 10^{21}$ tonnes

7) A) Write the number 34·5 million in Standard Form.
 	
B) Calculate (34·5 million)^2, expressing the answer in Standard Form.
	7)
	A) Method 1 Method 2
	    $34 \cdot 5 \times 1{,}000{,}000$ $34 \cdot 5 \times 10^6$
	    = $34{,}500{,}000$ = $3 \cdot 45 \times 10^1 \times 10^6$
	    = $3 \cdot 45 \times 10^7$ = $3 \cdot 45 \times 10^7$
	
	B) (34·5 million)^2
	    = $(3 \cdot 45 \times 10^7)^2$
	    = $(3 \cdot 45 \times 10^7) \times (3 \cdot 45 \times 10^7)$
	    = $11 \cdot 9025 \times 10^{14}$
	    = $1 \cdot 19025 \times 10^1 \times 10^{14}$
	    = $1 \cdot 19025 \times 10^{15}$

8) The mass of an electron is approx. ($9 \cdot 1 \times 10^{-28}$) grams.
 	
Calculate the total mass of 25,000 electrons, giving the answer in Standard Form.
	8) Total mass = 25,000 × Electron
	                   = $25{,}000 \times (9 \cdot 1 \times 10^{-28})$
	                   = $227{,}500 \times 10^{-28}$
	                   = $2 \cdot 275 \times 10^5 \times 10^{-28}$
	                   = $2 \cdot 275 \times 10^{-23}$ grams.

9) The sun is a sphere (approx.) and its radius is (7×10^5) km (approx.). Calculate the approximate volume of the sun, giving the answer in Standard Form. (let $\pi = 3 \cdot 14159...$)	9) V = Sphere $V = \frac{4}{3} \pi r^3$ $V = \frac{4}{3} \times \pi \times r \times r \times r$ $V = \frac{4}{3} \times 3 \cdot 14... \times (7 \times 10^5) \times (7 \times 10^5) \times (7 \times 10^5)$ $V = 1436 \cdot 755... \times 10^{15}$ $V = 1 \cdot 436755... \times 10^3 \times 10^{15}$ $V = 1 \cdot 436755... \times 10^{18}$ km^3

Surds 1 — Addition

Surds

A **surd** is the **positive square root of a natural number** (counting number) - but **not** of a square number.

It is written using the square root sign, ex., $\sqrt{2}, \sqrt{3}, \sqrt{5}, \sqrt{6}, \sqrt{7}, \sqrt{8}, \sqrt{10}, \ldots$

The square root of the number would be **irrational** -
  a non-terminating and non-recurring decimal number.   ex. $\sqrt{2} = 1.414\ldots$ ex. $\sqrt{6} = 2.449\ldots$

The positive square root of a square number is itself a natural number and is therefore **not** a surd.
  ex. $\sqrt{1} = 1$ ex. $\sqrt{4} = 2$ ex. $\sqrt{9} = 3$

A surd is an **exact** number and, when appropriate, an 'answer to a problem' can be written in surd form.

Ex.1 Calculate $\sqrt{3+5}$ Write the answer as a **surd**.   ⇨ $\sqrt{3+5} = \sqrt{8}$

*Note Here, the emphasis is on the **square root** - other roots are also considered surds.
  Ex. $\sqrt[3]{4}$ $\sqrt[4]{2}$ Ex. $\sqrt[3]{8} = 2$ so is **not** a surd.

Q	A	Q	A
1. Calculate - write answer as a **surd**.	1.	2. If $a = 2$ and $b = 3$ calculate - as a **surd**.	
1) $\sqrt{1+2}$	1) $\sqrt{3}$	1) $\sqrt{a+b}$	1) $\sqrt{2+3} = \sqrt{5}$
2) $\sqrt{9-4}$	2) $\sqrt{5}$	2) $\sqrt{10-a}$	2) $\sqrt{10-2} = \sqrt{8}$
3) $\sqrt{2 \times 3}$	3) $\sqrt{6}$	3) $\sqrt{5ab}$	3) $\sqrt{5 \times 2 \times 3} = \sqrt{30}$
4) $\sqrt{\dfrac{35}{5}}$	4) $\sqrt{7}$	4) $\sqrt{\dfrac{6}{b}}$	4) $\sqrt{\dfrac{6}{3}} = \sqrt{2}$
5) $\sqrt{4^2 - 1^2}$	5) $\sqrt{16-1} = \sqrt{15}$	5) $\sqrt{a^2 + b^2}$	5) $\sqrt{2^2 + 3^2} = \sqrt{4+9} = \sqrt{13}$

Addition

Consider the following -

(**not surds**) Ex.1 a) $\sqrt{9} + \sqrt{4}$ b) $\sqrt{9} + \sqrt{4}$
                = 3 + 2 $\neq \sqrt{9+4}$
                = 5 -------------------------- $\neq \sqrt{13}$

  The square roots have to be found   ... the numbers can **not** be added
  before the addition can take place ...   and **then** the square root found.

(**surds**) Ex.2 $\sqrt{5} + \sqrt{3}$

  Ex.1b) shows that the addition of <u>different</u> square roots can **not** be simplified.

(**not surds**) Ex.3 a) $\sqrt{9} + \sqrt{9}$ (**surds**) b) $\sqrt{11} + \sqrt{11}$
                = 3 + 3   = $2 \times \sqrt{11}$
                = 2 × 3   = $2\sqrt{11}$ This is in **surd form**.
                = $2 \times \sqrt{9}$ (the × sign
                = $2\sqrt{9}$ is omitted)

  Ex.3b) shows that the addition of <u>same</u> square roots **can** be simplified.

Q	A	Q	A
Simplify where possible.		5) $\sqrt{2} + \sqrt{2}$	5) $2\sqrt{2}$
1) $\sqrt{3} + \sqrt{5}$	1) $\sqrt{3} + \sqrt{5}$	6) $\sqrt{5} + \sqrt{5} + \sqrt{5}$	6) $3\sqrt{5}$
2) $\sqrt{7} + \sqrt{7}$	2) $2\sqrt{7}$	7) $2\sqrt{6} + \sqrt{6}$	7) $3\sqrt{6}$
3) $6 + \sqrt{8}$	3) $6 + \sqrt{8}$	8) $7\sqrt{2} + 2\sqrt{2}$	8) $9\sqrt{2}$
4) $\sqrt{5} + \sqrt{5}$	4) $2\sqrt{5}$	9) $4\sqrt{3} + \sqrt{3} + 3\sqrt{3}$	9) $8\sqrt{3}$

Surds 2 — Subtraction

Subtraction

Consider the following -

(not surds) Ex.1 a) $\sqrt{9} - \sqrt{4}$
                    $= 3 - 2$
                    $= 1$

b) $\sqrt{9} - \sqrt{4}$
     $\neq \sqrt{9-4}$
     $\neq \sqrt{5}$

The square roots have to be found **before** the subtraction can take place ...

... the numbers can **not** be subtracted and **then** the square root found.

(surds) Ex.2 $\sqrt{5} - \sqrt{3}$

Ex.1b) shows that the subtraction of <u>different</u> square roots can **not** be simplified.

(not surds) Ex.3 a) $\sqrt{9} - \sqrt{9}$
              $= 3 - 3$
              $= 0$

(surds) b) $\sqrt{11} - \sqrt{11}$
       $= 0$

(not surds) Ex.4 a) $2\sqrt{9} - \sqrt{9}$
              $= 2\sqrt{9} - 1\sqrt{9}$
              $= 2 \times 3 - 1 \times 3$
              $= 6 - 3$
              $= 3$
              $= \sqrt{9}$ $(2-1=1)$

(surds) b) $2\sqrt{11} - \sqrt{11}$
       $= \sqrt{11}$

(not surds) Ex.5 a) $7\sqrt{9} - 5\sqrt{9}$
              $= 7 \times 3 - 5 \times 3$
              $= 21 - 15$
              $= 6$
              $= 2 \times 3$
              $= 2\sqrt{9}$ $(7-5=2)$

(surds) b) $7\sqrt{11} - 5\sqrt{11}$
       $= 2\sqrt{11}$

Exs.3b) 4b) 5b) show that the subtraction of <u>same</u> square roots **can** be simplified.

Q	A	Q	A
Simplify where possible.			
1) $\sqrt{7} - \sqrt{5}$	1) $\sqrt{7} - \sqrt{5}$	5) $6\sqrt{7} - 5\sqrt{7}$	5) $\sqrt{7}$
2) $\sqrt{6} - \sqrt{6}$	2) 0	7) $7\sqrt{3} - 4\sqrt{3}$	6) $3\sqrt{3}$
3) $\sqrt{3} - 1$	3) $\sqrt{3} - 1$	7) $8\sqrt{2} - 3\sqrt{2}$	7) $5\sqrt{2}$
4) $2\sqrt{8} - \sqrt{8}$	4) $\sqrt{8}$	8) $15\sqrt{10} - 9\sqrt{10}$	8) $6\sqrt{10}$

Surds 3 — Multiplication

Multiplication

Consider the following -

(**not surds**) Ex.1
a) $\sqrt{4} \times \sqrt{9}$
$= 2 \times 3$
$= 6$

b) $\sqrt{4} \times \sqrt{9}$
$= \sqrt{4 \times 9}$
$= \sqrt{36}$
$= 6$

This shows that when multiplying 2 **square roots** there are 2 ways to produce the same answer -
either a) find the square root of each number and then multiply the results
or b) multiply the numbers and then find the square root of the result.

Now consider **multiplying surds** - using b) above [since a) can not be used with surds].

The values are in **surd form**.

Ex.2 $\sqrt{3} \times \sqrt{5}$
$= \sqrt{3 \times 5}$ ---- this line can be omitted
$= \sqrt{15}$
The answer is a **surd**.

Ex.3 $\sqrt{3} \times \sqrt{3}$
$= \sqrt{3 \times 3}$
$= \sqrt{9}$
$= 3$
The answer is a **natural number**.

Ex.4 $2\sqrt{3} \times 4\sqrt{5}$
$= 8\sqrt{3 \times 5}$
$= 8\sqrt{15}$
The answer is in **surd form**.

The square root of a **square number** may be multiplied by a **surd**.

Ex.4
a) $\sqrt{4} \times \sqrt{7}$
$= 2 \times \sqrt{7}$
$= 2\sqrt{7}$

b) $\sqrt{7} \times \sqrt{4}$
$= \sqrt{7} \times 2$
$= 2\sqrt{7}$

The answer is in **surd form**.

c) $\sqrt{4} \times \sqrt{7}$
$= \sqrt{4 \times 7}$
$= \sqrt{28}$

The answer is a **surd**.

Q	A	Q	A
Write as a surd / natural number / in surd form.		9) $(4\sqrt{5})^2$	9) $4\sqrt{5} \times 4\sqrt{5}$ $= 16\sqrt{25} = 16 \times 5 = 80$
1) $\sqrt{3} \times \sqrt{2}$	1) $\sqrt{6}$	10) $2\sqrt{7} \times 3\sqrt{7}$	10) $6\sqrt{49} = 6 \times 7 = 42$
2) $\sqrt{7} \times \sqrt{5}$	2) $\sqrt{35}$	11) $6\sqrt{2} \times 8\sqrt{11}$	11) $48\sqrt{22}$
3) $\sqrt{8} \times \sqrt{11}$	3) $\sqrt{88}$	12) $5\sqrt{5} \times 7\sqrt{7}$	12) $35\sqrt{35}$
4) $\sqrt{40} \times \sqrt{30}$	4) $\sqrt{1200}$	13) $\sqrt{4} \times \sqrt{5}$	13) $\sqrt{20}$ or $2\sqrt{5}$
5) $\sqrt{5} \times \sqrt{5}$	5) $\sqrt{25} = 5$	14) $\sqrt{9} \times \sqrt{2}$	14) $\sqrt{18}$ or $3\sqrt{2}$
6) $\sqrt{8} \times \sqrt{8}$	6) $\sqrt{64} = 8$	15) $\sqrt{7} \times \sqrt{81}$	15) $\sqrt{567}$ or $\sqrt{7} \times 9 = 9\sqrt{7}$
7) $(\sqrt{6})^2$	7) $\sqrt{6} \times \sqrt{6} = \sqrt{36} = 6$	16) $\sqrt{2} \times \sqrt{2} \times \sqrt{2}$	16) $\sqrt{8}$ or $\sqrt{4} \times \sqrt{2} = 2\sqrt{2}$
8) $(\sqrt{3})^2$	8) $\sqrt{3} \times \sqrt{3} = \sqrt{9} = 3$	17) $\sqrt{5} \times \sqrt{5} \times \sqrt{5}$	17) $\sqrt{125}$ or $\sqrt{25} \times \sqrt{5} = 5\sqrt{5}$

Surds 4 — Division

Division
Consider the following -

(**not surds**) Ex.1 a) $\dfrac{\sqrt{36}}{\sqrt{9}} = \dfrac{6}{3} = 2$ b) $\dfrac{\sqrt{36}}{\sqrt{9}} = \sqrt{\dfrac{36}{9}} = \sqrt{4} = 2$

This shows that when dividing 2 **square roots** there are 2 ways to produce the same answer -
either a) find the square root of each number and then divide the results
or b) divide the numbers and then find the square root of the result.

Now consider **dividing surds** - using b) above [since a) can not be used with surds]

Ex.2 $\dfrac{\sqrt{35}}{\sqrt{5}} = \sqrt{\dfrac{35}{5}} = \sqrt{7}$ Ex.3 $\dfrac{\sqrt{18}}{\sqrt{2}} = \sqrt{\dfrac{18}{2}} = \sqrt{9} = 3$

The answer is a **surd**. The answer is a **natural number**.

Ex.4 $\dfrac{8\sqrt{35}}{4\sqrt{5}} = 2\sqrt{\dfrac{35}{5}} = 2\sqrt{7}$ Ex.5 $\dfrac{3\sqrt{40}}{9\sqrt{8}} = \dfrac{1}{3}\sqrt{\dfrac{40}{8}} = \dfrac{1}{3}\sqrt{5}$

The answers are in **surd form**.

The square root of a **square number** may be divided by / into a **surd**.

Ex.6 $\dfrac{\sqrt{21}}{\sqrt{9}} = \dfrac{\sqrt{21}}{3}$ Ex.7 $\dfrac{\sqrt{49}}{\sqrt{8}} = \dfrac{7}{\sqrt{8}}$

The answers are in **surd form**.

Q	A	Q	A
Write as a **surd** / **natural number** / in surd form.			
1) $\dfrac{\sqrt{10}}{\sqrt{2}}$	1) $=\sqrt{\dfrac{10}{2}} = \sqrt{5}$	10) $\dfrac{2\sqrt{15}}{4\sqrt{5}}$	10) $=\dfrac{1}{2}\sqrt{\dfrac{15}{5}} = \dfrac{1}{2}\sqrt{3}$
2) $\dfrac{\sqrt{30}}{\sqrt{5}}$	2) $=\sqrt{\dfrac{30}{5}} = \sqrt{6}$	11) $\dfrac{6\sqrt{24}}{8\sqrt{3}}$	11) $=\dfrac{3}{4}\sqrt{\dfrac{24}{3}} = \dfrac{3}{4}\sqrt{8}$
3) $\dfrac{\sqrt{56}}{\sqrt{8}}$	3) $=\sqrt{\dfrac{56}{8}} = \sqrt{7}$	12) $\dfrac{\sqrt{19}}{\sqrt{4}}$	12) $=\dfrac{\sqrt{19}}{2}$
4) $\dfrac{\sqrt{48}}{\sqrt{3}}$	4) $=\sqrt{\dfrac{48}{3}} = \sqrt{16} = 4$	13) $\dfrac{\sqrt{55}}{\sqrt{81}}$	13) $=\dfrac{\sqrt{55}}{9}$
5) $\dfrac{\sqrt{63}}{\sqrt{7}}$	5) $=\sqrt{\dfrac{63}{7}} = \sqrt{9} = 3$	14) $\dfrac{\sqrt{36}}{\sqrt{8}}$	14) $=\dfrac{6}{\sqrt{8}}$
6) $\dfrac{\sqrt{800}}{\sqrt{8}}$	6) $=\sqrt{\dfrac{800}{8}} = \sqrt{100} = 10$	15) $\dfrac{\sqrt{25}}{\sqrt{47}}$	15) $=\dfrac{5}{\sqrt{47}}$
7) $\dfrac{\sqrt{5}}{\sqrt{5}}$	7) $=\sqrt{\dfrac{5}{5}} = \sqrt{1} = 1$	16) $\dfrac{\sqrt{5}\times\sqrt{6}}{\sqrt{2}}$	16) $=\sqrt{\dfrac{30}{2}} = \sqrt{15}$
8) $\dfrac{6\sqrt{42}}{2\sqrt{6}}$	8) $=3\sqrt{\dfrac{42}{6}} = 3\sqrt{7}$	17) $\dfrac{\sqrt{8}\times\sqrt{3}}{\sqrt{6}}$	17) $=\sqrt{\dfrac{24}{6}} = \sqrt{4} = 2$
9) $\dfrac{99\sqrt{110}}{9\sqrt{11}}$	9) $=11\sqrt{\dfrac{110}{11}} = 11\sqrt{10}$	18) $\dfrac{\sqrt{2}\times\sqrt{50}}{\sqrt{10}\times\sqrt{10}}$	18) $=\sqrt{\dfrac{100}{100}} = \sqrt{1} = 1$

Surds *5

Factors.

Factors

A number can be expressed as a product (multiplication) of 2 of its factors.
The order of the factors does not matter.

Ex. $6 = 1 \times 6$ or $6 = 6 \times 1$
$6 = 2 \times 3$ $6 = 3 \times 2$

Similarly, a **surd** can be expressed as a product (multiplication) of 2 of its factors -   **number × surd**
or   **surd × surd**.

Write each **surd** as the product (multiplication) of 2 factors.

Ex.1   a) $\sqrt{6} = \sqrt{1 \times 6}$ The **multiplication** of surds is **reversed**.   b) $\sqrt{6} = \sqrt{2 \times 3}$
$= \sqrt{1} \times \sqrt{6}$ $= \sqrt{2} \times \sqrt{3}$
$= 1 \times \sqrt{6}$ **surd × surd**

         number × surd   or $= \sqrt{3} \times \sqrt{2}$

The 1st. 'step' can be omitted.

Ex.2   a) $\sqrt{12} = \sqrt{1 \times 12}$ b) $\sqrt{12} = \sqrt{2 \times 6}$ c) $\sqrt{12} = \sqrt{3 \times 4}$ or $\sqrt{12} = \sqrt{4 \times 3}$
$= \sqrt{1} \times \sqrt{12}$ $= \sqrt{2} \times \sqrt{6}$ $= \sqrt{3} \times \sqrt{4}$ $= \sqrt{4} \times \sqrt{3}$
$= 1 \times \sqrt{12}$ **surd × surd** $= \sqrt{3} \times 2$ $= 2 \times \sqrt{3}$
       number × surd $= 2\sqrt{3}$ $= 2\sqrt{3}$
                                                                           number × surd

*Note 1   Factors of $12 = 1 \times 12$
$12 = 2 \times 6$
$12 = 3 \times 4$

*Note 2   The product of factors $1 \times$ **surd** is 'trivial' and is omitted in the following Q/A.

Q	A	Q	A
Write each **surd** as the product (multiplication) of 2 factors.	*Note The **minimum** work is shown.	9) $\sqrt{8}$	9) $\sqrt{4} \times \sqrt{2}$ $= 2\sqrt{2}$
1) $\sqrt{10}$	1) $\sqrt{2} \times \sqrt{5}$	10) $\sqrt{27}$	10) $\sqrt{9} \times \sqrt{3}$ $= 3\sqrt{3}$
2) $\sqrt{14}$	2) $\sqrt{2} \times \sqrt{7}$	11) $\sqrt{18}$	11) a) $\sqrt{9} \times \sqrt{2}$ $= 3\sqrt{2}$ b) $\sqrt{3} \times \sqrt{6}$
3) $\sqrt{15}$	3) $\sqrt{3} \times \sqrt{5}$		
4) $\sqrt{26}$	4) $\sqrt{2} \times \sqrt{13}$		
5) $\sqrt{35}$	5) $\sqrt{5} \times \sqrt{7}$	12) $\sqrt{20}$	12) a) $\sqrt{2} \times \sqrt{10}$ b) $\sqrt{4} \times \sqrt{5}$ $= 2\sqrt{5}$
6) $\sqrt{42}$	6) a) $\sqrt{2} \times \sqrt{21}$ b) $\sqrt{3} \times \sqrt{14}$ c) $\sqrt{6} \times \sqrt{7}$	13) $\sqrt{32}$	13) a) $\sqrt{4} \times \sqrt{8}$ $= 2\sqrt{8}$ b) $= \sqrt{16} \times \sqrt{2}$ $= 4\sqrt{2}$
7) $\sqrt{30}$	7) a) $\sqrt{2} \times \sqrt{15}$ b) $\sqrt{3} \times \sqrt{10}$ c) $\sqrt{5} \times \sqrt{6}$		
8) $\sqrt{66}$	8) a) $\sqrt{2} \times \sqrt{33}$ b) $\sqrt{3} \times \sqrt{22}$ c) $\sqrt{6} \times \sqrt{11}$	14) $\sqrt{50}$	14) a) $\sqrt{25} \times \sqrt{2}$ $= 5\sqrt{2}$ b) $\sqrt{5} \times \sqrt{10}$

Surds *6

Express Number × Surd as a Surd

Surds *5 Ex. 2c) shows how a **surd** can be expressed in the form of **number × surd**.
If the method is **reversed** then a **number × surd** can be written as a **surd**.

surd
$$\sqrt{12} = \sqrt{4 \times 3}$$
$$= \sqrt{4} \times \sqrt{3}$$
$$= 2 \times \sqrt{3}$$
$$= 2\sqrt{3}$$
number × surd

reverse

Ex. Express as a **surd**.
$$2\sqrt{3}$$
$$= 2 \times \sqrt{3}$$
$$= \sqrt{4} \times \sqrt{3}$$
$$= \sqrt{4 \times 3} \quad \text{(can be omitted)}$$
$$= \sqrt{12}$$

Q	A	Q	A
Express as a **surd**. 1) $2\sqrt{2}$	1) $\quad 2\sqrt{2}$ $= 2 \times \sqrt{2}$ $= \sqrt{4} \times \sqrt{2}$ $= \sqrt{8}$	5) $7\sqrt{6}$	5) $\quad 7\sqrt{6}$ $= 7 \times \sqrt{6}$ $= \sqrt{49} \times \sqrt{6}$ $= \sqrt{294}$
2) $3\sqrt{3}$	2) $\quad 3\sqrt{3}$ $= 3 \times \sqrt{3}$ $= \sqrt{9} \times \sqrt{3}$ $= \sqrt{27}$	6) $10\sqrt{8}$	6) $\quad 10\sqrt{8}$ $= 10 \times \sqrt{8}$ $= \sqrt{100} \times \sqrt{8}$ $= \sqrt{800}$
3) $5\sqrt{5}$	3) $\quad 5\sqrt{5}$ $= 5 \times \sqrt{5}$ $= \sqrt{25} \times \sqrt{5}$ $= \sqrt{125}$	7) $4\sqrt{11}$	7) $\quad 4\sqrt{11}$ $= 4 \times \sqrt{11}$ $= \sqrt{16} \times \sqrt{11}$ $= \sqrt{176}$
4) $6\sqrt{2}$	4) $\quad 6\sqrt{2}$ $= 6 \times \sqrt{2}$ $= \sqrt{36} \times \sqrt{2}$ $= \sqrt{72}$	8) $9\sqrt{20}$	8) $\quad 9\sqrt{20}$ $= 9 \times \sqrt{20}$ $= \sqrt{81} \times \sqrt{20}$ $= \sqrt{1620}$

Alternative Number × Surd

An alternative Number × Surd may be written if the surd number has a **square factor**.

Ex. $7\sqrt{8} = 7 \times \sqrt{4} \times \sqrt{2}$ (square)
$= 7 \times 2 \times \sqrt{2}$
$= 14\sqrt{2}$

Q	A	Q	A
Find an alternative Number × Surd 1) $2\sqrt{12}$	1) $= 2 \times \sqrt{4} \times \sqrt{3}$ $= 2 \times 2 \times \sqrt{3}$ $= 4\sqrt{3}$	4) $3\sqrt{27}$	4) $= 3 \times \sqrt{9} \times \sqrt{3}$ $= 3 \times 3 \times \sqrt{3}$ $= 9\sqrt{3}$
2) $5\sqrt{18}$	2) $= 5 \times \sqrt{9} \times \sqrt{2}$ $= 5 \times 3 \times \sqrt{2}$ $= 15\sqrt{2}$	5) $8\sqrt{125}$	5) $= 8 \times \sqrt{25} \times \sqrt{5}$ $= 8 \times 5 \times \sqrt{5}$ $= 40\sqrt{5}$
3) $6\sqrt{20}$	3) $= 6 \times \sqrt{4} \times \sqrt{5}$ $= 6 \times 2 \times \sqrt{5}$ $= 12\sqrt{5}$	6) $7\sqrt{96}$	6) $= 7 \times \sqrt{16} \times \sqrt{6}$ $= 7 \times 4 \times \sqrt{6}$ $= 28\sqrt{6}$

Surds *7 — Expand Brackets

Expand Brackets

A term **directly in front** of the brackets **multiplies** each term inside the brackets.

Ex. 1
$2(\sqrt{3} + \sqrt{5})$
$= 2\sqrt{3} + 2\sqrt{5}$

Ex. 2
$\sqrt{3}(2 - \sqrt{5})$
$= 2\sqrt{3} - \sqrt{3}\sqrt{5}$
$= 2\sqrt{3} - \sqrt{15}$

Ex. 3
$6(\sqrt{8} + \sqrt{6} + \sqrt{3})$
$= 6\sqrt{8} + 6\sqrt{6} + 6\sqrt{3}$

Ex. 4
$\sqrt{7}(4 + \sqrt{7})$
$= 4\sqrt{7} + \sqrt{7}\sqrt{7}$
$= 4\sqrt{7} + 7$

Q	A	Q	A
Expand the brackets.			
1) $5(\sqrt{2} + \sqrt{3})$	1) $= 5\sqrt{2} + 5\sqrt{3}$	5) $2(\sqrt{7} + \sqrt{5} - \sqrt{2})$	5) $= 2\sqrt{7} + 2\sqrt{5} - 2\sqrt{2}$
2) $7(\sqrt{6} - \sqrt{2})$	2) $= 7\sqrt{6} - 7\sqrt{2}$	6) $3(\sqrt{2} - \sqrt{3} + \sqrt{6})$	6) $= 3\sqrt{2} - 3\sqrt{3} + 3\sqrt{6}$
3) $\sqrt{3}(4 + \sqrt{8})$	3) $= 4\sqrt{3} + \sqrt{3}\sqrt{8}$ $= 4\sqrt{3} + \sqrt{24}$	7) $\sqrt{2}(9 + \sqrt{2})$	7) $= 9\sqrt{2} + \sqrt{2}\sqrt{2}$ $= 9\sqrt{2} + 2$
4) $\sqrt{5}(\sqrt{7} - 6)$	4) $= \sqrt{5}\sqrt{7} - 6\sqrt{5}$ $= \sqrt{35} - 6\sqrt{5}$	8) $\sqrt{8}(\sqrt{8} - 7)$	8) $= \sqrt{8}\sqrt{8} - 7\sqrt{8}$ $= 8 - 7\sqrt{8}$

Expand Brackets

Each term in the 1st bracket multiplies each term in the 2nd bracket.

Ex.1 $(\sqrt{2} + \sqrt{3})(\sqrt{5} + \sqrt{7})$
$= \sqrt{2}\sqrt{5} + \sqrt{2}\sqrt{7} + \sqrt{3}\sqrt{5} + \sqrt{3}\sqrt{7}$
$= \sqrt{10} + \sqrt{14} + \sqrt{15} + \sqrt{21}$

Ex.2 $(2 + \sqrt{3})(\sqrt{5} - \sqrt{3})$
$= 2\sqrt{5} - 2\sqrt{3} + \sqrt{3}\sqrt{5} - \sqrt{3}\sqrt{3}$
$= 2\sqrt{5} - 2\sqrt{3} + \sqrt{15} - 3$

Q	A
Expand the brackets.	
1) $(\sqrt{6} + \sqrt{5})(\sqrt{3} + \sqrt{8})$	1) $= \sqrt{6}\sqrt{3} + \sqrt{6}\sqrt{8} + \sqrt{5}\sqrt{3} + \sqrt{5}\sqrt{8}$ $= \sqrt{18} + \sqrt{48} + \sqrt{15} + \sqrt{40}$
2) $(\sqrt{5} - \sqrt{7})(\sqrt{6} - \sqrt{2})$	2) $= \sqrt{5}\sqrt{6} - \sqrt{5}\sqrt{2} - \sqrt{7}\sqrt{6} + \sqrt{7}\sqrt{2}$ $= \sqrt{30} - \sqrt{10} - \sqrt{42} + \sqrt{14}$
3) $(2 + \sqrt{3})(\sqrt{8} + \sqrt{3})$	3) $= 2\sqrt{8} + 2\sqrt{3} + \sqrt{3}\sqrt{8} + \sqrt{3}\sqrt{3}$ $= 2\sqrt{8} + 2\sqrt{3} + \sqrt{24} + 3$
4) $(\sqrt{8} - \sqrt{2})(\sqrt{8} + \sqrt{2})$	4) $= \sqrt{8}\sqrt{8} + \sqrt{8}\sqrt{2} - \sqrt{2}\sqrt{8} - \sqrt{2}\sqrt{2}$ $= 8 + \sqrt{16} - \sqrt{16} - 2$ $= 6$
5) $(\sqrt{5} + \sqrt{3})^2$	5) $= (\sqrt{5} + \sqrt{3})(\sqrt{5} + \sqrt{3})$ $= \sqrt{5}\sqrt{5} + \sqrt{5}\sqrt{3} + \sqrt{3}\sqrt{5} + \sqrt{3}\sqrt{3}$ $= 5 + \sqrt{15} + \sqrt{15} + 3$ $= 8 + 2\sqrt{15}$
6) $(\sqrt{7} - \sqrt{6})^2$	6) $= (\sqrt{7} - \sqrt{6})(\sqrt{7} - \sqrt{6})$ $= \sqrt{7}\sqrt{7} - \sqrt{7}\sqrt{6} - \sqrt{6}\sqrt{7} + \sqrt{6}\sqrt{6}$ $= 7 - \sqrt{42} - \sqrt{42} + 6$ $= 13 - 2\sqrt{42}$

Surds *8 — Fraction - Rationalise the Denominator.

Fraction - 'Rationalise the Denominator'
A fraction which contains a **surd in the denominator** may be 'simplified' by making
the **denominator an integer**.
This is done by '**rationalising the denominator**' - (in general, making it a **rational** number) -
multiplying both the numerator and denominator by the same term / expression ...
(as when finding equivalent fractions).
  2 kinds of example are explained ...
  1) If the denominator is either a **surd** or a <u>number multiplying a **surd**</u> -
     multiply numerator and denominator by the **same surd**.

Ex. Simplify

a) $\dfrac{1}{\sqrt{2}} \Rightarrow = \dfrac{1}{\sqrt{2}} \times \dfrac{\sqrt{2}}{\sqrt{2}} = \dfrac{\sqrt{2}}{2}$ (same surd)

b) $\dfrac{4}{\sqrt{5}} \Rightarrow = \dfrac{4}{\sqrt{5}} \times \dfrac{\sqrt{5}}{\sqrt{5}} = \dfrac{4\sqrt{5}}{5}$

c) $\dfrac{\sqrt{7}}{\sqrt{6}} \Rightarrow = \dfrac{\sqrt{7}}{\sqrt{6}} \times \dfrac{\sqrt{6}}{\sqrt{6}} = \dfrac{\sqrt{7}\sqrt{6}}{6} = \dfrac{\sqrt{42}}{6}$

d) $\dfrac{8}{7\sqrt{3}} \Rightarrow = \dfrac{8}{7\sqrt{3}} \times \dfrac{\sqrt{3}}{\sqrt{3}} = \dfrac{8\sqrt{3}}{7\times 3} = \dfrac{8\sqrt{3}}{21}$

In general $\times \dfrac{\sqrt{m}}{\sqrt{m}} = \times \dfrac{1}{1} = \times 1$ and does not change the value of the fraction multiplied.

Q	A	Q	A
Simplify 1) $\dfrac{1}{\sqrt{3}}$	1) $= \dfrac{1}{\sqrt{3}} \times \dfrac{\sqrt{3}}{\sqrt{3}} = \dfrac{\sqrt{3}}{3}$	7) $\dfrac{\sqrt{3}}{\sqrt{7}}$	7) $= \dfrac{\sqrt{3}}{\sqrt{7}} \times \dfrac{\sqrt{7}}{\sqrt{7}} = \dfrac{\sqrt{3}\sqrt{7}}{7} = \dfrac{\sqrt{21}}{7}$
2) $\dfrac{1}{\sqrt{8}}$	2) $= \dfrac{1}{\sqrt{8}} \times \dfrac{\sqrt{8}}{\sqrt{8}} = \dfrac{\sqrt{8}}{8}$	8) $\dfrac{1}{2\sqrt{3}}$	8) $= \dfrac{1}{2\sqrt{3}} \times \dfrac{\sqrt{3}}{\sqrt{3}} = \dfrac{\sqrt{3}}{2\times 3} = \dfrac{\sqrt{3}}{6}$
3) $\dfrac{4}{\sqrt{11}}$	3) $= \dfrac{4}{\sqrt{11}} \times \dfrac{\sqrt{11}}{\sqrt{11}} = \dfrac{4\sqrt{11}}{11}$	9) $\dfrac{9}{5\sqrt{2}}$	9) $= \dfrac{9}{5\sqrt{2}} \times \dfrac{\sqrt{2}}{\sqrt{2}} = \dfrac{9\sqrt{2}}{5\times 2} = \dfrac{9\sqrt{2}}{10}$
4) $\dfrac{2}{\sqrt{6}}$	4) $= \dfrac{2}{\sqrt{6}} \times \dfrac{\sqrt{6}}{\sqrt{6}} = \dfrac{^1 2\sqrt{6}}{{}_3 6} = \dfrac{\sqrt{6}}{3}$	10) $\dfrac{\sqrt{7}}{4\sqrt{6}}$	10) $= \dfrac{\sqrt{7}}{4\sqrt{6}} \times \dfrac{\sqrt{6}}{\sqrt{6}} = \dfrac{\sqrt{7}\sqrt{6}}{4\times 6} = \dfrac{\sqrt{42}}{24}$
5) $\dfrac{7}{\sqrt{7}}$	5) $= \dfrac{7}{\sqrt{7}} \times \dfrac{\sqrt{7}}{\sqrt{7}} = \dfrac{^1 7\sqrt{7}}{{}_1 7} = \sqrt{7}$	11) $\dfrac{\sqrt{8}}{3\sqrt{5}}$	11) $= \dfrac{\sqrt{8}}{3\sqrt{5}} \times \dfrac{\sqrt{5}}{\sqrt{5}} = \dfrac{\sqrt{8}\sqrt{5}}{3\times 5} = \dfrac{\sqrt{40}}{15}$
6) $\dfrac{\sqrt{5}}{\sqrt{2}}$	6) $= \dfrac{\sqrt{5}}{\sqrt{2}} \times \dfrac{\sqrt{2}}{\sqrt{2}} = \dfrac{\sqrt{5}\sqrt{2}}{2} = \dfrac{\sqrt{10}}{2}$	12) $\dfrac{9\sqrt{2}}{7\sqrt{7}}$	12) $= \dfrac{9\sqrt{2}}{7\sqrt{7}} \times \dfrac{\sqrt{7}}{\sqrt{7}} = \dfrac{9\sqrt{2}\sqrt{7}}{7\times 7} = \dfrac{9\sqrt{14}}{49}$

Fractions may be multiplied.
  Ex. Multiply

a) $\dfrac{1}{\sqrt{3}} \times \dfrac{8}{\sqrt{5}} \Rightarrow = \dfrac{8}{\sqrt{15}} = \dfrac{8}{\sqrt{15}} \times \dfrac{\sqrt{15}}{\sqrt{15}} = \dfrac{8\sqrt{15}}{15}$

b) $\dfrac{3}{\sqrt{5}} \times \dfrac{\sqrt{2}}{\sqrt{7}} \Rightarrow = \dfrac{3\sqrt{2}}{\sqrt{35}} = \dfrac{3\sqrt{2}}{\sqrt{35}} \times \dfrac{\sqrt{35}}{\sqrt{35}} = \dfrac{3\sqrt{70}}{35}$

Q	A	Q	A
Multiply 1) $\dfrac{1}{\sqrt{5}} \times \dfrac{3}{\sqrt{2}}$	1) $= \dfrac{3}{\sqrt{10}} = \dfrac{3}{\sqrt{10}} \times \dfrac{\sqrt{10}}{\sqrt{10}} = \dfrac{3\sqrt{10}}{10}$	4) $\dfrac{5}{\sqrt{7}} \times \dfrac{\sqrt{3}}{\sqrt{2}}$	4) $= \dfrac{5\sqrt{3}}{\sqrt{14}} = \dfrac{5\sqrt{3}}{\sqrt{14}} \times \dfrac{\sqrt{14}}{\sqrt{14}} = \dfrac{5\sqrt{42}}{14}$
2) $\dfrac{2}{\sqrt{7}} \times \dfrac{4}{\sqrt{3}}$	2) $= \dfrac{8}{\sqrt{21}} = \dfrac{8}{\sqrt{21}} \times \dfrac{\sqrt{21}}{\sqrt{21}} = \dfrac{8\sqrt{21}}{21}$	5) $\dfrac{2\sqrt{2}}{\sqrt{3}} \times \dfrac{2}{\sqrt{11}}$	5) $= \dfrac{4\sqrt{2}}{\sqrt{33}} = \dfrac{4\sqrt{2}}{\sqrt{33}} \times \dfrac{\sqrt{33}}{\sqrt{33}} = \dfrac{4\sqrt{66}}{33}$
3) $\dfrac{5}{\sqrt{2}} \times \dfrac{2}{\sqrt{6}}$	3) $= \dfrac{10}{\sqrt{12}} = \dfrac{10}{\sqrt{12}} \times \dfrac{\sqrt{12}}{\sqrt{12}} = \dfrac{10\sqrt{12}}{12}$ $= \dfrac{10\sqrt{4}\sqrt{3}}{12} = \dfrac{10\times 2\sqrt{3}}{12} \overset{\div 4}{=} \dfrac{5\sqrt{3}}{3}$	6) $\dfrac{\sqrt{8}}{\sqrt{2}} \times \dfrac{\sqrt{8}}{3\sqrt{5}}$	6) $= \dfrac{8}{3\sqrt{10}} = \dfrac{8}{3\sqrt{10}} \times \dfrac{\sqrt{10}}{\sqrt{10}} = \dfrac{8\sqrt{10}}{3\times 10}$ $= \dfrac{8\sqrt{10}}{30} = \dfrac{4\sqrt{10}}{15}$

Surds *9 — Fraction - Rationalise the Denominator.

Fraction - 'Rationalise the Denominator'

Q	A	Q	A
1) Calculate the area of the rectangle. $\frac{1}{\sqrt{2}}$ cm by $\frac{1}{\sqrt{3}}$ cm	1) $A = lb$ $= \frac{1}{\sqrt{2}} \times \frac{1}{\sqrt{3}} = \frac{1}{\sqrt{6}}$ $= \frac{1}{\sqrt{6}} \times \frac{\sqrt{6}}{\sqrt{6}} = \frac{\sqrt{6}}{6}$ cm²	4) Calculate the area of the triangle. Sides $\frac{1}{\sqrt{2}}$ m and $\frac{6}{\sqrt{7}}$ m	4) $A = \tfrac{1}{2}bh$ $= \tfrac{1}{2} \times \frac{\cancel{6}^{3}}{\sqrt{7}} \times \frac{1}{\sqrt{2}} = \frac{3}{\sqrt{14}}$ $= \frac{3}{\sqrt{14}} \times \frac{\sqrt{14}}{\sqrt{14}} = \frac{3\sqrt{14}}{14}$ m²
2) Calculate the area of the rectangle. $\frac{4}{\sqrt{3}}$ cm by $\frac{2}{\sqrt{5}}$ cm	2) $A = lb$ $= \frac{4}{\sqrt{3}} \times \frac{2}{\sqrt{5}} = \frac{8}{\sqrt{15}}$ $= \frac{8}{\sqrt{15}} \times \frac{\sqrt{15}}{\sqrt{15}} = \frac{8\sqrt{15}}{15}$ cm²	5) Calculate the volume of the sphere. Radius = $\frac{\sqrt{7}}{\sqrt{5}}$ km	5) $V = \tfrac{4}{3}\pi r^3$ $= \tfrac{4}{3} \times \pi \times \frac{\sqrt{7}}{\sqrt{5}} \times \frac{\sqrt{7}}{\sqrt{5}} \times \frac{\sqrt{7}}{\sqrt{5}}$ $= \tfrac{4}{3} \times \pi \times \frac{7}{5} \times \frac{\sqrt{7}}{\sqrt{5}}$ $= \tfrac{28}{15}\pi \times \frac{\sqrt{7}}{\sqrt{5}} \times \frac{\sqrt{5}}{\sqrt{5}}$ $= \tfrac{28}{15}\pi \times \frac{\sqrt{35}}{5}$ $= \frac{28\sqrt{35}}{75}\pi$ km³
3) Calculate the area of the triangle. Sides $\frac{1}{\sqrt{7}}$ m and $\frac{2}{\sqrt{3}}$ m	3) $A = \tfrac{1}{2}bh$ $= \tfrac{1}{\cancel{2}} \times \frac{\cancel{2}^{1}}{\sqrt{3}} \times \frac{1}{\sqrt{7}} = \frac{1}{\sqrt{21}}$ $= \frac{1}{\sqrt{21}} \times \frac{\sqrt{21}}{\sqrt{21}} = \frac{\sqrt{21}}{21}$ m²		

Fractions - Addition / Subtraction (same denominator)

Ex. Calculate

1) $\frac{1}{\sqrt{3}} + \frac{1}{\sqrt{3}}$ ⇨ 1) $= \frac{2}{\sqrt{3}} \times \frac{\sqrt{3}}{\sqrt{3}} = \frac{2\sqrt{3}}{3}$

2) $\frac{7\sqrt{5}}{\sqrt{2}} - \frac{4\sqrt{5}}{\sqrt{2}}$ ⇨ 2) $= \frac{3\sqrt{5}}{\sqrt{2}} \times \frac{\sqrt{2}}{\sqrt{2}} = \frac{3\sqrt{10}}{2}$

3) $\frac{7}{4\sqrt{5}} + \frac{\sqrt{3}}{4\sqrt{5}}$ ⇨ 3) $= \frac{7+\sqrt{3}}{4\sqrt{5}} \times \frac{\sqrt{5}}{\sqrt{5}} = \frac{7\sqrt{5}+\sqrt{15}}{4\times 5} = \frac{7\sqrt{5}+\sqrt{15}}{20}$

Q	A	Q	A
Calculate 1) $\frac{2}{\sqrt{7}} + \frac{3}{\sqrt{7}}$	1) $= \frac{5}{\sqrt{7}} \times \frac{\sqrt{7}}{\sqrt{7}} = \frac{5\sqrt{7}}{7}$	7) $\frac{4\sqrt{6}}{\sqrt{3}} + \frac{3\sqrt{6}}{\sqrt{3}}$	7) $= \frac{7\sqrt{6}}{\sqrt{3}} \times \frac{\sqrt{3}}{\sqrt{3}} = \frac{7\sqrt{18}}{3}$ $= \frac{7\sqrt{9}\sqrt{2}}{3} = \frac{7\times 3\sqrt{2}}{\cancel{3}} = 7\sqrt{2}$
2) $\frac{8}{\sqrt{6}} - \frac{4}{\sqrt{6}}$	2) $= \frac{4}{\sqrt{6}} \times \frac{\sqrt{6}}{\sqrt{6}} = \frac{4\sqrt{6}}{6} = \frac{2\sqrt{6}}{3}$		
3) $\frac{7}{3\sqrt{5}} - \frac{5}{3\sqrt{5}}$	3) $= \frac{2}{3\sqrt{5}} \times \frac{\sqrt{5}}{\sqrt{5}} = \frac{2\sqrt{5}}{3\times 5} = \frac{2\sqrt{5}}{15}$	8) $\frac{7}{4\sqrt{5}} + \frac{\sqrt{3}}{4\sqrt{5}}$	8) $= \frac{7+\sqrt{3}}{4\sqrt{5}} \times \frac{\sqrt{5}}{\sqrt{5}}$ $= \frac{7\sqrt{5}+\sqrt{15}}{4\times 5} = \frac{7\sqrt{5}+\sqrt{15}}{20}$
4) $\frac{1}{5\sqrt{8}} + \frac{2}{5\sqrt{8}}$	4) $= \frac{3}{5\sqrt{8}} \times \frac{\sqrt{8}}{\sqrt{8}} = \frac{3\sqrt{8}}{5\times 8} = \frac{3\sqrt{8}}{40}$		
5) $\frac{9\sqrt{2}}{\sqrt{7}} - \frac{3\sqrt{2}}{\sqrt{7}}$	5) $= \frac{6\sqrt{2}}{\sqrt{7}} \times \frac{\sqrt{7}}{\sqrt{7}} = \frac{6\sqrt{14}}{7}$	9) $\frac{\sqrt{5}}{2\sqrt{2}} - \frac{5}{2\sqrt{2}}$	9) $= \frac{\sqrt{5}-5}{2\sqrt{2}} \times \frac{\sqrt{2}}{\sqrt{2}}$ $= \frac{\sqrt{10}-5\sqrt{2}}{2\times 2} = \frac{\sqrt{10}-5\sqrt{2}}{4}$
6) $\frac{4\sqrt{5}}{\sqrt{6}} - \frac{\sqrt{5}}{\sqrt{6}}$	6) $= \frac{3\sqrt{5}}{\sqrt{6}} \times \frac{\sqrt{6}}{\sqrt{6}} = \frac{3\sqrt{30}}{6} = \frac{\sqrt{30}}{2}$		

Surds *10 — Fraction - Rationalise the Denominator.

Fractions - Addition / Subtraction (different denominators)
'Rationalise the Denominator' of each fraction.

Ex. Calculate

a) $\dfrac{5}{\sqrt{2}} + \dfrac{1}{\sqrt{3}}$ ⇨ a) $= \dfrac{5}{\sqrt{2}}\dfrac{\sqrt{2}}{\sqrt{2}} + \dfrac{1}{\sqrt{3}}\dfrac{\sqrt{3}}{\sqrt{3}}$

$= \times 3\left(\dfrac{\frac{5\sqrt{2}}{2} + \frac{\sqrt{3}}{3}}{\frac{15\sqrt{2}}{6} + \frac{2\sqrt{3}}{6}}\right)\times 2$

$= \dfrac{15\sqrt{2} + 2\sqrt{3}}{6}$

b) $\dfrac{7}{\sqrt{8}} - \dfrac{3}{\sqrt{2}}$ ⇨ b) $= \dfrac{7}{\sqrt{8}}\dfrac{\sqrt{8}}{\sqrt{8}} - \dfrac{3}{\sqrt{2}}\dfrac{\sqrt{2}}{\sqrt{2}}$

$= \times 1\left(\dfrac{\frac{7\sqrt{8}}{8} - \frac{3\sqrt{2}}{2}}{\frac{7\sqrt{8}}{8} - \frac{12\sqrt{2}}{8}}\right)\times 4$

$= \dfrac{7\sqrt[2]{4}\sqrt{2}}{8} - \dfrac{12\sqrt{2}}{8}$

$= \dfrac{14\sqrt{2}}{8} - \dfrac{12\sqrt{2}}{8}$

$= \dfrac{2\sqrt{2}}{8}$

$= \dfrac{\sqrt{2}}{4}$

Q	A	Q	A
Calculate 1) $\dfrac{2}{\sqrt{5}} + \dfrac{4}{\sqrt{7}}$	1) $= \dfrac{2}{\sqrt{5}}\dfrac{\sqrt{5}}{\sqrt{5}} + \dfrac{4}{\sqrt{7}}\dfrac{\sqrt{7}}{\sqrt{7}}$ $= \times 7\left(\dfrac{\frac{2\sqrt{5}}{5} + \frac{4\sqrt{7}}{7}}{\frac{14\sqrt{5}}{35} + \frac{20\sqrt{7}}{35}}\right)\times 5$ $= \dfrac{14\sqrt{5} + 20\sqrt{7}}{35}$	4) $\dfrac{5}{\sqrt{2}} + \dfrac{2}{\sqrt{6}}$ Calculate and express in terms of $\sqrt{2}$.	4) $= \dfrac{5}{\sqrt{2}}\dfrac{\sqrt{2}}{\sqrt{2}} + \dfrac{2}{\sqrt{6}}\dfrac{\sqrt{6}}{\sqrt{6}}$ $= \times 3\left(\dfrac{\frac{5\sqrt{2}}{2} + \frac{2\sqrt{6}}{6}}{\frac{15\sqrt{2}}{6} + \frac{2\sqrt{6}}{6}}\right)\times 1$ $= \dfrac{15\sqrt{2} + 2\sqrt{3}\sqrt{2}}{6}$ $= \dfrac{(15 + 2\sqrt{3})\sqrt{2}}{6}$
2) $\dfrac{7}{\sqrt{3}} - \dfrac{5}{\sqrt{8}}$	2) $= \dfrac{7}{\sqrt{3}}\dfrac{\sqrt{3}}{\sqrt{3}} - \dfrac{5}{\sqrt{8}}\dfrac{\sqrt{8}}{\sqrt{8}}$ $= \times 8\left(\dfrac{\frac{7\sqrt{3}}{3} - \frac{5\sqrt{8}}{8}}{\frac{56\sqrt{3}}{24} - \frac{15\sqrt{8}}{24}}\right)\times 3$ $= \dfrac{56\sqrt{3} - 15\sqrt{8}}{24}$	5) $\dfrac{11}{\sqrt{12}} - \dfrac{13}{\sqrt{48}}$ Calculate and express in terms of $\sqrt{3}$.	5) $= \dfrac{11}{\sqrt{12}}\dfrac{\sqrt{12}}{\sqrt{12}} - \dfrac{13}{\sqrt{48}}\dfrac{\sqrt{48}}{\sqrt{48}}$ $= \times 4\left(\dfrac{\frac{11\sqrt{12}}{12} - \frac{13\sqrt{48}}{48}}{\frac{44\sqrt{12}}{48} - \frac{13\sqrt{48}}{48}}\right)\times 1$ $= \dfrac{44\sqrt[2]{4}\sqrt{3}}{48} - \dfrac{13\sqrt[4]{16}\sqrt{3}}{48}$ $= \dfrac{88\sqrt{3}}{48} - \dfrac{52\sqrt{3}}{48}$ $= \dfrac{36\sqrt{3}}{48}$ $= \dfrac{3\sqrt{3}}{4}$
3) $\dfrac{1}{\sqrt{6}} + \dfrac{9}{\sqrt{5}}$	3) $= \dfrac{1}{\sqrt{6}}\dfrac{\sqrt{6}}{\sqrt{6}} + \dfrac{9}{\sqrt{5}}\dfrac{\sqrt{5}}{\sqrt{5}}$ $= \times 5\left(\dfrac{\frac{1\sqrt{6}}{6} + \frac{9\sqrt{5}}{5}}{\frac{5\sqrt{6}}{30} + \frac{54\sqrt{5}}{30}}\right)\times 6$ $= \dfrac{5\sqrt{6} + 54\sqrt{5}}{30}$		

Surds *11 — Fraction - Rationalise the Denominator.

Fraction - 'Rationalise the Denominator'

2) If the denominator is the addition / subtraction of either a <u>surd</u> **and** a <u>number</u> or **2 surds** - multiply numerator and denominator by the **same terms** but **change the addition/ subtraction sign**.

Ex. Simplify $\dfrac{1}{\sqrt{8}+2}$ ⇨ $= \dfrac{1}{\sqrt{8}+2} \times \dfrac{\sqrt{8}-2}{\sqrt{8}-2}$ — same as — $(\sqrt{8}+2)(\sqrt{8}-2)$

same terms / change sign

$$= \dfrac{\sqrt{8}-2}{\sqrt{8}\sqrt{8}-2\sqrt{8}+2\sqrt{8}-2\times 2}$$

$$= \dfrac{\sqrt{8}-2}{8 \qquad -4}$$

$$= \dfrac{\sqrt{8}-2}{4}$$

Q	A	Q	A
Simplify 1) $\dfrac{1}{\sqrt{2}+1}$	1) $= \dfrac{1}{\sqrt{2}+1} \times \dfrac{\sqrt{2}-1}{\sqrt{2}-1}$ $= \dfrac{\sqrt{2}-1}{\sqrt{2}\sqrt{2}-1\sqrt{2}+1\sqrt{2}-1\times 1}$ $= \dfrac{\sqrt{2}-1}{2 \quad -1}$ $= \dfrac{\sqrt{2}-1}{1}$ $= \sqrt{2}-1$	4) $\dfrac{\sqrt{6}}{3-\sqrt{8}}$	4) $= \dfrac{\sqrt{6}}{3-\sqrt{8}} \times \dfrac{3+\sqrt{8}}{3+\sqrt{8}}$ $= \dfrac{\sqrt{6}\times 3 + \sqrt{6}\times\sqrt{8}}{3\times 3 + 3\sqrt{8} - 3\sqrt{8} - \sqrt{8}\sqrt{8}}$ $= \dfrac{3\sqrt{6}+\sqrt{6}\sqrt{8}}{9 \quad -8}$ $= \dfrac{3\sqrt{6}+\sqrt{48}}{1}$ $= 3\sqrt{6}+\sqrt{16}\sqrt{3}$ $= 3\sqrt{6}+4\sqrt{3}$
2) $\dfrac{1}{\sqrt{6}-5}$	2) $= \dfrac{1}{\sqrt{6}-5} \times \dfrac{\sqrt{6}+5}{\sqrt{6}+5}$ $= \dfrac{\sqrt{6}+5}{\sqrt{6}\sqrt{6}+5\sqrt{6}-5\sqrt{6}-5\times 5}$ $= \dfrac{\sqrt{6}+5}{6 \quad -25}$ $= \dfrac{\sqrt{6}+5}{-19}$	5) $\dfrac{9\sqrt{2}}{\sqrt{7}+\sqrt{2}}$	5) $= \dfrac{9\sqrt{2}}{\sqrt{7}+\sqrt{2}} \times \dfrac{\sqrt{7}-\sqrt{2}}{\sqrt{7}-\sqrt{2}}$ $= \dfrac{9\sqrt{2}\times\sqrt{7}-9\sqrt{2}\times\sqrt{2}}{\sqrt{7}\sqrt{7}-\sqrt{7}\sqrt{2}+\sqrt{2}\sqrt{7}-\sqrt{2}\sqrt{2}}$ $= \dfrac{9\sqrt{14}-9\times 2}{7 \quad -2}$ $= \dfrac{9\sqrt{14}-18}{5}$
3) $\dfrac{2}{4+\sqrt{7}}$	3) $= \dfrac{2}{4+\sqrt{7}} \times \dfrac{4-\sqrt{7}}{4-\sqrt{7}}$ $= \dfrac{2\times 4 - 2\times\sqrt{7}}{4\times 4 - 4\sqrt{7}+4\sqrt{7}-\sqrt{7}\sqrt{7}}$ $= \dfrac{8-2\sqrt{7}}{16 \quad -7}$ $= \dfrac{8-2\sqrt{7}}{9}$	6) $\dfrac{\sqrt{5}+\sqrt{3}}{\sqrt{5}-\sqrt{3}}$	6) $= \dfrac{\sqrt{5}+\sqrt{3}}{\sqrt{5}-\sqrt{3}} \times \dfrac{\sqrt{5}+\sqrt{3}}{\sqrt{5}+\sqrt{3}}$ $= \dfrac{\sqrt{5}\sqrt{5}+\sqrt{5}\sqrt{3}+\sqrt{3}\sqrt{5}+\sqrt{3}\sqrt{3}}{\sqrt{5}\sqrt{5}+\sqrt{5}\sqrt{3}-\sqrt{3}\sqrt{5}-\sqrt{3}\sqrt{3}}$ $= \dfrac{5+2\sqrt{3}\sqrt{5}+3}{5 \quad -3}$ $= \dfrac{{}^4 8 + {}^1 2\sqrt{15}}{{}^1 2} = 4+\sqrt{15}$

Trial and Improvement 1 — Square Root.

*Note see also **Algebra**. **Quadratic Equations** *19.

'Trial and Improvement'
This is a term used to describe a method of finding a solution to a problem -
usually when an **exact** answer is not possible.
We **try an estimate**, use it to carry out a calculation(s) and compare the answer with the required answer
- depending on the difference between the two / the accuracy required, if necessary,
we **improve the estimate** and re-calculate ... until the required answer / accuracy is achieved.

For example, the Square Root of a number can be found simply by using a calculator -
if not exact, then a given answer is to a very high degree of accuracy.
We can use the method to show how the same answer can be found - a calculator may be used!

(Estimates may be different than shown here.)

Ex. Use the method of 'trial and improvement' to find $\sqrt{3}$ (positive value)
  a) correct to 1 decimal place.
  b) correct to 2 decimal places.

  *Note
  We need to find a number so that **the number** × **the number** = 3.
                    or **the number**^{2} = 3.

a) correct to 1 decimal place.   (Only the table is required -
                        the notes help explain.)

Try	Calculate		Answer	Compare with 3
1	1^2	=	1	Smaller
2	2^2	=	4	Bigger
1·5	$1·5^2$	=	2·25	Smaller
1·6	$1·6^2$	=	2·56	Smaller
1·7	$1·7^2$	=	2·89	Smaller
1·8	$1·8^2$	=	3·24	Bigger
1·75	$1·75^2$	=	3·0625	Bigger

(a 'reasonable' estimate is made to start)
The answer required is between 1 and 2.

('half way' between values is often chosen)

The answer required is either 1·**7** or 1·**8**
- the ·7 or ·8 is the '1 decimal place' number.

The answer required is now between 1·7 and 1·75
and any of these values round to 1·7
'to 1 decimal place'.

So 1·7 is the required answer.

$\sqrt{3}$ is nearer to 1·7   3·   3·24 $(1·8)^2$
         than 1·8   − 2·89 $(1·7)^2$ − 3·
      'check difference' ...   0·11   0·24

⇩ (continue)

b) correct to 2 decimal places.

1·74	$1·74^2$	=	3·0276	Bigger
1·73	$1·73^2$	=	2·9929	Smaller
1·735	$1·735^2$	=	3·010225	Bigger

The answer required is either 1·**74** or 1·**73**
- the ·74 or ·73 are the '2 decimal places' numbers.

The answer required is now between 1·73 and 1·735
and any of these values round to 1·73
'to 2 decimal places'.

So 1·73 is the required answer.

         $\sqrt{3}$ is nearer to 1·73   3·   3·0276 $(1·74)^2$
A calculator shows $\sqrt{3}$ = 1·7320...   than 1·74   − 2·9929 $(1·73)^2$ − 3·
      'check difference' ...   0·0071   0·0276

Trial and Improvement 2

Square Root. Cube Root.

Q | A

1) Use the method of **'trial and improvement'** to find $\sqrt{5}$ (positive value)

a) correct to 1 decimal place.
b) correct to 2 decimal places.

Use a calculator to find $\sqrt{5}$.

1)

	Try	Calculate	Answer	Compare with 5	
a)	2	2^2 =	4	Smaller	answer is between 2 and 3.
	3	3^2 =	9	Bigger	
	2·5	$2·5^2$ =	6·25	Bigger	
	2·4	$2·4^2$ =	5·76	Bigger	
	2·3	$2·3^2$ =	5·29	Bigger	2·3 or 2·2 between 2·2 and 2·25 round to 2·2
	2·2	$2·2^2$ =	4·84	Smaller	
	2·25	$2·25^2$ =	5·0625	Bigger	
b)	2·24	$2·24^2$ =	5·0176	Bigger	2·24 or 2·23 between 2·24 and 2·235 round to 2·24
	2·23	$2·23^2$ =	4·9729	Smaller	
	2·235	$2·235^2$ =	4·995225	Smaller	

a) 2·2   ($\sqrt{5}$ is nearer to 2·2 than 2·3)

b) 2·24 ($\sqrt{5}$ is nearer to 2·24 than 2·23)

Calculator $\sqrt{5} = 2·2360...$

---

2) Use the method of **'trial and improvement'** to find $\sqrt[3]{60}$

a) correct to 1 decimal place.
b) correct to 2 decimal places.

Use a calculator to find $\sqrt[3]{60}$.

2)

	Try	Calculate	Answer	Compare with 60	
a)	3	3^3 =	27	Smaller	answer is between 3 and 4.
	4	4^3 =	64	Bigger	
	3·5	$3·5^3$ =	42·875	Smaller	4·0 or 3·9
	3·9	$3·9^3$ =	59·319	Smaller	between 3·9 and 3·95 round to 3·9
	3·95	$3·95^3$ =	61·629875	Bigger	
b)	3·92	$3·92^3$ =	60·236288	Bigger	3·92 or 3·91 between 3·91 and 3·915 round to 3·91
	3·91	$3·91^3$ =	59·776471	Smaller	
	3·915	$3·915^3$ =	60·006...	Bigger	

a) 3·9   ($\sqrt[3]{60}$ is nearer to 3·9 than 4·0)

b) 3·91 ($\sqrt[3]{60}$ is nearer to 3·91 than 3·92)

Calculator $\sqrt[3]{60} = 3·914...$

Measure 1   Discrete / Continuous.

Measure
Discrete Measure and Continuous Measure
There are 2 kinds of measure - Discrete and Continuous.

Discrete
A **Discrete** measure is **counted**.
The possible values are **discrete** or **separate** from each other and there are no values in between them.

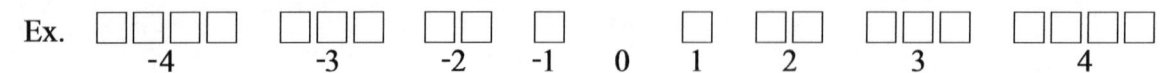

Ex. ☐☐☐☐   ☐☐☐   ☐☐   ☐   ☐   ☐☐   ☐☐☐   ☐☐☐☐
       -4   -3   -2   -1   0   1   2   3   4

The values are **exact**.
Ex. The number of sweets in a box.
Ex. The number of cars passing a junction during a day.
Ex. The number of people in the world.
Ex. The number of £'s owed to a bank (can be referred to as a negative number).

Continuous
A **Continuous** measure is **read from an instrument**.
The possible values are **continuous** or **not separate** from each other and so a measurement can **never be exact**.

Ex. -15, -2·8, $-\frac{3}{7}$, 0, 4·999, 5, $10\frac{39}{100}$

The **accuracy** of a reading depends on the accuracy of the instrument being used ... and the person using it!

Ex. The **exact** reading on the **continuous scale** is not possible -
      it appears to be 2·5 but could be 2·49 or 2·496 or 2·5087 or 2·51 or ...

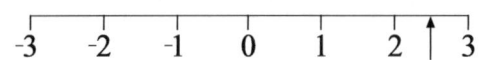

         -3   -2   -1   0   1   2 ↑ 3

	Instrument used
Ex. The **length** of a table -	**ruler**.
Ex. The **weight** of a person -	**scales**.
Ex. The **capacity** of a liquid -	**scaled container ex. jug**.
Ex. The **time** taken to walk a mile -	**watch**.
Ex. The **temperature** of the weather -	**thermometer**.

A measurement is often stated with the degree of accuracy - ex. ' 5mm **to the nearest mm**'.
(This is considered further in **Measure 2** and **Rounding Numbers 1-5**.)

Q	A	Q	A
State whether each measure is **discrete** or **continuous**.			
1) The number of people in a room.	1) discrete	9) The area of a field.	9) continuous
2) The length of a pencil.	2) continuous	10) The number of cars in a city.	10) discrete
3) The number of people in a crowd.	3) discrete	11) The capacity of a container.	11) continuous
4) The height of a building.	4) continuous	12) The time taken to boil an egg.	12) continuous
5) The weight of a baby.	5) continuous	13) The number of fish in the sea.	13) discrete
6) The number of bricks in a wall.	6) discrete	14) The number of coins in a box.	14) discrete
7) The number of pages in a book.	7) discrete	15) The amount of money in a box.	15) discrete
8) The temperature in a room.	8) continuous	16) The speed of a bicycle.	16) continuous

*Note
It is useful to gain '**real life experience**' of dealing with each kind of measure to gain a better understanding.

Measure 2

Continuous Measure - Accuracy.

Continuous Measure - Accuracy

Since an **exact** measure on a **continuous scale** is not possible, it can be stated to the **nearest unit**.
This means that the measure has to fall within a **range** which allows it to be **rounded** to that unit -
the 'minimum' / 'greatest **lower** bound' and the 'maximum' / 'least **upper** bound'.
(The range can be determined **mentally**.)

Ex.1 Consider the following continuous scale -
the arrow marks a measurement of '2 units, to the **nearest 1**'.

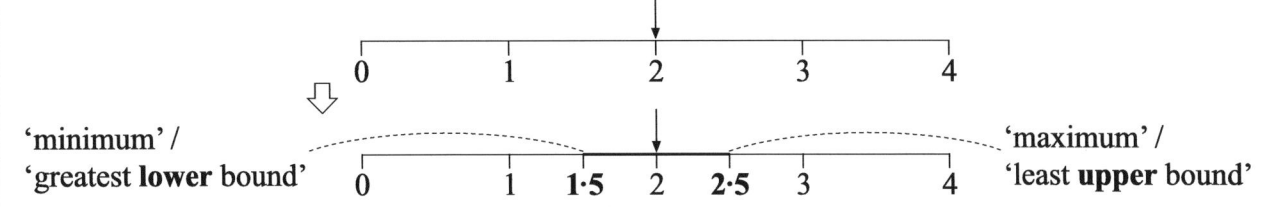

```
              ↓
  0    1    2    3    4
```

'minimum' /
'greatest **lower** bound'   0   1   **1·5**   2   **2·5**   3   4   'maximum' / 'least **upper** bound'

Even though we can **see** that the measurement is **close** to 2,
we still have to allow for it to be in a **range**   as low as   1·5 (minimum) which **rounds up** to 2
                             and   as high as   2·5 (maximum) which **rounds down** to 2.

The measurement can be said to lie in a range which is 'half a unit' below it and 'half a unit' above it.

Here, the unit is **1**.   'Half a unit' is $\frac{1}{2}$ of 1 = 0·5   ⇐ $\begin{array}{r} 0{\cdot}5 \\ 2\overline{)1{\cdot}0} \end{array}$

Ex.2 Consider the following continuous scale -
the arrow marks a measurement of '2 units, to the **nearest 0·1**'.

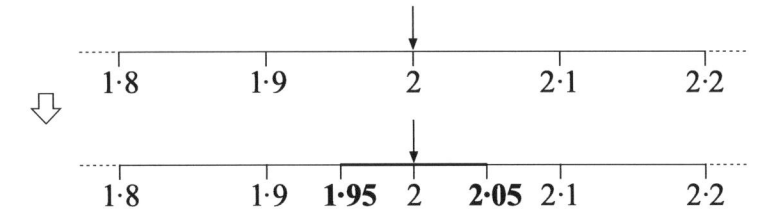

```
         ↓
  1·8   1·9   2   2·1   2·2
```

  1·8   1·9   **1·95**   2   **2·05**   2·1   2·2

Even though we can **see** that the measurement is **close** to 2,
we still have to allow for it to be in a **range**   as low as   1·95 (minimum) which **rounds up** to 2
                             and   as high as   2·05 (maximum) which **rounds down** to 2.

The measurement can be said to lie in a range which is 'half a unit' below it and 'half a unit' above it.
Here, the unit is **0·1**.   'Half a unit' is $\frac{1}{2}$ of 0·1 = 0·05   ⇐ $\begin{array}{r} 0{\cdot}05 \\ 2\overline{)0{\cdot}10} \end{array}$

Ex.3 Consider the following continuous scale -
the arrow marks a measure of '2 units, to the **nearest 0·01**'.

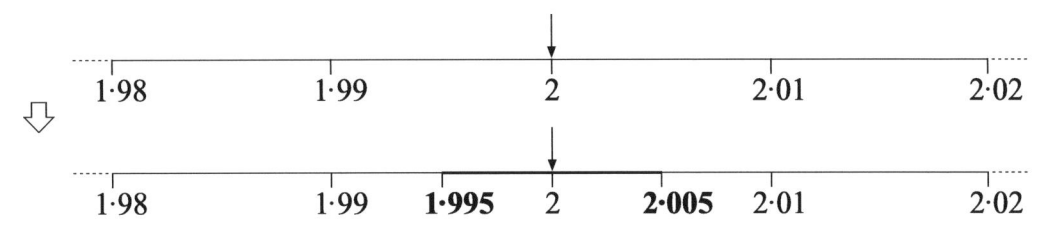

```
          ↓
  1·98   1·99   2   2·01   2·02
```

  1·98   1·99   **1·995**   2   **2·005**   2·01   2·02

Even though we can **see** that the measurement is **close** to 2,
we still have to allow for it to be in a **range**   as low as   1·995 (minimum) which **rounds up** to 2
                             and   as high as   2·005 (maximum) which **rounds down** to 2.

The measurement can be said to lie in a range which is 'half a unit' below it and 'half a unit' above it.

Here, the unit is **0·01**.   'Half a unit' is $\frac{1}{2}$ of 0·01 = 0·005   ⇐ $\begin{array}{r} 0{\cdot}005 \\ 2\overline{)0{\cdot}010} \end{array}$

Measure 3

Continuous Measure - Accuracy.

Q
Show on each continuous scale the measurement, which is given to the '**nearest 1**' unit and its **minimum** and **maximum** value.

1) 3cm

   |—|—|—|—|—|
   0   1   2   3   4cm

2) 6g

   |—|—|—|—|—|
   5   6   7   8   9   10g

3) 99ml

   |—|—|—|
   97   98   99   100ml

4) 50sec

   |—|—|—|
   49   50   51   52sec

A
'Half a unit' is $\frac{1}{2}$ of $1 = 0.5$ ⇐ $2\overline{)1.0} = 0.5$

1) ↓ at 3
   0   1   2   **2·5**   3   **3·5**   4cm

2) ↓ at 6
   5   **5·5**   6   **6·5**   7   8   9   10g

3) ↓ at 99
   97   98   **98·5**   99   **99·5**   100ml

4) ↓ at 50
   49   **49·5**   50   **50·5**   51   52sec

Q
Show on each continuous scale the measurement, which is given to the '**nearest 0·1**' unit and its **minimum** and **maximum** value.

1) 0·1 °C

   0   0·1   0·2   0·3   0·4 °C

2) 77·4 amps

   77·3   77·4   77·5 amps

3) 4·5m

   4·3   4·4   4·5   4·6 m

4) 8·0kg

   7·9   8   8·1 kg

A
'Half a unit' is $\frac{1}{2}$ of $0.1 = 0.05$ ⇐ $2\overline{)0.10} = 0.05$

1) ↓ at 0·1
   0   **0·05**   0·1   **0·15**   0·2   0·3   0·4 °C

2) ↓ at 77·4
   77·3   **77·35**   77·4   **77·45**   77·5 amps

3) ↓ at 4·5
   4·3   4·4   **4·45**   4·5   **4·55**   4·6 m

4) ↓ at 8
   7·9   **7·95**   8   **8·05**   8·1 kg

Q
Show on each continuous scale the measurement, which is given to the '**nearest 0·01**' unit and its **minimum** and **maximum** value.

1) 2·07 *l*

   2·06   2·07   2·08 *l*

2) 1·10 mg

   1·09   1·10   1·11   1·12 mg

3) Angle 93·92°

   93·91   93·92   93·93°

4) 5·55 mph

   5·54   5·55   5·56   5·57 mph

A
'Half a unit' is $\frac{1}{2}$ of $0.01 = 0.005$ ⇐ $2\overline{)0.010} = 0.005$

1) ↓ at 2·07
   2·06   **2·065**   2·07   **2·075**   2·08 *l*

2) ↓ at 1·10
   1·09   **1·095**   1·10   **1·105**   1·11   1·12 mg

3) The line is an arc of a very large circle! ↙
   93·91   **93·915**   93·92   **93·925**   93·93°

4) ↓ at 5·55
   5·54   **5·545**   5·55   **5·555**   5·56   5·57 mph

Measure 4 Continuous Measure - Accuracy.

Continuous Measure - Accuracy
Combining Measures

When the accuracy of a measurement - minimum / maximum - is to be found,
a value may be stated in one unit of measure and its accuracy in a **smaller** unit of measure.

It is usually easier to convert the value to the **smaller** unit of measure ($\times 10, \times 100, \times 1000$)
to determine the minimum / maximum (these can be **mental** calculations)
and then convert these back to the larger unit of measure ($\div 10, \div 100, \div 1000$).

Ex. The length of a rod is 8·3**cm**, correct to the nearest **mm**.
State the minimum and maximum length of the rod, in **cm**.

$$\overset{\times 10}{8\cdot 3} = 83\text{mm} \qquad \text{'Half a unit' is } \tfrac{1}{2} \text{ of } 1 = 0\cdot 5 \qquad \Leftarrow \quad 2\overline{)1\cdot 0}^{\,0\cdot 5}$$

minimum = $\overset{\div 10}{82\cdot 5}$mm = 8·25**cm**

maximum = $\overset{\div 10}{83\cdot 5}$mm = 8·35**cm**

Q	A
1) The height of a toy is 4·7**cm**, correct to the nearest **mm**. State the minimum and maximum height of the toy, in **cm**.	1) $\overset{\times 10}{4\cdot 7} = 47$mm 'Half a unit' = 0·5mm minimum = $\overset{\div 10}{46\cdot 5}$mm = 4·65**cm** maximum = $\overset{\div 10}{47\cdot 5}$mm = 4·75**cm**
2) The length of a room is 5·32**m**, correct to the nearest **cm**. State the minimum and maximum length of the room, in **m**.	2) $\overset{\times 100}{5\cdot 32} = 532$cm 'Half a unit' = 0·5cm minimum = $\overset{\div 100}{531\cdot 5}$cm = 5·315**m** maximum = $\overset{\div 100}{532\cdot 5}$cm = 5·325**m**
3) The weight of a box is 1·069**kg**, correct to the nearest **g**. State the minimum and maximum weight of the box, in **kg**.	3) $\overset{\times 1000}{1\cdot 069} = 1069$g 'Half a unit' = 0·5g minimum = $\overset{\div 1000}{1068\cdot 5}$g = 1·0685**kg** maximum = $\overset{\div 1000}{1069\cdot 5}$g = 1·0695**kg**
4) A volume of water is 2·38 l, correct to the nearest **ml**. State the minimum and maximum volume of water, in l.	4) $\overset{\times 1000}{2\cdot 380} = 2380$ml 'Half a unit' = 0·5ml minimum = $\overset{\div 1000}{2379\cdot 5}$ml = 2·3795 l maximum = $\overset{\div 1000}{2380\cdot 5}$ml = 2·3805 l
5) The length of a plane is 10**m**, correct to the nearest **cm**. State the minimum and maximum length of the plane, in **m**.	5) $\overset{\times 100}{10\cdot 00} = 1000$cm 'Half a unit' = 0·5cm minimum = $\overset{\div 100}{999\cdot 5}$cm = 9·995**m** maximum = $\overset{\div 100}{1000\cdot 5}$cm = 10·005**m**

Measure 5

Continuous Measure - Accuracy.

Continuous Measure - Accuracy: 'Problems'

Q	A

1) The lengths of the sides of a triangle are 5cm, 7cm and 9cm, correct to the nearest **cm**.
 	
Calculate the perimeter of the triangle using
a) the given lengths of the sides.
b) the minimum lengths of the sides.
c) the maximum lengths of the sides.
	1) 'Half a unit' = 0·5cm
	
	  a) given b) minimum c) maximum
	    5 4·5 5·5
	    7 6·5 7·5
	+ 9 + 8·5 + 9·5
	21cm 19·5cm 22·5cm

2) The lengths of the sides of a rectangle are 8m and 3m, correct to the nearest **m**.
 	
Calculate the area of the rectangle using
a) the given lengths of the sides.
b) the minimum lengths of the sides.
c) the maximum lengths of the sides.
	2) 'Half a unit' = 0·5m
	
	  a) given b) minimum c) maximum
	    A = lb A = lb A = lb
	      = 8×3 = 7·5×2·5 = 8·5×3·5
	      = 24m² = 18·75m² = 29·75m²

3) The length of the side of a cube is 2mm, correct to the nearest **mm**.
 	
Calculate the volume of the cube using
a) the given length of the side.
b) the minimum length of the side.
c) the maximum length of the side.
	3) 'Half a unit' = 0·5mm
	
	a) given b) minimum c) maximum
	V = l^3 V = l^3 V = l^3
	  = 2³ = 1·5³ = 2·5³
	  = 2×2×2 = 1·5×1·5×1·5 = 2·5×2·5×2·5
	  = 8mm³ = 3·375mm³ = 15·625mm³

4) A sphere weighs 4g, correct to the nearest **g**.
 	
Calculate the weight of 100 spheres using
a) the given weight of the sphere.
b) the minimum weight of the sphere.
c) the maximum weight of the sphere.
	4) 'Half a unit' = 0·5g
	
	  a) given b) minimum c) maximum
	    4g 3·5g 4·5g
	× 100 spheres × 100 spheres × 100 spheres
	  400g 350g 450g

5) An oil tank holds 20 l, correct to the nearest l. The oil is poured equally into 10 containers. Calculate the amount of oil in a container using a) the given volume in the tank. b) the minimum volume in the tank. c) the maximum volume in the tank.	5) 'Half a unit' = 0·5 l a) given $\dfrac{20\, l}{10 \text{ containers}} = 2\, l$ b) minimum $\dfrac{19\cdot 5\, l}{10 \text{ containers}} = 1\cdot 95\, l$ c) maximum $\dfrac{20\cdot 5\, l}{10 \text{ containers}} = 2\cdot 05\, l$

Measure 6

Continuous Measure - Accuracy.

Q	A

6) The lengths of the sides of a rectangle are 8mm and $6\frac{1}{2}$mm, correct to the nearest $\frac{1}{2}$**mm**.
 	
Calculate the perimeter of the rectangle using
a) the given lengths of the sides.
b) the minimum lengths of the sides.
c) the maximum lengths of the sides.
 	
[rectangle: 8mm by $6\frac{1}{2}$mm]
	6) 'Half a unit' = $\frac{1}{2}$ of $\frac{1}{2}$ = $\frac{1}{4}$ mm
	
	a) given b) minimum c) maximum
	   8 $7\frac{3}{4}$ $8\frac{1}{4}$
	   8 $7\frac{3}{4}$ $8\frac{1}{4}$
	   $6\frac{1}{2}$ $6\frac{1}{4}$ $6\frac{3}{4}$
	+ $6\frac{1}{2}$ + $6\frac{1}{4}$ + $6\frac{3}{4}$
	29mm 28mm 30mm

7) The radius of a circle is 5·1cm, correct to the nearest **0·1cm**. Calculate the area of the circle using a) the given length of the radius. b) the minimum length of the radius. c) the maximum length of the radius. (let $\pi = 3.14$) [circle: radius 5·1cm]	7) 'Half a unit' is $\frac{1}{2}$ of $0.1 = 0.05$cm a) given $A = \pi r^2$ $= \pi \times r \times r$ $= 3.14 \times 5.1 \times 5.1$ $= 81.6714$cm^2 b) minimum c) maximum $A = \pi r^2$ $A = \pi r^2$ $= \pi \times r \times r$ $= \pi \times r \times r$ $= 3.14 \times 5.05 \times 5.05$ $= 3.14 \times 5.15 \times 5.15$ $= 80.07785$cm^2 $= 83.28065$cm^2

8) A gold bar weighs 0·09kg, correct to the nearest **0·01kg**.
 	
Calculate the weight of 50 gold bars using
a) the given weight of the bar.
b) the minimum weight of the bar.
c) the maximum weight of the bar.
	8) 'Half a unit' is $\frac{1}{2}$ of $0.01 = 0.005$kg
	
	a) given b) minimum c) maximum
	  0·09kg 0·085kg 0·095kg
	× 50 bars × 50 bars × 50 bars
	  4·5kg 4·25kg 4·75kg

9) An athlete runs 30m in 3·7s, correct to the nearest $\frac{1}{100}$**s**. State a) the athlete's minimum time. b) the athlete's maximum time.	9) 'Half a unit' is $\frac{1}{2}$ of $0.01 \Big\} \frac{1}{100} = 0.005$s a) minimum = 3·695s b) maximum = 3·705s

10) The height of Student X is 164cm and Student Y is 160cm, both correct to the nearest **cm**.
Calculate
a) the difference in heights using the given values.
b) the **smallest** difference in heights.
c) the **largest** difference in heights.
	10) 'Half a unit' is 0·5cm
	
	a) given
	X 164
	Y − 160
	       4cm
	
	b) c)
	minimum X 163·5 maximum X 164·5
	maximum Y − 160·5 minimum Y − 159·5
	                  3cm 5cm

Measure 7 — Error Interval.

Error Interval

A measurement may be larger or smaller by a given Error Interval.
For example, a 10% Error Interval means that a measurement may be up to 10% larger or 10% smaller than the measurement given.

Ex. 100g chocolate bars are made with a 10% error interval.
Calculate
a) the largest possible weight of a bar. ⇨ a) Calculate 10% of 100g

$$\frac{10}{100} \times \frac{100}{1} = 10g \qquad \begin{array}{r} 100 \\ +\ 10 \\ \hline 110g \end{array}$$

b) the smallest possible weight of a bar.

b) $\quad \begin{array}{r} 100 \\ -\ 10 \\ \hline 90g \end{array}$

c) state the possible weights as an inequality.

c) $90g \leqslant \text{weight} \leqslant 110g$

Q	A
1) 700ml bottles of milk have a 5% error interval. Calculate a) the maximum possible capacity of milk in a bottle. b) the minimum possible capacity of milk in a bottle. c) state the possible capacity as an inequality.	1) a) Calculate 5% of 700ml $\frac{5}{100} \times \frac{700}{1} = 35ml \qquad \begin{array}{r} 700 \\ +\ 35 \\ \hline 735ml \end{array}$ b) $\quad \begin{array}{r} 700 \\ -\ 35 \\ \hline 665ml \end{array}$ c) $665ml \leqslant \text{capacity} \leqslant 735ml$
2) A machine makes 20mm nails with a 2% error interval. Calculate a) the largest possible length of a nail. b) the smallest possible length of a nail. c) state the possible lengths as an inequality.	2) a) Calculate 2% of 20mm $\frac{2}{100} \times \frac{20}{1} = \frac{40}{100} = 0.4mm \qquad \begin{array}{r} 20.0 \\ +\ 0.4 \\ \hline 20.4mm \end{array}$ b) $\quad \begin{array}{r} 20.0 \\ -\ 0.4 \\ \hline 19.6mm \end{array}$ c) $19.6mm \leqslant \text{length} \leqslant 20.4mm$
3) The temperature of a process is set with an error interval which allows for a minimum possible temperature of 38°C and a maximum possible temperature of 44°C. Calculate a) the set temperature of the process. b) the error interval as a percentage. c) state the possible temperatures as an inequality.	3) a) Calculate the average possible temperature $\frac{38 + 44}{2} = \frac{82}{2} = 41°C$ b) Calculate the error interval value $44°C - 41°C = 3°C$ or $41°C - 38°C = 3°C$ $\frac{\text{error interval}}{\text{set temperature}} \qquad \frac{3}{41} \times 100\% = \frac{300}{41} = 7.31...\%$ c) $38°C \leqslant \text{temperature} \leqslant 44°C$

Measure 8 — Absolute Error. Percentage Error.

Absolute Error. Percentage Error

The **absolute error** is the difference between the **measured value** and the **actual value** of a quantity.
This is **always positive** so the [highest − lowest] value is calculated.
(the 'actual value' can be considered to be its 'expected value'.)

The **percentage error** may then be calculated $= \dfrac{\text{absolute error}}{\text{actual value}} \times 100\%$

Ex. The **actual weight** of a chocolate bar is 100g.
(it is 'expected' to be this weight)
Its **measured weight** is 103g.

Calculate a) the absolute error of the weight.
b) the percentage error of the weight.

a) measured weight 103g
    − actual weight   100g
    absolute error   3g

b) $\dfrac{\text{absolute error}}{\text{actual value}} \times 100\%$

$\dfrac{3}{100} \times 100\% = \dfrac{300}{100} = 3\%$

Q	A
1) A can of juice has an actual capacity of 500ml. It is measured to have 495ml. Calculate a) the absolute error of the capacity. b) the percentage error of the capacity.	1) a) actual capacity 500ml − measured capacity 495ml absolute error 5ml b) $\dfrac{\text{absolute error}}{\text{actual value}} \; \dfrac{5}{500} \times 100\% = \dfrac{500}{500} = 1\%$
2) A pot of jam, labelled 250g is measured to weigh 264g. Calculate a) the absolute error of the weight. b) the percentage error of the weight.	2) a) measured weight 264g − actual weight 250g absolute error 14g b) $\dfrac{\text{absolute error}}{\text{actual value}} \; \dfrac{14}{250} \times 100\% = \dfrac{1400}{250} = 5\cdot 6\%$
3) The actual length of a path, 38m, is measured to be 37m. Calculate a) the absolute error of the length. b) the percentage error of the length.	3) a) actual length 38m − measured length 37m absolute error 1m b) $\dfrac{\text{absolute error}}{\text{actual value}} \; \dfrac{1}{38} \times 100\% = \dfrac{100}{38} = 2\cdot 63...\%$
4) In a time trial, the actual time between cyclists starting is 40secs. The measured time between 2 cyclists starting is 41·5secs. Calculate a) the absolute error of the time. b) the percentage error of the time.	4) a) measured time 41·5secs − actual time 40 secs absolute error 1·5secs b) $\dfrac{\text{absolute error}}{\text{actual value}} \; \dfrac{1\cdot 5}{40} \times 100\% = \dfrac{150}{40} = 3\cdot 75\%$
5) The actual length of a steel rod is 8cm, to the nearest cm. Calculate a) the maximum length of the rod. b) the minimum length of the rod. c) the maximum absolute error of the length. d) the percentage error of the length.	5) a) maximum = 8·5cm { 'Half a unit' = 0·5cm b) minimum = 7·5cm c) measured length 8·5cm − actual length 8 cm absolute error 0·5cm or actual length 8 cm − measured length 7·5cm absolute error 0·5cm d) $\dfrac{\text{absolute error}}{\text{actual value}} \; \dfrac{0\cdot 5}{8} \times 100\% = \dfrac{50}{8} = 6\cdot 25\%$

Measure - Metric System 1

Metric Units - Length / Weight / Capacity.

Metric System
Here, the **metric system** for Length, Weight and Capacity is used with **decimals** and not fractions.
*Note Here, Area and Volume are also included.
The following units and conversions are used -

Length

mm	= millimetre	10 mm = 1 cm
cm	= centimetre	100 cm = 1 m
m	= metre	1000 mm = 1 m
km	= kilometre	1000 m = 1 km

Weight

mg	= milligram	1000 mg = 1 g
g	= gram	1000 g = 1 kg
kg	= kilogram	1000 kg = 1 tonne
tonne	= tonne	

Capacity

			*Note
ml	= millilitre	1000 ml = 1 l = 1000 cm^3	1 ml = 1 cm^3
cl	= centilitre	10 ml = 1 cl = 10 cm^3	
l	= litre	100 cl = 1 l	1 ml of **water** weighs 1g

$\frac{1}{1000}$ = $\frac{1}{100}$

$\frac{1}{1000}$ = $\frac{1}{100}$ of a metre

$\frac{1}{1000}$ = $\frac{1}{100}$ of a litre

$\frac{1}{1000}$

kilo = 1000
1 **kilo**metre = 1000 metres
1 **kilo**gram = 1000 grams

*Note
Decimals ... Division / Multiplication by 10, 100 and 1000 are needed to convert between the units.
A brief reminder is given here.

The decimal point (if needed) is on the RIGHT of a whole number.   Exs.   5 = 5·   409 = 409·

÷ 10   Move the point **1 place LEFT**

÷ 100   Move the point **2 places LEFT**

÷ 1000   Move the point **3 places LEFT**

Ex.   a)   44 ÷ 10   b)   809·2 ÷ 100   c)   76·5 ÷ 1000
         = 4·4   = 8·092   = 0·0765

× 10   Move the point **1 place RIGHT**

× 100   Move the point **2 places RIGHT**

× 1000   Move the point **3 places RIGHT**

Ex.   a)   12·3 × 10   b)   0·584 × 100   c)   0·97 × 1000
         = 123   = 58·4   = 970
(whole number - omit point)

Measure - Metric System 2

Measure / Draw Line (cm / mm).

Length
Length is a **continuous** measure.

The Metric Units of Length
The following units are used here -

    10 mm = 1 cm   [mm = millimetre]
                            [cm = centimetre]

Measure / Draw a line
The end points of a line are usually labelled with capital letters.   Ex.   A————B

The line is referred to as **'the line AB'**.

When measuring/drawing the length of a line, starting from 0 on the ruler is usually easier.

Ex.   X————————Y   XY = 3cm 5mm = 35mm

Q	A
Measure and state the length of each line in a) **cm** and **mm** b) **mm**	
1) A — B	1) a) 0cm 4mm b) 4mm
2) C ——— D	2) a) 2cm 7mm b) 27mm
3) E ———————— F	3) a) 5cm 0mm b) 50mm
4) G ———————————————— H	4) a) 8cm 8mm b) 88mm
5) I ———————————————————— J	5) a) 10cm 5mm b) 105mm
6) K ———————————————————————— L	6) a) 11cm 1mm b) 111mm

Q	A
Draw each line.	
1) MN 0cm 9mm	1) M —— N
2) OP 3cm 2mm	2) O ———————— P
3) QR 60mm	3) Q ———————————————— R
4) ST 7cm 6mm	4) S ———————————————————— T
5) UV 95mm	5) U ———————————————————————— V
6) WX 123mm	6) W ———————————————————————————————— X

Measure - Metric System 3
Convert Units - Length.

Length

The Metric Units of Length

The following units and conversions are used -

    mm = millimetre   10 mm = 1 cm
    cm = centimetre   100 cm = 1 m
    m = metre   1000 m = 1 km
    km = kilometre

Convert Units

Measurements can be converted between the different units.

To change		To change	
mm to cm	÷ **10**	cm to mm	× **10**
cm to m	÷ **100**	m to cm	× **100**
m to km	÷ **1000**	km to m	× **1000**

Ex. Convert 63mm to cm.

$$63 \div 10$$
$$= 6.3 \text{ cm}$$

Ex. Convert 5.08m to cm.

$$5.08 \times 100$$
$$= 508 \text{ cm}$$

Q	A	Q	A
Convert			
1) 80mm to cm	1) ÷ 10 = 8cm	5) 90cm to mm	5) × 10 = 900mm
2) 6mm to cm	2) ÷ 10 = 0.6cm	6) 5.5cm to mm	6) × 10 = 55mm
3) 17mm to cm	3) ÷ 10 = 1.7cm	7) 0.3cm to mm	7) × 10 = 3mm
4) 521mm to cm	4) ÷ 10 = 52.1cm	8) 74.74cm to mm	8) × 10 = 747.4mm

Q	A	Q	A
Convert			
1) 200cm to m	1) ÷ 100 = 2m	5) 10m to cm	5) × 100 = 1000cm
2) 76cm to m	2) ÷ 100 = 0.76m	6) 8.4m to cm	6) × 100 = 840cm
3) 953cm to m	3) ÷ 100 = 9.53m	7) 0.19m to cm	7) × 100 = 19cm
4) 5cm to m	4) ÷ 100 = 0.05m	8) 5.885m to cm	8) × 100 = 588.5cm

Q	A	Q	A
Convert			
1) 3000m to km	1) ÷ 1000 = 3km	5) 59.648km to m	5) × 1000 = 59648m
2) 8250m to km	2) ÷ 1000 = 8.25km	6) 1.3km to m	6) × 1000 = 1300m
3) 167m to km	3) ÷ 1000 = 0.167km	7) 0.02km to m	7) × 1000 = 20m
4) 49m to km	4) ÷ 1000 = 0.049km	8) 0.007km to m	8) × 1000 = 7m

Measure - Metric System 4
'Problems' - Length.

Measure - Length
'Problems' often involve converting units, as in the following Q/A.

Q	A
1) A piece of wood is 789**mm** long. a) Express the length in **m**. b) Find the total length of 5 such pieces i) in **mm** ii) in **m**	1) a) $\quad\div 1000$ $\qquad 789\text{mm} = 0.789\text{m}$ b) i) $\quad 789$ $\qquad \times\ \ \ 5$ $\qquad \overline{3945}\text{mm}\ \Rightarrow$ ii) $\div 1000$ $\qquad 3945\text{mm} = 3.945\text{m}$
2) Each brick in a pile has a length of 30**cm**. a) State the length in **mm**. b) Find the total length of 60 bricks i) in **cm** ii) in **m**	2) a) $\quad\xrightarrow{\times 10}$ $\qquad 30\text{cm} = 300\text{mm}$ b) i) $60\times 30\text{cm} = 1800\text{cm}$ $\qquad\qquad\Downarrow\ \xrightarrow{\div 100}$ ii) $\qquad 1800\text{cm} = 18\text{m}$
3) The inside lane of a running track is measured at 400**m**. a) Express the distance in **km**. b) An athlete runs round the track 25 times. How far did the athlete run i) in **m** ii) in **km**	3) a) $\quad\div 1000$ $\qquad 400\text{m} = 0.4\text{km}$ b) i) $25\times 400\text{m} = 10{,}000\text{m}$ $\qquad\qquad\Downarrow\ \xrightarrow{\div 1000}$ ii) $\qquad 10{,}000\text{m} = 10\text{km}$
4) How many pieces of wire, each 9**cm** long, can be cut from a reel holding 18**m**.	4) **Method 1** $\xrightarrow{\times 100}$ $18\text{m} = 1800\text{cm}\ \Rightarrow\ \dfrac{1800\text{cm}}{9\text{cm}} = 200\text{pieces}$ **Method 2** $\div 100$ $\qquad 9\text{cm} = 0.09\text{m}\ \Rightarrow\ \dfrac{18\text{m}}{0.09\text{m}} = 200\text{pieces}$
5) A cyclist travels 30·9**km**, rests, and then travels a further 8707**m**. Calculate the total distance travelled i) in **km** ii) in **m**	5) i) $\div 1000$ $\quad 8707\text{m} = 8.707\text{km}\ \Rightarrow\ \begin{array}{r}30.9\ \text{km}\\ +\ 8.707\text{km}\\ \hline 39.607\text{km}\end{array}$ $\qquad\qquad\qquad\times 1000$ ii) $\quad 39.607\text{km} = 39607\text{m}$
6) The diameter of pipe A is 2·1**cm**, pipe B is 19**mm** and pipe C is 23·5**mm**. a) Find the difference, in **mm**, between the diameters of pipes A and B. b) Find the difference, in **cm**, between the diameters of pipes A and C.	6) a) $\times 10$ $\quad 2.1\text{cm} = 21\text{mm}\ \Rightarrow\ \begin{array}{r}21\text{mm}\ \ \text{A}\\ -19\text{mm}\ \ \text{B}\\ \hline 2\text{mm}\end{array}$ b) $\div 10$ $\quad 23.5\text{mm} = 2.35\text{cm}\ \Rightarrow\ \begin{array}{r}2.35\text{cm}\ \ \text{C}\\ -2.1\ \ \text{cm}\ \ \text{A}\\ \hline 0.25\text{cm}\end{array}$
7) Convert 7**km** to **cm**.	7) $\xrightarrow{\times 1000}\ \xrightarrow{\times 100}$ $\quad 7\text{km} = 7{,}000\text{m} = 700{,}000\text{cm}$

Measure - Metric System 5　　　　　　　　　　　　Convert Units - Weight.

Weight
Weight is a **continuous** measure.

The Metric Units of Weight
The following units and conversions are used -

    mg　= milligram　　　　1000 mg　= 1 g
    g　= gram　　　　　　 1000 g　= 1 kg
    kg　= kilogram　　　　 1000 kg　= 1 tonne
    tonne = tonne

Convert Units
Measurements can be converted between the different units.

To change　　　　　　　　　　　　　　　　To change
    mg　to　g　　÷ **1000**　　　　　　　g　to　mg　× **1000**
    g　to　kg　　÷ **1000**　　　　　　　kg　to　g　× **1000**
    kg　to　tonnes　÷ **1000**　　　　　tonnes　to　kg　× **1000**

Ex.　Convert　1234 mg to g.　　　　　　　　Ex.　Convert　5·67 kg to g.
　　　　　　1 2 3 4　÷ 1000　　　　　　　　　　　　　5·6 7　× 1000
　　　　　= 1·2 3 4 g　　　　　　　　　　　　　　　 = 5 6 7 0 g

Q	A	Q	A
Convert			
1) 7000mg to g	1) ÷ 1000 = 7g	4) 4·624g to mg	4) × 1000 = 4624mg
2) 5839mg to g	2) ÷ 1000 = 5·839g	5) 1·5g to mg	5) × 1000 = 1500mg
3) 105mg to g	3) ÷ 1000 = 0·105g	6) 0·09g to mg	6) × 1000 = 90mg

Q	A	Q	A
Convert			
1) 4000g to kg	1) ÷ 1000 = 4kg	4) 8·023kg to g	4) × 1000 = 8023g
2) 12345g to kg	2) ÷ 1000 = 12·345kg	5) 5·1kg to g	5) × 1000 = 5100g
3) 769g to kg	3) ÷ 1000 = 0·769kg	6) 0·004kg to g	6) × 1000 = 4g

Q	A	Q	A
Convert			
1) 85000kg to tonnes	1) ÷ 1000 = 85tonnes	4) 200·7 tonnes to kg	4) × 1000 = 200700kg
2) 54300kg to tonnes	2) ÷ 1000 = 54·3tonnes	5) 0·05tonnes to kg	5) × 1000 = 50kg
3) 990kg to tonnes	3) ÷ 1000 = 0·99tonnes	6) 0·0016tonnes to kg	6) × 1000 = 1·6kg

Measure - Metric System 6

'Problems' - Weight.

Measure - Weight

'Problems' often involve converting units, as in the following Q/A.

Q	A

1) A pill weighs 475**mg**.
   a) Express the weight in **g**.
 	
   b) Find the total weight of 32 pills
      i) in **mg**
      ii) in **g**
	1) a) ÷1000
	   475mg = 0·475g
	
	   b) i) 475
	        × 32
	        ─────
	          950
	       +14250
	        ─────
	        15200mg ⇒ ii) ÷1000
	                        15200 = 15·2g

2) A tube of sweets weighs 0·123**kg**.
   a) Express the weight in **g**.
 	
   b) Find the total weight of 40 tubes
      i) in **kg**
      ii) in **g**
	2) a) ×1000
	   0·123kg = 123g
	
	   b) i) 0·123
	        × 40
	        ──────
	        4·920kg
	      = 4·92 kg ⇒ ii) ×1000
	                          4·92 kg = 4920g

3) 600 equal size paving stones weigh 8·4**tonnes** in total.
   a) Express the weight in **kg**.
 	
   b) Calculate the weight of 1 paving stone
      i) in **kg**
      ii) in **tonnes**
	3) a) ×1000
	   8·4 tonnes = 8400kg
	
	   b) i) 8400kg / 600 = 14kg
	
	      ii) ÷1000
	          14kg = 0·014tonnes

4) 97**mg** of a chemical is added to 2·9**g** of another. Find the total weight, in **mg**, of the chemicals.	4) ×1000
	   2·9 g = 2900mg ⇒ 2900g
	                      + 97g
	                      ──────
	                       2997g

5) A cook uses 462**g** of flour from a 1·5**kg** bag. What weight of flour, in **kg**, is left in the bag?	5) ÷1000
	   462g = 0·462kg ⇒ 1·500kg
	                      − 0·462kg
	                      ────────
	                        1·038kg

6) How many 550**g** bags of dog biscuits can be filled from a sack weighing 12·1**kg**?	6) **Method 1**
	   ÷1000
	   550g = 0·55kg ⇒ 12·1kg / 0·55kg = 22bags
	
	   Method 2
	   ×1000
	   12·1 kg = 12100g ⇒ 12100g / 550g = 22bags

7) A lorry, weighing 15·777 **tonnes**, carries a load of 653 metal pipes each weighing 6·1**kg**. What is the combined weight, in **tonnes**, of the lorry and its load?	7) 653
	     × 6·1
	     ──────
	       653
	    +3 9 1 8 0
	    ─────────
	     3,9 8 3·3kg = 3·9833tonnes ⇒ 15·777 tonnes
	     ÷1000 + 3·9833tonnes
	                                       ──────────────
	                                        19·7603tonnes

8) Convert 3**tonnes** to **g**.	8) ×1000 ×1000 3tonnes = 3000kg = 3,000,000g

Measure - Metric System 7

Convert Units - Capacity.

Capacity
Capacity is a **continuous** measure.

The Metric Units of Capacity
The following units and conversions are used -

ml = millilitre
cl = centilitre
l = litre

10 ml = 1 cl
100 cl = 1 l

1000 ml = 1 l ⬅ *Note
This is more commonly used

*Note
The following units are also used -

1ml = 1cm^3
1cl = 10ml = 10cm^3
1 l = 1000ml = 1000cm^3

Convert Units
Measurements can be converted between the units.

To change
ml to cl ÷ **10**
cl to l ÷ **100**
ml to l ÷ **1000**

To change
cl to ml × **10**
l to cl × **100**
l to ml × **1000**

Ex. Convert 4321ml to l.

4 3 2 1 ÷ 1000
= 4·3 2 1 l

Ex. Convert 6·58cl to ml.

6·58 × 10
= 6 5·8 ml

Q	A	Q	A
Convert			
1) 6000ml to l	1) ÷ 1000 = 6 l	5) 3·815 l to ml	5) × 1000 = 3815ml
2) 95000ml to l	2) ÷ 1000 = 95 l	6) 0·29 l to ml	6) × 1000 = 290ml
3) 2468ml to l	3) ÷ 1000 = 2·468 l	7) 1·1 l to ml	7) × 1000 = 1100ml
4) 70ml to l	4) ÷ 1000 = 0·07 l	8) 0·006 l to ml	8) × 1000 = 6ml

Q	A	Q	A
Convert			
1) 20ml to cl	1) ÷ 10 = 2cl	5) 400cl to l	5) ÷ 100 = 4 l
2) 507ml to cl	2) ÷ 10 = 50·7cl	6) 730cl to l	6) ÷ 100 = 7·3 l
3) 8·5cl to ml	3) × 10 = 85ml	7) 6·2 l to cl	7) × 100 = 620cl
4) 0·9cl to ml	4) × 10 = 9ml	8) 0·38 l to cl	8) × 100 = 38cl

Q	A	Q	A
Convert			
1) 300cm^3 to ml	1) = 300ml	4) 9·7ml to cm^3	4) = 9·7cm^3
2) 50cm^3 to cl	2) ÷ 10 = 5 cl	5) 4·9cl to cm^3	5) × 10 = 49cm^3
3) 8642cm^3 to l	3) ÷ 1000 = 8·642 l	6) 0·25 l to cm^3	6) × 1000 = 250cm^3

Measure - Metric System 8

'Problems' - Capacity.

Measure - Capacity

'Problems' often involve converting units, as in the following Q/A.

Q	A

1) A bottle holds 900**ml** of milk.
  a) Express the capacity in **cl**.
 	
  b) What is the total amount of milk in 9 bottles
    i) in **ml**
    ii) in **cl**
	1) a) $\xrightarrow{\div 10}$
	   900ml = 90cl
	
	   b) i) 900
	        × 9
	        ―――
	        8100ml $\Rightarrow$ ii) $\xrightarrow{\div 10}$ 8100 = 810cl

2) A jar has a capacity of 220**ml**.
  a) What is the capacity in **l**.
 	
  b) Find the total capacity of 18 jars
    i) in **ml**
    ii) in **l**
	2) a) $\xrightarrow{\div 1000}$
	   220ml = 0·22 l
	
	   b) i) 220
	        × 18
	        ―――
	        1760
	        +2200
	        ―――
	        3960ml $\Rightarrow$ ii) $\xrightarrow{\div 1000}$ 3960ml = 3·96 l

3) The volume of a can is 0·34 **l**.
  a) State the volume in **ml**.
 	
  b) Calculate the total volume of 50 cans
    i) in **l**
    ii) in **ml**
	3) a) $\xrightarrow{\times 1000}$
	   0·34 l = 340ml
	
	   b) i) 0·34
	        × 50
	        ―――
	        17·00 l
	        = 17 l $\Rightarrow$ ii) 17 ×1000 = 17000ml

4) 9·2 **l** of water are in a tank.
  When 645**ml** are added, how much water is in the tank
  a) in **l**
  b) in **ml**
	4) a) $\xrightarrow{\div 1000}$ 9·2 l ---- tank
	   645ml = 0·645 l $\Rightarrow$ +0·645 l ---- added
	                                 ―――――
	                                 9·845 l ---- total
	                                   $\Downarrow$
	   b) × 1000
	        $\xrightarrow{}$
	        9·845 l = 9845 ml

5) How many 40**ml** tins of oil can be filled from a can holding 80**cl** ?	5) Method 1
	   $\xrightarrow{\div 10}$
	   40ml = 4cl $\Rightarrow$ $\dfrac{80cl}{4cl}$ = 20 tins
	   --- 	
	   Method 2
	   $\xrightarrow{\times 10}$
	   80cl = 800ml $\Rightarrow$ $\dfrac{800ml}{40ml}$ = 20 tins

6) The volumes of 2 containers are 85·9 **l** and 77,750**cm³**.
   Calculate the difference between the 2 volumes in **cm³**.
	6) $\xrightarrow{\times 1000}$
	   85·9 l = 85900cm³ $\Rightarrow$ 85900cm³
	                                 −77750cm³
	                                 ―――――
	                                  8150cm³

7) Bottle X has a capacity of 6**ml**.
   Bottle Y has a capacity of 5 times Bottle X.
   Bottle Z has a capacity of 4 times Bottle Y.
 	
   a) Find the capacity of Bottle Y in **cl**.
 	
   b) Find the capacity of Bottle Z in **l**.
	7) a) Y = 5×X b) Z = 4×Y
	      = 5×6ml = 4×30ml
	      = 30ml = 120ml
	        $\Downarrow$ $\Downarrow$
	      $\xrightarrow{\div 10}$ $\xrightarrow{\div 1000}$
	      30ml = 3cl 120ml = 0·12 l

Measure - Metric System 9 — Ascending / Descending order.

Measurement Order
Measurements can be arranged in **ascending** order (from smallest to largest)
or **descending** order (from largest to smallest).

Given several numbers to arrange in order, it is useful to list them in a **column** in **descending** order from top to bottom (even if **ascending** order is required).
It is easier to compare numbers if the digits are in their correct columns.
It is easy to then list the numbers in a **row**, in **either** order, if required.

It may be possible to arrange the numbers in their given form - but if we are not sure,
it is usual to change the necessary values to the **same unit of measure**, arrange the numbers,
then write them in their given form.
It is best if the **same unit of measure** is the **smallest unit** of those given since this generally ensures the easier task of comparing **integers** (positive/negative whole numbers) and not decimals.

Ex.1
List the following lengths in **ascending** order.

    4·2cm   15mm   0·03m

×10 → 4·2cm   15mm   ×1000 → 0·03m
= 42**mm**   15**mm**   30**mm**

ascending ↑ 42mm   4·2cm
            30mm ⇨ 0·03m
            15mm   15mm

Ex.2
List the following weights in **descending** order.

    5·99kg   6487g   920g   0·8kg

×1000 → 5·99kg   6487g   920g   ×1000 → 0·8kg
= 5 990**g**   6487**g**   920**g**   800**g**

descending ↓ 6487g   6487g
             5990g   5·99kg
             920g ⇨ 920g
             800g   0·8kg

Q

1) List in **ascending** order

    500mm   0·6m   80cm

A

1)   ×1000   ×10
    500mm   0·6m   80cm
= 500**mm**   600**mm**   800**mm**

ascending ↑ 800mm   80cm
             600mm ⇨ 0·6m
             500mm   500mm

---

2) List in **descending** order

    0·4kg   36g   0·17kg   45g

2)   ×1000   ×1000
    0·4kg   36g   0·17kg   45g
=   400**g**   36**g**   170**g**   45**g**

descending ↓ 400g   0·4kg
             170g   0·17kg
             45g ⇨ 45g
             36g   36g

---

3) List in **ascending** order

    88m*l*   0·5 *l*   762m*l*   93c*l*

3)   ×1000   ×10
    88m*l*   0·5 *l*   762m*l*   93c*l*
= 88**m*l***   500**m*l***   762**m*l***   930**m*l***

ascending ↑ 930m*l*   93c*l*
             762m*l*   762m*l*
             500m*l* ⇨ 0·5 *l*
             88m*l*   88m*l*

Measure - Metric System 10

Ascending / Descending order.

Q	A
4) List in **descending** order 7327m 7·23km 7·3km 7227m	4) ×1000 ×1000 7327m 7·23km 7·3km 7227m = 7327**m** 7 230**m** 7 300**m** 7227**m** **descending** 7327m 7300m 7230m 7227m ⇨ 7327m 7·3km 7·23km 7227m
5) List in **ascending** order 2·98g 3g 3456mg 2892mg	5) ×1000 ×1000 2·98g 3g 3456mg 2892mg = 2 980**mg** 3000**mg** 3456mg 2892**mg** **ascending** 3456mg 3000mg 2980mg 2892mg ⇨ 3456mg 3g 2·98g 2892mg
6) List in **descending** order 6450cm³ 654 c*l* 6554m*l* 6·405 *l*	6) ×10 ×1000 6450cm³ 654 c*l* 6554m*l* 6·405 *l* = 6450**m*l*** 6540**m*l*** 6554m*l* 6 405**m*l*** **descending** 6554m*l* 6540m*l* 6450m*l* 6405m*l* ⇨ 6554m*l* 654c*l* 6450cm³ 6·405 *l*
7) With sea level and Town A situated at 0m, the heights of 4 other towns are situated as follows: B 80m, C −0·03km, D 0·1km, E −25m. List the towns in **ascending** order of height.	7) ×1000 ×1000 0m 80m −0·03km 0·1km −25**m** = 0m 80m −30m 100**m** −25**m** A B C D E **ascending** 100m 80m 0m −25m −30m ⇨ D 0·1km B 80m A 0m E −25m C −0·03km
8) List in **descending** order 4·3tonnes 4299·9kg 4310kg	8) ×1000 4·3tonnes 4299·9kg 4310kg = 4 300**kg** 4299·9kg 4310**kg** **descending** 4310 kg 4300 kg 4299·9kg ⇨ 4310kg 4·3tonnes 4299·9kg
9) List in **ascending** order 70·7 *l* 70070m*l* 70707cm³ 7007·7 c*l*	9) ×1000 ×10 70·7 *l* 70070m*l* 70707cm³ 7007·7 c*l* = 7 0700**m*l*** 70070m*l* 70707**m*l*** 7007 7**m*l*** **ascending** 70707m*l* 70700m*l* 70077m*l* 70070m*l* ⇨ 70707cm³ 70·7 *l* 7007·7c*l* 70070m*l*

240

Measure - Metric System 11

Estimation.

Estimation
Length / Weight / Capacity

It is useful to be able to make a 'reasonable' estimation of a measure in 'everyday life'.
This should be based on actual experience and / or ' general knowledge'.

Ex. The weight of a can of soup could be found by using scales or reading its label.
Ex. The weight of a passenger plane would rely on information from its manufacturer.

The size of either object may vary considerably so any estimation would need to take this into account.

The following Q/A give just 2 'estimates' - the more 'reasonable' estimate should be **chosen**.

Length

Q	A	Q	A
Estimate			
1) Length of a nail. a) 2cm b) 60cm	1) a)	4) Height of a house. a) 7m b) 70m	4) a)
2) Width of a cereal box. a) 80cm b) 30cm	2) b)	5) Length of a sports pitch. a) 1km b) 110m	5) b)
3) Height of an adult. a) 1·7m b) 2·7m	3) a)	6) Length of a car. a) 30m b) 3m	6) b)

Weight

Q	A	Q	A
Estimate			
1) Weight of A4 sheet of paper. a) 200mg b) 5g	1) b)	4) Weight of a can of peas. a) 30g b) 300g	4) b)
2) Weight of a chocolate bar. a) 70g b) 7mg	2) a)	5) Weight of a bicycle. a) 14kg b) 90kg	5) a)
3) Weight of a teenager. a) 50kg b) 8kg	3) a)	6) Weight of a car. a) 18 tonne b) 1 tonne	6) b)

Capacity

Q	A	Q	A
Estimate			
1) Capacity of a teaspoon. a) 4ml b) 25ml	1) a)	4) Capacity of a bucket. a) 2000ml b) 16 l	4) b)
2) Capacity of a cup. a) 250ml b) 950ml	2) a)	5) Capacity of a petrol tanker. a) 30,000 l b) 700 l	5) a)
3) Capacity of a milk carton. a) 10 l b) 2 l	3) b)	6) Capacity of a swimming pool. a) 40,000 l b) 2million l	6) b)

Measure - Metric System 12

Convert Units - Area

The following units and conversions are used -

Length units	Hectare
1 cm = 10 mm	**1 hectare = 10,000m^2**
1 m = 100 cm	(1 way to remember it ...)
1 km = 1000 m	100m □ 100m
	100m × 100m = 10000m^2

*Note It is useful to draw a square and label the lengths as here.

Q	A
1) Convert a) 1cm^2 to mm^2 b) 6cm^2 to mm^2	1) □ 1cm, 10mm / 1cm, 10mm a) 1cm^2 = 1cm × 1cm = 10mm × 10mm = 100mm^2 b) 6cm^2 = 6 × 1cm × 1cm = 6 × 10mm × 10mm = 600mm^2
2) Convert a) 1m^2 to cm^2 b) 4m^2 to cm^2	2) □ 1m, 100cm / 1m, 100cm a) 1m^2 = 1m × 1m = 100cm × 100cm = 10,000cm^2 b) 4cm^2 = 4 × 1m × 1m = 4 × 100cm × 100cm = 40,000cm^2
3) Convert a) 1km^2 to m^2 b) 3km^2 to m^2.	3) □ 1km, 1000m / 1km, 1000m a) 1km^2 = 1km × 1km = 1000m × 1000m = 1,000,000m^2 b) 3km^2 = 3 × 1km × 1km = 3 × 1000m × 1000m = 3,000,000m^2
4) Convert a) 1hectare to m^2. b) 7hectares to m^2.	4) hectare □ 100m / 100m a) 1 hectare = 100m × 100m = 10,000m^2 b) 7 hectares = 7 × 100m × 100m = 70,000m^2
5) Convert 500mm^2 to cm^2	5) □ 1cm, 10mm / 1cm, 10mm 1cm^2 = 1cm × 1cm = 10mm × 10mm = 100mm^2 $\frac{500mm^2}{100mm^2} = 5cm^2$
6) Convert 80,000cm^2 to m^2.	6) □ 1m, 100cm / 1m, 100cm 1m^2 = 1m × 1m = 100cm × 100cm = 10,000cm^2 $\frac{80,000cm^2}{10,000cm^2} = 8m^2$
7) Convert 29,000,000m^2 to km^2.	7) □ 1km, 1000m / 1km, 1000m 1km^2 = 1km × 1km = 1000m × 1000m = 1,000,000m^2 $\frac{29,000,000m^2}{1,000,000m^2} = 29km^2$
8) Convert 500,000m^2 to hectares.	8) hectare □ 100m / 100m 1 hectare = 100m × 100m = 10,000m^2 $\frac{500,000m^2}{10,000m^2} = 50hectare$

Measure - Metric System 13

Convert Units - Volume.

Convert Units - Volume
The following units and conversions are used -

Length units
1 cm = 10 mm
1 m = 100 cm
1 km = 1000 m

*Note It is useful to draw a cube and label the lengths as here.

Q	A
1) Convert a) $1cm^3$ to mm^3. b) $8cm^3$ to mm^3.	1) *(cube: 1cm = 10mm on each side)* a) $1cm^3$ $= 1cm \times 1cm \times 1cm$ $= 10mm \times 10mm \times 10mm$ $= 1,000mm^3$ b) $8cm^3$ $= 8 \times 1cm \times 1cm \times 1cm$ $= 8 \times 10mm \times 10mm \times 10mm$ $= 8,000mm^3$
2) Convert a) $1m^3$ to cm^3. b) $3m^3$ to cm^3.	2) *(cube: 1m = 100cm on each side)* a) $1m^3$ $= 1m \times 1m \times 1m$ $= 100cm \times 100cm \times 100cm$ $= 1,000,000cm^3$ b) $3m^3$ $= 3 \times 1m \times 1m \times 1m$ $= 3 \times 100cm \times 100cm \times 100cm$ $= 3,000,000cm^3$
3) Convert a) $1km^3$ to m^3. b) $5km^3$ to m^3.	3) *(cube: 1km = 1000m on each side)* a) $1km^3$ $= 1km \times 1km \times 1km$ $= 1000m \times 1000m \times 1000m$ $= 1,000,000,000m^3$ b) $5km^3$ $= 5 \times 1km \times 1km \times 1km$ $= 5 \times 1000m \times 1000m \times 1000m$ $= 5,000,000,000m^3$
4) Convert $7,000mm^3$ to cm^3.	4) *(cube: 1cm = 10mm on each side)* $1cm^3$ $= 1cm \times 1cm \times 1cm$ $= 10mm \times 10mm \times 10mm$ $= 1,000mm^3$ $\dfrac{7,000mm^3}{1,000mm^3} = 7cm^3$
5) Convert $4,000,000cm^3$ to m^3.	5) *(cube: 1m = 100cm on each side)* $1m^3$ $= 1m \times 1m \times 1m$ $= 100cm \times 100cm \times 100cm$ $= 1,000,000cm^3$ $\dfrac{4,000,000cm^3}{1,000,000cm^3} = 4m^3$
6) Convert $29,000,000,000m^3$ to km^3.	6) *(cube: 1km = 1000m on each side)* a) $1km^3$ $= 1km \times 1km \times 1km$ $= 1000m \times 1000m \times 1000m$ $= 1,000,000,000m^3$ $\dfrac{29,000,000,000m^3}{1,000,000,000m^3} = 29km^3$

Measure - Imperial System 1　　　　　　　　　　　Convert Units - Length.

The **Imperial System** is a system of measure for Length, Weight and Capacity.
*Note Here, Area and Volume are also included.

Length

The system for length is based on the **inch**.
On a ruler, **fractions of an inch** are used for more accurate measure, usually -
1/2, 1/4, 1/8, 1/16, 1/32 or 1/10.

Imperial System	unit abbreviations	
1 inch	in	
12 inches = 1 foot	ft	
3 feet = 1 yard	yd	(3 × 12 in = 36 in)
36 inches = 1 yard		
1760 yards = 1 mile	mi	(1760 × 3 ft = 5280 ft)
5280 feet = 1 mile		

*Note The abbreviation for **miles** may be written **mls**, and **miles per hour** written **mph**.

Larger units can be converted to smaller units, and vice versa.

Ex.1　　　　　　　　　　　　　　　　　Ex.2
Convert 4yd to in.　　　　　　　　　　　Convert 5000yd to mi and yd.

　　1yd = 36in　　　　　　　　　　　　　　1760yd = 1mile
　　4yd = 36in × 4 = 144in　　　　　　　　5000yd = $\frac{5000yd}{1760}$ = 2mi 1480yd

Q	A	
1) Convert 6ft to in.	1)	1ft = 12in 6ft = 12in × 6 = 72in
2) Convert 5yd to ft.	2)	1yd = 3ft 5yd = 3ft × 5 = 15ft
3) Convert 3yd to in.	3)	1yd = 36in 3yd = 36in × 3 = 108in
4) Convert 9mi to yd.	4)	1mi = 1760yd 9mi = 1760yd × 9 = 15840yd
5) Convert 2mi to ft.	5)	1mi = 5280ft 2mi = 5280ft × 2 = 10560ft
6) Convert 45in to ft and in.	6)	12in = 1ft 45in = $\frac{45}{12}$ = 3ft 9in
7) Convert 22ft to yd and ft.	7)	3ft = 1yd 22ft = $\frac{22}{3}$ = 7yd 1ft
8) Convert 10,000 yd to mi and yd.	8)	1760yd = 1mi 10000yd = $\frac{10000yd}{1760}$ = 5mi 1200yd
9) Convert 20,000 ft to mi and ft.	9)	5280ft = 1mi 20000ft = $\frac{20000ft}{5280}$ = 3mi 4160ft

Measure - Imperial System 2 — Convert Units - Weight.

Weight

The system for weight is based on the **ounce**.

Imperial System	unit abbreviations
1 ounce	oz
16 ounces = 1 pound	lb (lb is from the Latin 'libra' -
14 pounds = 1 stone	st meaning 'balances' or 'scales')
160 stones = 1 ton	t
2240 pounds = 1 ton	

Larger units can be converted to smaller units, and vice versa.

Ex.1
Convert 3 lb to oz.

$1 \text{ lb} = 16 \text{oz}$
$3 \text{ lb} = 16 \text{oz} \times 3 = 48 \text{oz}$

Ex.2
Convert 153 lb to st and lb.

$14 \text{ lb} = 1 \text{st}$
$153 \text{ lb} = \frac{153 \text{ lb}}{14} = 10 \text{st } 13 \text{ lb}$

Q	A
1) Convert 7 lb to oz.	1) $1 \text{ lb} = 16 \text{oz}$ $7 \text{ lb} = 16 \text{oz} \times 7 = 112 \text{oz}$
2) Convert 4st to lb.	2) $1 \text{st} = 14 \text{ lb}$ $4 \text{st} = 14 \text{ lb} \times 4 = 56 \text{ lb}$
3) Convert 0·2st to lb.	3) $1 \text{st} = 14 \text{ lb}$ $0·2 \text{st} = 14 \text{ lb} \times 0·2 = 2·8 \text{ lb}$
4) Convert 5t to st.	4) $1 \text{t} = 160 \text{st}$ $5 \text{t} = 160 \text{st} \times 5 = 800 \text{st}$
5) Convert 3t to lb.	5) $1 \text{t} = 2240 \text{ lb}$ $3 \text{t} = 2240 \text{ lb} \times 3 = 6720 \text{ lb}$
6) Convert 25oz to lb and oz.	6) $16 \text{oz} = 1 \text{ lb}$ $25 \text{oz} = \frac{25 \text{ oz}}{16} = 1 \text{ lb } 9 \text{oz}$
7) Convert 80 lb to st and lb.	7) $14 \text{ lb} = 1 \text{st}$ $80 \text{ lb} = \frac{80 \text{ lb}}{14} = 5 \text{st } 10 \text{ lb}$
8) Convert 1000st to t and st.	8) $160 \text{st} = 1 \text{t}$ $1000 \text{st} = \frac{1000 \text{st}}{160} = 6 \text{t } 40 \text{st}$
9) Convert 30,000 lb to t, st and lb.	9) $2240 \text{ lb} = 1 \text{t}$ $30000 \text{ lb} = \frac{30000 \text{ lb}}{2240} = 13 \text{t } \mathbf{880 \text{ lb}}$ $14 \text{ lb} = 1 \text{st}$ $880 \text{ lb} = \frac{880 \text{ lb}}{14} = 62 \text{st } 12 \text{ lb}$ $30000 \text{ lb} = = 13 \text{t } 62 \text{st } 12 \text{ lb}$

Measure - Imperial System 3

Convert Units - Capacity.

Capacity

The system for capacity is based on the **fluid ounce**.

Imperial System   unit abbreviations
1 fluid ounce   fl oz

20 fluid ounces = 1 pint   pt
8 pints   = 1 gallon gal

Larger units can be converted to smaller units, and vice versa.

Ex.1
Convert 4gal to pt.

1gal = 8pt
4gal = 8pt × 4 = 32pt

Ex.2
Convert 50pt to gal and pt.

8pt = 1gal
50pt = $\frac{50pt}{8}$ = 6gal 2pt

Q	A
1) Convert 7pt to fl oz.	1) 1pt = 20fl oz 7pt = 20fl oz × 7 = 140fl oz
2) Convert 9gal to pt.	2) 1gal = 8pt 9gal = 8pt × 9 = 72pt
3) Convert 3gal to fl oz.	3) 1gal = 8pt 3gal = 8pt × 3 = 24pt 1 pt = 20fl oz 24 pt = 20fl oz × 24 = 480fl oz
4) Convert 0·6pt to fl oz.	4) 1pt = 20fl oz 0·6pt = 20fl oz × 0·6 = 12fl oz
5) Convert 2·2gal to pt.	5) 1gal = 8pt 2·2gal = 8pt × 2·2 = 17·2pt
6) Convert 27fl oz to pt.	6) 20fl oz = 1pt 27fl oz = $\frac{27\,oz}{20}$ = $1\frac{7}{20}$ pt = 1·35pt
7) Convert 48pt to gal.	7) 8pt = 1gal 48pt = $\frac{48pt}{8}$ = 6gal
8) Convert 5·5pt to gal.	8) 8pt = 1gal 5·5pt = $\frac{5\cdot5pt}{8}$ = 0·6875gal
9) Convert 2000fl oz to gal and pt.	9) 20fl oz = 1pt 2000fl oz = $\frac{2000fl\,oz}{20}$ = 100pt 8pt = 1gal 100pt = $\frac{100pt}{8}$ = 12gal 4pt

Measure - Imperial System 4

Convert Units - Area.

Area

The system for area is based on the **inch²** ('square inch').
This itself is based on the units of **length** - since area = length × breadth.
Only the following conversion units of length are used here.

Imperial System	Length unit abbreviations	Area unit abbreviations
1 inch	in	in²
12 inches = 1 foot	ft	ft²
3 feet = 1 yard	yd	yd²
1760 yards = 1 mile	mi	mi²

Acre

1 acre = 4840yd²

(1 way to remember it ...)

22yd □ 220yd

22yd × 220yd = 4840yd²

*Note It is useful to draw a square and label the lengths as here.

Q	A
1) Convert a) 1ft² to in². b) 5ft² to in².	1) □ 1ft / 12in / 1ft / 12in a) 1ft² = 1ft × 1ft = 12in × 12in = 144in² b) 5ft² = 5 × 1ft × 1ft = 5 × 12in × 12in = 720in²
2) Convert a) 1yd² to ft². b) 6yd² to ft².	2) □ 1yd / 3ft / 1yd / 3ft a) 1yd² = 1yd × 1yd = 3ft × 3ft = 9ft² b) 6yd² = 6 × 1yd × 1yd = 6 × 3ft × 3ft = 54ft²
3) Convert a) 1mi² to yd² b) 8mi² to yd²	3) □ 1mi / 1760yd / 1mi / 1760yd a) 1mi² = 1mi × 1mi = 1760yd × 1760yd = 3,097,600yd² b) 8mi² = 8 × 1mi × 1mi = 8 × 1760yd × 1760yd = 24,780,800yd²
4) Convert 360in² to ft².	4) □ 1ft / 12in / 1ft / 12in 1ft² = 1ft × 1ft = 12in × 12in = 144in² $\dfrac{360\text{in}^2}{144\text{in}^2} = 2\dfrac{72}{144}\text{ft}^2 = 2\tfrac{1}{2}\text{ft}^2$
5) Convert 9,000ft² to yd²	5) □ 1yd / 3ft / 1yd / 3ft 1yd² = 1yd × 1yd = 3ft × 3ft = 9ft² $\dfrac{9{,}000\text{ft}^2}{9\text{ft}^2} = 1{,}000\text{yd}^2$
6) Convert 15,488,000yd² to mi².	6) □ 1mi / 1760yd / 1mi / 1760yd 1mi² = 1mi × 1mi = 1760yd × 1760yd = 3,097,600yd² $\dfrac{15{,}488{,}000\text{yd}^2}{3{,}097{,}600\text{yd}^2} = 5\text{mi}^2$
7) Convert a) 5acres to yd² b) 220acres to yd²	7) 1 acre = 4840yd² a) 5acres = 5 × 4840yd² = 24,200yd² b) 220acres = 220 × 4840yd² = 1,064,800yd²
8) Convert a) 48,400yd² to acres. b) 121,000yd² to acres.	8) 1 acre = 4840yd² a) $\dfrac{48{,}400\text{yd}^2}{4840\text{yd}^2} = 10$ acres b) $\dfrac{121{,}000\text{yd}^2}{4840\text{yd}^2} = 25$ acres

Measure - Imperial System 5

Convert Units - Volume.

Volume

The system for volume is based on the **inch³** ('cubic inch').
This itself is based on the units of **length** - since volume = length × breadth × height.
Only the following conversion units of length are used here.

	Imperial System	Length unit abbreviations	Volume unit abbreviations
	1 inch	in	in³
	12 inches = 1 foot	ft	ft³
	3 feet = 1 yard	yd	yd³
	1760 yards = 1 mile	mi	mi³

*Note It is useful to draw a cube and label the lengths as here.

Q	A

1) Convert
a) 1ft³ to in³.
b) 7ft³ to in³.
	1) [cube: 1ft/12in on each side] a) 1ft³ = 1ft × 1ft × 1ft b) 7ft³ = 7 × 1ft × 1ft × 1ft
	         = 12in × 12in × 12in = 7 × 12in × 12in × 12in
	         = 1,728in³ = 12,096in³

2) Convert
a) 1yd³ to ft³.
b) 4yd³ to ft³.
	2) [cube: 1yd/3ft on each side] a) 1yd³ = 1yd × 1yd × 1yd b) 4yd³ = 4 × 1yd × 1yd × 1yd
	         = 3ft × 3ft × 3ft = 4 × 3ft × 3ft × 3ft
	         = 27ft³ = 108ft³

3) Convert 8,640in³ to ft³.	3) [cube: 1ft/12in] 1ft³ = 1ft × 1ft × 1ft
	      = 12in × 12in × 12in $\dfrac{8640\text{in}^3}{1728\text{in}^3} = 5\text{ft}^3$
	      = 1,728in³

4) Convert 540ft³ to yd³.	4) [cube: 1yd/3ft] 1yd³ = 1yd × 1yd × 1yd
	      = 3ft × 3ft × 3ft $\dfrac{540\text{ft}^3}{27\text{ft}^3} = 20\text{yd}^3$
	      = 27ft³

5) Convert 3mi³ to yd³.	5) [cube: 1mi/1760yd] 3mi³ = 3 × 1mi × 1mi × 1mi
	       = 3 × 1760 × 1760 × 1760
	       = 16,355,328,000yd³

6) Convert 27,258,880,000yd³ to mi³.	6) [cube: 1mi/1760yd] 1mi³ = 1mi × 1mi × 1mi
	       = 1760 × 1760 × 1760 $\dfrac{27,258,880,000\text{yd}^3}{5,451,776,000\text{yd}^3} = 5\text{mi}^3$
	       = 5,451,776,000yd³

Measure Conversion 1

Metric / Imperial - Length.

Measure Conversion - Metric / Imperial.
The **Metric System** and the **Imperial System** are 2 systems of measure for Length, Weight and Capacity.
It is useful to be able to **convert** (**change**) ...
<u>Metric units to Imperial units</u> and <u>Imperial units to Metric units</u>.

Whole numbers of Imperial units do not convert to whole numbers of Metric units, and vice versa, so the degree of accuracy needed when converting will depend on the 'particular problem'.

<u>Length</u>

Metric system	**Imperial System**
10 mm = 1 cm	12 in = 1 ft
100 cm = 1 m	3 ft = 1 yd
1000 m = 1 km	36 in = 1 yd
	1760 yd = 1 mi
	5280 ft = 1 mi

The following give <u>approximate</u> conversions from ...

Metric units to **Imperial units**	**Imperial units** to **Metric units**
1 cm = 0·4 in	1 in = 2·5 cm
1 m = 39 in	1 ft = 30 cm
1 m = 3ft 3in	1 yd = 0·9 m
1 m = 1·1yd	1 yd = 90 cm
1 km = 0·6 mi	1 mi = 1·6 km

*Note 8 km = 5 mi
is often stated.

---

*Note The following Exs. and Q/A use the above <u>approximate</u> conversions.

<u>Ex.1</u>
Convert 50cm to ft and in.

first, convert 50cm to in ...
　1cm = 0·4in
　50cm = 0·4in × 50 = 20in

then, convert 20 ins to ft and in ...
　12in = 1ft
　20in = $\frac{20}{12}$ = 1ft 8in

<u>Ex.2</u>
Convert 7 mi to m.

first, convert 7mi to km ...
　1 mi = 1·6km
　7 mi = 1·6km × 7 = 11·2km ⎫
　　　　　　　　　　　　　　　⎬ ×1000
then, convert 11·2 km to m ... ⎭
　　　　　　　　= 11,200m

Q	A
1) Convert 25cm to in.	1) 　1cm = 0·4in 　25cm = 0·4in × 25 = 10in
2) Convert 77cm to in.	2) 　1cm = 0·4in 　77cm = 0·4in × 77 = 30·8in 　　　　　　　　　　　≈ 31in
3) Convert 130cm to ft and in.	3) 　1cm = 0·4in 　130cm = 0·4in × 130 = 52in 　12in = 1ft 　52in = $\frac{52}{12}$ = 4ft 4in

Measure Conversion 2

Metric / Imperial - Length.

Metric system	Metric to Imperial	Imperial to Metric	Imperial System
10 mm = 1 cm	1 cm = 0·4 in	1 in = 2·5 cm	12 in = 1 ft
100 cm = 1 m	1 m = 39 in	1 ft = 30 cm	3 ft = 1 yd
1000 m = 1 km	1 m = 3ft 3in	1 yd = 0·9 m	36 in = 1 yd
	1 m = 1·1 yd	1 yd = 90 cm	1760 yd = 1 mi
	1 km = 0·6 mi	1 mi = 1·6 km	5280 ft = 1 mi

Q	A
4) Convert 8m to yd, ft and in.	4) 1m = 1·1yd 8m = 1·1yd × 8 = 8·**8**yd 1 yd = 3ft 0·8yd = 3ft × 0·8 = 2·**4**ft 1 ft = 12in 0·4ft = 12in × 0·4 = 4·8in ≈ 5in 8m = 8yd 2ft 5in
5) Convert 10m to ft and in.	5) 1m = 3ft 3in 10m = 3ft 3in × 10 = 30ft **30**in 12in = 1ft 30in = $\frac{30}{12}$ = 2ft 6in 10m = 32ft 6in
6) Convert 162km to mi and yd.	6) 1km = 0·6 mi 162km = 0·6 mi × 162 = 97·**2** mi 1 mi = 1760yd 0·2 mi = 1760yd × 0·2 = 352yd 162km = 97miles 352yd
7) Convert 4in to cm.	7) 1in = 2·5cm 4in = 2·5cm × 4 = 10cm
8) Convert 7ft to cm.	8) 1ft = 30cm 7ft = 30cm × 7 = 210cm
9) Convert 55yd to m.	9) 1yd = 0·9m 55yd = 0·9m × 55 = 49·5m
10) Convert 2yds 2ft 2in to m and cm.	10) 1yd = 0·9m 2yd = 0·9m × 2 = 1·8m = 1m 80cm 1ft = 30cm 2ft = 30cm × 2 = 60cm 1in = 2·5cm 2in = 2·5cm × 2 = 5cm 2yd 2ft 2in = 2m 45cm
11) Convert 2 mi to km.	11) 1 mi = 1·6km 2 mi = 1·6km × 2 = 3·2km
12) Convert 11 mi to m.	12) 1 mi = 1·6km 11 mi = 1·6km × 11 = 17·6km ×1000 = 17,600m

Measure Conversion 3
Metric / Imperial - Weight.

Weight	
Metric system	**Imperial System**
1000 mg = 1 g	16 oz = 1 lb
1000 g = 1 kg	14 lb = 1 st
1000 kg = 1 tonne	2240 lb = 1 t

The following give <u>approximate</u> conversions from ...

Metric units to **Imperial units**	**Imperial units** to **Metric units**
1 g = 0·03 oz	1 oz = 30 g
1 kg = 2·2 lb	1 lb = 450 g
1 kg = 2 lb 3 oz	1 lb = 0·45 kg
1 tonne = 2200 lb	1 st = 6·5 kg
1 tonne = 0·98 t	1 t = 1·02 tonne

*Note The following Exs. and Q/A use the above <u>approximate</u> conversions.

Ex.1
Convert 7kg to lb.

1kg = 2·2 lb
7kg = 2·2 lb × 7 = 15·4 lb

Ex.2
Convert 8st to kg.

1st = 6·5kg
8st = 6·5kg × 8 = 52kg

Q	A
1) Convert 800g to oz.	1) 1g = 0·03oz 800g = 0·03oz × 800 = 24oz
2) Convert 4kg to lb and oz.	2) 1kg = 2·2 lb 4kg = 2·2 lb × 4 = 8·**8** lb 1 lb = 16oz 0·8 lb = 16oz × 0·8 = 12·8oz 4kg = 8 lb 12·8oz (small difference) or 1kg = 2 lb 3oz × 4 4kg = 8 lb 12oz
3) Convert 3·4tonne to lb.	3) 1tonne = 2200 lb 3·4tonne = 2200 lb × 3·4 = 7480 lb
4) Convert 6·95tonne to t.	4) 1tonne = 0·98t 6·95tonne = 0·98t × 6·95 = 6·811t
5) Convert 9oz to g.	5) 1oz = 30g 9oz = 30g × 9 = 270g
6) Convert 70 lb to kg.	6) 1 lb = 0·45kg 70 lb = 0·45kg × 70 = 31·5kg
7) Convert 2st 13 lb to kg.	7) 1st = 6·5kg 2st = 6·5kg × 2 = 13kg 1 lb = 0·45kg 13 lb = 0·45kg × 13 = 5·85kg 2st 13 lb = 18·85kg
8) Convert 500t to tonne.	8) 1t = 1·02tonne 500t = 1·02tonne × 500 = 510tonne

Measure Conversion 4

Metric / Imperial - Capacity.

Capacity

Metric system	Imperial System
$1\ l = 1000\ ml$ $1\ l = 1000\ cm^3$	$1\ gal = 8\ pt$

The following give <u>approximate</u> conversions from ...

Metric units to **Imperial units**	**Imperial units** to **Metric units**
$1\ ml = 0 \cdot 0018\ pt$ $1\ cm^3 = 0 \cdot 0018\ pt$ $1\ l = 1 \cdot 8\ pt$ $1\ l = 0 \cdot 22\ gal$	$1\ pt = 600\ ml$ $1\ pt = 600\ cm^3$ $1\ pt = 0 \cdot 6\ l$ $1\ gal = 4 \cdot 5\ l$

*<u>Note</u> The following Exs. and Q/A use the above <u>approximate</u> conversions.

<u>Ex.1</u>
Convert $4\ l$ to pt.

$1\ l = 1 \cdot 8\ pt$
$4\ l = 1 \cdot 8\ pt \times 4 = 7 \cdot 2\ pt$

<u>Ex.2</u>
Convert 5pts to l.

$1\ pt = 0 \cdot 6\ l$
$5\ pts = 0 \cdot 6\ l \times 5 = 3\ l$

Q	A
1) Convert 700ml to pt.	1) $1\ ml = 0 \cdot 0018\ pt$ $700\ ml = 0 \cdot 0018\ pt \times 700 = 1 \cdot 26\ pt$
2) Convert 2500cm^3 to pt.	2) $1\ cm^3 = 0 \cdot 0018\ pt$ $2500\ cm^3 = 0 \cdot 0018\ pt \times 2500 = 4 \cdot 5\ pt$
3) Convert $9 \cdot 3\ l$ to pt.	3) $1\ l = 1 \cdot 8\ pt$ $9 \cdot 3\ l = 1 \cdot 8\ pt \times 9 \cdot 3 = 16 \cdot 74\ pt$
4) Convert $70\ l$ to gal and pt.	4) $1\ l = 0 \cdot 22\ gal$ $70\ l = 0 \cdot 22\ gal \times 70 = 15 \cdot \underline{4}\ gal$ $1\ gal = 8\ pt$ $0 \cdot 4\ gal = 8\ pt \times 0 \cdot 4 = 3 \cdot 2\ pt$ $70\ l = \qquad 15\ gal\ \ 3 \cdot 2\ pt$
5) Convert 6pt to ml.	5) $1\ pt = 600\ ml$ $6\ pt = 600\ ml \times 6 = 3600\ ml$
6) Convert 15pt to cm^3.	6) $1\ pt = 600\ cm^3$ $15\ pt = 600\ cm^3 \times 15 = 9000\ cm^3$
7) Convert 4pt to l.	7) $1\ pt = 0 \cdot 6\ l$ $4\ pt = 0 \cdot 6\ l \times 4 = 2 \cdot 4\ l$
8) Convert 11pt to l and ml.	8) $1\ pt = 0 \cdot 6\ l$ $11\ pt = 0 \cdot 6\ l \times 11 = 6 \cdot \underline{6}\ l$ $1\ l = 1000\ ml$ $0 \cdot 6\ l = 1000\ ml \times 0 \cdot 6 = 600\ ml$ $11\ pt = \qquad 6\ l\ \ 600\ ml$
9) Convert 3000gal to l.	9) $1\ gal = 4 \cdot 5\ l$ $3000\ gal = 4 \cdot 5\ l \times 3000 = 13500\ l$

Measure Conversion 5

Metric / Imperial - Area.

Area

The following give <u>approximate</u> conversions from ...

Metric units to **Imperial units**	**Imperial units** to **Metric units**
1 cm = 0·4 in	1 in = 2·5 cm
1 m = 1·1 yd	1 ft = 30 cm
1 km = 0·6 mi	1 yd = 0·9 m
	1 mi = 1·6 km

*Note 1 The following Q/A use the above approximate conversions.
*Note 2 It is useful to draw a square and label the lengths as here.

Q	A
1) Convert a) $1cm^2$ to in^2. b) $50cm^2$ to in^2.	1) [square: 1cm / 0·4in sides] a) $1cm^2 = 1cm \times 1cm$ $= 0·4in \times 0·4in$ $= 0·16in^2$ b) $50cm^2 = 50 \times 1cm \times 1cm$ $= 50 \times 0·4in \times 0·4in$ $= 8in^2$
2) Convert a) $1m^2$ to yd^2. b) $3m^2$ to yd^2.	2) [square: 1m / 1·1yd sides] a) $1m^2 = 1m \times 1m$ $= 1·1yd \times 1·1yd$ $= 1·21yd^2$ b) $3m^2 = 3 \times 1m \times 1m$ $= 3 \times 1·1yd \times 1·1yd$ $= 3·63yd^2$
3) Convert a) $1km^2$ to mi^2. b) $80km^2$ to mi^2.	3) [square: 1km / 0·6mi sides] a) $1km^2 = 1km \times 1km$ $= 0·6mi \times 0·6mi$ $= 0·36mi^2$ b) $80km^2 = 80 \times 1km \times 1km$ $= 80 \times 0·6mi \times 0·6mi$ $= 28·8mi^2$
4) Convert a) $1in^2$ to cm^2. b) $7in^2$ to cm^2.	4) [square: 1in / 2·5cm sides] a) $1in^2 = 1in \times 1in$ $= 2·5cm \times 2·5cm$ $= 6·25cm^2$ b) $7in^2 = 7 \times 1in \times 1in$ $= 7 \times 2·5cm \times 2·5cm$ $= 43·75cm^2$
5) Convert a) $1ft^2$ to cm^2. b) $6ft^2$ to cm^2.	5) [square: 1ft / 30cm sides] a) $1ft^2 = 1ft \times 1ft$ $= 30cm \times 30cm$ $= 900cm^2$ b) $6ft^2 = 6 \times 1ft \times 1ft$ $= 6 \times 30cm \times 30cm$ $= 5,400cm^2$
6) Convert a) $1yd^2$ to m^2. b) $25yd^2$ to m^2.	6) [square: 1yd / 0·9m sides] a) $1yd^2 = 1yd \times 1yd$ $= 0·9m \times 0·9m$ $= 0·81m^2$ b) $25yd^2 = 25 \times 1yd \times 1yd$ $= 25 \times 0·9m \times 0·9m$ $= 20·25m^2$
7) Convert a) $1mi^2$ to km^2. b) $4mi^2$ to km^2.	7) [square: 1mi / 1·6km sides] a) $1mi^2 = 1mi \times 1mi$ $= 1·6km \times 1·6km$ $= 2·56km^2$ b) $4mi^2 = 4 \times 1mi \times 1mi$ $= 4 \times 1·6km \times 1·6km$ $= 10·24km^2$

Measure Conversion 6 Metric / Imperial - Volume.

<u>Volume</u>

The following give <u>approximate</u> conversions from ...

Metric units to **Imperial units**
- 1 cm = 0·4 in
- 1 m = 1·1 yd

Imperial units to **Metric units**
- 1 in = 2·5 cm
- 1 ft = 30 cm
- 1 yd = 0·9 m

*Note 1 The following Q/A use the above approximate conversions.
*Note 2 It is useful to draw a cube and label the lengths as here.

Q	A

1) Convert
a) $1cm^3$ to in^3.
b) $20cm^3$ to in^3.
	1) [cube: 1cm / 0·4in on each side]
	a) $1cm^3 = 1cm \times 1cm \times 1cm$
	     $= 0·4in \times 0·4in \times 0·4in$
	     $= 0·064in^3$
	b) $20cm^3 = 20 \times 1cm \times 1cm \times 1cm$
	     $= 20 \times 0·4in \times 0·4in \times 0·4in$
	     $= 1·28in^3$

2) Convert
a) $1m^3$ to yd^3.
b) $90m^3$ to yd^3.
	2) [cube: 1m / 1·1yd on each side]
	a) $1m^3 = 1m \times 1m \times 1m$
	     $= 1·1yd \times 1·1yd \times 1·1yd$
	     $= 1·331yd^3$
	b) $90m^3 = 90 \times 1m \times 1m \times 1m$
	     $= 90 \times 1·1yd \times 1·1yd \times 1·1yd$
	     $= 119·79yd^3$

3) Convert
a) $1in^3$ to cm^3
b) $5in^3$ to cm^3.
	3) [cube: 1in / 2·5cm on each side]
	a) $1in^3 = 1in \times 1in \times 1in$
	     $= 2·5cm \times 2·5cm \times 2·5cm$
	     $= 15·625cm^3$
	b) $5in^3 = 5 \times 1in \times 1in \times 1in$
	     $= 5 \times 2·5cm \times 2·5cm \times 2·5cm$
	     $= 78·125cm^3$

4) Convert
a) $1ft^3$ to cm^3.
b) $3ft^3$ to cm^3.
	4) [cube: 1ft / 30cm on each side]
	a) $1ft^3 = 1ft \times 1ft \times 1ft$
	     $= 30cm \times 30cm \times 30cm$
	     $= 27,000cm^3$
	b) $3ft^3 = 3 \times 1ft \times 1ft \times 1ft$
	     $= 3 \times 30cm \times 30cm \times 30cm$
	     $= 81,000cm^3$

5) Convert
a) $1yd^3$ to m^3.
b) $47yd^3$ to m^3.
	5) [cube: 1yd / 0·9m on each side]
	a) $1yd^3 = 1yd \times 1yd \times 1yd$
	     $= 0·9m \times 0·9m \times 0·9m$
	     $= 0·729m^3$
	b) $47yd^3 = 47 \times 1yd \times 1yd \times 1yd$
	     $= 47 \times 0·9m \times 0·9m \times 0·9m$
	     $= 34·263m^3$

Density - Mass - Volume 1   Formulas.

Density - Mass - Volume
The Density of a substance is defined as the Mass per unit of Volume.
It is a compound measure - a combination of other measures (like Speed = Distance per unit of Time).

The Metric units of Mass shown here are mg, g, kg. (these are the same units for weight)

The units of Volume are mm^3, cm^3, m^3.

Formulas
Density - Mass - Volume can be presented in terms of formulas.

$$\text{Density} = \frac{\text{Mass}}{\text{Volume}}$$

$$D = \frac{M}{V}$$

$$\text{Density} = \frac{\text{Mass}}{\text{Volume}} \quad \text{'swop'}$$

$$\text{Volume} = \frac{\text{Mass}}{\text{Density}}$$

$$V = \frac{M}{D}$$

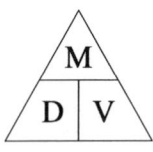

$$\text{Density} = \frac{\text{Mass}}{\text{Volume}}$$

$$\text{Density} \times \text{Volume} = \text{Mass} \quad \text{'swop'}$$

$$\text{Mass} = \text{Density} \times \text{Volume}$$

$$M = DV$$

The triangle is useful to help remember the 3 formulas ...

In the triangle, in turn ...
choose the letter on the **left** of each formula -
this leaves the 2 letters on the **right** of each formula AND in their correct 'position'.

Ex.1
A piece of metal has a mass of 200g and a volume of $50cm^3$. Calculate its **density**.

$$D = \frac{M}{V} = \frac{200g}{50cm^3} = 4\text{g per } cm^3$$

4gpcm³
4g/cm³
4gcm⁻³

— means 'for each cm^3'
any of these terms may be used

*Note

$$4g/cm^3 = \frac{4g}{cm^3} = 4gcm^{-3}$$

same

(see **Indices / Powers 8**)

Ex.2
A machine part has a mass of 2·1kg and a density of $7000kg/m^3$. Calculate its **volume**.

$$V = \frac{M}{D} = \frac{2 \cdot 1\text{kg}}{7000\text{kg}/m^3} = 0 \cdot 0003 m^3$$

The units must agree - ex. **kg**

Ex.3
A pipe has a density of $8mg/mm^3$ and a volume of $50,000mm^3$. Calculate its **mass**.

$$M = DV$$
$$= 8mg/mm^3 \times 50,000mm^3$$
$$= 400,000mg$$

The units must agree - ex. mm^3

Density - Mass - Volume 2

Formulas.

Q	A
1) A lead cube has a mass of 90·4g and a volume of 8cm³. Calculate its **density**.	1) $$D = \frac{M}{V} = \frac{90\cdot4g}{8cm^3} = 11\cdot3g/cm^3$$
2) 9 Plastic cuboids have a total mass of 45kg and a density of 900kg/m³. Calculate their **volume**.	2) $$V = \frac{M}{D} = \frac{45kg}{900kgpm^3} = 0\cdot05m^3$$
3) A silver ring has a density of 10·5mg/mm³ and a volume of 70mm³. Calculate its **mass**.	3) $$M = DV$$ $$= 10\cdot5mg/mm^3 \times 70mm^3 = 735mg$$
4) A rod of plutonium has a mass of 1·98kg and a volume of 0·0001m³. Calculate its **density**.	4) $$D = \frac{M}{V} = \frac{1\cdot98kg}{0\cdot0001m^3} = 19800kgm^{-3}$$
5) A sphere has a mass of 6,000mg and a density of 3mg/mm³. Calculate its **volume**.	5) $$V = \frac{M}{D} = \frac{6000mg}{3mg/mm^3} = 2,000mm^3$$
6) A metal sculpture has a density of 2·5gcm⁻³ and a volume of 248cm³. Calculate its **mass**.	6) $$M = DV$$ $$= 2\cdot5gcm^{-3} \times 248cm^3 = 620g$$
7) A solid pole has a mass of 300g and a volume of 100cm³. Calculate its **density**.	7) $$D = \frac{M}{V} = \frac{300g}{100cm^3} = 3g/cm^3$$
8) A load of concrete has a mass of 9,500kg and a density of 2,000kg/m³. Calculate its **volume**.	8) $$V = \frac{M}{D} = \frac{9500kg}{2000kg/m^3} = 4\cdot75m^3$$
9) A chocolate bar has a density of 1·25g/cm³ and a volume of 80cm³. Calculate its **mass**.	9) $$M = DV$$ $$= 1\cdot25g/cm^3 \times 80cm^3 = 100g$$

Density - Mass - Volume 3

Formulas. Change of Units.

Change of Units
Density - Mass - Volume units may need to be changed as part of a calculation.

Q	A

1) A mass of 8,000kg of sulphur has a volume of 4m³.
 	
Calculate its **density** a) in **kg/m3**
 	
b) in **g/cm3**.
	1)
	a) $D = \dfrac{M}{V} = \dfrac{8000kg}{4m^3} = $ **2,000kg/m³**
	
	b) 1kg ──── ×1000g
	                 2,000,000g/m³
	    1m³ ──── ÷1,000,000cm³
	                 2gm/cm³
	
	or
	
	               ×1000
	$D = \dfrac{M}{V} = \dfrac{8,000kg}{4m^3} = \dfrac{8,000,000gm}{4,000,000cm^3} = $ **2gm/cm³**
	               ×1,000,000

2) A reel of copper wire has a mass of 6kg and a density of 8·9g/cm³. Calculate its **volume**.	2) ×1000 $V = \dfrac{M}{D} = \dfrac{6kg}{8 \cdot 9g/cm^3} = \dfrac{6000g}{8 \cdot 9g/cm^3} = $ **674·15... cm³**
3) A gold bar has a density of 19·3gcm⁻³ and a volume of 300cm³. Calculate its **mass** in kg.	3) $M = DV$ $= 19 \cdot 3$**gcm⁻³** $\times 300$**cm³** $= 5790g = $ **5·79kg**

4) A quantity of mercury has a mass of 135g and a volume of 10cm³.
 	
Calculate its **density** a) in **g/cm³**
 	
b) in **kg/m³**.
	4)
	a) $D = \dfrac{M}{V} = \dfrac{135g}{10cm^3} = $ **13·5g/cm³**
	
	b) 1m³ ──── ×1,000,000cm³
	                 13,500,000g/m³
	    1kg ──── ÷1000g
	             = **13,500kg/m³**
	
	or
	
	               ×1000
	$D = \dfrac{M}{V} = \dfrac{135g}{10cm^3} = \dfrac{0 \cdot 135kg}{0 \cdot 00001m^3} = $ **13,500kg/m³**
	              ÷1,000,000

5) A tin has a mass of 365g and a density of 7300kg/m³. Calculate the **volume** of the tin in m³.	5) ÷1000 $V = \dfrac{M}{D} = \dfrac{365g}{7300kg/m^3} = \dfrac{0 \cdot 365kg}{7300kg/m^3} = $ **0·00005m³**
6) A cylinder has a density of 5gpcm³ and a volume of 50cm³. Calculate its **mass** in mg.	6) $M = DV$ ×1000 $= 5$**gpcm³** $\times 50$**cm³** $= 250g = $ **250,000mg**

Pressure - Force - Area 1   Formulas.

Pressure - Force - Area

Pressure is the force applied to a unit of area.

It is a compound measure - a combination of other measures (like Speed = Distance per unit of Time).

The units of Force are **Newtons (N)**.

The units of Area are mm^2, cm^2, m^2.

Formulas

Pressure - Force - Area can be presented in terms of formulas.

$$\text{Pressure} = \frac{\text{Force}}{\text{Area}}$$

$$P = \frac{F}{A}$$

$$\text{Pressure} = \frac{\text{Force}}{\text{Area}} \quad \text{'swop'}$$

$$\text{Area} = \frac{\text{Force}}{\text{Pressure}}$$

$$A = \frac{F}{P}$$

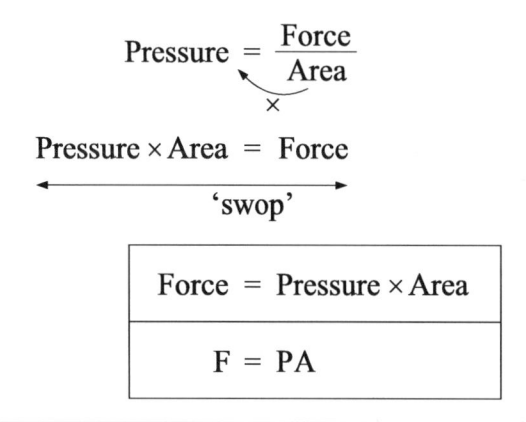

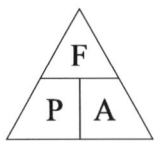

$$\text{Pressure} = \frac{\text{Force}}{\text{Area}} \times$$

$$\text{Pressure} \times \text{Area} = \text{Force} \quad \text{'swop'}$$

$$\text{Force} = \text{Pressure} \times \text{Area}$$

$$F = PA$$

The triangle is useful to help remember the 3 formulas ...

In the triangle, in turn ...
choose the letter on the **left** of each formula - this leaves the 2 letters on the **right** of each formula AND in their correct 'position'.

Ex.1
A force of 50 Newtons is applied to an area of 5m^2. Calculate the **pressure**.

$$P = \frac{F}{A} = \frac{50N}{5m^2} = 10 \text{ N per m}^2$$

10 N pm^2
10 N/m^2
10 Nm^{-2}

— means 'for each m^2'

any of these terms may be used

*Note

$$10 \text{ N/m}^2 = \frac{10 \text{ N}}{m^2} = 10 \text{ Nm}^{-2}$$

same

(see **Indices / Powers 8**)

Ex.2
A force of 35 N produces a pressure of 14 N/m^2 when applied to an area. Calculate the **area**.

$$A = \frac{F}{P} = \frac{35 \text{ N}}{14 \text{ N/m}^2} = 2 \cdot 5 \text{m}^2$$

Ex.3
A force produces a pressure of 20 N/m^2 when applied to an area of 4m^2. Calculate the **force**.

$$F = PA$$
$$= 20 \text{ N/m}^2 \times 4\text{m}^2$$
$$= 80 \text{ N}$$

The units must agree - ex. **m^2**

Pressure - Force - Area 2

Formulas.

Q	A
1) A force of 60 Newtons is applied to an area of 3m². Calculate the **pressure** produced.	1) $P = \dfrac{F}{A} = \dfrac{60N}{3m^2} = 20$ N per m²
2) A force of 96 N produces a pressure of 12 N/m² when applied to an area. Calculate the **area**.	2) $A = \dfrac{F}{P} = \dfrac{96 \text{ N}}{12 \text{ N/m}^2} = 8m^2$
3) When applied to an area of 4m² a force produces a pressure of 9·5 Nm⁻². Calculate the **force**.	3) $F = PA$ $= 9\cdot5$ Nm⁻² $\times$ 4m² $= 38$ N
4) A force of 90·6 N is applied to an area of 20·5cm². Calculate the **pressure** produced.	4) $P = \dfrac{F}{A} = \dfrac{90\cdot6 \text{ N}}{20\cdot5 \text{cm}^2} = 4\cdot419...$ N/cm²
5) A pressure of 10 Ncm⁻² is produced by a force of 200 N when applied to an area. Calculate the **area**.	5) $A = \dfrac{F}{P} = \dfrac{200 \text{ N}}{10 \text{ Ncm}^{-2}} = 20$cm²
6) A force produces a pressure of 7 N pcm² when applied to an area of 7cm². Calculate the **force**.	6) $F = PA$ $= 7$ N pcm² $\times$ 7cm² $= 49$ N
7) A force of 7,500 Newtons is applied to an area of 10m². Calculate the **pressure** produced.	7) $P = \dfrac{F}{A} = \dfrac{7500N}{10m^2} = 750$ N/m²
8) A force of 3,600 N produces a pressure of 45 N/m² when applied to an area. Calculate the **area**.	8) $A = \dfrac{F}{P} = \dfrac{3600 \text{ N}}{45 \text{ N/m}^2} = 80m^2$
9) When applied to an area of 8m² a force produces a pressure of 80 Nm⁻². Calculate the **force**.	9) $F = PA$ $= 80$ Nm⁻² $\times$ 8m² $= 640$ N

Scale 1 Read Scale.

Scale
Read a Measurement on a Scale
The unit and marks on a scale, indicating the measurements, is accurate enough for its purpose.
A reading is 'acceptable' if it is close enough to a mark for that purpose.
If more accuracy is needed, then a smaller unit / more marks are needed - and so a different scale!

The **difference** between 2 given values may have to be **divided by** the **number of spaces** in between them to give the value of each **mark** - and so allow a measure to be read.

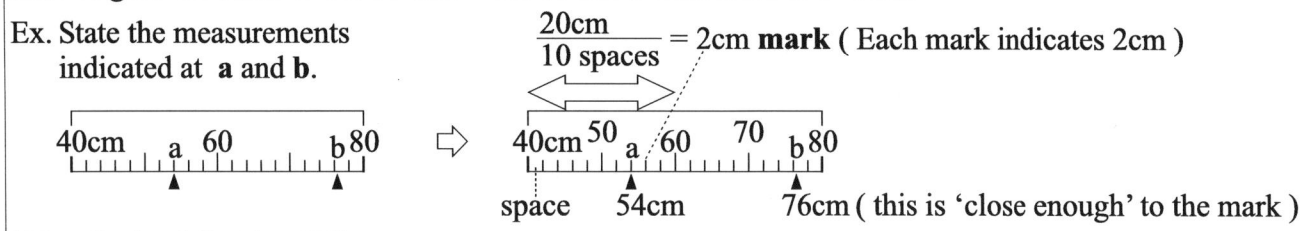

Ex. State the measurements indicated at **a** and **b**.

$\dfrac{20cm}{10 \text{ spaces}} = 2cm$ **mark** (Each mark indicates 2cm)

space   54cm   76cm (this is 'close enough' to the mark)

*Note In the following Q/A ...
1) the calculation of the '**mark**' is shown but is usually done mentally.
2) each 'part' scale starts from 0 (even if not shown).
3) any unit could be written on any scale.

Q	A	Q	A
State each measurement indicated by an arrow.		12)	12) 0.19cl $\dfrac{0.1cl}{10} = 0.01cl$ mark
1)	1) 3cm		
2)	2) 1.5cm $\dfrac{1cm}{2} = 0.5cm$ mark		
3)	3) 2.2cm $\dfrac{1cm}{10} = 0.1cm$ mark		
4)	4) 59mm $\dfrac{10mm}{10} = 1mm$ mark	13)	13) 740 l $\dfrac{200\,l}{20} = 10\,l$ mark
5)	5) 0.1m		
6)	6) 47.5km $\dfrac{10km}{4} = 2.5km$ mark		
7)	7) 30mg $\dfrac{20mg}{2} = 10mg$ mark		
8)	8) 3.7g $\dfrac{0.2g}{2} = 0.1g$ mark	14)	14) 17 kmph $\dfrac{5kmph}{5} = 1$ kmph mark
9)	9) 275kg $\dfrac{100kg}{4} = 25kg$ mark		
10)	10) 4.25 kg $\dfrac{1kg}{4} = 0.25$ kg mark	15)	15) 1.8 Amps $\dfrac{1Amp}{5} = 0.2$ Amp mark
11)	11) 68ml $\dfrac{10ml}{5} = 2ml$ mark		

260

Scale 2   Estimate a Scale Measure - Read.

Scale
Estimating a Measurement on a Scale - **Read** a value

Any '**reasonable**' estimate is acceptable.

Here, to **estimate** the value of a given **Point** on a scale, the method is based on that used in **Scales 1**.

By **estimating** its position, we **draw** the **midpoint** of the 2 given values on either side of the given **Point** (if it is not already drawn) - then **calculate** its value. This may be 'close enough' to the given **Point** and not require any further work! If not 'close enough' (or we are not sure) ...

The d**ifference** between the 2 given values is **divided by 10** - to create 10 spaces in between them and to give the value of each mark. (10 is an 'easy number' to divide by.)

By **estimation** again, we **draw 4 marks** for **5** equal spaces on the side of the midpoint which has the given **Point** - (the 5 spaces on the other side of the midpoint are **not needed**).

State the measurement of the line **closest to** the **Point** - it is usually a '**reasonable**' estimate.

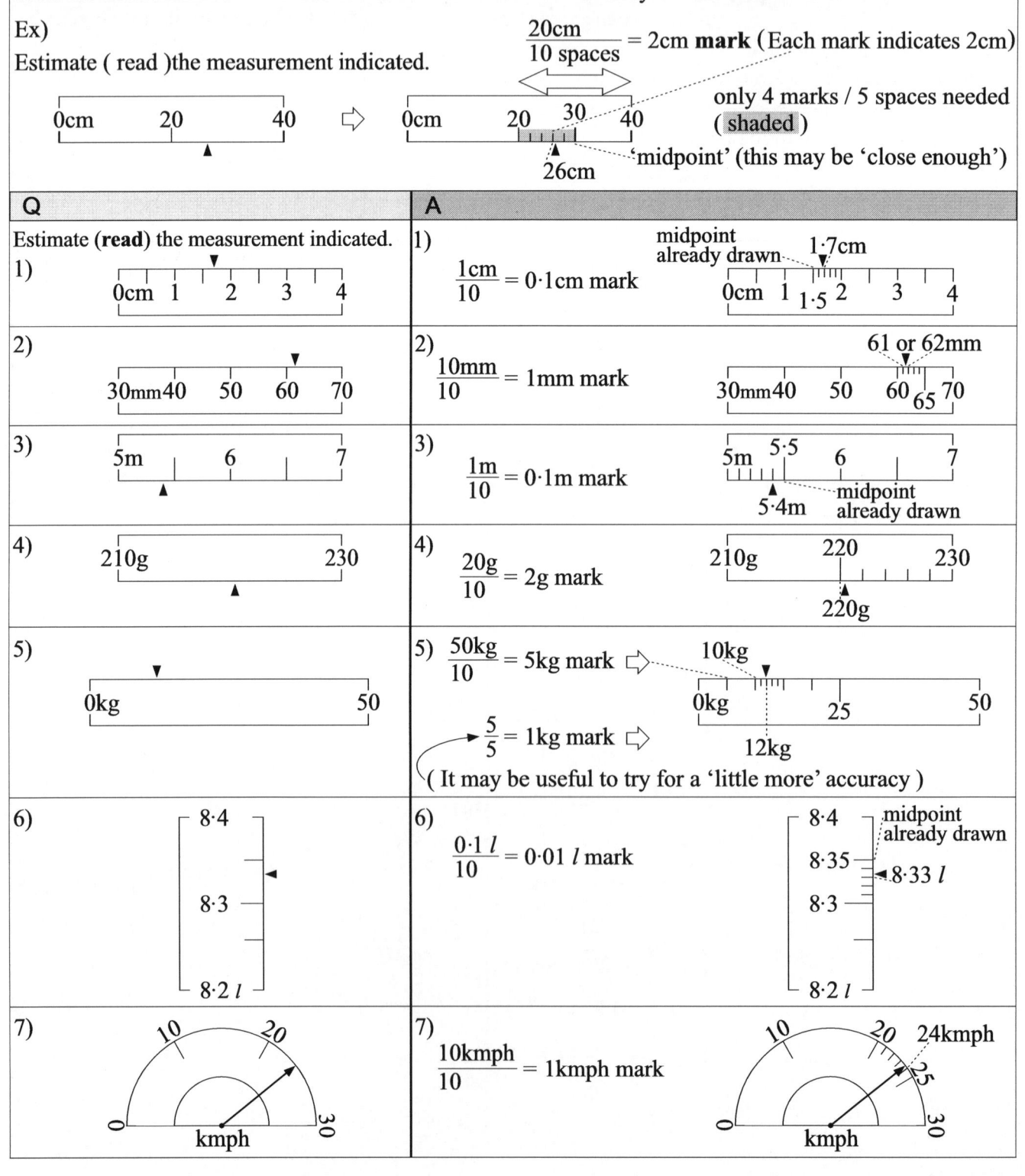

Ex) Estimate (read) the measurement indicated.

$\dfrac{20\text{cm}}{10 \text{ spaces}} = 2\text{cm mark}$ (Each mark indicates 2cm)

only 4 marks / 5 spaces needed (shaded)

'midpoint' (this may be 'close enough')

26cm

Q	A
Estimate (**read**) the measurement indicated. 1) scale 0cm 1 2 3 4 with pointer	1) $\dfrac{1\text{cm}}{10} = 0\cdot1\text{cm mark}$ midpoint already drawn 1·7cm 0cm 1 1·5 2 3 4
2) scale 30mm 40 50 60 70	2) $\dfrac{10\text{mm}}{10} = 1\text{mm mark}$ 61 or 62mm 30mm 40 50 60 65 70
3) scale 5m 6 7	3) $\dfrac{1\text{m}}{10} = 0\cdot1\text{m mark}$ 5m 5·5 6 7 5·4m midpoint already drawn
4) scale 210g 230	4) $\dfrac{20\text{g}}{10} = 2\text{g mark}$ 210g 220 230 220g
5) scale 0kg 50	5) $\dfrac{50\text{kg}}{10} = 5\text{kg mark}$ ⇨ 10kg 0kg 25 50 $\dfrac{5}{5} = 1\text{kg mark}$ ⇨ 12kg (It may be useful to try for a 'little more' accuracy)
6) vertical scale 8·4, 8·3, 8·2 *l*	6) $\dfrac{0\cdot1\,l}{10} = 0\cdot01\,l\text{ mark}$ 8·4 midpoint already drawn 8·35 ◄ 8·33 *l* 8·3 8·2 *l*
7) dial 0, 10, 20, 30 kmph	7) $\dfrac{10\text{kmph}}{10} = 1\text{kmph mark}$ 24kmph dial showing 25 kmph

Scale 3

Estimate a Scale Measure - Mark.

Scale
Estimating a Measurement on a Scale - **Mark** a value
Scale 2 ... **read** a value ... is also the method used to **mark** a given value.

Ex) Estimate and mark the measurement of 17cm.

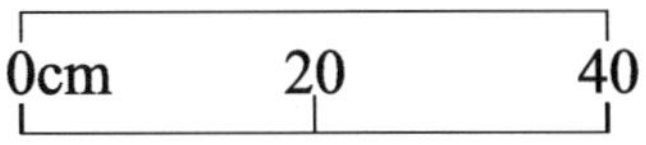

$\dfrac{20\text{cm}}{10 \text{ spaces}} = 2\text{cm}$ **mark** (Each mark indicates 2cm)

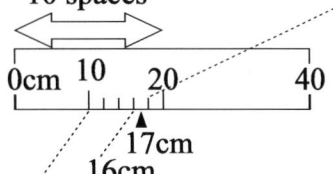

'midpoint' (this may be 'close enough')

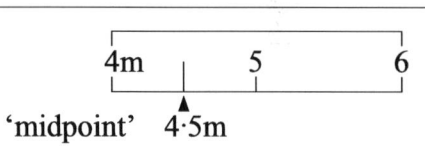

Q	A
Estimate (**mark**) the measurement indicated. 1) 3·3cm	1) $\dfrac{1}{10} = 0{\cdot}1\text{cm}$ mark
2) 66mm 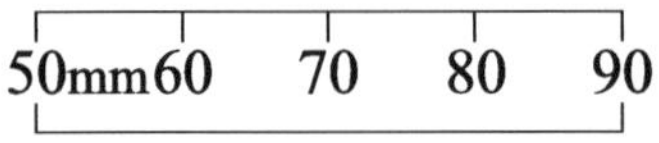	2) $\dfrac{10}{10} = 1\text{mm}$ mark 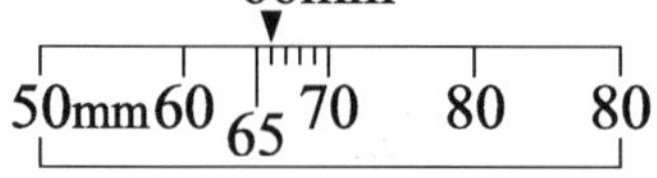
3) 4·5m 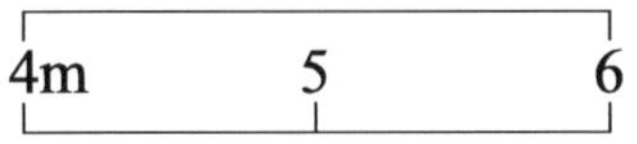	3) 'midpoint' 4·5m
4) 109g 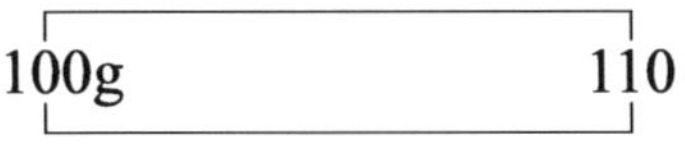	4) $\dfrac{10}{10} = 1\text{g}$ mark 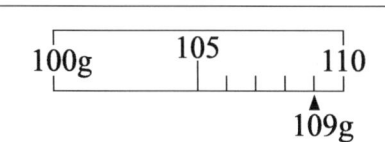
5) 7·77 l 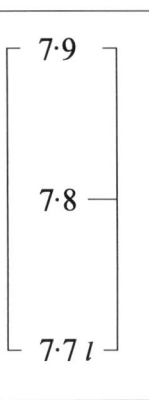	5) $\dfrac{0{\cdot}1}{10} = 0{\cdot}01\ l$ mark 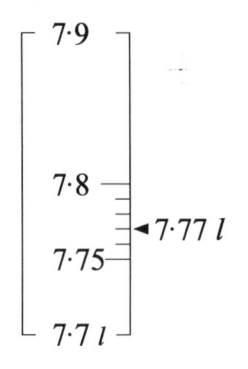
6) 12 Amps 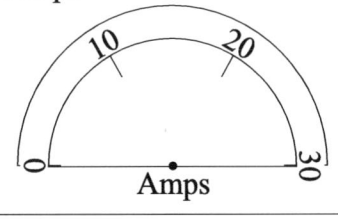	6) $\dfrac{10}{10} = 1$ Amp mark 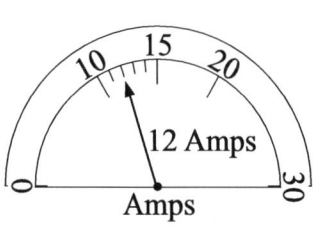 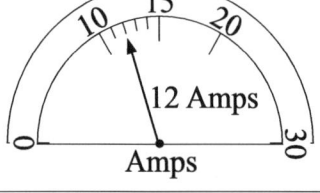
7) 19kg 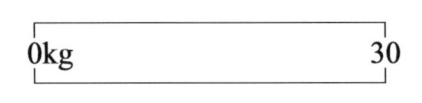	7) *Note Dividing the scale into 3 parts is more useful here. $\dfrac{30}{3} = 10\text{kg}$ mark $\dfrac{10}{10} = 1\text{kg}$ mark 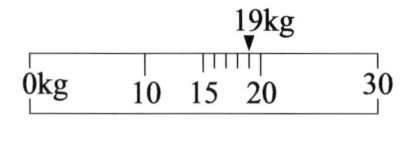

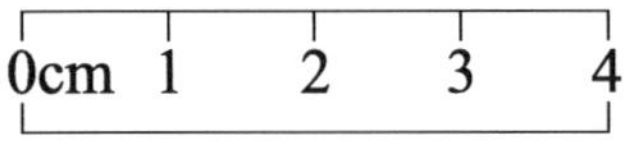

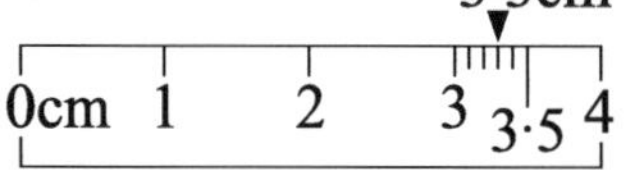

Time 1 — Units of Time.

Time

The units of time are:

60 seconds = 1 minute	'at least 4 weeks' = 1 month
60 minutes = 1 hour	'from 28 to 31 days' = 1 month
24 hours = 1 day	3 months = 1 quarter (of a year)
7 days = 1 week	12 months = 1 year

Q	A
1) State the number of days in each month. January February March April May June July August September October November December	1) January 31 February 28 or 29 in a leap year March 31 April 30 May 31 June 30 July 31 August 31 September 30 October 31 November 30 December 31
2) State the total number of days in a year.	2) 365
3) State the total number of days in a leap year.	3) 366
4) Explain when a leap year occurs.	4) A leap year occurs every 4 years. The year divides exactly by 4. Since the thousands and hundreds of a given year divide exactly by 4, ... $\frac{1000}{4} = 250 \qquad \frac{100}{4} = 25$ ... the **tens and units** together need to divide by 4.
5) State if each year is a leap year or not. A) 1700 B) 1944 C) 2010 D) 2016 E) 2031 F) 2052	5) A) $\frac{1700}{4} = 425$ or $\frac{00}{4} = 0$ leap year B) $\frac{1944}{4} = 486$ or $\frac{44}{4} = 11$ leap year C) $\frac{2010}{4} = 502\ r\ 2$ or $\frac{10}{4} = 2\ r\ 2$ **not** leap year D) $\frac{2016}{4} = 504$ or $\frac{16}{4} = 4$ leap year E) $\frac{2031}{4}$ **odd** number can not divide by 4 exactly - **any odd** number year is **not** a leap year. F) $\frac{2052}{4} = 513$ or $\frac{52}{4} = 13$ leap year

Time 2   12-Hour Clock.

12-Hour Clock Time
The time on a clock can be **read** / **stated** in more than 1 way.

Analogue Clock - **hours** and **minutes** are marked on a **dial** (clock **face**) and 2 hands turn through 360°.
The **small hand** indicates the **hour** (out of 12) - the **large hand** indicates the **minutes** (out of 60).
Each hour mark is also a '5 minute mark' - these are shown on the **outside** of the clocks A and B.

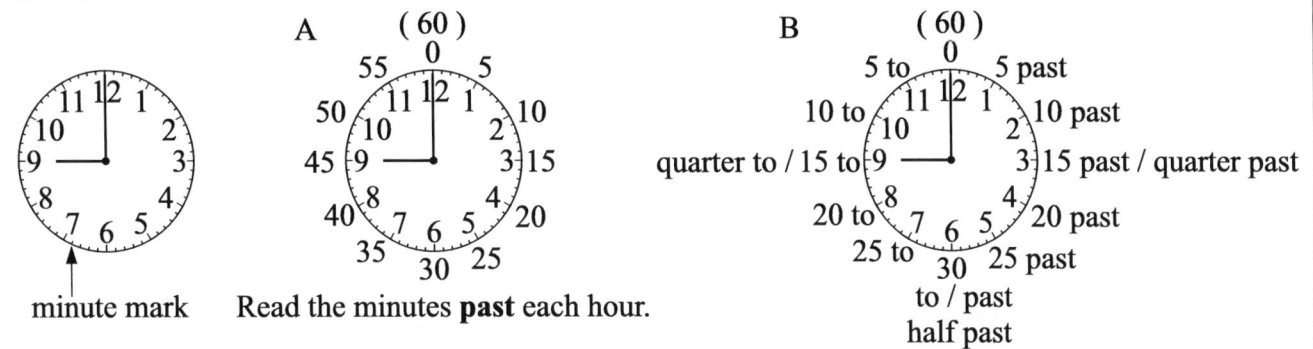

A — Read the minutes **past** each hour.

B — Read the minutes (up to 30) **past** each hour - then read the minutes **to** the next hour.

Digital Clock - the **hour** and **minutes past the hour** are shown in **3** or **4 digit** / **number** form.

Ex. 12-hour time **9:10** AM   Ex. 24-hour time **09:10** (see **Time 5**)

Q	A	Q	A
State each time.			
1)	1) 4 o'clock	7)	7) 3 past 7 3 minutes past 7 7.03
2)	2) 20 past 2 2.20 *Note This is **not** a **decimal point**.	8)	8) 12 past 10 12 minutes past 10 10.12
3)	3) 25 to 12 35 past 11 11.35	9)	9) 19 to 8 19 minutes to 8 7.41
4)	4) 15 past 9 a quarter past 9 9.15	10)	10) 29 past 3 29 minutes past 3 3.29
5)	5) 10 to 5 50 past 4 4.50	11)	11) 8 to 9 8 minutes to 9 8.52
6)	6) 5 past 6 6.05	12)	12) 1 to 1 1 minute to 1 12.59

Time 3 — 12-Hour Clock.

12-Hour Clock
Indicate Hand Positions

For any given time - to the nearest minute - the **exact** position of the **minute hand** can be indicated.
The **exact** position or a **good estimate** of the position of the **hour hand** is possible - and can be found quite quickly.

Q	A
1) Consider the **hour hand only** moving from 1 o'clock to 2 o'clock. Calculate a) how many minutes it takes to reach **each minute mark**. b) how many minutes it takes to reach **half way** between each minute mark.	1)a) 1 hour = 60 minutes The **hour hand** has to reach <u>5</u> **minute marks**. $$\frac{60 \text{ minutes}}{5 \text{ minute marks}} = 12 \text{ minutes}$$ b) $\frac{12 \text{ minutes}}{2} = 6$ minutes The **hour hand** takes 12 minutes to reach **each minute mark** ... 6 minutes to reach **half way** between each minute mark.

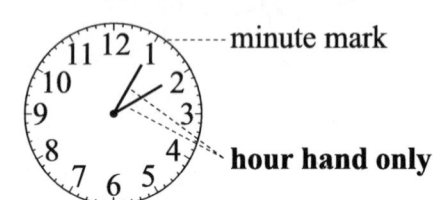

From the above Q/A, it can be seen that to determine the position of the **hour hand** for a given time -

Divide the minutes <u>past the hour</u> of a given time by 12 to give the number of 1 minute marks the **hour hand** moves past the hour.

The movement for any 'remainder' minutes can be **estimated** using $\frac{1}{2}$ marks also.

The calculation is avoided when the time is an exact hour - ex. 5 o'clock, since we **know** that the position of the **hour hand** is **exactly** in line with that hour.

It may also be avoided when the time is 'half past' / '30 minutes past' the hour - since we **know** that the position of the **hour hand** is **exactly half way** between the hours.

Ex. Indicate each time on the clock.

1) 12 minutes past 1
   1.12

   ⇨ Position the **minute hand**. ⇨

   $\frac{12 \text{ mins}}{12} = 1$

   The **hour hand** moves **1 minute mark** past 1.

   The 'arrow' emphasises the **hour hand** position.

2) 20 to 11
   10.40

   ⇨ Position the **minute hand**. ⇨

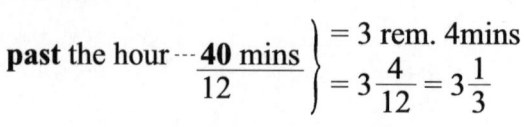

   past the hour — $\frac{40 \text{ mins}}{12}$ = 3 rem. 4 mins
   $= 3\frac{4}{12} = 3\frac{1}{3}$

   The **hour hand** moves $3\frac{1}{3}$ **minute marks** past 10.

   The 'arrow' emphasises the **hour hand** position.

Time 4

12-Hour Clock.

Q	A	Q	A
Indicate each time. 1) 5 o'clock	1)	7) 5 to 12 11.55	7) $\frac{55m}{12} = 4\frac{7}{12}$ min. marks past 11
2) 24 past 1 1.24	2) $\frac{24m}{12} = 2$ min. marks past 1	8) Quarter past 3 3.15	8) $\frac{15m}{12} = 1\frac{3}{12} = 1\frac{1}{4}$ min. marks past 3
3) 24 minutes to 9 8.36	3) $\frac{36m}{12} = 3$ min. marks past 8	9) Quarter to 7 6.45	9) $\frac{45m}{12} = 3\frac{9}{12} = 3\frac{3}{4}$ min. marks past 7
4) 12 minutes to 6 5.48	4) $\frac{48m}{12} = 4$ min. marks past 5	10) 6 minutes past 9 9.06	10) $\frac{6m}{12} = \frac{1}{2}$ min. mark past 9
5) Half past 10 10.30	5) $\frac{30m}{12} = 2\frac{6}{12} = 2\frac{1}{2}$ min. marks past 10	11) 20 past 12 12.20	11) $\frac{20m}{12} = 1\frac{8}{12} = 1\frac{2}{3}$ min. marks past 12
6) 10 past 4 4.10	6) $\frac{10m}{12} = \frac{5}{6}$ min. mark past 4	12) 11 minutes past 2 2.11	12) $\frac{11m}{12}$ min. mark past 2

Time 5

12 / 24 - Hour Clock.

12 / 24-Hour Clock Time
Time can be presented in either **12-hour clock time** or **24-hour clock time**.

12-Hour Clock Time
The time starts from 12 midnight and each hour is then indicated by the numbers 1, 2, 3, ...11, 12.

To show that a time is between **12 midnight** and **12 noon (midday)** it is followed by **a.m.**
a.m. is short for <u>a</u>nte <u>m</u>eridiem - which is Latin for 'before midday'.   Exs. 0.30 a.m.   5.00 a.m.

To show that a time is between **12 noon** and **12 midnight** it is followed by **p.m.**
p.m. is short for <u>p</u>ost <u>m</u>eridiem - which is Latin for 'after midday'.   Exs. 4.00 p.m.   10.22 p.m.

24-Hour Clock Time.
The time starts from **midnight 00.00** and each hour is then indicated by the numbers 01, 02, 03, ... 23 until the time reaches midnight again (the hour number 24 is **not** used).
Each time is written with 4 digits.   Exs. 00.50   06.21   17.48 ⎱ Both forms are used /acceptable
                                    or Exs. 0050   0621   1748 ⎰ in the following Exs. and Q/A.

Ex. Write in 24-hour clock time.

a) 3.30 a.m. ⇨ 0330

b) 3.30 p.m. ⇨ 3.30
              +12
              ─────
              15.30

12 hours are **added** to p.m. times to give 24 hour times.

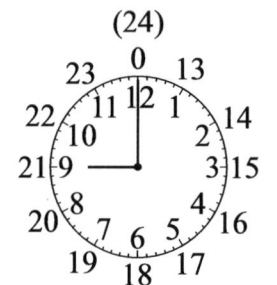

Ex. Write in 12-hour clock time.

a) 09.26 ⇨ 9.26 a.m.

b) 1748   ⇨ 1748
             −12
             ─────
             5.48 p.m.

12 hours are **subtracted** for times over 1200 to give p.m. times.

Q	A	Q	A
1. Write in 24-hour clock time.	1.	2. Write in 12-hour clock time.	2.
1) 12 midnight.	1) 00.00	1) 0000	1) 12 midnight.
2) 0.27 a.m.	2) 00.27	2) 00.15	2) 0.15 a.m.
3) 6.40 a.m.	3) 0640	3) 0456	3) 4.56 a.m.
4) 11.18 a.m.	4) 11.18	4) 09.00	4) 9.00 a.m.
5) 12 noon (midday)	5) 1200	5) 12.00	5) 12 noon (midday)
6) 12.09 p.m.	6) 12.09	6) 12.20	6) 12.20 p.m.
7) 1.00 p.m.	7) 13.00	7) 1300	7) 1.00 p.m.
8) 1.55 p.m.	8) 13.55	8) 14.37	8) 2.37 p.m.
9) 3.21 p.m.	9) 1521	9) 16.45	9) 4.45 p.m.
10) 7.00 p.m.	10) 19.00	10) 1809	10) 6.09 p.m.
11) 10.10 p.m.	11) 22.10	11) 20.30	11) 8.30 p.m.
12) 11.59 p.m.	12) 2359	12) 2222	12) 10.22 p.m.

Time 6

Converting Time Units.

Converting Time - Higher Units to Lower Units
To convert higher time units to lower value units - we **multiply** as follows:

weeks ⇨ days ⇨ hours ⇨ minutes ⇨ seconds
 ×7   ×24   ×60   ×60

Ex. Convert 3 hours 57 minutes to minutes.

$$3 \times 60 \text{ minutes} = \begin{array}{r} 180 \text{ mins} \\ + 57 \text{ mins} \\ \hline 237 \text{ mins} \end{array}$$

Q	A	Q	A
Convert each time.		6) 5 minutes 2 seconds to seconds.	6) 5×60 mins = 300 secs + 2 secs 302 secs
1) 9 weeks 5 days to days.	1) 9×7 days = 63 days + 5 days 68 days	7) 7 hours 48 minutes to minutes.	7) 7×60 mins = 420 mins + 48 mins 468 mins
2) 8 days to hours.	2) 8×24 hrs = 192 hrs	8) 1 hour to seconds.	8) 1×60 mins = 60 mins × 60 secs 3600 secs
3) 3 days 19 hours to hours.	3) 3×24 hrs = 72 hrs + 19 hrs 91 hrs	9) 2 hrs 3 mins 4 secs to seconds.	9) 2×60 mins = 120 mins + 3 mins 123 mins × 60 secs 7380 secs + 4 secs 7384 secs
4) 4 hours 35 minutes to minutes.	4) 4×60 mins = 240 mins + 35 mins 275 mins		
5) 6 minutes to seconds.	5) 6×60 secs = 360 secs		

Converting Time - Lower units to Higher units
To convert lower time units to higher value units - we **divide** as follows:

seconds ⇨ minutes ⇨ hours ⇨ days ⇨ weeks
 ÷60   ÷60   ÷24   ÷7

Ex. Convert 63 hours to days and hours.

$$\frac{63 \text{ hours}}{24} = 2 \text{ days } 15 \text{ hours}$$

Q	A	Q	A
Convert each time.		9) 7200 seconds to hours.	9) $\frac{7200 \text{ secs}}{60} = 120$ mins $\frac{120 \text{ mins}}{60} = 2$ hrs
1) 180 seconds to minutes.	1) $\frac{180 \text{ secs}}{60} = 3$ mins		
2) 444 seconds to mins and secs.	2) $\frac{444 \text{ secs}}{60} = 7$ mins 24 secs	10) 4000 seconds to hrs mins secs.	10) $\frac{4000 \text{ secs}}{60} = 66$ mins 40 secs $\frac{66 \text{ mins}}{60} = 1$ hr 6 mins **1 hr 6 mins 40 secs**
3) 600 minutes to hours.	3) $\frac{600 \text{ mins}}{60} = 10$ hrs		
4) 333 minutes to hrs and mins.	4) $\frac{333 \text{ mins}}{60} = 5$ hrs 33 mins	11) 5000 minutes to days hrs mins.	11) $\frac{5000 \text{ mins}}{60} = 83$ hrs 20 mins $\frac{83 \text{ hrs}}{24} = 3$ days 11 hrs **3 days 11hrs 20mins**
5) 96 hours to days.	5) $\frac{96 \text{ hrs}}{24} = 4$ days		
6) 280 hours to days and hrs.	6) $\frac{280 \text{ hrs}}{24} = 11$ days 16 hrs	12) 2409 hours to wks days hrs.	12) $\frac{2409 \text{ hrs}}{24} = 100$ days 9 hrs $\frac{100 \text{ days}}{7} = 14$ wks 2 days **14 wks 2 days 9 hrs**
7) 147 days to weeks.	7) $\frac{147 \text{ days}}{7} = 21$ wks		
8) 60 days to wks and days.	8) $\frac{60 \text{ days}}{7} = 8$ wks 4 days		

Converting Time - Decimal / Fraction Units to Units

To convert decimal /fraction time units to units - we **multiply** as follows:

weeks ⇨ days ⇨ hours ⇨ minutes ⇨ seconds
×7 ×24 ×60 ×60

Ex.1 Convert 0·7hr to minutes.

$0{\cdot}7 \times 60 \text{ mins} = 12 \text{ mins}$

Ex.2 Convert $\frac{5}{8}$ day to hours.

$\frac{5}{\cancel{8}_1} \times \frac{\cancel{24}^{3}\text{hrs}}{1} = \frac{15}{1} = 15 \text{ hrs}$

Q	A	Q	A
Convert each time. 1) 0·2 min to secs.	1) $0{\cdot}2 \times 60$ secs = 12 secs	7) $\frac{2}{5}$ hr to mins.	7) $\frac{2}{\cancel{5}_1} \times \frac{\cancel{60}^{12}\text{m}}{1} = \frac{24}{1} = 24$ mins
2) 0·75 day to hours.	2) $0{\cdot}75 \times 24$ hrs = 18 hrs		
3) 0·3 wk to days.	3) $0{\cdot}3 \times 7$ days = 2·1 days	8) $\frac{3}{8}$ min to secs.	8) $\frac{3}{\cancel{8}_2} \times \frac{\cancel{60}^{15}\text{s}}{1} = \frac{45}{2} = 22\frac{1}{2}$ secs
4) 0·89 min to secs.	4) $0{\cdot}89 \times 60$ secs = 53·4 secs		
5) 0·12 hr to mins and secs.	5) $0{\cdot}12 \times 60$ mins = **7**·2 mins $0{\cdot}2 \times 60$ secs = **12** secs **7** mins **12** secs	9) $\frac{1}{9}$ hr to mins and secs.	9) $\frac{1}{\cancel{9}_3} \times \frac{\cancel{60}^{20}\text{m}}{1} = \frac{20}{3} = \mathbf{6}\frac{2}{3}$ mins $\frac{2}{\cancel{3}_1} \times \frac{\cancel{60}^{20}\text{s}}{1} = \frac{40}{1} = \mathbf{40}$ secs **6** mins **40** secs
6) $\frac{1}{4}$ wk to days.	6) $\frac{1}{4} \times \frac{7\text{days}}{1} = \frac{7}{4} = 1\frac{3}{4}$ days		

Converting Time - units to Fraction / Decimal units.

To convert units to Fraction / Decimal time units - we **divide** as follows:

seconds ⇨ minutes ⇨ hours ⇨ days ⇨ weeks
÷60 ÷60 ÷24 ÷7

Ex. Convert 45 mins to hours.

$\frac{45}{60} = \frac{3}{4} \text{ hr} = 0{\cdot}75 \text{ hr}$

Q	A	Q	A
Convert each time. 1) 10 mins to hrs.	1) $\frac{10}{60} = \frac{1}{6}$ hr = 0·1666... hr	5) 14 hrs to days.	5) $\frac{14}{24} = \frac{7}{12}$ day = 0·583... day
2) 48 mins to hrs.	2) $\frac{48}{60} = \frac{4}{5}$ hr = 0·8 hr	6) 32 secs to mins.	6) $\frac{32}{60} = \frac{8}{15}$ min = 0·53... min
3) 2 days to weeks.	3) $\frac{2}{7}$ wk = 0·285... wk	7) 18 mins to hrs.	7) $\frac{18}{60} = \frac{3}{10}$ hr = 0·3 hr
4) 6 hrs to days.	4) $\frac{6}{24} = \frac{1}{4}$ day = 0·25 day	8) 1 sec to mins.	8) $\frac{1}{60}$ min = 0·0166... min

Time 8

Addition.

Addition
Add the smallest time units first - if necessary, convert a time unit to involve a higher time unit.

Ex.1
```
   days  hrs         days  hrs
     2    7            2    7
   + 3    9   ⇨      + 3    9
                       5   16
```

Ex.2
```
   days  hrs         days  hrs           days  hrs
     4   22            4   22              4   22  ⎫
   + 1    8   ⇨      + 1    8    ⇨      + 1₁   8  ⎬ +
                                ⎫             6    6 ⎭
                              30⎭                  30
                                                   24
                   ( convert 30 hours to        = 1 r 6
                       days and hours )
```

Q	A	Q	A

Add the following times.
1) wks days
     2   3
   + 1   2
	1) wks days
	     2 3
	   + 1 2
	     3 5
		7) mins secs
		     24 16
		   + 24 20
			7) mins secs
			     24 16
			   + 24 20
			     48 36

2) wks days
     1   4
   + 1   6
	2) wks days
	     1 4
	   + 1₁ 6 ⎫+
	     3 3 ⎬=
	         10
	          7
	       = 1 r 3
		8) mins secs
		     19 58
		   + 39 49
			8) mins secs
			     19 58 ⎫+
			   + 39₁ 49 ⎬
			     59 47 ⎭=
			          107
			           60
			       = 1 r 47

3) days hrs
     1   8
   + 2   15
	3) days hrs
	     1 8
	   + 2 15
	     3 23
		9) hrs mins secs
		     5 20 38
		   +14 37 18
			9) hrs mins secs
			     5 20 38
			   +14 37 18
			    19 57 56

4) days hrs
     4   21
   + 1   17
	4) days hrs
	     4 21
	   + 1₁ 17 ⎫+
	     6 14 ⎬=
	          38
	          24
	       = 1 r 14
		10) hrs mins secs
		     12 15 46
		    + 3 44 53
			10) hrs mins secs
			     12 15 46 ⎫+
			    + 3₁ 44₁ 53 ⎬=
			     16 0 39
			          60 99
			          60 60
			         =1r0 =1r39

5) hrs mins
    10   44
   +11   9
	5) hrs mins
	    10 44
	   +11 9
	    21 53
		11) days hrs mins
		     3 1 10
		   + 3 22 20
			11) days hrs mins
			     3 1 10
			   + 3 22 20
			     6 23 30

6) hrs mins
     6   37
     4   58
   + 5   40
	6) hrs mins
	    16 37 ⎫
	    14 58 ⎬+
	   + 5₂ 40 ⎭+
	    37 15 ⎬=
	         135
	          60
	       = 2 r 15
	
	(it is not necessary to convert the hours to days)
		12) days hrs mins
		     2 20 24
		   + 2 19 36
			12) days hrs mins
			     2 20 24 ⎫+
			   + 2₁ 19₁ 36 ⎬=
			     5 16 0
			          40 60
			          24 60
			         =1r16 =1r0

		13) days hrs mins
		     3 7 5
		     1 7 42
		   + 2 5 8
			13) days hrs mins
			     3 7 5
			     1 7 42
			   + 2 5 8
			     6 19 55

Time 9 — Subtraction

Subtraction
When the values in each column of time allow a direct subtraction ... we simply subtract.

Ex.
```
    days  hrs         days  hrs
     10   21           10   21
  -   7    9        -   7    9
                         3   12
```

Q	A	Q	A

Subtract the following times.
1) days hrs
    36   18
−11   2
	1) days hrs
	    36 18
	−11 2
	    25 16
		2) hrs mins secs
		  22 44 57
		−12 8 33
			2) hrs mins secs
			  22 44 57
			−12 8 33
			  10 36 24

Subtraction
When the values in a column of time do **not** allow a direct subtraction ...
3 Methods are considered.

Method 1
Subtract the smaller units first -
to allow the subtraction we have to
'convert' units to smaller units.

Ex.
```
    hrs  mins              hrs   mins
                           21    60 }75
     22   15               2̶2̶    15
   -  8   40            -   8    40
                           13    35
```
(**'convert'** 1 hr to 60 mins)

Method 2
Subtract the smaller units first -
to allow the subtraction we have to
'borrow and pay back' between the units.

Ex.
```
    hrs  mins              hrs    mins
                                   60 }75
     22   15               22     15
   -  8   40           -   8₁     40
                           13     35
```
(**'borrow'** 60 mins and **'pay back'** 1 hr)

Q	A		
Subtract the following times.	Method 1		Method 2

1) hrs mins
  19 00
−16 10
	1) hrs mins
	  18 60 }60
	  1̶9̶ 00
	−16 10
	   2 50
			hrs mins
			      60 }60
			  19 00
			−16₁ 10
			   2 50

2) mins secs
  46 25
−31 40
	2) mins secs
	  45 60 }85
	  4̶6̶ 25
	−31 40
	  14 45
			mins secs
			      60 }85
			  46 25
			−31₁ 40
			  14 45

3) days hrs
  7   8
−5 21
	3) days hrs
	  6 24 }32
	  7̶ 8
	−5 21
	  1 11
			days hrs
			    24 }32
			  7 8
			−5₁ 21
			  1 11

4) days hrs mins
  9   6 13
−4 21 59
	4) days hrs mins
	      24 }29
	  8 5 60 }73
	  9̶ 6̶ 13
	−4 21 59
	  4 8 14
			days hrs mins
			    24 }30 60 }73
			  9 6 13
			−4₁ 21₁ 59
			  4 8 14

	(**'convert'** 1 hr to 60 mins) (**'convert'** 1 day to 24 hrs)		(**'borrow'** 60 mins and **'pay back'** 1 hr) (**'borrow'** 24 hrs and **'pay back'** 1 day)

Time 10
Subtraction.

Method 3
Given 2 units of time ...
Starting from the **lowest time**, find the **difference** to the **next whole unit** of time.
From this unit of time, find the **difference** to the **highest time**.
(These differences are usually calculated **mentally**.)
Add the differences to find the total difference between the given times.

Given 3 or more units of time ...
Starting from the **lowest time**, find the **difference** to the **next whole unit** of time.
From this unit of time, find the difference to the **next whole unit** of time (and so on ...).
Eventually, find the **difference** to the **highest time**.
Add the differences.

Ex. 1
```
 hrs  mins  ⇨                          hrs  mins
  8    00      4 - 25 to 5 - 00 =      0    35
 −4    25      5 - 00 to 8 - 00 =     +3    00
                                       3    35
```

Ex. 2
```
 hrs  mins  ⇨                           hrs  mins
  22   19     16 - 37 to 17 - 00 =       0    23
 −16   37     17 - 00 to 22 - 19 =      +5    19
                                         5    42
```

Q	A
Subtract the following times.	
1) mins secs 29 00 −18 53	1) mins secs 18 - 53 to 19 - 00 = 0 7 19 - 00 to 29 - 00 = +10 00 10 7
2) mins secs 46 25 −31 40	2) mins secs 31 - 40 to 32 - 00 = 0 20 32 - 00 to 46 - 25 = +14 25 14 45
3) hrs mins 58 5 −24 29	3) hrs mins 24 - 29 to 25 - 00 = 0 31 25 - 00 to 58 - 05 = +33 05 33 36
4) days hrs 12 00 − 7 18	4) days hrs 7 18 to 8 00 = 0 6 8 00 to 12 00 = + 4 00 4 6
5) days hrs 7 8 − 5 21	5) days hrs 5 21 to 6 00 = 0 3 6 00 to 7 08 = +1 8 1 11
6) hrs mins secs 21 15 32 − 9 48 47	6) h m s h m s hrs mins secs 9 - 48 - 47 to 9 - 49 - 00 = 0 0 13 9 - 49 - 00 to 10 - 00 - 00 = 0 11 0 10 - 00 - 00 to 21 - 15 - 32 = +11 15 32 11 26 45
7) days hrs mins 14 6 13 − 1 19 51	7) d h m d h m days hrs mins 1 19 51 to 1 20 00 = 0 0 9 1 20 00 to 2 00 00 = 0 4 0 2 00 00 to 14 6 13 = +12 6 13 12 10 22

Time 11

Multiplication

Multiply the smallest time units first - if necessary, convert a time unit to involve a higher time unit.

Ex.1
```
  days  hrs          days  hrs
    3    9             3    9
  ×       2    ⇨     ×       2
                       6   18
```

Ex.2
```
  days  hrs          days  hrs          days  hrs
    6   10             6   10    ⎫        6   10
  ×       5         ×       5    ⎬ ×    × 2    5
                            50   ⎭       32    2
                                              50
                                              24
( convert 50 hours to                       = 2 r 2
    days and hours )
```

Q	A	Q	A
Multiply each time.			

1) wks days
  7   2
×   3
	1) wks days
	  7 2
	× 3
	 21 6
		7) mins secs
		  9 7
		× 8
			7) mins secs
			  9 7
			× 8
			 72 56

2) wks days
  4   6
×   6
	2) wks days
	  4 6 ⎫×
	× 5 6 ⎬=
	  29 1
	      36
	       7
	    = 5 r 1
		8) mins secs
		  20 45
		× 9
			8) mins secs
			  20 45 ⎫×
			× 6 9 ⎬=
			 186 45
			      405
			       60
			    = 6 r 45

3) days hrs
  9   4
×   5
	3) days hrs
	  9 4
	× 5
	 45 20
		9) hrs mins secs
		  6 22 17
		× 2
			9) hrs mins secs
			  6 22 17
			× 2
			 12 44 34

4) days hrs
  2 18
×   3
	4) days hrs
	  2 18 ⎫×
	× 2 3 ⎬=
	  8 6
	     54
	     24
	   = 2 r 6
		10) hrs mins secs
		  11 40 33
		× 7
			10) hrs mins secs
			  11 40 33 ⎫×
			× 4 3 7 ⎬=
			  81 43 51
			      283 231
			       60 60
			    =4r43 =3r51

5) hrs mins
  5 12
×   4
	5) hrs mins
	  5 12
	× 4
	 20 48
		11) days hrs mins
		  8 3 6
		× 6
			11) days hrs mins
			  8 3 6
			× 6
			 48 18 36

6) hrs mins
  7 30
× 10
	6) hrs mins
	  7 30 ⎫×
	× 5 10 ⎬=
	 75 0
	     300
	      60
	   = 5 r 0
		12) days hrs mins
		  4 4 4
		× 20
			12) days hrs mins
			  4 4 4 ⎫×
			× 3 1 20 ⎬=
			  83 9 20
			      81 80
			      24 60
			    =3r9 =1r20

Time 12

'Problems': +, −, ×.

'Problems'
The following 'problems' involve addition, subtraction, multiplication.

Q	A

1) The times of a team's 3 runners are recorded in a relay race - 5m 45s, 5m 39s and 5m 54s.
 	
Calculate the total time of the 3 runners.
	1)
	    mins secs
	     5 45
	     5 39 } +
	 + 5₂ 54
	   17 18) =
	         138
	          60
	      = 2 r 18

2) A weather forecast expects it to be 'fine' for 2 weeks and 5 days before 'unsettled' for 1 week and 4 days.
 	
What is the weather forecast period?
	2)
	    wks days
	     2 5 } +
	 + 1₁ 4
	    4 2) =
	        9
	        7
	    = 1 r 2

3) 2 cyclists complete a course in 18m 32s and 20m 16s.
 	
Calculate the difference in their times.
	3)
	
	Method 1
	  m s
	 20 60 } 76
	 2̶1̶ 16
	−18 32
	  2 44
	
	Method 2
	  m s
	       60 } 76
	 21 16
	−18₁ 32
	  2 44
	
	Method 3
	             m s
	18-32 to 19-00 = 0 28
	19-00 to 21-16 = +2 16
	                     2 44

4) A passenger arrives at a station at 7.47 a.m. and catches a train at 9.08 a.m.
 	
How long was the passenger waiting at the station?
	4)
	
	Method 1
	 h m
	 8 60 } 68
	 9̶ 08
	−7 47
	 1 21
	
	Method 2
	 h m
	     60 } 68
	 9 08
	−7₁ 47
	 1 21
	
	Method 3
	             h m
	7-47 to 8-00 = 0 13
	8-00 to 9-08 = +1 08
	                  1 21

5) It takes an average of 2hrs 35 mins to make a gate.
 	
Find the total time that it takes to make 9 gates.
	5)
	    hrs mins
	     2 35 } ×
	 × 5, 9
	   23 15) =
	         315
	          60
	      = 5 r 15

6) The time for a scientific experiment is recorded as 8hrs 23mins 51secs.
 	
If the experiment is repeated 8 times find the total time recorded.
	6)
	    hrs mins secs
	     8 23 51 } ×
	 × 3, 6, 8
	   67 10 48) =
	         190 408
	          60 60
	      =3r10 =6r48

Time 13

Division.

Division - by a Whole Number

1) No remainder

Division of units of time by a whole number may divide exactly with no 'remainder'.

Ex. 42 hours work is shared equally between 6 people. ⇨ $\frac{42 \text{ hours}}{6 \text{ people}} = 7$ hours work each.
How many hours work does each person have?

Q	A	Q	A
1) Share 55 hours work between 5 people.	1) $\frac{55 \text{ hours}}{5 \text{ people}} = 11$ hours.	4) If it is 16 days 12 hours to a year end, find $\frac{1}{4}$ of this time.	4) days hrs 4 3 4 ⟌ 16 12
2) A process lasts 21 mins. Calculate $\frac{1}{3}$ of the time.	2) $\frac{21 \text{ mins}}{3} = 7$ mins.	5) 9km of cable are made in 72 hrs 45 mins. In what time is 1km made?	5) hrs mins 8 5 9 ⟌ 72 45
3) 8 pins are made in 64 secs. In what time is 1 pin made?	3) $\frac{64 \text{ secs}}{8 \text{ pins}} = 8$ secs.	6) 2 laps of a course are run in 20 mins 32 secs. What is the average lap time?	6) mins secs 10 16 2 ⟌ 20 32

2) Remainder

Division of units of time by a whole number may divide to leave a '**remainder**' ('**rem.**' or '**r.**').

How we deal with the remainder depends on how we want / need to present the answer.
  A) The remainder can be left as it is.
  B) The remainder can be written as a fraction or decimal of a unit.

A)
  Ex.1 It takes 2 hours to make a table.
      How many complete tables are made in 9 hours?

      4 tables rem. 1hr
      2 ⟌ 9

B)
  Ex.1 It takes 9 hours to make 2 tables.
      How long does it take to make 1 table?

      $\frac{9}{2} = 4\frac{1}{2}$ hrs = 4·5 hrs

Q	A	Q	A
1) It takes 3 hours to make a chair. How many chairs are made in 11 hours?	1) 2 chairs rem. 2 hrs 3 ⟌ 11	4) 8 litres of water leak out of a tank in 77 hours. How long does it take for 1 litre to leak out?	4) $\frac{77}{8} = 9\frac{5}{8}$ hrs = 9·625 hrs (·625×60 = 37·5) = 9 hrs 37·5 mins
2) A process takes 7 mins. Calculate $\frac{1}{4}$ of the time.	2) $\frac{7}{4} = 1\frac{3}{4}$ mins = 1·75 mins (·75×60 = 45) = 1 min 45 secs	5) An athlete takes 7 mins to complete a training circuit. How many circuits are completed in 1 hour?	5) 8 circuits r. 4 mins 7 ⟌ 60 mins (= 1 hour)
3) If a leaflet is printed in 5 seconds, how many can be printed in 56 seconds?	3) 11 leaflets r. 1 sec 5 ⟌ 56	6) A clock gains 20 seconds in 3 hours. How much time is gained per hour?	6) $\frac{20}{3} = 6\frac{2}{3}$ secs = 6·666... secs

Time 14

Division.

Remainder

C) It may be possible to change the remainder to smaller units and continue dividing.

Ex. It takes 9 mins 30 secs to make 2 hats.
How long does it take to make 1 hat?

```
         mins    secs
           4      45
       2 │ 9      30
          -8     +60
           1      90
        ×60secs  -90
         60secs   0
```

*Note
The work is shown as 'long division' but can be shown in other ways.

2 is divided into **90**

*Note
To divide into minutes and seconds
1) divide into the minutes
2) change the remainder into seconds (×60) before adding to the seconds column
3) divide into the seconds

Q	A	Q	A
1) It takes 7 mins to make 4 flags. How long does it take to make 1 flag?	``` mins secs 1 45 4 │ 7 0 -4 +180 3 180 ×60secs -180 180secs 0 ```	6) A journey takes 19 days 4 hours. How long is each stage of the journey if there are 5 approximately equal stages?	``` days hrs 3 20 5 │ 19 4 -15 + 96 4 100 ×24hrs -100 96hrs 0 ```
2) A process takes 8 hours. Calculate $\frac{1}{3}$ of the time.	``` hrs mins 2 40 3 │ 8 0 -6 +120 2 120 ×60mins -120 120mins 0 ```	7) In a week, an athlete completes 7 laps of a course in a total time of 12 hrs 28 mins 53 secs. Calculate the average time per lap.	``` hrs mins secs 1 46 59 7 │ 12 28 53 - 7 +300 +360 5 328 413 ×60m -322 -413 300m 6 0 × 60s 360s ```
3) A space craft travels for 11 days. Find an $\frac{1}{8}$ of this time.	``` days hrs 1 9 8 │ 11 0 - 8 +72 3 72 ×24hrs -72 72hrs 0 ```	8) A machine makes 10 items in 56 mins 21secs. How long does it take to make 1 item?	``` mins secs 5 38 10 │ 56 21 -50 +360 6 381 ×60secs -380 360secs 1 ``` ⇩ There is still a remainder - the answer can be shown as ``` mins secs 5 38$\frac{1}{10}$ ``` or ``` mins secs 5 38·1 ```
4) A car journey took 23 hrs 15 mins. If 5 people shared the driving equally, how long did each drive for?	``` hrs mins 4 39 5 │ 23 15 -20 +180 3 195 ×60mins -195 180mins 0 ```		
5) It takes 35 mins 25 secs to make 11 brushes. How long does it take to make 1 brush?	``` mins secs 3 25 11 │ 37 35 -33 +240 4 275 ×60secs -275 240secs 0 ```		

Time 15

Timetable.

Timetable

A **Timetable** may be presented in many ways.
It is more usual for times to be shown using the 24-hour clock rather than the 12-hour clock.

The following Timetable relates to 4 trains and a route involving Stations A to G.
For example, Train T1 starts from Station A, stops at Stations B, C, D, E, F and finally arrives at G.

A train's arrival and departure time at a Station along the route is shown as the **same time**, since the difference between them is small and a single time is clearer.

If a train does not stop at a Station then a time is not given - ex. Train T2 does not stop at B.

		Train			
		T1	T2	T3	T4
	A	0600	1100	1530	
	B	0615		1545	1915
	C	0630	1130	1600	1930
Station	D	0700	1200	1630	2000
	E	0720		1650	2020
	F	0735	1235	1705	2035
	G	0800	1300	1730	

The following Q/A refer to this Timetable.
Calculations may be set out differently than shown here.

Q	A
1) At what time does T1 leave A?	1) 0600
2) At what time does T1 arrive at F?	2) 0735
3) How long is the T1 journey from B to C?	3) 0630 – 0615 = 15 minutes
4) How long is the T1 journey from A to G?	4) 0800 – 0600 = 2 hours
5) After leaving A, at which Stations does T2 stop?	5) C, D, F, G.
6) A family arrives at E at 1200 and boards a train. What is the earliest time the family can arrive at F?	6) 1705 (boards Train T3 at 1650)
7) What is the time difference between T2 and T3 at A?	7) 1530 – 1100 = 4 hours 30 minutes
8) If T3 is 30 minutes late, when does it leave B?	8) 1545 + 30mins = 1615
9) How long is the T3 journey from B to E?	9) 1650 – 1545 = 1 hour 5 minutes
10) Between which 2 stations is the T3 time 25 minutes?	10) F and G
11) What is the total time for the T4 journey?	11) 2035 – 1915 = 1 hour 20 minutes
12) Complete the Timetable for T5. The times between Stations are the same as T1. 　　　　T5 　A　2145 　B 　C 　D 　E 　F 　G	12) 　　　　T5 　A　2145　⎫ +15 mins 　B　2200　⎬ +15 　C　2215　⎬ +30 　D　2245　⎬ +20 　E　2305　⎬ +15 　F　2320　⎬ +25 　G　2345

Time 16

Calender. Estimation.

Calendar

A **Calendar** may be presented in many ways.

The month can be written before or after the number in the date, for ex., May 3rd or 3rd May.

The following Q/A refer just to the months of May and June (for no particular year).

```
          May                              June
Monday    1  8 15 22 29         Monday       5 12 19 26
Tuesday   2  9 16 23 30         Tuesday      6 13 20 27
Wednesday 3 10 17 24 31         Wednesday    7 14 21 28
Thursday  4 11 18 25            Thursday   1 8 15 22 29
Friday    5 12 19 26             Friday     2 9 16 23 30
Saturday  6 13 20 27            Saturday   3 10 17 24
Sunday    7 14 21 28            Sunday     4 11 18 25
```

Q		A	
1)	What day is 23rd May?	1)	Tuesday
2)	What dates are the Fridays in May?	2)	5th, 12th, 19th, 26th
3)	What date is the last Sunday in May?	3)	28th
4)	What date is 9 days after 12th May?	4)	21st May
5)	What is the date 2 weeks after May 8th?	5)	May 22nd
6)	What is the date 13 days after 25th May?	6)	7th June
7)	What is the date 4 weeks after May 18th?	7)	June 15th
8)	How many days are there between May 8th and May19th - not including the 2 dates?	8)	10 days (9th, 10th, ... 18th)
9)	What day is 8th June?	9)	Thursday
10)	What dates are the Wednesdays in June?	10)	7th, 14th, 21st, 28th
11)	What date is the second Saturday in June?	11)	10th
12)	What date is 5 days after 25th June?	12)	30th June
13)	What is the date 3 weeks after June 4th?	13)	June 25th
14)	What is the date 15 days before 2nd June?	14)	18th May
15)	What is the date 4 weeks before June 21st?	15)	May 24th
16)	How many days are there between May 27th and June 6th - not including the 2 dates?	16)	9 days (May 28th, 29th, ... June 5th)

Time - Estimation

It is useful to be able to make a 'reasonable' estimation of a measure of time in 'everyday life'.

This should be based on actual experience and / or ' general knowledge'.

The time of an 'event' may vary considerably so any estimation would need to take this into account.

The following Q/A give just 2 'estimates' - the more 'reasonable' estimate should be **chosen**.

Q	A	Q	A
Estimate			

1) Time to run 100m.
   a) 50sec   b) 13sec
	1) b)	4) Time to build a house.
		   a) 12hr b) 6wk
			4) b)

2) Time to cycle 1km.
   a) 40min   b) 4min
	2) b)	5) Time a plane flies 400miles.
		   a) 1hr b) 10hr
			5) a)

3) Time to boil a kettle of water.
   a) 3min   b) 1hr
	3) a)	6) Time to write my own name.
		   a) 5sec b) 45sec
			6) a)

Time - Speed - Distance 1 Formulas.

Time - Speed - Distance

The units of Time are usually seconds (sec, s), minutes (min, m), hours (hr, h).

Speed is the Distance travelled (by an object) in 1 unit of Time.
Speed is taken to be **constant** - it stays the same throughout the time.

The Metric units of Distance (length) are mm, cm, m, km.

Formulas

Time - Speed - Distance can be presented in terms of formulas.

$$\text{Speed} = \frac{\text{Distance}}{\text{Time}}$$

$$S = \frac{D}{T}$$

$$\text{Speed} = \frac{\text{Distance}}{\text{Time}} \quad \text{'swop'}$$

$$\text{Time} = \frac{\text{Distance}}{\text{Speed}}$$

$$T = \frac{D}{S}$$

$$\text{Speed} = \frac{\text{Distance}}{\text{Time}}$$

$$\text{Speed} \times \text{Time} = \text{Distance} \quad \text{'swop'}$$

$$\text{Distance} = \text{Speed} \times \text{Time}$$

$$D = ST$$

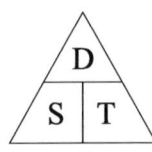

The triangle is useful to help remember the 3 formulas ...

In the triangle, in turn ...
choose the letter on the **left** of each formula -
this leaves the 2 letters on the **right** of each formula AND in their correct 'position'.

Ex.1
A girl walks 24km in 3 hours.
Calculate her **speed**.

$$S = \frac{D}{T} = \frac{24\text{km}}{3\text{hours}} = 8\text{km per hour}$$

— means 'for each hour'

8kmph
8km/h
8kmh^{-1}

any of these terms may be used

*Note

$$8\text{km/h} = \frac{8\text{km}}{\text{h}} = 8\text{kmh}^{-1}$$

same

(see **Indices / Powers 8**)

Ex.2
A robot travels 150m at 30mpm.
Calculate the **time** it takes.

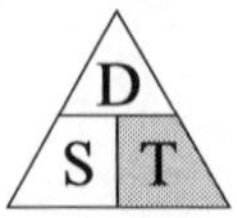

$$T = \frac{D}{S} = \frac{150\text{m}}{30\text{mpm}} = 5 \text{ minutes}$$

The units must agree - ex. **m**

Ex.3
An object travels at 80cm/sec for 7seconds.
Calculate the **distance** it moves.

$$D = ST$$
$$= 80\text{cm/sec} \times 7\text{seconds}$$
$$= 560\text{cm}$$

The units must agree - ex. **sec**

Time - Speed - Distance 2

Formulas. Change of Units.

Q	A
1) A boy runs 20km in 2 hours. Calculate his **speed**.	1) $S = \dfrac{D}{T} = \dfrac{20km}{2hours} = 10kmph$
2) A girl runs 1,500m at 300m/min. Calculate the **time** it takes her.	2) $T = \dfrac{D}{S} = \dfrac{1500m}{300m/min} = 5$ minutes
3) An object travels at 40cm/s for 4 seconds. Calculate the **distance** it travels.	3) $D = ST$ $= 40cm/s \times 4seconds = 160cm$
4) A cyclist travels 3,721m in 8 minutes. Calculate the cyclist's **speed**.	4) $S = \dfrac{D}{T} = \dfrac{3721m}{8mins} = 465\dfrac{1}{8}$ m/min $= 465 \cdot 125$ m/min
5) An object moves 9·8mm at 7mm/s. Calculate the **time** it takes.	5) $T = \dfrac{D}{S} = \dfrac{9 \cdot 8mm}{7mm/s} = 1 \cdot 4$ seconds
6) A car travels at 70kmh^{-1} for 2·5 hours. Calculate the **distance** it travels.	6) $D = ST$ $= 70kmh^{-1} \times 2 \cdot 5 hours = 175km$

Change of Units

Time - Speed - Distance units may need to be changed as part of a calculation.

Ex.1 A bus travels 45km in 1·5 hours. Calculate the **speed** a) in **kmph**
b) in **m/min**.

a) $S = \dfrac{D}{T} = \dfrac{45km}{1 \cdot 5h} = 30$ **kmph**

b)
$\times 1000$
30,000 **mph**
$\div 60$ min
500 **m/min**

or

$S = \dfrac{D}{T} = \dfrac{45km}{1 \cdot 5h} = \dfrac{45000m}{90min} = 500$ **m/min**
(×1000 / ×60)

Ex.2 An object moves 5m at 25cm/min. Calculate the **time** it takes.

$T = \dfrac{D}{S} = \dfrac{5m}{25cm/min} = \dfrac{500cm}{25cm/min} = 20$ min
(×100)

Ex.3 An aircraft flies at 420kmh^{-1} for 35 minutes. Calculate the **distance** it flies.

$D = ST$
$= 420kmh^{-1} \times \dfrac{35}{60}$ hour $= 245km$

Time - Speed - Distance 3

Formulas. Change of Units.

Q	A
1) A car travels 90km in 2 hours. Calculate the **speed** a) in **kmph** b) in **m/min**.	1) a) $S = \dfrac{D}{T} = \dfrac{90km}{2h} = 45kmph$ b) $\xrightarrow{\times 1000}$ 45,000mph $\xrightarrow{\div 60min}$ 750m/min or $S = \dfrac{D}{T} = \dfrac{90km}{2h} \xrightarrow{\times 1000}_{\times 60} \dfrac{90000m}{120min} = 750$m/min
2) An object moves 77cm at 11mm/s. Calculate the **time** it takes.	2) $T = \dfrac{D}{S} = \dfrac{77cm}{11mm/s} \xrightarrow{\times 10} \dfrac{770mm}{11mm/s} = 70$ sec
3) An object travels at 80cm/s for 3 minutes. Calculate the **distance** it travels, in metres.	3) $D = ST$ 3min × 60 ÷1000 $= 80$cm/s × 180s $= 14400$cm $= 14{\cdot}4$m
4) A girl runs 100m in 20s. Calculate her **speed** a) in **mps** b) in **kmph**.	4) a) $S = \dfrac{D}{T} = \dfrac{100m}{20s} = 5$mps b) (How far will she run in **1 hour**?) ×60sec → 300m in 1 min ×60min → 18000m in 1 hour ÷1000 = **18kmph**
5) A man walks 900m at 5km/h. Calculate the **time** it takes, in minutes.	5) $T = \dfrac{D}{S} = \dfrac{900m}{5km/h} \xrightarrow{\div 1000} \dfrac{0{\cdot}9km}{5km/h} = 0{\cdot}18$h ×60 min → 10·8 min
6) An aircraft flies at 240kmh⁻¹ for 50 minutes. Calculate the **distance** it flies.	6) $D = ST$ $= 240$kmh$^{-1} \times \dfrac{50}{60}$ hour $= 200$km
7) A train travels 600km in 4 hours. Calculate the **speed** a) in **kmph** b) in **m/min**.	7) a) $S = \dfrac{D}{T} = \dfrac{600km}{4h} = 150$kmph b) ×1000 → 150,000mph ÷60min → 2,500m/min or $S = \dfrac{D}{T} = \dfrac{600km}{4h} \xrightarrow{\times 1000}_{\times 60} \dfrac{600,000m}{240min} = 2,500$m/min
8) An object moves 3m at 5cm/s. Calculate the **time** it takes.	8) $T = \dfrac{D}{S} = \dfrac{3m}{5cm/s} \xrightarrow{\times 100} \dfrac{300cm}{5cm/s} = 60$ sec
9) An cyclist rides at 550m/min for 2 hours. Calculate the **distance** travelled, in km.	9) $D = ST$ 2hrs × 60 ÷1000 $= 550$m/min × 120min $= 66,000$m $= 66$km

Time - Speed - Distance 4

Average Speed.

Average Speed

Here, Average Speed is referred to as the average of 2 speeds which are **noticeably different**.

$$\text{Average Speed} = \frac{\text{Total Distance}}{\text{Total Time}}$$

$$S = \frac{D}{T}$$

*Note
The **speed** formula is still used.

Ex.1 A car travels 60km from A to B in 2 hours, then 50km from B to C in 1 hour. Calculate

a) the speed from A to B.

a) $S = \dfrac{D}{T} = \dfrac{60\text{km}}{2\text{hours}} = 30\text{kmph}$

b) the speed from B to C.

b) $S = \dfrac{D}{T} = \dfrac{50\text{km}}{1\text{hour}} = 50\text{kmph}$

c) the **average speed** from A to C.

c) $S = \dfrac{D}{T} = \dfrac{60\text{km} + 50\text{km}}{2\text{h} + 1\text{h}} = \dfrac{110\text{km}}{3\text{h}} = 36\dfrac{2}{3}\text{kmph}$
$= 36.666...\text{ kmph}$

*Note
Since **more time** was spent travelling at a **lower speed**, the average speed is nearer the lower speed.

The Distance and / or the Time may have to be calculated before the speed formula is used.

Ex.2 A plane flies from X to Y at 500kmph for 4h, then from Y to Z at 400kmph for 5h.
Calculate its average speed from X to Z.

For each speed, the **time** is given - the **distance** is needed.

the **distance** from X to Y.

$D = ST$
$= 500\text{kmph} \times 4\text{h}$
$= 2,000\text{km}$

the **distance** from Y to Z.

$D = ST$
$= 400\text{kmph} \times 5\text{h}$
$= 2,000\text{km}$

$\text{Average Speed} = \dfrac{\text{Total Distance}}{\text{Total Time}}$

$= \dfrac{2000\text{km} + 2000\text{km}}{4\text{h} + 5\text{h}}$

$= \dfrac{4000\text{km}}{9\text{h}} = 444\dfrac{4}{9}\text{ kmph}$

$= 444.444...\text{ kmph}$

Ex.3 A boat sails from F to G, 80km at 40km/h, then from G to H, 75km at 25km/h.
Calculate its average speed.

For each speed, the **distance** is given - the **time** is needed.

the **time** from F to G. $T = \dfrac{D}{S} = \dfrac{80\text{km}}{40\text{km/h}} = 2\text{ hours}$

the **time** from G to H. $T = \dfrac{D}{S} = \dfrac{75\text{km}}{25\text{km/h}} = 3\text{ hours}$

$\text{Average Speed} = \dfrac{\text{Total Distance}}{\text{Total Time}}$

$= \dfrac{80\text{km} + 75\text{km}}{2\text{h} + 3\text{h}} = \dfrac{155\text{km}}{5\text{h}} = 31\text{km/h}$

Time - Speed - Distance 5

Average Speed.

Q	A
1) A girl walks 10km from X to Y in 2 hours, then 18km from Y to Z in 3 hours. Calculate a) her speed from X to Y. b) her speed from Y to Z. c) her **average speed** from X to Z.	1) a) $S = \dfrac{D}{T} = \dfrac{10km}{2h} = 5kmph$ b) $S = \dfrac{D}{T} = \dfrac{18km}{3h} = 6kmph$ c) $S = \dfrac{D}{T} = \dfrac{10km + 18km}{2h + 3h} = \dfrac{28km}{5h} = 5\dfrac{3}{5}kmph$ $= 5\cdot 6kmph$ *Note Since **more time** was spent travelling at a **higher speed**, the average speed is nearer the higher speed.
2) A cyclist rides 40km from G to H in 2 hrs, then 36km from H to K in 2 hrs. Calculate a) the speed from G to H. b) the speed from H to K. c) the **average speed** from G to K.	2) a) $S = \dfrac{D}{T} = \dfrac{40km}{2h} = 20kmh^{-1}$ b) $S = \dfrac{D}{T} = \dfrac{36km}{2h} = 18kmh^{-1}$ c) $S = \dfrac{D}{T} = \dfrac{40km + 36km}{2h + 2h} = \dfrac{76km}{4h} = 19kmh^{-1}$ *Note Since the **same time** was spent travelling at **both speeds**, the average speed is 'in the middle' of both speeds.
3) A boy runs 900m from M to N in 3 mins, then 900m from N to P in 6 mins. Calculate a) his speed from M to N. b) his speed from N to P. c) his **average speed** from M to P.	3) a) $S = \dfrac{D}{T} = \dfrac{900m}{3min} = 300m/min$ b) $S = \dfrac{D}{T} = \dfrac{900m}{6min} = 150m/min$ c) $S = \dfrac{D}{T} = \dfrac{900m + 900m}{3min + 6min} = \dfrac{1800m}{9min} = 200m/min$
4) A train travels 180km from A to B in 1·5 hrs, then 100km from B to C in 2hrs. Calculate a) the speed from A to B. b) the speed from B to C. c) the **average speed** from A to C.	4) a) $S = \dfrac{D}{T} = \dfrac{180km}{1\cdot 5hrs} = 120km/h$ b) $S = \dfrac{D}{T} = \dfrac{100km}{2hrs} = 50km/h$ c) $S = \dfrac{D}{T} = \dfrac{180km + 100km}{1\cdot 5hrs + 2hrs} = \dfrac{280km}{3\cdot 5hrs} = 80km/h$

Time - Speed - Distance 6

Average Speed.

Q	A
5) An object moves at 8cm/s for 5s, then at 20cm/s for 15s. Calculate its average speed.	5) For each speed, the **distance** moved is needed. $$D = ST \qquad\qquad D = ST$$ $$= 8\text{cm/s} \times 5\text{s} \qquad = 20\text{cm/s} \times 15\text{s}$$ $$= 40\text{cm} \qquad\qquad = 300\text{cm}$$ Average Speed $= \dfrac{\text{Total Distance}}{\text{Total Time}}$ $$= \dfrac{40\text{cm} + 300\text{cm}}{5\text{s} + 15\text{s}} = \dfrac{340\text{cm}}{20\text{s}} = 17\text{cm/s}$$
6) A boat sails 60km at 15km/h, then 80km at 20km/h. Calculate its average speed.	6) For each speed, the **time** sailed is needed. $$T = \dfrac{D}{S} = \dfrac{60\text{km}}{15\text{km/h}} = 4 \text{ hours}$$ $$T = \dfrac{D}{S} = \dfrac{80\text{km}}{20\text{km/h}} = 4 \text{ hours}$$ Average Speed $= \dfrac{\text{Total Distance}}{\text{Total Time}}$ $$= \dfrac{60\text{km} + 80\text{km}}{4\text{h} + 4\text{h}} = \dfrac{140\text{km}}{8\text{h}} = 17 \cdot 5\text{km/h}$$
7) A plane flies at 700kmph for 2h, then at 550kmph for 4h. Calculate its average speed.	7) For each speed, the **distance** flown is needed. $$D = ST \qquad\qquad D = ST$$ $$= 700\text{kmph} \times 2\text{h} \qquad = 550\text{kmph} \times 4\text{h}$$ $$= 1400\text{km} \qquad\qquad = 2200\text{km}$$ Average Speed $= \dfrac{\text{Total Distance}}{\text{Total Time}}$ $$= \dfrac{1400\text{km} + 2200\text{km}}{2\text{h} + 4\text{h}}$$ $$= \dfrac{3600\text{km}}{6\text{h}} = 600\text{kmph}$$
8) An athlete runs 90m at 9m/s, then 120m at 8m/s. Calculate the athlete's average speed.	8) For each speed, the **time** travelled is needed. $$T = \dfrac{D}{S} = \dfrac{90\text{m}}{9\text{m/s}} = 10\text{s}$$ $$T = \dfrac{D}{S} = \dfrac{120\text{m}}{8\text{m/s}} = 15\text{s}$$ Average Speed $= \dfrac{\text{Total Distance}}{\text{Total Time}}$ $$= \dfrac{90\text{m} + 120\text{m}}{10\text{s} + 15\text{s}} = \dfrac{210\text{m}}{25\text{s}} = 8\dfrac{10}{25} \text{ m/s}$$ $$= 8 \cdot 4\text{m/s}$$
9) A bus journey from A to B takes 3 hours at 80km/h. The bus makes the return journey from B to A at 48km/h. Calculate the average speed of the bus over the 2 journeys together.	9) For A to B, the **distance** is needed. $$D = ST$$ $$= 80\text{kmph} \times 3\text{h} = 240\text{km}$$ For B to A, (AB=BA), the **time** is needed. $$T = \dfrac{D}{S} = \dfrac{240\text{km}}{48\text{km/h}} = 5 \text{ hours}$$ Average Speed $= \dfrac{\text{Total Distance}}{\text{Total Time}}$ $$= \dfrac{240\text{km} + 240\text{km}}{3\text{h} + 5\text{h}} = \dfrac{480\text{km}}{8\text{h}} = 60\text{km/h}$$

Velocity - Time - Acceleration 1 Formulas.

Velocity - Time - Acceleration

Velocity is the **speed** (of an object) in 1 unit of Time **in a particular direction**.
The units of Time are usually seconds (sec, s), minutes (min, m), hours (hr, h).
(Here, for simplicity, the **direction** is not important and so not considered.)
Acceleration is the **rate of change of Velocity** with respect to Time.
Negative Acceleration is **Deceleration**.
The Metric units of Distance (length) are mm, cm, m, km.
*Note Velocity and acceleration are accepted as **average velocity** and **average acceleration**.

Formulas

Velocity - Time - Acceleration can be presented in terms of formulas.

$$\text{Acceleration} = \frac{\text{Change in Velocity}}{\text{Time}}$$

$$\text{Acc} = \frac{\text{Ch.V}}{\text{T}}$$

'swop'

$$\text{Acceleration} = \frac{\text{Change in Velocity}}{\text{Time}}$$

$$\text{Time} = \frac{\text{Change in Velocity}}{\text{Acceleration}}$$

$$\text{T} = \frac{\text{Ch.V}}{\text{Acc}}$$

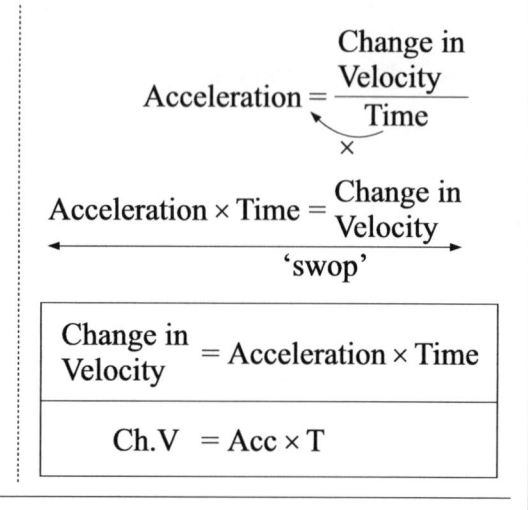

$$\text{Acceleration} = \frac{\text{Change in Velocity}}{\text{Time}}$$

$$\text{Acceleration} \times \text{Time} = \text{Change in Velocity}$$

'swop'

$$\frac{\text{Change in Velocity}}{} = \text{Acceleration} \times \text{Time}$$

$$\text{Ch.V} = \text{Acc} \times \text{T}$$

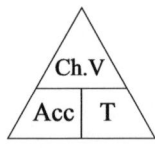

The triangle is useful to help remember the 3 formulas ...

In the triangle, in turn ...
choose the term on the **left** of each formula -
this leaves the 2 terms on the **right** of each formula AND in their correct 'position'.

Triangle: Ch.V on top, Acc and T on bottom.

Ex.1
The change in velocity of an object is 8m per second in 2 seconds. Calculate the **acceleration**.

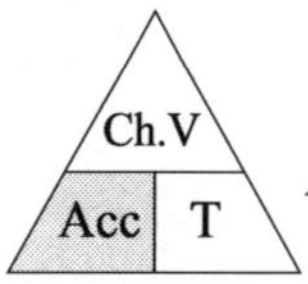

$$\text{Acc} = \frac{\text{Ch.V}}{\text{T}} = \frac{8\text{m/s}}{2\text{s}} = \textbf{4m per second per second}$$
$$= \textbf{4m/s per second}$$
$$= \textbf{4m/s}^2$$
$$= \textbf{4ms}^{-2}$$

any of these terms may be used

The units must agree - ex. **s**

'**4m per second per second**'
means that the velocity increases by **4m per second every second**.

For example, if the velocity is 15mpsec, it increases to 15+4 = 19mpsec in 1 second, then to 19+4 = 23mpsec in the next second.

*Note $\frac{8\text{m/s}}{2\text{s}} = \frac{8\text{m}}{\text{s}} \times \frac{1}{2\text{s}} = \frac{8\text{m}}{2\text{s}^2} = \frac{4\text{m}}{\text{s}^2} = 4\text{ms}^{-2}$

same

(see **Indices / Powers 8**)

Ex.2 The change in velocity of an object is 30m/s.
The acceleration is 5m/s^2.
Calculate the **time** it takes.

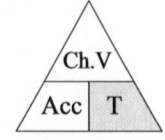

$$\text{T} = \frac{\text{Ch.V}}{\text{Acc}} = \frac{30\text{m/s}}{5\text{m/s}^2} = 6\text{s}$$

The units must agree - ex. **m/s**

Ex.3 The acceleration of an object is 10m/s^2 for 7 seconds.
Calculate the **change in velocity**.

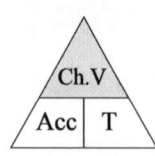

$$\text{Ch.V} = \text{Acc} \times \text{T}$$
$$= 10\text{m/s}^2 \times 7\text{s}$$
$$= 70\text{m/s}$$

The units must agree - ex. **s**

Velocity - Time - Acceleration 2

Formulas.

Q	A
1) The change in velocity of an object is 40m per second in 5 seconds. Calculate the **acceleration**.	1) $Acc = \dfrac{Ch.V}{T} = \dfrac{40m/s}{5s} = 8m/s^2$
2) The change in velocity of an object is 60m/s. The acceleration is 20m/s². Calculate the **time** it takes.	2) $T = \dfrac{Ch.V}{Acc} = \dfrac{60m/s}{20m/s^2} = 3s$
3) The acceleration of an object is 9cm/s² for 4 seconds. Calculate the **change in velocity**.	3) $Ch.V = Acc \times T$ $= 9cm/s^2 \times 4s$ $= 36cm/s$
4) A car changes its velocity from 10ms⁻¹ to 40 ms⁻¹ in 24 seconds. Calculate the car's **acceleration**.	4) $Ch.V = 40 - 10 = 30ms^{-1}$ $Acc = \dfrac{Ch.V}{T} = \dfrac{30ms^{-1}}{24s} = 1\dfrac{6}{24} = 1\dfrac{1}{4} ms^{-2}$
5) The change in velocity of a plane is 90km/h. The acceleration is 45km/h². Calculate the **time** it takes.	5) $T = \dfrac{Ch.V}{Acc} = \dfrac{90km/h}{45km/h^2} = 2h$
6) A cyclist accelerates at 50m/min² for 7 mins. Calculate the **change in velocity**.	6) $Ch.V = Acc \times T$ $= 50m/min^2 \times 7mins$ $= 350m/min$
7) A car's velocity changes by 24m/s in 0·2 minutes. Calculate the car's **acceleration** in m/s².	7) $0{\cdot}2mins = 0{\cdot}2 \times 60secs = 12secs$ $Acc = \dfrac{Ch.V}{T} = \dfrac{24m/s}{12s} = 2m/s^2$
8) The change in velocity of a plane is 3km/min. The acceleration is 360km/h². Calculate the **time** it takes in hours.	8) $Ch.V = 3km/min \times 60mins = 180km/hr$ $T = \dfrac{Ch.V}{Acc} = \dfrac{180km/h}{360km/h^2} = \dfrac{1}{2} hr$
9) A cyclist accelerates at 600m/min² for 45 secs. Calculate the **change in velocity** in m/min.	9) $45secs = \dfrac{45}{60} = \dfrac{3}{4} = 0{\cdot}75mins$ $Ch.V = Acc \times T$ $= 600m/min^2 \times 0{\cdot}75mins$ $= 450m/min$
10) A car changes its velocity from 40ms⁻¹ to 20 ms⁻¹ in 5 seconds. Calculate the car's **acceleration**.	10) $Ch.V = 20 - 40 = -20ms^{-1}$ $Acc = \dfrac{Ch.V}{T} = \dfrac{-20ms^{-1}}{5s} = -4ms^{-2}$ (= **deceleration**)
11) The acceleration of an object is -7cm/s² for 8 seconds. Calculate the **change in velocity**.	11) $Ch.V = Acc \times T$ $= -7cm/s^2 \times 8s$ $= -56cm/s$ (= **slowing down**)

Velocity - Time - Acceleration 3

Kinematics Formulas.

Kinematics Formulas

The following 3 Velocity - Time - Acceleration Formulas are referred to as the **kinematics formulas**.

$v = u + at$

$s = ut + \dfrac{1}{2}at^2$

$v^2 = u^2 + 2as$

a is constant acceleration
u is initial velocity
v is final velocity
s is displacement (distance) from the position when t = 0
t is time

Q	A
1) $v = u + at$ Calculate v when $u = 5$m/s, $a = 3$m/s^2, $t = 6$s	1) $v = u + at$ $v = 5 + 3\times 6$ $v = 5 + 18$ $v = 23$m/s
2) $v = u + at$ Calculate v when $u = 7$m/s, $a = -2$m/s^2, $t = 5$s	2) $v = u + at$ $v = 7 + -2\times 5$ $v = 7 + -10$ $v = -3$m/s
3) $s = ut + \dfrac{1}{2}at^2$ Calculate s when $u = 0$cm/s, $a = 9$cm/s^2, $t = 4$s	3) $s = ut + \dfrac{1}{2}at^2$ $s = 0\times 4 + \dfrac{1}{2}\times 9\times 4^2$ $s = 0 + \dfrac{1}{2}\times 9\times 16$ $s = 0 + 72$ $s = 72$cm
4) $v^2 = u^2 + 2as$ Calculate v when $u = 6$m/s, $a = 4$m/s^2, $s = 8$m	4) $v^2 = u^2 + 2as$ $v^2 = 6^2 + 2\times 4\times 8$ $v^2 = 36 + 64 = 100$ $v = \sqrt{100}$ $v = 10$m/s
5) Calculate the final velocity of a particle which accelerates from 7cm/s at 5cm/s^2 for 2s.	5) $v = u + at$ $v = 7 + 5\times 2$ $v = 7 + 10$ $v = 17$cm/s
6) Calculate the distance travelled by a car which starts from rest and accelerates at 3m/s^2 for 10s.	6) $s = ut + \dfrac{1}{2}at^2$ $s = 0\times 10 + \dfrac{1}{2}\times 3\times 10^2$ $s = 0 + \dfrac{1}{2}\times 3\times 100$ $s = 0 + 150$ $s = 150$m
7) Calculate the final velocity of a car travelling at 20m/s which accelerates at 5m/s^2 for 50m.	7) $v^2 = u^2 + 2as$ $v^2 = 20^2 + 2\times 5\times 50$ $v^2 = 400 + 500 = 900$ $v = \sqrt{900}$ $v = 30$m/s

Calculator

Contents

Page	Title		Details
	Calculator		
	Introduction		
	Calculator -		
1	**Direct Input**	1	Direction. Shift. Mode. Clear. Error. +/−.
2		2	Addition. Subtraction. Multiplication. Division.
3		3	Prime Factors. Indices / Powers.
4		4	Roots. Percentages. Factorial.
5		5	Fractions: Input. To Decimal. Equivalent. Improper / Mixed Number.
6		6	+, −, ×.
7		7	÷. Decimal to Fraction.
8		8	Answer Key.
			Combining Operations. Brackets / = Sign.
9		9	
10		10	
11		11	Indices / Powers and Roots.
12		12	π (Pi). Reciprocal. Random Numbers. Trigonometry.
13		13	Standard Form to/from Number.
14		14	Calculations. Multiplication.
	Calculator -		
15	**Indirect Input**	1	Shift. Clear. Error. +/−. Memory.
16		2	Addition. Subtraction. Multiplication. Division.
17		3	Constant. Indices / Powers.
18		4	Roots. Percentages. Factorial.
19		5	Fractions: Input. To/from Decimal. Equivalent. Improper / Mixed Number.
20		6	+, −, ×, ÷.
21		7	Combining Operations. Brackets / = Sign.
22		8	
23		9	Indices / Powers and Roots.
24		10	Standard Form to/from Number. Calculations.
25		11	π (Pi). Reciprocal. Random Numbers. Trigonometry.

Calculator - Introduction   Direct / Indirect Input Calculator.

<u>Calculator</u>
This section provides a guide to using the calculator.

There are generally 2 types of calculator -

<u>'Direct Input'</u>
This requires 'input' in the **order as it appears on the page**.
A calculation with its numbers and operations are shown together on the screen as they are 'input'.
It is more 'sophisticated' than the 'Indirect Input' calculator.

<u>'Indirect Input'</u>
This requires 'input' in the **order which is not always as it appears on the page**.
A number on the screen disappears when followed by an operation / another number of a calculation.
It is less 'sophisticated' than the 'Direct Input' calculator and operations are easier to generalise.

Here, the main calculations required for the course are generalised as well as possible -
but each depends on the make / model.
The maker's 'user manual' should be studied accordingly!

Generally, when using a calculator in Maths...

It is useful to practice with smaller numbers - that can be checked **mentally** - to gain confidence.
It is better to work in 'steps' - writing answers down - rather than rely on the calculator to do 'too much'.
It is also useful to carry out calculations at least twice in order to avoid errors -
and even in a different order when possible.

The 'Display' on a calculator is usually up to 10 digits.
A calculation may not show all the digits of a calculation - but involve some degree of 'rounding'.
Very large / small numbers may automatically be shown in **Standard Form**.

<u>*Note</u>
Here, for clarity, a large / infinite decimal display is indicated with a few digits and 3 dots, ex. 0.123...

Calculator - Direct Input 1

Direction. Shift. Mode. Clear. Error. +/−.

'Direct Input' Calculator

Direction

When required ... A '4-directions' key allows the **screen cursor** to be moved to allow for input and changes (delete / insert).

This is shown as 4 individual arrows to indicate the direction required. △ ▷ ▽ ◁

Shift / Inv / 2nd F

Depending on calculator ...

- shift
- Inv 'Inverse'
- 2nd F '2nd Function'

... this key is used to allow some keys to perform a 2nd task - written **over** the keys.

Here, such keys are shown blank ☐ since they can differ between calculators.

s-I-2 Here, this symbol represents each of the 3 keys.

Calculator Mode / Set Up

s-I-2   MODE / SET UP ☐

Screen: Natural Display / Degrees

choose 'Natural Display' shows input / results 'as it appears on the page'.

choose 'Degrees' for angle calculations (Trigonometry).

Clear

Press Key Screen

C ☐

At the end of a calculation, a new one may be started without clearing the screen.

Clears the screen ready for a new calculation.

Clear Entry / Error

5 + 2

 error
5 + 3 DEL 2 =   Screen: 7

Clears a **digit / symbol / operation**.

6 + 19

 error error
6 − 9 DEL DEL + 1 9 =   25

Clear / Change / Insert

1 + 3

 clear
1 + 2 3 ◁ DEL =   4

Clear / Change / Insert **digit / symbol / operation**.

25 × 4

 change
2 7 × 4 ◁ ◁ DEL 5 =   100

96 ÷ 8

 insert
9 6 8 ◁ ÷ =   12

The screen cursor may be moved back to the end of the calculation before pressing the = sign or to extend the calculation shown.

Error

Screen: Error

Change Positive / Negative Answer

7 to -7   7 (−) =   -7

-4 to 4   -4 (−) =   4

A calculation may produce a positive or negative number which can be changed in order to continue a calculation.

Calculator - Direct Input 2   Addition. Subtraction. Multiplication. Division.

Addition

4·6 + 3   [4][·][6][+][3][=]   [7.6]   *Note The decimal point is usually at the base of the numbers.

-8 + 9   [−][8][+][9][=]   [1]

7 + -5   [7][+][−][5][=]   [2]

-2 + -1   [−][2][+][−][1][=]   [-3]

Subtraction

17 − 3   [1][7][−][3][=]   [14]

2 − 8   [2][−][8][=]   [-6]

-1 − 6   [−][1][−][6][=]   [-7]

-4 − -5   [−][4][−][−][5][=]   [1]

Multiplication

8 × 11   [8][×][1][1][=]   [88]

1·3 × 0·2   [1][·][3][×][0][·][2][=]   [0.26]

7 × -5   [7][×][−][5][=]   [-35]

-9 × -6   [−][9][×][−][6][=]   [54]

Division

*Note When a decimal answer occurs, the fraction form (in lowest terms) is displayed first.

To change a **fraction display** to a decimal, press [F⇔D] (or similar key).
(This does not work for all fractions and often only provides an answer in Standard Form!)

To change the **same decimal** back to the **fraction display**, press [F⇔D] again.

This is only shown in the first example.

9 ÷ 4 $\frac{9}{4}$ [9][÷][4][=] $\frac{9}{4}$ [F⇔D] [2.25] [F⇔D] $\frac{9}{4}$

1 ÷ 3 $\frac{1}{3}$ [1][÷][3][=] $\frac{1}{3}$ [F⇔D] [0.333333...] or [$0.\dot{3}$]   3 recurs infinitely

2 ÷ 7 $\frac{2}{7}$ [2][÷][7][=] $\frac{2}{7}$ [F⇔D] [0.285714...] or [$0.\dot{2}8571\dot{4}$]   285714 recurs infinitely

60 ÷ 20 $\frac{60}{20}$ [6][0][÷][2][0][=]   [3]

5 ÷ 0 $\frac{5}{0}$ [5][÷][0][=]   [Error]   Division by 0 is **not allowed**.

-8 ÷ 2 $\frac{-8}{2}$ [−][8][÷][2][=]   [-4]

-9 ÷ -3 $\frac{-9}{-3}$ [−][9][÷][−][3][=]   [3]

Calculator - Direct Input 3

Prime Factors. Indices / Powers.

Prime Factors

Number	Input		Result
6	[6] [=]	[6] [s-I-2]^FACT	[2×3]
8	[8] [=]	[8] [s-I-2]^FACT	[2^3]
45	[4][5] [=]	[45] [s-I-2]^FACT	[$3^2 × 5$]

Indices / Powers

3^2 ⎫   [3] [x^2] [=]   [9]

3^2 ⎭   [3] [$x^\square$] [2] [=]   [9]

2^5 [2] [$x^\square$] [5] [=]   [32]

$(-4)^2$ [(] [−] [4] [)] [x^2] [=]   [16]   = -4 × -4   [-4 is squared]

-4^2 [−] [4] [x^2] [=]   [-16]   = -(4 × 4) = -(16)   [4 is squared]

$(-4)^3$ [(] [−] [4] [)] [$x^\square$] [3] [=]   [-64]

$1·2^5$ [1] [·] [2] [$x^\square$] [5] [=]   [2.48832]

$1·234^5$ [1] [·] [2] [3] [4] [$x^\square$] [5] [=]   [2.861381721]   The number is 'rounded' -
                                                         it should have 15 decimal places.

100^5 [1] [0] [0] [$x^\square$] [5] [=]   [$1 × 10^{10}$]   = 10,000,000,000
                                                    In Standard Form - the decimal point
                                                    is moved 10 places to the **right**.

88^8 [8] [8] [$x^\square$] [8] [=]   [$3.596... × 10^{15}$]   The number is 'rounded'.
                                                         In Standard Form - the decimal point
                                                         is moved 15 places to the **right**.

2^{-1} $= \dfrac{1}{2^1}$ [2] [$x^\square$] [−] [1] [=]   [0.5]

5^{-2} $= \dfrac{1}{5^2}$ [5] [$x^\square$] [−] [2] [=]   [0.04]

3^{-4} $= \dfrac{1}{3^4}$ [3] [$x^\square$] [−] [4] [=]   [0.0123...]

$(-4)^{-1}$ $= \dfrac{1}{(-4)^1}$ [(] [−] [4] [)] [$x^\square$] [−] [1] [=]   [-0.25]

$(-2)^{-6}$ $= \dfrac{1}{(-2)^6}$ [(] [−] [2] [)] [$x^\square$] [−] [6] [=]   [0.015625]

Calculator - Direct Input 4

Roots. Percentages. Factorial.

Roots

*Note
For an even number root, ex., $\sqrt[2]{}$, $\sqrt[4]{}$, $\sqrt[6]{}$, ... the answer is $+$ or $-$. Ex. $\sqrt[2]{9} = +3$ or -3.
The calculator only displays the +ve answer.

2 is usually omitted

$\sqrt[2]{9}$ → [√☐] [9] [=] → 3

$\sqrt{2}$ → [√☐] [2] [=] → 1.4142...

$\sqrt[3]{8}$ → [s-I-2] (³√☐) [8] [=] → 2

$\sqrt[3]{8}$ → [3] [s-I-2] (√☐) [8] [=] → 2

$\sqrt[4]{625}$ → [4] [s-I-2] (√☐) [6] [2] [5] [=] → 5

$\sqrt[3]{-8}$ → [3] [s-I-2] (√☐) [−] [8] [=] → -2

$\sqrt[7]{9}$ → [7] [s-I-2] (√☐) [9] [=] → 1.368...

Percentages

$\left(5\% = \frac{5}{100} = 0{\cdot}05 \right)$

5% → [5] [s-I-2] (%) [=] → 0.05   Gives the decimal form of a percentage.

5% of 600 → [5] [s-I-2] (%) [×] [6] [0] [0] [=] → 30

→ [6] [0] [0] [×] [5] [s-I-2] (%) [=] → 30

Increase 600 by 5% → [6] [0] [0] [+] [5] [s-I-2] (%) [×] [6] [0] [0] [=] → 630

Decrease 600 by 5% → [6] [0] [0] [−] [5] [s-I-2] (%) [×] [6] [0] [0] [=] → 570

Factorial

[5] [s-I-2] (x!) or (n!) [=] → 120.

5! is read as '5 **factorial**' and multiplies numbers as shown.

$5! = 5 \times 4 \times 3 \times 2 \times 1 = 120$
$4! = 4 \times 3 \times 2 \times 1 = 24$

Calculator - Direct Input 5

Fractions: Input. To Decimal. Equivalent. Improper / Mixed Number.

Fractions

A fraction answer to a calculation is shown **in its lowest terms**.

Input a Fraction   (ready for further input)

$\frac{1}{2}$ [▤][1][▽][2][▷]······ $\frac{1}{2}$

$3\frac{4}{7}$ [▥][3][▷][4][▽][7][▷] $3\frac{4}{7}$

or ···

$\frac{1}{2}$ [1][▤][2][▷] $\frac{1}{2}$ This method is a little quicker and preferred - it is used in the rest of the examples.

$3\frac{4}{7}$ [3][▥][4][▽][7][▷] $3\frac{4}{7}$

Change a Fraction to a Decimal

change decimal back to same fraction

$\frac{1}{2}$ [1][▤][2][=][F⇔D]   0.5   [F⇔D] $\frac{1}{2}$

$3\frac{4}{7}$ [3][▥][4][▽][7][=][F⇔D]   3.571...   [F⇔D] $3\frac{4}{7}$

Equivalent Fraction - Lowest Terms (Cancel)

$\frac{2}{4}$ [2][▤][4][=] $\frac{1}{2}$

$\frac{8}{6}$ [8][▤][6][=] $\frac{4}{3}$

Change Improper Fraction to Mixed Number

$\frac{9}{2}$ [9][▤][2][=][s-I-2][F⇔D] $4\frac{1}{2}$

$\frac{8}{6}$ [8][▤][6][=] $\frac{4}{3}$ [s-I-2][F⇔D] $1\frac{1}{3}$

Gives lowest terms first.

Change Mixed Number to Improper Fraction

$1\frac{3}{5}$ [1][s-I-2][▥][3][▽][5][=] $\frac{8}{5}$

$5\frac{4}{6}$ [5][s-I-2][▥][4][▽][6][=] $\frac{17}{3}$ ([s-I-2][F⇔D] $5\frac{2}{3}$)

Answer $\frac{34}{6}$ given in lowest terms.

Lowest terms.

Calculator - Direct Input 6

Fractions: $+, -, \times$.

Addition

$\dfrac{1}{2} + \dfrac{1}{2}$ [1][▫/▫][2][▷][+][1][▫/▫][2][=] 1

$\dfrac{2}{7} + \dfrac{4}{7}$ [2][▫/▫][7][▷][+][4][▫/▫][7][=] $\dfrac{6}{7}$

$\dfrac{1}{4} + \dfrac{3}{5}$ [1][▫/▫][4][▷][+][3][▫/▫][5][=] $\dfrac{17}{20}$

$3\dfrac{1}{2} + 5\dfrac{1}{4}$ [3][s-I-2 ▫▫/▫][1][▽][2][▷][+][5][s-I-2 ▫▫/▫][1][▽][4][=] $\dfrac{35}{4}$

                                 [s-I-2][F⇔D] $8\dfrac{3}{4}$

Subtraction

$\dfrac{4}{5} - \dfrac{1}{5}$ [4][▫/▫][5][▷][−][1][▫/▫][5][=] $\dfrac{3}{5}$

$\dfrac{7}{8} - \dfrac{3}{4}$ [7][▫/▫][8][▷][−][3][▫/▫][4][=] $\dfrac{1}{8}$

$7\dfrac{2}{9} - 1\dfrac{5}{6}$ [7][s-I-2 ▫▫/▫][2][▽][9][▷][−][1][s-I-2 ▫▫/▫][5][▽][6][=] $\dfrac{97}{18}$

                                   [s-I-2][F⇔D] $5\dfrac{7}{18}$

$\dfrac{1}{7} - \dfrac{6}{7}$ [1][▫/▫][7][▷][−][6][▫/▫][7][▷][=] $-\dfrac{5}{7}$

Multiplication

$\dfrac{1}{2} \times \dfrac{3}{4}$ [1][▫/▫][2][▷][×][3][▫/▫][4][=] $\dfrac{3}{8}$

$\dfrac{7}{6} \times \dfrac{9}{5}$ [7][▫/▫][6][▷][×][9][▫/▫][5][=] $\dfrac{21}{10}$ Answer $\dfrac{63}{30}$ given in lowest terms.

                                   [s-I-2][F⇔D] $2\dfrac{1}{10}$

$2\dfrac{4}{9} \times 8\dfrac{1}{5}$ [2][s-I-2 ▫▫/▫][4][▽][9][▷][×][8][s-I-2 ▫▫/▫][1][▽][5][=] $\dfrac{902}{45}$

                                   [s-I-2][F⇔D] $20\dfrac{2}{45}$

$-\dfrac{3}{8} \times \dfrac{3}{7}$ or   [−][▫/▫][3][▽] / [8][▷][×][3][▫/▫][7][=] $-\dfrac{9}{56}$

                       [−][3][▫/▫]

Calculator - Direct Input 7

Fractions: ÷ . Decimal to Fraction.

Division

$\dfrac{1}{3} \div 2$ → $\dfrac{\frac{1}{3}}{2}$ → [1][▤][3][▷][÷][2][=] → $\dfrac{1}{6}$

$7 \div \dfrac{4}{9}$ → $\dfrac{7}{\frac{4}{9}}$ → [7][÷][4][▤][9][=] → $\dfrac{63}{4}$ [s-I-2][F⇔D] → $15\dfrac{3}{4}$

$\dfrac{5}{6} \div \dfrac{1}{2}$ → $\dfrac{\frac{5}{6}}{\frac{1}{2}}$ → [5][▤][6][▷][÷][1][▤][2][=] → $\dfrac{5}{3}$ [s-I-2][F⇔D] → $1\dfrac{2}{3}$

$9\dfrac{3}{4} \div 1\dfrac{4}{5}$ → $\dfrac{9\frac{3}{4}}{1\frac{4}{5}}$ → [9][s-I-2][ ][3][▽][4][▷][÷][1][s-I-2][ ][4][▽][5][=] → $\dfrac{65}{12}$

[s-I-2][F⇔D] → $5\dfrac{5}{12}$

$\dfrac{-2}{7} \div \dfrac{-7}{8}$ → $\dfrac{\frac{-2}{7}}{\frac{-7}{8}}$ → [−][▤][2][▽][7][▷][÷][▤][−][7][▽][8][=] → $\dfrac{16}{49}$

Change Decimal to Fraction

changes fraction back to same decimal

0·1 → [0][.][1][= or =,F⇔D] → $\dfrac{1}{10}$ → [F⇔D] → 0.1

0·5 → [0][.][5][= or =,F⇔D] → $\dfrac{1}{2}$

3·7 → [3][.][7][= or =,F⇔D] → $\dfrac{37}{10}$ [s-I-2][F⇔D] → $3\dfrac{7}{10}$

0·99 → [0][.][9][9][= or =,F⇔D] → $\dfrac{99}{100}$

8·25 → [8][.][2][5][= or =,F⇔D] → $\dfrac{33}{4}$ [s-I-2][F⇔D] → $8\dfrac{1}{4}$

Calculator - Direct Input 8

Answer Key. Combining Operations. Brackets / = Sign.

Ans Answer Key

The Answer to the **last calculation** is stored and can be obtained by pressing this key.
This allows the **last answer** ... to be retrieved if cleared ...

$1 + 2$ | 1 | + | 2 | = | 3 | C | Ans | = | Ans 3

... or used in the **next calculation** ...

$1 + 2$ | 1 | + | 2 | = | 3
Continue calculation ...
$+ 4$ | + | 4 | = | 7
Continue calculation ...
$\times 5$ | × | 5 | = | 35
Continue calculation ...
$2 \times$ **last answer**. | 2 | × | Ans | = | 70

Combining Operations
Brackets / = Sign

When combining operations, calculations are carried out in an order of 'importance' -
Brackets, **Indices**, **Division** / **Multiplication**, **Addition** / **Subtraction** (often referred to as **BIDMAS**).

In each of the following examples, the operations are the same or have the same 'importance'.

$1 + 2 + 3$ | 1 | + | 2 | + | 3 | = | 6.
$8 - 4 - 1$ | 8 | − | 4 | − | 1 | = | 3.
$5 + 4 - 2$ | 5 | + | 4 | − | 2 | = | 7.
$3 \times 1 \times 6$ | 3 | × | 1 | × | 6 | = | 18.
$6 \div 3 \div 2$ | 6 | ÷ | 3 | ÷ | 2 | = | 1.
$8 \times 5 \div 4$ | 8 | × | 5 | ÷ | 4 | = | 10.

In each of the following examples, the value of each bracket is calculated separately.

$7 - (3 + 2)$ | 7 | − | (| 3 | + | 2 |) | = | 2.
$= 7 - (\ 5\ )$

$7 - (3 - 2)$ | 7 | − | (| 3 | − | 2 |) | = | 6.
$= 7 - (\ 1\ )$

$(6 + 5) \times (4 + 1)$ | (| 6 | + | 5 |) | × | (| 4 | + | 1 |) | = | 55.
$= (\ 11\ ) \times (\ 5\ )$

$(6 - 5) \times (4 - 1)$ | (| 6 | − | 5 |) | × | (| 4 | − | 1 |) | = | 3.
$= (\ 1\ ) \times (\ 3\ )$

Calculator - Direct Input 9

Combining Operations. Brackets / = Sign.

The = Sign can often be used to complete a calculation as if it was in brackets.

Multiplication **with** Addition / Subtraction

$7 \times 9 + 1$
$= 63 + 1$

| 7 | × | 9 | + | 1 | = | 64. | × is carried out first.

The use of brackets changes ... the order of the 'importance' / the order of calculations.

$7 \times (9 + 1)$
$= 7 \times 10$

| 7 | × | (| 9 | + | 1 |) | = | 70. The value of the brackets ( ) is calculated first, then $7 \times (\ )$.

| 9 | + | 1 | = | × | 7 | = | 70. The 1st.= Sign calculates $9 + 1$ first.

$1 + 2 \times 3$
$= 1 + 6$

| 1 | + | 2 | × | 3 | = | 7. Even though + is first, × is carried out first.

The use of brackets changes ... the order of the 'importance' / the order of calculations.

$(1 + 2) \times 3$
$= 3 \times 3$

| (| 1 | + | 2 |) | × | 3 | = | 9. The value of the brackets ( ) is calculated first, then ×3.

| 1 | + | 2 | = | × | 3 | = | 9. The 1st.= Sign calculates $1 + 2$ first.

$6 \times 5 - 4$
$= 30 - 4$

| 6 | × | 5 | − | 4 | = | 26. × is carried out first.

The use of brackets changes ... the order of the 'importance' / the order of calculations.

$6 \times (5 - 4)$
$= 6 \times 1$

| 6 | × | (| 5 | − | 4 |) | = | 6. The value of the brackets ( ) is calculated first, then $6 \times (\ )$.

| 5 | − | 4 | = | × | 6 | = | 6. The 1st.= Sign calculates $5 - 4$ first.

$9 - 1 \times 2$
$= 9 - 2$

| 9 | − | 1 | × | 2 | = | 7. Even though − is first, × is carried out first.

The use of brackets changes ... the order of the 'importance' / the order of calculations.

$(9 - 1) \times 2$
$= 8 \times 2$

| (| 9 | − | 1 |) | × | 2 | = | 16. The value of the brackets ( ) is calculated first, then ×2.

| 9 | − | 1 | = | × | 2 | = | 16. The 1st.= Sign calculates $9 - 1$ first.

Calculator - Direct Input 10

Combining Operations. Brackets / = Sign.

Multiplication with Addition / Subtraction

The 'Direct Input' calculator should allow the × sign to be omitted when multiplying a bracket ...

$7 \times (9 + 1)$ } $7(9 + 1)$
$= 7 \times 10$ } $= 7(10)$

| 7 | (| 9 | + | 1 |) | = | | 70 |

$6 \times (5 - 4)$ } $6(5 - 4)$
$= 6 \times 1$ } $= 6(1)$

| 6 | (| 5 | − | 4 |) | = | | 6 |

$(8 + 2) \times (3 + 1)$
$= (10) \times (4)$

| (| 8 | + | 2 |) | (| 3 | + | 1 |) | = | | 40 |

$(8 - 2)(3 - 1)$
$= (6)(2)$

| (| 8 | − | 2 |) | (| 3 | − | 1 |) | = | | 12 |

Division with Addition / Subtraction
(... is similar to multiplication ...)

$6 \div 2 + 1$
$= 3 + 1$

| 6 | ÷ | 2 | + | 1 | = | | 4. | ÷ is carried out first.

The use of brackets changes ... the order of the 'importance' / the order of calculations.

$6 \div (2 + 1)$
$= 6 \div 3$

| 6 | ÷ | (| 2 | + | 1 |) | = | | 2. | The value of the brackets ( ) is calculated first, then $6 \div (\ )$.

$\dfrac{8 + 6}{2}$ }

| (| 8 | + | 6 |) | ÷ | 2 | = | | 7. | $\dfrac{8+6}{2}$ ⇐ The ÷ line acts like brackets ( ).

$= \dfrac{14}{2}$ }

| 8 | + | 6 | = | ÷ | 2 | = | | 7. |

$\dfrac{8 + 6}{2}$

| 8 | + | 6 | ÷ or □/□ | 2 | = | | 11. |

$= 8 + 3$

$\dfrac{7 + 3}{9 - 4}$ }

| (| 7 | + | 3 |) | ÷ | (| 9 | − | 4 |) | = | | 2. |

$= \dfrac{10}{5}$ }

| 7 | + | 3 | = | ÷ | (| 9 | − | 4 |) | = | | 2. |

Division with Multiplication

$\dfrac{8}{1 \times 2}$ }

| 8 | ÷ | (| 1 | × | 2 |) | = | | 4. |

| 8 | ÷ | 1 | ÷ | 2 | = | | 4. | 8 is **divided** by 1 AND is **divided** by 2.

$\dfrac{8}{1} \times 2$

| 8 | ÷ | 1 | × | 2 | = | | 16. |

Calculator - Direct Input 11

Combining Operations - Indices / Powers and Roots.

Indices / Powers and Roots

$\sqrt{4^3}$
$4^{\frac{3}{2}}$ }   [√☐] [4] [x^☐] [3] [=] $\boxed{8}$

$(\sqrt{4})^3$
$4^{\frac{3}{2}}$ }   [(] [√☐] [4] [▷] [)] [x^☐] [3] [=] $\boxed{8}$

$4^{\frac{3}{2}}$ [4] [x^☐] [3] [▭/▭] [2] [=] $\boxed{8}$

$4^{\frac{3}{2}} = 4^{1.5}$ [4] [x^☐] [1] [·] [5] [=] $\boxed{8}$

$8^{-\frac{2}{3}} = \dfrac{1}{(\sqrt[3]{8})^2}$ [8] [x^☐] [−] [2] [▭/▭] [3] [=] $\boxed{\dfrac{1}{4}}$

$\sqrt[6]{7^5}$ [6] [s-I-2] [ᵡ√☐] [7] [x^☐] [5] [=] $\boxed{5.061...}$

$\sqrt{3^2 + 4^2}$ [√☐] [3] [x²] [+] [4] [x²] [=] $\boxed{5}$

$\sqrt[3]{2^2 \times 5^2}$ [∛☐] or [∛☐ s-I-2] [2] [x²] [×] [5] [x²] [=] $\boxed{4.641...}$

$\sqrt[5]{5^5 - 3^6}$ [5] [s-I-2] [ᵡ√☐] [5] [x^☐] [5] [▷] [−] [3] [x^☐] [6] [=] $\boxed{4.741...}$

$\dfrac{1}{2^2}$ [1] [▭/▭ or ÷] [2] [x²] [=] $\boxed{\dfrac{1}{4}}$

$\dfrac{1}{90^6}$ [1] [▭/▭ or ÷] [9] [0] [x^☐] [6] [=] $\boxed{1.881...\ 10^{-12}}$ The number is 'rounded'. In Standard Form - the decimal point is moved 12 places to the **left**.

$\dfrac{4^5}{5^4}$ [4] [x^☐] [5] [▷] [▭/▭ or ÷] [5] [x^☐] [4] [=] $\boxed{\dfrac{1024}{625}}$

$\dfrac{\sqrt[7]{9}}{2^7}$ [7] [s-I-2] [ᵡ√☐] [9] [▷] [▭/▭ or ÷] [2] [x^☐] [7] [=] $\boxed{0.0106...}$

$3^{-2} \times 6^4$
$= \dfrac{1}{3^2} \times 6^4$ [3] [x^☐] [−] [2] [▷] [×] [6] [x^☐] [4] [=] $\boxed{144}$

Calculator - Direct Input 12

π (Pi). Reciprocal. Random Numbers. Trigonometry.

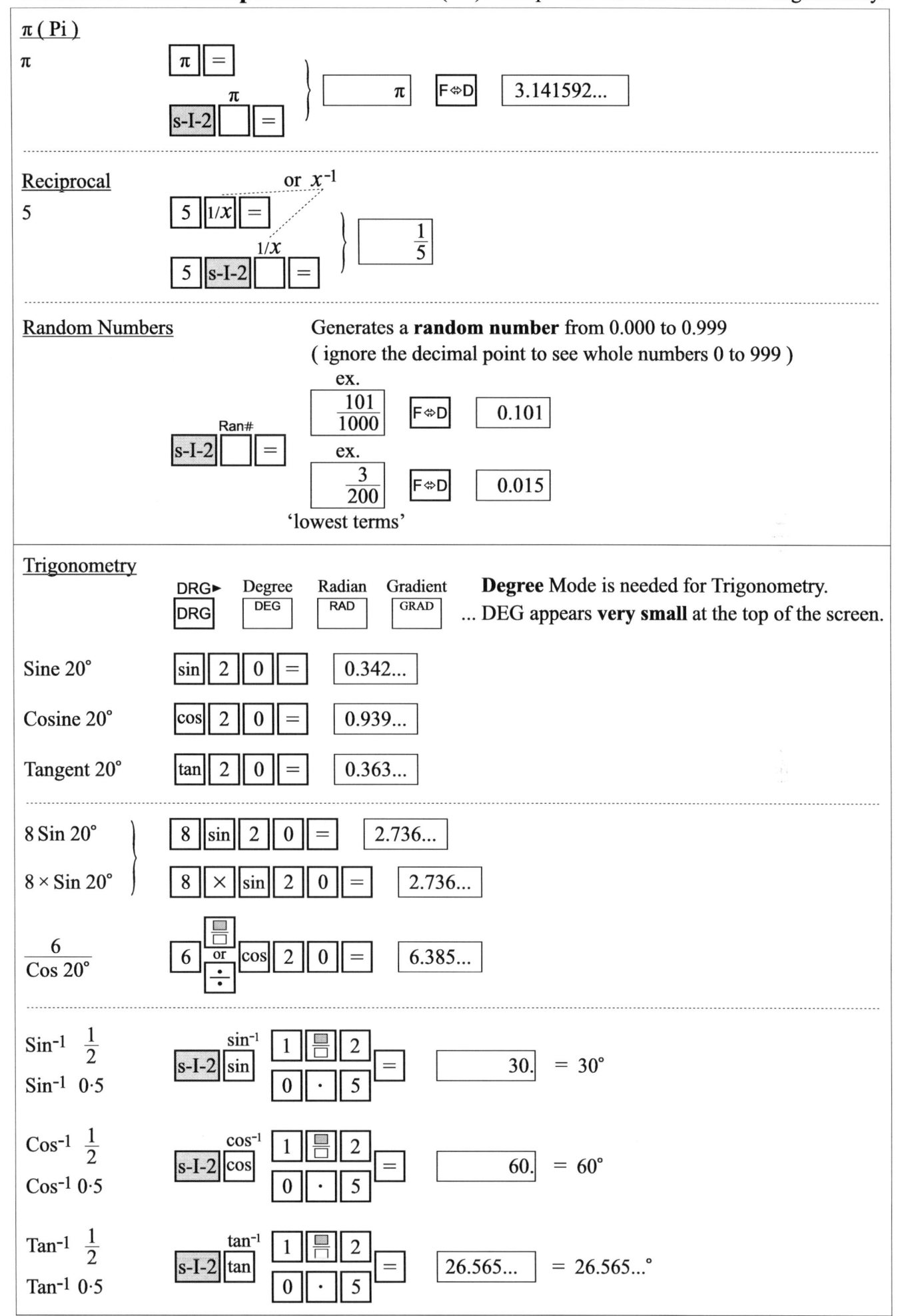

π (Pi)

π

[π][=]
[s-I-2][π][ ][=]   }   [π]   [F⇔D]   [3.141592...]

Reciprocal

5

[5][1/x][=]   or x^{-1}
[5][s-I-2][1/x][ ][=]   }   [$\frac{1}{5}$]

Random Numbers

Generates a **random number** from 0.000 to 0.999
(ignore the decimal point to see whole numbers 0 to 999)

ex.
[s-I-2][Ran#][ ][=]   [$\frac{101}{1000}$]   [F⇔D]   [0.101]

ex.
[$\frac{3}{200}$]   [F⇔D]   [0.015]
'lowest terms'

Trigonometry

[DRG►] Degree Radian Gradient   **Degree** Mode is needed for Trigonometry.
[DRG]   [DEG]   [RAD]   [GRAD]   ... DEG appears **very small** at the top of the screen.

Sine 20°   [sin][2][0][=]   [0.342...]

Cosine 20°   [cos][2][0][=]   [0.939...]

Tangent 20°   [tan][2][0][=]   [0.363...]

8 Sin 20°   [8][sin][2][0][=]   [2.736...]

8 × Sin 20°   [8][×][sin][2][0][=]   [2.736...]

$\frac{6}{\cos 20°}$ [6][▢/▢ or ÷][cos][2][0][=]   [6.385...]

Sin⁻¹ $\frac{1}{2}$
Sin⁻¹ 0·5   [s-I-2][sin⁻¹/sin][1][▢/▢][2] or [0][·][5][=]   [30.]   = 30°

Cos⁻¹ $\frac{1}{2}$
Cos⁻¹ 0·5   [s-I-2][cos⁻¹/cos][1][▢/▢][2] or [0][·][5][=]   [60.]   = 60°

Tan⁻¹ $\frac{1}{2}$
Tan⁻¹ 0·5   [s-I-2][tan⁻¹/tan][1][▢/▢][2] or [0][·][5][=]   [26.565...]   = 26.565...°

Calculator - Direct Input 13

Standard Form to/from Number.

Standard Form

Calculator Mode / Set Up

[s-I-2] MODE / SET UP [ ] → Scientific   choose 'Scientific' gives **answer** in Standard Form.

↓

Sci 0~9   choose 0 allows for infinite decimal. (1-9 significant figures allowed)

*Note
'Scientific Mode' is changed back to 'Natural Display' when not required.
This may involve restoring the calculator to its 'default settings' -

[s-I-2] CLEAR [ ] All   choose 'All'

'Natural Display' may have to be re-selected.

Change Number to Standard Form   (allows $Number \times 10^{-9}$ to $Number \times 10^9$ to appear on screen)

123·4   [1][2][3][·][4][=][F⇔D] 1.234×10^2

0·0097   [0][·][0][0][9][7][=][F⇔D] 9.7×10^{-3}

Change Standard Form to Number   (allows $Number \times 10^{-9}$ to $Number \times 10^9$ to appear on screen)
('Natural Display' needed.)

[s-I-2] CLEAR [ ]   Clear

[s-I-2] MODE / SET UP [ ]   Natural Display

1.234×10^2 [1][·][2][3][4][×10^x][2][=][F⇔D]   123·4

9.7×10^{-2} [9][·][7][×10^x][−][2][=][F⇔D]   0·097

*Note
$Number \times 10^{-9}$ to $Number \times 10^{-3}$ may only provide answer in Standard Form ... and not as a decimal!

9.7×10^{-6} [9][·][7][×10^x][−][6][=][F⇔D] 9.7×10^{-6} (no change !)

Calculator - Direct Input 14

Standard Form - Calculations. Multiplication.

Standard Form
Calculations

Input Number Form - answer in Standard Form.

54×87 [5][4][×][8][7][=] $4 \cdot 698 \times 10^3$

Input Standard Form - answer in Standard Form.

$6 \cdot 8 \times 10^4 + 3 \cdot 079 \times 10^2$

[6][·][8][×10^x][4][+][3][·][0][7][9][×10^x][2][=] [F⇔D] 6.83079×10^4

$1 \cdot 4 \times 10^3 - 1 \cdot 4 \times 10^{-3}$

[1][·][4][×10^x][3][−][1][·][4][×10^x][−][3][=] 1.3999986×10^3

$3 \cdot 5 \times 10^6 \times 2 \cdot 2 \times 10^9$

[3][·][5][×10^x][6][×][2][·][2][×10^x][9][=] 7.7×10^{15}

$5 \cdot 7 \times 10^{-1} \div 9 \cdot 5 \times 10^8$

[5][·][7][×10^x][−][1][÷][9][·][5][×10^x][8][=] 6×10^{-10}

$(4 \cdot 8 \times 10^5)^2$

[(][4][·][8][×10^x][5][)][x^2][=] 2.304×10^{11}

$\sqrt[5]{1 \cdot 6 \times 10^4}$

[5][s-I-2][$\sqrt[\square]{\square}$][1][·][6][×10^x][4][=] $6.931... \times 10^0$

Multiplication

The multiplication sign may be omitted in several calculations.

$4(3 - 1)$ [4][(][3][−][1][)][=]   8
$= 4(\ 2 \ )$

$(1 + 2)(3 + 4)$ [(][1][+][2][)][(][3][+][4][)][=]   21
$= (\ 3 \ )(\ 7 \ )$

9π [9][π][=]   28.274...

$5\sqrt{4}$ [5][$\sqrt{\square}$][4][=]   10

$6\sin 7°$ [6][sin][7][=]   0.731...

Calculator - Indirect Input 1

Shift. Clear. Error. +/−. Memory.

'Indirect Input' Calculator

Shift / Inv / 2nd F

Depending on calculator ... |shift| ... this key is used to allow some keys to perform a 2nd task - written **over** the keys.

|Inv| 'Inverse'

|2nd F| '**2nd** Function'

Here, such keys are shown blank □ since they can differ between calculators.

⇩

|s-I-2| Here, this symbol represents each of the 3 keys.

Clear

	Press Key	Screen	
	|C|	|0.|	

At the end of a calculation, a new one may be started without clearing the screen.

Clears the screen ready for a new calculation.

Clear Entry / Error

5 + 2 |5||+||3| error |CE||2||=| |7.| Clears the last **number** (1,2,3, ... digits) input.

6 + 19 |6||+||1||4| error |CE||1||9||=| |25.| *Note The decimal point is usually at the base of the numbers.

8 + 1 |8||−| error |CE||+||1||=| |9.| **May** clear the last +,−,×,÷ operation input.

Error

Screen |E   0.| E is an Error. 0. appears also but it is **not** the answer.

Change Positive / Negative

7 to -7 |7.||+/−| |-7.| 1) A positive number is made negative and vice versa.

-4 to 4 |-4.||+/−| |4.| 2) A calculation may produce a positive or negative number which can be changed in order to continue a calculation.

Memory

9 |9| |9.| |x→M| |M   9.| Transfers the number on screen to Memory. **M** appears on screen when the Memory is in use.

1 + 2 |1||+||2||=| |M   3.| Other calculations may be carried out before MR is used.

         |MR| |M   9.| 'Memory Recall' displays the number in Memory. 9 is still in the Memory.

5 + MR |5||+||MR||=| |M   14.| |x→M| |M   14.| 14 replaces 9 in the Memory.

Clear Memory |C|

4 × cos 8° |4||×||8||cos||=| |3.961...| |x→M| |M   3.961...|

Calculator - Indirect Input 2

Addition. Subtraction. Multiplication. Division.

Addition

5 + 2	[5][+][2][=]	[7.]
4·6 + 3	[4][·][6][+][3][=]	[7.6]
-8 + 9	[8][+/−][+][9][=]	[1.]
7 + -5	[7][+][5][+/−][=]	[2.]
4 + -9	[4][+][9][+/−][=]	[-5.]
-2 + -1	[2][+/−][+][1][+/−][=]	[-3.]

Subtraction

17 − 3	[1][7][−][3][=]	[14.]
9·5 − 6·4	[9][·][5][−][6][·][4][=]	[3.1]
2 − 8	[2][−][8][=]	[-6.]
-1 − 6	[1][+/−][−][6][=]	[-7.]
-4 − -5	[4][+/−][−][5][+/−][=]	[1.]
-7 − -6	[7][+/−][−][6][+/−][=]	[-1.]

Multiplication

8 × 11	[8][×][1][1][=]	[88.]
1·3 × 0·2	[1][·][3][×][0][·][2][=]	[0.26]
7 × -5	[7][×][5][+/−][=]	[-35.]
-4 × 4	[4][+/−][×][4][=]	[-16.]
-9 × -6	[9][+/−][×][6][+/−][=]	[54.]

Division

60 ÷ 20	$\frac{60}{20}$	[6][0][÷][2][0][=]	[3.]
9 ÷ 4	$\frac{9}{4}$	[9][÷][4][=]	[2.25]
1 ÷ 3	$\frac{1}{3}$	[1][÷][3][=]	[0.333333...] 3 recurs infinitely
2 ÷ 7	$\frac{2}{7}$	[2][÷][7][=]	[0.285714...] 285714 recurs infinitely
5 ÷ 0	$\frac{5}{0}$	[5][÷][0][=]	[E 0.] Error - division by 0 is **not allowed**.
-8 ÷ 2	$\frac{-8}{2}$	[8][+/−][÷][2][=]	[-4.]
-9 ÷ -3	$\frac{-9}{-3}$	[9][+/−][÷][3][+/−][=]	[3.]

Calculator - Indirect Input 3

Constant. Indices / Powers.

Constant

The = Sign can be used to 'automatically' $+, -, \times, \div$ a constant.

$5 + 5$	[5][+][=]	10.	3×3	[3][×][=]	9.
$5 + 5 + 5$	[5][+][=][=]	15.	$3 \times 3 \times 3$	[3][×][=][=]	27.
$5 + 5 + 5 + 5$	[5][+][=][=][=]	20.	$3 \times 3 \times 3 \times 3$	[3][×][=][=][=]	81.
$4 - 4$	[4][−][=]	0.	$2 \div 2$	[2][÷][=]	1.
$4 - 4 - 4$	[4][−][=][=]	-4.	$2 \div 2 \div 2$	[2][÷][=][=]	0.5
$4 - 4 - 4 - 4$	[4][−][=][=][=]	-8.	$2 \div 2 \div 2 \div 2$	[2][÷][=][=][=]	0.25

Indices / Powers

3^2 — [3][x^2] 9.

3^2 — [3][y^x][2][=] 9.

2^5 [2][y^x][5][=] 32.

$(-4)^2$ [4][+/−][x^2] 16.

$(-4)^3$ [4][+/−][y^x][3][=] -64.

$1 \cdot 2^5$ [1][.][2][y^x][5][=] 2.48832

$1 \cdot 234^5$ [1][.][2][3][4][y^x][5][=] 2.861381721   The number is 'rounded' - it should have 15 decimal places.

100^5 [1][0][0][y^x][5][=] 1. 10 $= 10,000,000,000$
1×10^{10} In Standard Form - the decimal point is moved 10 places to the **right**.

88^8 [8][8][y^x][8][=] 3.5963452 15   The number is 'rounded'.
$3.596... \times 10^{15}$ In Standard Form - the decimal point is moved 15 places to the **right**.

$2^{-1} = \dfrac{1}{2^1}$ [2][y^x][1][+/−][=] 0.5

$5^{-2} = \dfrac{1}{5^2}$ [5][y^x][2][+/−][=] 0.04

$3^{-4} = \dfrac{1}{3^4}$ [3][y^x][4][+/−][=] 0.0123...

$(-4)^{-1} = \dfrac{1}{(-4)^1}$ [4][+/−][y^x][1][+/−][=] -0.25

$(-2)^{-6} = \dfrac{1}{(-2)^6}$ [2][+/−][y^x][6][+/−][=] 0.015625

Calculator - Indirect Input 4

Roots. Percentages. Factorial.

Roots

*Note
For an even number root, ex., $\sqrt[2]{}$, $\sqrt[4]{}$, $\sqrt[6]{}$, ... the answer is $+$ or $-$.   Ex. $\sqrt[2]{9} = +3$ or -3.
The calculator only displays the +ve answer.

₂ is usually omitted

$\sqrt[2]{9}$ ⎫   [9] [√]   [3.]
$\sqrt{9}$ ⎭   [9] [s-I-2] [√]   [3.]

$\sqrt{2}$ [2] [√]   [1.4142...]

$\sqrt{0.16}$ [0] [·] [1] [6] [√]   [0.4]

$\sqrt[3]{8}$ ⎫ [8] [s-I-2] [$\sqrt[3]{}$]   [2.]

$\sqrt[3]{8}$ ⎭ [8] [s-I-2] [$\sqrt[x]{y}$] [3] [=]   [2.]

$\sqrt[4]{625}$ [6] [2] [5] [s-I-2] [$\sqrt[x]{y}$] [4] [=]   [5.]

$\sqrt[3]{-8}$ [8] [+/−] [s-I-2] [$\sqrt[3]{}$]   [-2.]

$\sqrt[7]{9}$ [9] [s-I-2] [$\sqrt[x]{y}$] [7] [=]   [1.368...]

Percentages

($5\% = \frac{5}{100} = 0.05$)

5%   [5] [s-I-2] [%]   [0.05]   Gives the decimal form of a percentage.

5% of 600 ⎫ [5] [s-I-2] [%] [×] [6] [0] [0] [=]   [30.]

          ⎭ [6] [0] [0] [×] [5] [s-I-2] [%] [=]   [30.]

Increase   [6] [0] [0] [+] [5] [s-I-2] [%] increase [30.] [=] [630.]
600 by 5%

Decrease   [6] [0] [0] [−] [5] [s-I-2] [%] decrease [30.] [=] [570.]
600 by 5%

Factorial

                [5] [s-I-2] [x!] or [n!] [=]   [120.]

5! is read as '5 **factorial**' and multiplics numbers as shown.

$5! = 5 \times 4 \times 3 \times 2 \times 1 = 120$
$4! = 4 \times 3 \times 2 \times 1 = 24$

Calculator - Indirect Input 5

Fractions: Input. To/from Decimal. Equivalent. Improper / Mixed Number.

Fractions
A calculator **may** have a fraction key - ex. [a b/c]

Input a Fraction

$\frac{1}{2}$	[1] [a b/c] [2]	1⌟2.
$3\frac{4}{7}$	[3] [a b/c] [4] [a b/c] [7]	3_4⌟7.

Change Fraction to Decimal

change decimal back to same fraction

$\frac{1}{2}$	[1] [a b/c] [2] [=]	1⌟2.	[a b/c]	0.5	[a b/c]	1⌟2.
$3\frac{4}{7}$	[3] [a b/c] [4] [a b/c] [7]	3_4⌟7.	[a b/c]	0.5	[a b/c]	3_4⌟7.

Change Decimal to Fraction (Calculator **may** do this)

change fraction back to same decimal

0·1	[0] [·] [1] [a b/c]	1⌟10.	$\frac{1}{10}$	[a b/c]	0.1
5·9	[5] [·] [9] [a b/c]	5_9⌟10.	$5\frac{9}{10}$	[a b/c]	5.9

Equivalent Fraction - Lowest Terms (Cancel)

$\frac{2}{4}$	[2] [a b/c] [4] [=]	1⌟2.	$\frac{1}{2}$	
$\frac{8}{6}$	[8] [a b/c] [6] [s-I-2] [a b/c]	4⌟3.	$\frac{4}{3}$	Remains an improper fraction.
$5\frac{6}{9}$	[5] [a b/c] [6] [a b/c] [9] [=]	5_2⌟3.	$5\frac{2}{3}$	

Change Improper Fraction to Mixed Number

$\frac{9}{2}$	[9] [a b/c] [2] [=]	4_1⌟2.	$4\frac{1}{2}$			
$\frac{8}{6}$	[8] [a b/c] [6] [=]	1_1⌟3.	$1\frac{1}{3}$	Answer $1\frac{2}{6}$ given in lowest terms.		
$\frac{8}{6}$	[8] [a b/c] [6] [s-I-2] [a b/c]	4⌟3.	[=]	1_1⌟3.	$1\frac{1}{3}$	Gives lowest terms first.

Change Mixed Number to Improper Fraction

$1\frac{3}{5}$	[1] [a b/c] [3] [a b/c] [5] [s-I-2] [a b/c]	8⌟5.	$\frac{8}{5}$			
$5\frac{4}{6}$	[5] [a b/c] [4] [a b/c] [6] [s-I-2] [a b/c]	17⌟3.	$\frac{17}{3}$	Answer $\frac{34}{6}$ given in lowest terms.		
$5\frac{4}{6}$	[5] [a b/c] [4] [a b/c] [6] [=]	5_2⌟3.	[s-I-2] [a b/c]	17⌟3.	$\frac{17}{3}$	Gives lowest terms first.

Calculator - Indirect Input 6

Fractions: $+, -, \times, \div$.

Addition

$\frac{1}{2} + \frac{1}{2}$ ⟶ $\boxed{1}\,\boxed{a\frac{b}{c}}\,\boxed{2}\,\boxed{+}\,\boxed{1}\,\boxed{a\frac{b}{c}}\,\boxed{2}\,\boxed{=}$ $\boxed{1.}$ 1.

$\frac{2}{7} + \frac{4}{7}$ ⟶ $\boxed{2}\,\boxed{a\frac{b}{c}}\,\boxed{7}\,\boxed{+}\,\boxed{4}\,\boxed{a\frac{b}{c}}\,\boxed{7}\,\boxed{=}$ $\boxed{6\lrcorner 7.}$ $\frac{6}{7}$

$\frac{1}{4} + \frac{3}{5}$ ⟶ $\boxed{1}\,\boxed{a\frac{b}{c}}\,\boxed{4}\,\boxed{+}\,\boxed{3}\,\boxed{a\frac{b}{c}}\,\boxed{5}\,\boxed{=}$ $\boxed{17\lrcorner 20.}$ $\frac{17}{20}$

$3\frac{1}{2} + 5\frac{1}{4}$ ⟶ $\boxed{3}\,\boxed{a\frac{b}{c}}\,\boxed{1}\,\boxed{a\frac{b}{c}}\,\boxed{2}\,\boxed{+}\,\boxed{5}\,\boxed{a\frac{b}{c}}\,\boxed{1}\,\boxed{a\frac{b}{c}}\,\boxed{4}\,\boxed{=}$ $\boxed{8\lrcorner 3\lrcorner 4.}$ $8\frac{3}{4}$

Subtraction

$\frac{4}{5} - \frac{1}{5}$ ⟶ $\boxed{4}\,\boxed{a\frac{b}{c}}\,\boxed{5}\,\boxed{-}\,\boxed{1}\,\boxed{a\frac{b}{c}}\,\boxed{5}\,\boxed{=}$ $\boxed{3\lrcorner 5.}$ $\frac{3}{5}$

$\frac{7}{8} - \frac{3}{4}$ ⟶ $\boxed{7}\,\boxed{a\frac{b}{c}}\,\boxed{8}\,\boxed{-}\,\boxed{3}\,\boxed{a\frac{b}{c}}\,\boxed{4}\,\boxed{=}$ $\boxed{1\lrcorner 8.}$ $\frac{1}{8}$

$7\frac{2}{9} - 1\frac{5}{6}$ ⟶ $\boxed{7}\,\boxed{a\frac{b}{c}}\,\boxed{2}\,\boxed{a\frac{b}{c}}\,\boxed{9}\,\boxed{-}\,\boxed{1}\,\boxed{a\frac{b}{c}}\,\boxed{5}\,\boxed{a\frac{b}{c}}\,\boxed{6}\,\boxed{=}$ $\boxed{5\lrcorner 7\lrcorner 18.}$ $5\frac{7}{18}$

$\frac{1}{7} - \frac{6}{7}$ ⟶ $\boxed{1}\,\boxed{a\frac{b}{c}}\,\boxed{7}\,\boxed{-}\,\boxed{6}\,\boxed{a\frac{b}{c}}\,\boxed{7}\,\boxed{=}$ $\boxed{-5\lrcorner 7.}$ $-\frac{5}{7}$

Multiplication

$\frac{1}{2} \times \frac{3}{4}$ ⟶ $\boxed{1}\,\boxed{a\frac{b}{c}}\,\boxed{2}\,\boxed{\times}\,\boxed{3}\,\boxed{a\frac{b}{c}}\,\boxed{4}\,\boxed{=}$ $\boxed{3\lrcorner 8.}$ $\frac{3}{8}$

$\frac{7}{6} \times \frac{9}{5}$ ⟶ $\boxed{7}\,\boxed{a\frac{b}{c}}\,\boxed{6}\,\boxed{\times}\,\boxed{9}\,\boxed{a\frac{b}{c}}\,\boxed{5}\,\boxed{=}$ $\boxed{2\lrcorner 1\lrcorner 10.}$ $2\frac{1}{10}$

$2\frac{4}{9} \times 8\frac{1}{5}$ ⟶ $\boxed{2}\,\boxed{a\frac{b}{c}}\,\boxed{4}\,\boxed{a\frac{b}{c}}\,\boxed{9}\,\boxed{\times}\,\boxed{8}\,\boxed{a\frac{b}{c}}\,\boxed{1}\,\boxed{a\frac{b}{c}}\,\boxed{5}\,\boxed{=}$ $\boxed{20\lrcorner 2\lrcorner 45.}$ $20\frac{2}{45}$

$-\frac{3}{8} \times \frac{3}{7}$ ⟶ $\boxed{3}\,\boxed{+/-}\,\boxed{a\frac{b}{c}}\,\boxed{8}\,\boxed{\times}\,\boxed{3}\,\boxed{a\frac{b}{c}}\,\boxed{7}\,\boxed{=}$ $\boxed{-9\lrcorner 56.}$ $-\frac{9}{56}$

Division

$\frac{1}{3} \div 2$ $\frac{\frac{1}{3}}{2}$ ⟶ $\boxed{1}\,\boxed{a\frac{b}{c}}\,\boxed{3}\,\boxed{\div}\,\boxed{2}\,\boxed{=}$ $\boxed{1\lrcorner 6.}$ $\frac{1}{6}$

$7 \div \frac{4}{9}$ $\frac{7}{\frac{4}{9}}$ ⟶ $\boxed{7}\,\boxed{\div}\,\boxed{4}\,\boxed{a\frac{b}{c}}\,\boxed{9}\,\boxed{=}$ $\boxed{15\lrcorner 3\lrcorner 4.}$ $15\frac{3}{4}$

$\frac{5}{6} \div \frac{1}{2}$ $\frac{\frac{5}{6}}{\frac{1}{2}}$ ⟶ $\boxed{5}\,\boxed{a\frac{b}{c}}\,\boxed{6}\,\boxed{\div}\,\boxed{1}\,\boxed{a\frac{b}{c}}\,\boxed{2}\,\boxed{=}$ $\boxed{1\lrcorner 2\lrcorner 3.}$ $1\frac{2}{3}$

$9\frac{3}{4} \div 1\frac{4}{5}$ $\frac{9\frac{3}{4}}{1\frac{4}{5}}$ ⟶ $\boxed{9}\,\boxed{a\frac{b}{c}}\,\boxed{3}\,\boxed{a\frac{b}{c}}\,\boxed{4}\,\boxed{\div}\,\boxed{1}\,\boxed{a\frac{b}{c}}\,\boxed{4}\,\boxed{a\frac{b}{c}}\,\boxed{5}\,\boxed{=}$ $\boxed{5\lrcorner 5\lrcorner 12.}$ $5\frac{5}{12}$

$-\frac{2}{7} \div \frac{-7}{8}$ $\frac{-\frac{2}{7}}{\frac{-7}{8}}$ ⟶ $\boxed{2}\,\boxed{+/-}\,\boxed{a\frac{b}{c}}\,\boxed{7}\,\boxed{\div}\,\boxed{7}\,\boxed{+/-}\,\boxed{a\frac{b}{c}}\,\boxed{8}\,\boxed{=}$ $\boxed{16\lrcorner 49.}$ $\frac{16}{49}$

Calculator - Indirect Input 7

Combining Operations. Brackets / = Sign.

Combining Operations
Brackets / = Sign

When combining operations, calculations are carried out in an order of 'importance' -
Brackets, Indices, Division / Multiplication, Addition / Subtraction (often referred to as **BIDMAS**).

In each of the following examples, the operations are the same or have the same 'importance'.

$1 + 2 + 3$ $\boxed{1}\boxed{+}\boxed{2}\boxed{+}\boxed{3}\boxed{=}$ $\boxed{6.}$

$8 - 4 - 1$ $\boxed{8}\boxed{-}\boxed{4}\boxed{-}\boxed{1}\boxed{=}$ $\boxed{3.}$

$5 + 4 - 2$ $\boxed{5}\boxed{+}\boxed{4}\boxed{-}\boxed{2}\boxed{=}$ $\boxed{7.}$

$3 \times 1 \times 6$ $\boxed{3}\boxed{\times}\boxed{1}\boxed{\times}\boxed{6}\boxed{=}$ $\boxed{18.}$

$6 \div 3 \div 2$ $\boxed{6}\boxed{\div}\boxed{3}\boxed{\div}\boxed{2}\boxed{=}$ $\boxed{1.}$

$8 \times 5 \div 4$ $\boxed{8}\boxed{\times}\boxed{5}\boxed{\div}\boxed{4}\boxed{=}$ $\boxed{10.}$

In each of the following examples, the value of each bracket is calculated separately.

$7 - (3 + 2)$
$= 7 - (\ 5 \ )$ $\boxed{7}\boxed{-}\boxed{(}\boxed{3}\boxed{+}\boxed{2}\boxed{)}\boxed{=}$ $\boxed{2.}$

$7 - (3 - 2)$
$= 7 - (\ 1 \ )$ $\boxed{7}\boxed{-}\boxed{(}\boxed{3}\boxed{-}\boxed{2}\boxed{)}\boxed{=}$ $\boxed{6.}$

$(6 + 5) \times (4 + 1)$
$= (\ 11 \ ) \times (\ 5 \ )$ $\boxed{(}\boxed{6}\boxed{+}\boxed{5}\boxed{)}\boxed{\times}\boxed{(}\boxed{4}\boxed{+}\boxed{1}\boxed{)}\boxed{=}$ $\boxed{55.}$

$(6 - 5) \times (4 - 1)$
$= (\ 1 \ ) \times (\ 3 \ )$ $\boxed{(}\boxed{6}\boxed{-}\boxed{5}\boxed{)}\boxed{\times}\boxed{(}\boxed{4}\boxed{-}\boxed{1}\boxed{)}\boxed{=}$ $\boxed{3.}$

The = Sign can often be used to complete a calculation as if it was in brackets.

Multiplication with Addition / Subtraction

$7 \times 9 + 1$
$= \ 63 \ + 1$ $\boxed{7}\boxed{\times}\boxed{9}\boxed{+}\boxed{1}\boxed{=}$ $\boxed{64.}$ $\times$ is carried out first.

The use of brackets changes ... the order of the 'importance' / the order of calculations.

$7 \times (9 + 1)$
$= 7 \times \ 10$ $\boxed{7}\boxed{\times}\boxed{(}\boxed{9}\boxed{+}\boxed{1}\boxed{)}\boxed{=}$ $\boxed{70.}$ The value of the brackets ( ) is calculated first, then $7 \times (\ )$.

$\boxed{9}\boxed{+}\boxed{1}\boxed{=}\boxed{\times}\boxed{7}\boxed{=}$ $\boxed{70.}$ The 1st.= Sign calculates $9 + 1$ first.

$1 + 2 \times 3$
$= 1 + \ 6$ $\boxed{1}\boxed{+}\boxed{2}\boxed{\times}\boxed{3}\boxed{=}$ $\boxed{7.}$ Even though $+$ is first, $\times$ is carried out first.

The use of brackets changes ... the order of the 'importance' / the order of calculations.

$(1 + 2) \times 3$
$= \ 3 \ \times 3$ $\boxed{(}\boxed{1}\boxed{+}\boxed{2}\boxed{)}\boxed{\times}\boxed{3}\boxed{=}$ $\boxed{9.}$ The value of the brackets ( ) is calculated first, then $\times 3$.

$\boxed{1}\boxed{+}\boxed{2}\boxed{=}\boxed{\times}\boxed{3}\boxed{=}$ $\boxed{9.}$ The 1st.= Sign calculates $1 + 2$ first.

Calculator - Indirect Input 8

Combining Operations. Brackets / = Sign.

$6 \times 5 - 4$
$= 30 - 4$

| 6 | × | 5 | − | 4 | = | | 26. | × is carried out first.

The use of brackets changes ... the order of the 'importance' / the order of calculations.

$6 \times (5 - 4)$
$= 6 \times \ \ 1$

| 6 | × | (| 5 | − | 4 |) | = | | 6. | The value of the brackets ( ) is calculated first, then $6 \times$ ( ).

| 5 | − | 4 | = | × | 6 | = | | 6. | The 1st.= Sign calculates $5 - 4$ first.

$9 - 1 \times 2$
$= 9 - \ \ 2$

| 9 | − | 1 | × | 2 | = | | 7. | Even though − is first, × is carried out first.

The use of brackets changes ... the order of the 'importance' / the order of calculations.

$(9 - 1) \times 2$
$= \ \ 8 \ \ \times 2$

| (| 9 | − | 1 |) | × | 2 | = | | 16. | The value of the brackets ( ) is calculated first, then ×2.

| 9 | − | 1 | = | × | 2 | = | | 16. | The 1st.= Sign calculates $9 - 1$ first.

Division with Addition / Subtraction
(... is similar to multiplication ...)

$6 \div 2 + 1$
$= \ \ 3 \ \ + 1$

| 6 | ÷ | 2 | + | 1 | = | | 4. | ÷ is carried out first.

The use of brackets changes ... the order of the 'importance' / the order of calculations.

$6 \div (2 + 1)$
$= 6 \div \ \ 3$

| 6 | ÷ | (| 2 | + | 1 |) | = | | 2. | The value of the brackets ( ) is calculated first, then $6 \div$ ( ).

$\dfrac{8 + 6}{2}$
$= \dfrac{14}{2}$

| (| 8 | + | 6 |) | ÷ | 2 | = | | 7. | $\dfrac{8+6}{2}$ ⇐ The ÷ line acts like brackets ( ).

| 8 | + | 6 | = | ÷ | 2 | = | | 7. |

$8 + \dfrac{6}{2}$
$= 8 + 3$

| 8 | + | 6 | ÷ | 2 | = | | 11. |

$\dfrac{7 + 3}{9 - 4}$
$= \dfrac{10}{5}$

| (| 7 | + | 3 |) | ÷ | (| 9 | − | 4 |) | = | | 2. |

| 7 | + | 3 | = | ÷ | (| 9 | − | 4 |) | = | | 2. |

Division with Multiplication

$\dfrac{8}{1 \times 2}$

| 8 | ÷ | (| 1 | × | 2 |) | = | | 4. |

| 8 | ÷ | 1 | ÷ | 2 | = | | 4. | 8 is **divided** by 1 **AND** is **divided** by 2.

$\dfrac{8}{1} \times 2$

| 8 | ÷ | 1 | × | 2 | = | | 16. |

Calculator - Indirect Input 9

Combining Operations - Indices / Powers and Roots.

Indices / Powers and Roots

$\sqrt{4^3}$
$4^{\frac{3}{2}}$ } | [4] [y^x] [3] [=] [$\sqrt{}$ or s-I-2 $\sqrt{}$] [8.]

$(\sqrt{4})^3$
$4^{\frac{3}{2}}$ } | [(] [4] [$\sqrt{}$ or s-I-2 $\sqrt{}$] [)] [y^x] [3] [=] [8.]

$4^{\frac{3}{2}}$ | [4] [y^x] [3] [$a\frac{b}{c}$] [2] [=] [8.]

$4^{\frac{3}{2}} = 4^{1·5}$ | [4] [y^x] [1] [·] [5] [=] [8.]

$\sqrt[6]{7^5}$ | [7] [y^x] [5] [=] [s-I-2 $\sqrt[x]{y}$] [6] [=] [5.061...]

$\sqrt{3^2 + 4^2}$ | [3] [x^2] [+] [4] [x^2] [=] [$\sqrt{}$ or s-I-2 $\sqrt{}$] [5.]

$\sqrt[3]{2^2 \times 5^2}$ | [2] [x^2] [×] [5] [x^2] [=] [$\sqrt[3]{}$ or s-I-2 $\sqrt[3]{}$] [4.641...]

$\sqrt[5]{5^5 - 3^6}$ | [5] [y^x] [5] [−] [3] [y^x] [6] [=] [s-I-2 $\sqrt[x]{y}$] [5] [=] [4.741...]

$\dfrac{1}{2^2}$ | [1] [÷] [2] [x^2] [=] [0.25]

$\dfrac{1}{90^6}$ | [1] [÷] [9] [0] [y^x] [6] [=] [1.8816764 -12]   The number is 'rounded'.
$1·881... \times 10^{-12}$ In Standard Form - the decimal point is moved 12 places to the **left**.

$\dfrac{4^5}{5^4}$ | [4] [y^x] [5] [÷] [5] [y^x] [4] [=] [1.6384]

$\dfrac{\sqrt[7]{9}}{2^7}$ | [9] [s-I-2 $\sqrt[x]{y}$] [7] [÷] [2] [y^x] [7] [=] [0.0106...]

$3^{-2} \times 6^4$
$= \dfrac{1}{3^2} \times 6^4$ | [3] [y^x] [2] [+/−] [×] [6] [y^x] [4] [=] [144.]

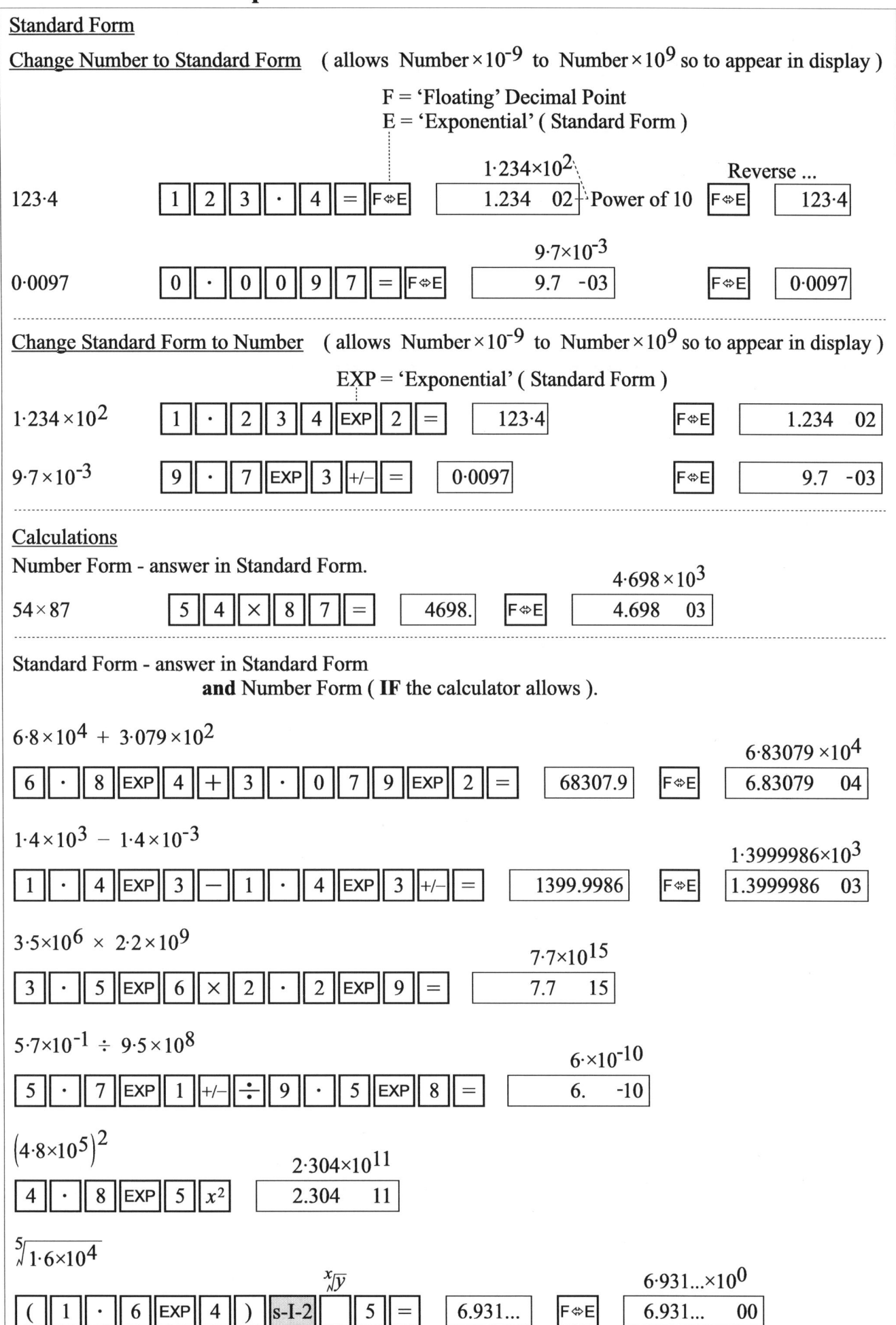

Calculator - Indirect Input 10
Standard Form to/from Number. Calculations.

Standard Form

Change Number to Standard Form   (allows Number×10^{-9} to Number×10^9 so to appear in display)

F = 'Floating' Decimal Point
E = 'Exponential' (Standard Form)

123·4   [1] [2] [3] [·] [4] [=] [F⇔E]   1.234 02 ─ Power of 10 $1·234×10^2$ Reverse ...   [F⇔E]   123·4

0·0097   [0] [·] [0] [0] [9] [7] [=] [F⇔E]   9.7 -03 $9·7×10^{-3}$ [F⇔E]   0·0097

Change Standard Form to Number   (allows Number×10^{-9} to Number×10^9 so to appear in display)

EXP = 'Exponential' (Standard Form)

$1·234×10^2$ [1] [·] [2] [3] [4] [EXP] [2] [=]   123·4   [F⇔E]   1.234 02

$9·7×10^{-3}$ [9] [·] [7] [EXP] [3] [+/−] [=]   0·0097   [F⇔E]   9.7 -03

Calculations

Number Form - answer in Standard Form.

54×87   [5] [4] [×] [8] [7] [=]   4698.   [F⇔E]   4.698 03 $4·698×10^3$

Standard Form - answer in Standard Form and Number Form (**IF** the calculator allows).

$6·8×10^4 + 3·079×10^2$

[6] [·] [8] [EXP] [4] [+] [3] [·] [0] [7] [9] [EXP] [2] [=]   68307.9   [F⇔E]   6.83079 04 $6·83079×10^4$

$1·4×10^3 − 1·4×10^{-3}$

[1] [·] [4] [EXP] [3] [−] [1] [·] [4] [EXP] [3] [+/−] [=]   1399.9986   [F⇔E]   1.3999986 03 $1·3999986×10^3$

$3·5×10^6 × 2·2×10^9$

[3] [·] [5] [EXP] [6] [×] [2] [·] [2] [EXP] [9] [=]   7.7 15 $7·7×10^{15}$

$5·7×10^{-1} ÷ 9·5×10^8$

[5] [·] [7] [EXP] [1] [+/−] [÷] [9] [·] [5] [EXP] [8] [=]   6. -10 $6·×10^{-10}$

$(4·8×10^5)^2$

[4] [·] [8] [EXP] [5] [x^2]   2.304 11 $2·304×10^{11}$

$\sqrt[5]{1·6×10^4}$

[(] [1] [·] [6] [EXP] [4] [)] [$\sqrt[x]{y}$ s-I-2] [5] [=]   6.931...   [F⇔E]   6.931... 00 $6·931...×10^0$

Calculator - Indirect Input 11   π (Pi). Reciprocal. Random Numbers. Trigonometry.

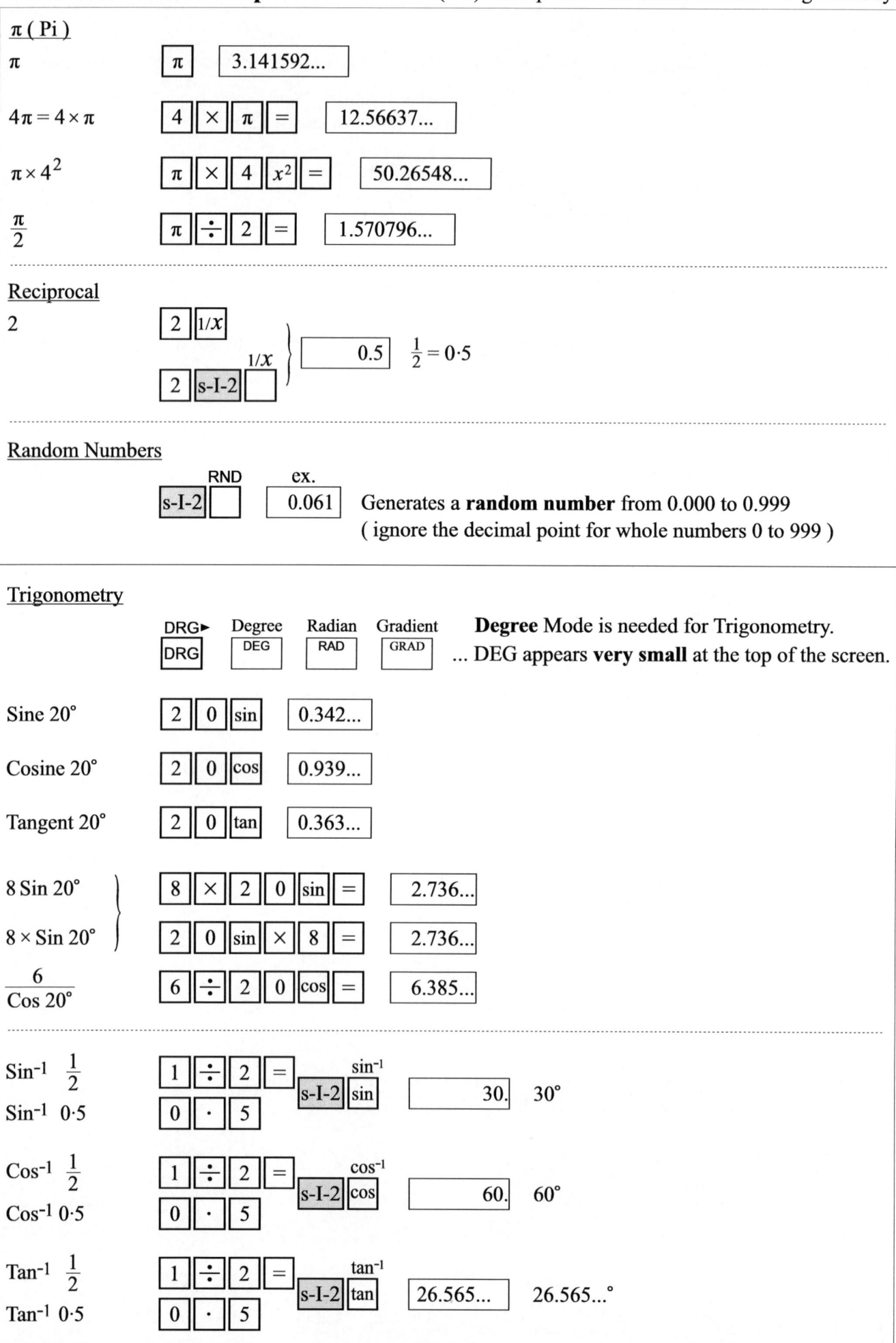

π (Pi)

π | π | 3.141592... |

$4\pi = 4 \times \pi$ | 4 | × | π | = | 12.56637... |

$\pi \times 4^2$ | π | × | 4 | x² | = | 50.26548... |

$\dfrac{\pi}{2}$ | π | ÷ | 2 | = | 1.570796... |

Reciprocal

2 | 2 | 1/x |

               1/x
| 2 | s-I-2 | | 0.5 $\dfrac{1}{2} = 0·5$

Random Numbers

     RND   ex.
| s-I-2 | | 0.061   Generates a **random number** from 0.000 to 0.999
                                                (ignore the decimal point for whole numbers 0 to 999)

Trigonometry

        DRG►   Degree   Radian   Gradient   **Degree** Mode is needed for Trigonometry.
        DRG   DEG   RAD   GRAD   ... DEG appears **very small** at the top of the screen.

Sine 20° | 2 | 0 | sin | 0.342... |

Cosine 20° | 2 | 0 | cos | 0.939... |

Tangent 20° | 2 | 0 | tan | 0.363... |

8 Sin 20° | 8 | × | 2 | 0 | sin | = | 2.736... |
8 × Sin 20° | 2 | 0 | sin | × | 8 | = | 2.736... |

$\dfrac{6}{\text{Cos } 20°}$ | 6 | ÷ | 2 | 0 | cos | = | 6.385... |

Sin⁻¹ $\dfrac{1}{2}$ | 1 | ÷ | 2 | = | s-I-2 | sin | 30.   30°
Sin⁻¹ 0·5 | 0 | · | 5 |

Cos⁻¹ $\dfrac{1}{2}$ | 1 | ÷ | 2 | = | s-I-2 | cos | 60.   60°
Cos⁻¹ 0·5 | 0 | · | 5 |

Tan⁻¹ $\dfrac{1}{2}$ | 1 | ÷ | 2 | = | s-I-2 | tan | 26.565...   26.565...°
Tan⁻¹ 0·5 | 0 | · | 5 |

Printed in Great Britain
by Amazon